LIFE: Conquest of Energy

LIFE: Conquest of Energy

RICHARD M. TULLAR
Los Angeles Pierce College

Holt, Rinehart and Winston, Inc.
New York Chicago San Francisco Atlanta Dallas Montreal Toronto London Sydney

ISBN: 0-03-081198-8
Library of Congress Catalog Card Number: 74-136776
Printed in the United States of America
2 3 4 5 038 9 8 7 6 5 4 3 2 1

Dedicated to My Fathers

Arthur Gilman Tullar, M.D.
who gave me life

and

Loye Holmes Miller, Ph.D.
who inspired me to its study

PREFACE

Insofar as their areas of interest are concerned, college students enrolled in introductory classes in general biology are a diverse group. Some may continue on for additional biological training in any one of a number of specialized fields. For many others, however, one formal exposure to general biology is all they ever receive. It is out of a sense of obligation to the latter group that this book was conceived. A proper study of general biology has much to offer that is relevant to the lives of all of us, regardless of the direction in which our interests lie. The nature of scientific proof and truth, the fundamental unity that pervades the diversity of life, the biological bases of behavior, the implications of organic evolution and interaction to man's future are of real concern to incipient historians, artists, philosophers, and poets as well as future engineers, businessmen, and scientists. These things are also a part of that proper study of general biology which introduces the student to the fascinating phenomenon of life.

As the reader proceeds through this book she or he will note that discussions within chapters as well as the chapters themselves are bound together in a logical sequence of concept built upon concept as the textual material gradually unfolds and enlarges upon its central theme. Interwoven throughout the text, this unifying theme is the title of the book: life is a conquest of energy.

The book is written in three sections. Section I, called "Science, Biology, and Life Itself," introduces the reader to the science of biology and considers various levels of organization from the physicochemical to the biological. The last chapter in this section (Chapter 3) is devoted to a discussion of life. It is the longest chapter in the book and serves as both an abstract and a prologue of everything that follows. Section II ("The Self-Directed Organism: Unit of Life") is an organismic approach to biology and consists of four chapters dealing with the organism's use of energy, and with behavior, growth, and reproduction. The last section discusses life as a populational phenomenon and is called "The Responsive Population: Source of Stability and Change." This section consists of four chapters which are concerned with the fundamental principles of heredity, organic evolution, organic interaction (ecology), and organic revolution (man's interrelationships with himself and with other living things). A list of questions follows each chapter, which serves as both a review and a summary of the chapter's contents. Following the questions are a number of recommended readings for those readers who may wish to pursue a given topic in greater depth or breadth. At the end of the book an appendix will be found containing an illustrated classification of organisms and a glossary. An Instructor's Manual written by Patricia Enkema of Los Angeles Pierce College is also available.

In the discussions that follow, scientific minutiae and terminology will be kept to a minimum, particularly in the areas of biochemistry and organic structure. Emphasis will be placed on concept and process. Readability has been a constant aim throughout the text because an uninteresting book about life is obviously not doing its subject justice. This does not mean that the readers of this book will necessarily have easy going. It does mean, however, that an attempt has been made to eliminate discussions which, to the general reader, are irrelevant.

For whatever success this book has in attaining these aims, I am indebted to the efforts of many people. Noel Korn of Los Angeles Valley College provided the initial encouragement to write. Continued support and assistance were received from A. Lee Haines and Sol A. Karlin of Los Angeles Pierce College, Anthony J. Gaudin of San Fernando Valley State College (California), and Fanny-Fern Davis (retired) of Okaloosa-Walton Junior College in Florida. Roberta Beatty, instructor in English (and my neighbor), has been a veritable oracle in matters of grammatical usage. The consulting editors of Holt, Rinehart and Winston, notably James D. Ebert (Carnegie Institution of Washington), Richard S. Miller (Yale University), and Howard A. Schneiderman (University of California at Irvine), have offered many helpful suggestions. Joseph C. Byers (now of Dryden Press) was both friend and counselor throughout the entire preparation of the book. Dorothy Garbose, senior project editor, was of constant assistance in coordinating the efforts of the many people responsible for getting a book together and in maintaining effective communication between us despite our being scattered from California to Italy. Many individuals and institutions have graciously provided illustrative materials for the book. Their permission to use these materials is gratefully acknowledged in the legends for these illustrations. The quality of the original artwork is due largely to the fine craftmanship of Tino Di Palmo and J & R Services. Howard Leiderman, design supervisor, also rendered helpful assistance.

Finally I would like to acknowledge the continuous encouragement proffered by a very good friend who not only shielded me from the perennial interruptions of worldly affairs but also took on the tedium and details of permissions correspondence and indexing, and with whom it is my extremely good fortune to be quite closely associated—my wife, Shirley Scott Tullar.

August 1971
Woodland Hills, California

Richard Montgomery Tullar

CONTENTS

SECTION I Science, Biology, and Life Itself

CHAPTER 1 Introducing the Science of Biology

WHAT BIOLOGY IS

A hummingbird flies 500 miles across the Gulf of Mexico with a "gas tank" 2 grams full. A majestic pine lifts water to its crown by microscopic "pumps." A human heart during a restful day performs enough work to lift a 150-pound man over 500 feet into the air. A population of people can become resistant to malaria just as a population of insects becomes tolerant to an insecticide. A human brain can contemplate itself!

These facts of life are a few of similar thousands born of the science of life or biology. By definition biology is the discipline that is devoted to the study of living organisms of all kinds.

An inventory of all the living things on this planet would yield an astronomically large number of individuals. A teaspoonful of rich soil, for example, may contain 1 billion bacteria, 20 million microscopic plants, and 200,000 single-celled animals (not to mention quite a number of larger ones)—all alive and growing vigorously. Any one of the groups of organisms represented in this soil—bacteria, plants, or animals—exist in the world in almost endless variety. Animals, for instance, include about 1 million kinds. A kind or species is defined as a population of individuals that may interbreed among themselves but is usually unable to breed with other species. All domestic breeds of dogs count therefore as only one species. Each one of these million species claims a roster of organisms numbering, except in rare cases, from tens of thousands to countless trillions of individuals. They run the gamut from single-celled microscopic animals to such varied types as sponges, corals, tapeworms, starfishes, insects, frogs, birds, and mammals including man himself. This proliferation in number and variety could similarly be traced among the half million species of bacteria and plants.

In observing the great diversity of organisms and their interrelationships with one another and the world around them, biologists have pondered an impressive number of major problems about life and living things. Biological subsciences and related disciplines have sprung up around these problems as questions were answered and answers became assembled into organized bodies of knowledge (see Fig. 1-1). How are living organisms built (anatomy)? How does an organism carry on its activities (physiology)? How do living things interact with one another and their environment (ecology, evolution)? What are the biological characteristics of a particular group of organisms (bacteriology, ornithology)? How can biological knowledge further man's welfare (medicine, agriculture)?

One penetrating problem, however, embodies the very foundation of all biological science—What is the "life" in life science? It is not difficult to write *about* life nor is it a particular chore to list the properties that all living things share. We know that organisms can reproduce their own kind. They are capable of a highly organized kind of growth, and are responsive to their surroundings, and so on. But none of these things sheds much light on what life *really* is. No

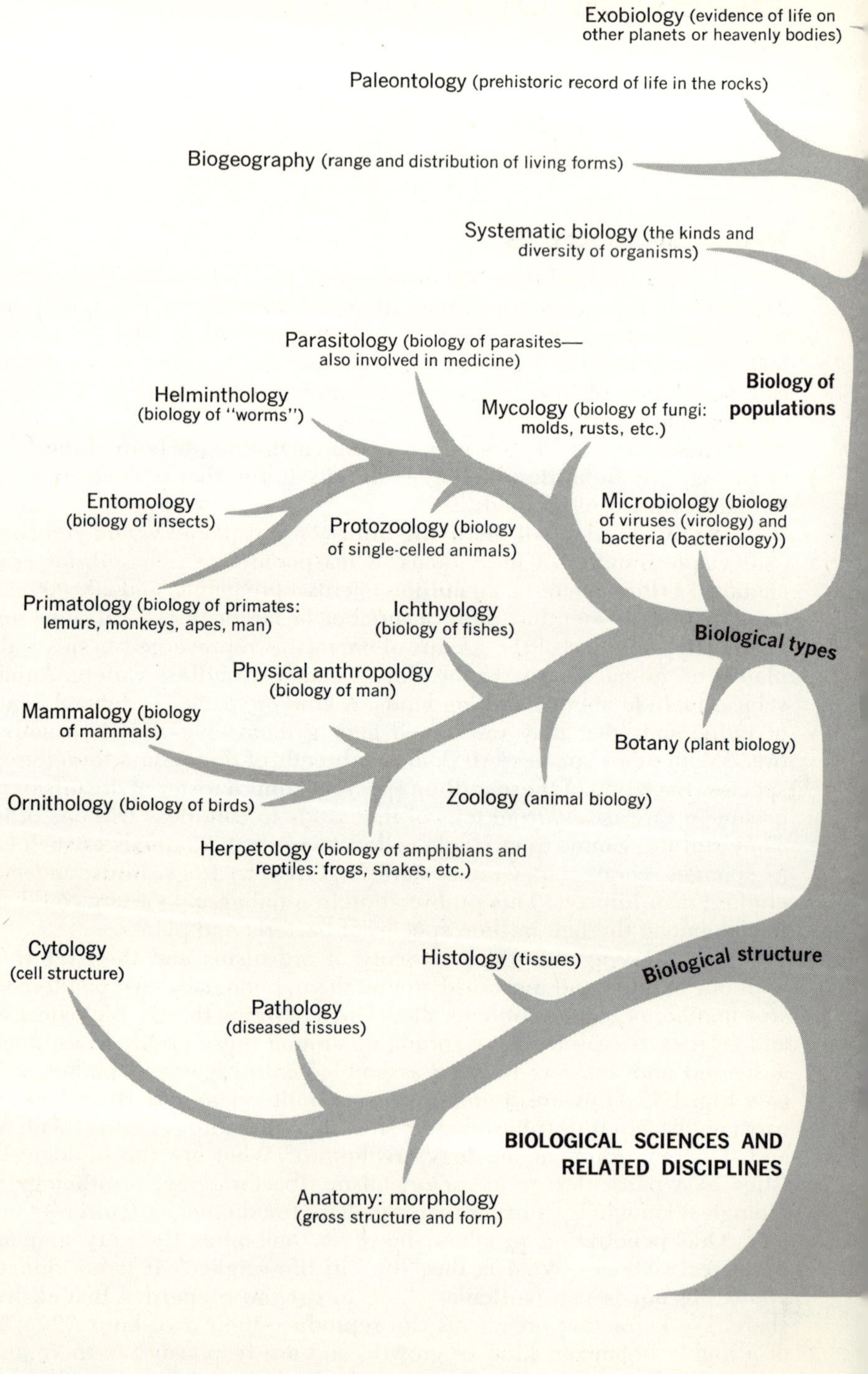

Figure 1-1 Biological sciences and related disciplines.

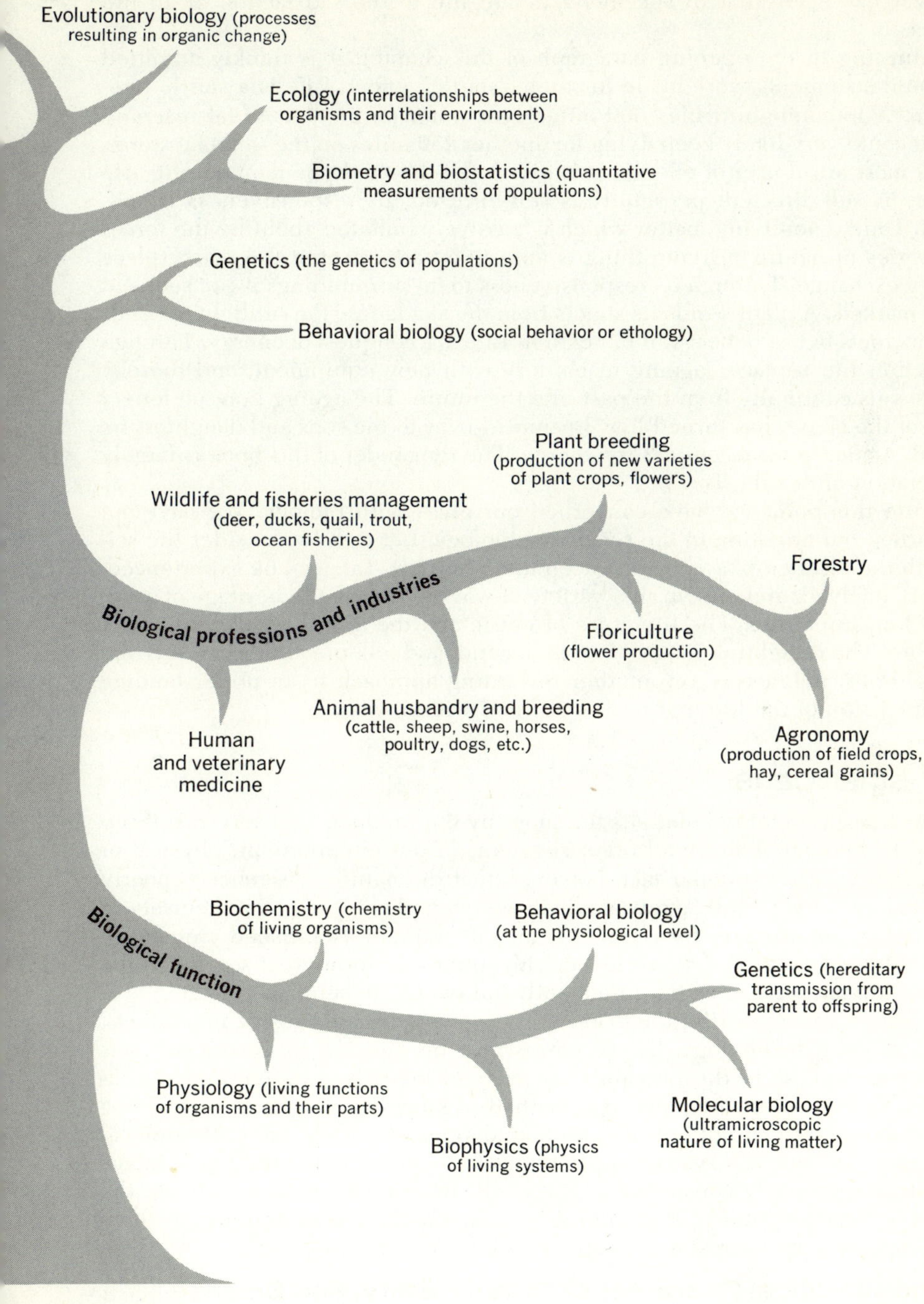
Evolutionary biology (processes resulting in organic change)
Ecology (interrelationships between organisms and their environment)
Biometry and biostatistics (quantitative measurements of populations)
Genetics (the genetics of populations)
Behavioral biology (social behavior or ethology)
Plant breeding (production of new varieties of plant crops, flowers)
Wildlife and fisheries management (deer, ducks, quail, trout, ocean fisheries)
Forestry
Biological professions and industries
Floriculture (flower production)
Animal husbandry and breeding (cattle, sheep, swine, horses, poultry, dogs, etc.)
Human and veterinary medicine
Agronomy (production of field crops, hay, cereal grains)
Biochemistry (chemistry of living organisms)
Behavioral biology (at the physiological level)
Biological function
Genetics (hereditary transmission from parent to offspring)
Physiology (living functions of organisms and their parts)
Molecular biology (ultramicroscopic nature of living matter)
Biophysics (physics of living systems)

one as yet has been able to succinctly define life to the satisfaction of all life scientists.

Returning to the opening paragraph of this chapter, it is frankly admitted that several statements were made for their "startle" effect. Life is a startle phenomenon. A hummingbird flies 500 miles over water on 2 grams of fat reserves. In fact, it could very likely keep flying for another 250 miles on the same fat stores. Life is a most artful user of energy. It is also a self-directed phenomenon: its use of energy is self-directed; perpetuity is self-directed; its responsiveness is self-directed. Unlike nonliving matter which is passively buffeted about by the forces and energies of nature, a living thing is an active participant at the marketplace of energy exchange. Through its responsiveness to its surroundings it can seek out the best markets. A plant bends its shoots from the shade into the sunlight; it sends most of its roots to the richest soil beneath it. Life is a conquest of energy. Through reproduction life replaces ageing machinery with new equipment, and thereby guarantees its continuity from the past into the future. The ageing lion, no longer capable of the chase, has turned his "business" over to the sons and daughters he has sired. Again, *life is a conquest of energy.* The remainder of this book is largely an elaboration of this thesis.

Up to this point we have concerned ourselves with biology. We have not really turned our attention to the *science* of biology. Let us now consider life science with the focal emphasis on science rather than life. Life can be experienced, described, enjoyed, and known in a variety of ways as rich as the heritage of accumulated human culture. The literature of music and the arts abounds with reference to life. The delightful poetry of myth, legend, and folklore alludes to a myriad of living forms. Science is yet another rewarding approach to an understanding and appreciation of the life that is about us and within us.

WHAT SCIENCE IS

Our lives are molded and shaped more by the products and services of science and technology than by all other segments of our environment, physical or cultural. It is a rather singular fact, however, that the nature of science is poorly understood by some people, a few of whom are even playing scientist themselves. This is rather unfortunate, for science is an integral and established element of human culture. Scientific discoveries and inventions perform great service in the interest of human welfare as the nineteenth and twentieth centuries attest.

Science's basic contribution to culture, however, does not reside in products, services, or even technology. These efforts may be directed to destructive acts of warfare as well as to the humanitarian ends of medicine. Its cultural significance lies in the fact that science is a method of knowing, a system for learning truth. Fundamentally, if not entirely, the foundations of science are philosophical to their core. Science is a system of principles for explaining natural phenomena. This system is logically consistent within itself. Its criteria for truth validate the premises on which it stands. It is unique because the basic assumptions on which its philosophic premises rest are essentially biologic in nature.

The Biological Character of Truth, Reality, and Science

Idealist philosophers from Plato on have proclaimed that the senses are deceptive: appearances deceive. A straight stick held in the water appears to be bent at the water's surface. Removing the pencil from behind your ear does not destroy the sensation that it is still there. Human perception therefore (runs the argument) cannot be relied on to reveal the truth about, or the nature of, reality. The idea of straight sticks is real; perceiving a straight stick is illusionary. At this point the scientist objects. How did the idealist demonstrate that the bent stick was an illusion? He ran his hand down the stick and it felt straight. He trusted the sense of touch in this instance to prove an error in the sense of vision. If perception is unequivocally not to be trusted, is this quite cricket? How can deceptive "tools" prove their own deception? If all 12-inch rulers were really 13 inches long, would we be aware of the inherent deceptiveness of rulers?

As a matter of fact perception is by no means delusive or out of touch with reality. One million species of animals are living testimony to the pertinence of perception to the real world. The young ape who swung for an imaginary limb millions of years ago not only missed being one of our ancestors but adulthood and parenthood as well! The very existence and continued survival of animals depend on the *relevance* of the sensory clues they receive from the real world in which they live.

Another biological element that lies at the heart of scientific knowledge and truth is that science is a *human* creation. Science is man's inquiry into his world. At first glance this appears so obvious as to be unworthy of mention, much less of any considered discussion. A moment's reflection, however, will reveal that this axiom has some rather far-reaching ramifications. There is a human (biological) component in all science. Scientific objectivity in the traditional sense—removing the observer from the scientific observation—is obviously impossible. We can construct equipment to respond automatically to whatever natural phenomena we may wish to investigate. But this in no way eliminates the human observer. He now merely observes (perceives) the responses of the machine. For example, an X-ray machine makes a visual picture from invisible X rays. The instrument was manmade to respond in a sensory department (in this example, vision) that man is capable of perceiving. To ask what something *is*, by itself and unrelated to human observation, is utterly meaningless. Everything we know about the world is in terms of how it *appears* to us, how it *sounds* to us, how it *feels* and *smells* and *tastes* to us. Everything we know about that which is real exists in terms of how it affects us, not in terms of what it is all by itself and unrelated to us. Scientific knowledge is based entirely on our sensory interaction with the world around us. Having perceived certain things about the world, we respond by making certain assumptions about what the world is really like. But these assumptions are strictly on our own terms, as we perceive and imagine them to be. All knowledge is a man-reacts-to-something kind of thing.

Our only avenues of sensory reaction with reality are (1) those sensory departments responsive to our external environment (sight, hearing, smell, taste, pressure, warmth, cold, and pain) and (2) those whose sensitivities are tuned to internal, visceral, and emotional states (the feel of body movement, pleasure, sexual response, anxiety, nausea, thirst, and so on). Using these sources of raw data, man is capable, through abstraction and reason, of performing feats of intellectual majesty: relativity theory, organic evolution, philosophic systems. But such abstractions must be finally predicated on the originating source of knowledge; human sensory experience is the ultimate sovereign of understanding. When we find ourselves discussing "an atom," "a force," or "a moral value" in terms that are unrelated to touch, push, heat, cold, light, sound, taste, or smell, or not founded in something akin to elation, anger, fear, or depression, we find that we are vacuously describing—nothing.

Some Philosophic Premises of Science

Now that we have presented certain aspects of the nature of scientific knowledge, we can turn our attention to some of the assumptions scientists make concerning the world around them. These assumptions might be compared to a policy statement about the nature of the world and the relative validity of knowledge gained about it. Some of these constitute a base from which all definitive scientific exploration must begin. Others embody criteria that guide the course of scientific inquiry and set the standards for interpreting the results of inquiry. As such these assumptions may be said to be the basic premises of science. That they are essentially philosophical in nature will become evident as the discussion develops.

THE LAW OF PARSIMONY AND UNIFORMITARIANISM

The reader should be immediately reassured that this caption is not nearly so formidable as it looks. The concept is the important issue, and there is fundamentally but one concept here.

The law of parsimony states that the number or complexity of causes for a phenomenon should not be unnecessarily magnified. The minimum number of reasons that adequately account for an event should be the maximum number. Given two different possible explanations, the simpler and the more plausible of the two is preferable. The simplest adequate hypothesis is the best hypothesis.

The doctrine of uniformitarianism is an excellent and far-reaching example of the law of parsimony in action. This doctrine states simply (and broadly) that all *process* in the universe is uniform. From the standpoint of the uniformity of process the universe is a very orderly place. The properties of elemental matter and energy as well as their potential for interaction have not been subject to exceptional or mystical modifications. Throughout all time they have been just as they are today. The principles of thermodynamics, the process of erosion, the oxidation-reduction reactions of energy exchange in living organisms do not change. Process is timeless: it has no history. Stars may grow old and disintegrate. Mountains may be worn by erosion to rounded hills. Organisms reproduce, die, and evolve into a myriad of forms. The products of process are framed in time; their changing parade is historically endless. The processes themselves and the elemental properties of nature are changeless. They are today as they have always been.

The astronomer, the geologist, and the biologist subscribe to this doctrine. Their success in explaining natural phenomena is eminent testimony of its fertility and usefulness. It is the simplest hypothesis capable of explaining the cosmos, the earth, life, or indeed anything that has a history. Living things, as we shall soon see, cannot be understood without a knowledge of their history. They are what they are *because* of their history. If biologists did not subscribe to the doctrine of uniformitarianism, there could be no real *science* of biology.

TRUTH, PROOF, AND DEGREES OF CONFIDENCE

A scientific truth may be defined as a judgment relative to which universal agreement can be reached. It is a statement that bears a definite relevance to reality because it has been proven. The nature of this proof and what it constitutes are as axiomatic to science as to any philosophic system.

The manner in which scientific proof becomes established can best be shown by example. It has long been known that fish travel hundreds, even thousands of miles sometimes, in order to lay their eggs where they themselves were originally hatched. In the case of salmon and some trout these migrations begin in the ocean; from here the fish find their way into a large river and ultimately into the small stream in which they began their lives. How do they find their way on these extended trips? Let us suppose that an hypothesis is proposed, based on a number of carefully controlled observations, that these fish use the sense of smell in an important way to guide them during migration, at least toward the end of their long journey. Scientific support of this would be established through repeated observations, preferably by several observers using both similar and different techniques. As the number of confirming observations increases (in the absence of refuting observations) so increases the confidence the scientist could place in the original hypothesis. The relative amount of confidence that is so engendered through repetitive observation is often referred to as the degree of confidence. It is obvious that the repetition of observations can extend into the indefinite future, utilizing thousands of observers and scores of investigative techniques. The evidence can therefore never be completely gathered; the evidence is never all in. As a result, confidence can never be unlimited; there are varying degrees of confidence. The scientist, however, has the potential of reaching high degrees of confidence by utilizing observations that would *disprove* his hypothesis. Refutation or disproof is often possible. In our example, the sensory pathway for smell could be surgically blocked or severed in a number of fish, thus destroying their sense of smell. The inability of these fish to orient themselves properly (failure of disproof) would certainly add validity to the original premise. It would, of course, be necessary to show that it was the absence of smell, not the surgery, that produced this incapacity. This could be accomplished by performing similar surgery *without severing the sensory pathways* in an equal number of fish and observing their behavior. If an hypothesis has not only withstood a large number of consistently confirming observations but has also persistently resisted all attempts to disprove it, confidence in it will be high. Such an hypothesis will then be called a *theory*. Most, if not all, scientific theories have considerably more evidence to support them than those elements commonly referred to as "facts."

Truth for the scientist, then, is dynamic, not absolute. Absolute proof is never possible. Our original statement would more properly read, "A scientific truth is a judgment relative to which universal agreement can be reached, *subject to whatever modifications may be necessary in the light of future study.*" Truth at its best can be brought up only to the present; it is not something that is ever fully attained. It is a direction, not a goal. It is a criterion, not an end. It lies in the act of searching and in the adventure of seeking.

EXPLANATION, PREDICTION, AND INDETERMINACY

The chief and proper exercise of science comes from man's driving curiosity to understand and interpret the world about him. "How?," "Why?," "What for?," and "How come?" are the questions that motivate the scientist in his quest for truth. The vigor of his drive is matched only by the rigor of his proof. All these questions share a common theme: what are the causes (or the reasons for) a given phenomenon? Still closer examination shows that these questions are dealing not with one but with two kinds of causes or explanations.

When the word "how" is used or implied, the questioner has an operational or functional problem in mind. How does water run downhill? How does a heart pump blood? How does a plant manufacture sugars? Figure 1-1 shows that there are a number of biological sciences dealing with structure and function. The problems posed by these disciplines are functional in nature. By what physical and chemical processes does a given structure (organ, tissue, or cell) perform some vital function? The emphasis may be on structure or it may be on process, but in the last analysis the answers to questions must include the structure(s) in operation.

When the word "why" is used or implied, the questioner is asking a totally different kind of question. Why does water run downhill (that is, what *purpose* is served)? The physical principle of gravitation explains how, but not why. The question is meaningless unless we suppose that water flows downhill so that man can harness its power. This is rather silly because water was flowing downhill for hundreds of thousands of years before man started to build dams, and, indeed, hundreds of millions of years prior to the existence of man. Explanations which assume that the utility of an event is also the cause of that event (a carryover from Aristotle and other early Greeks) are not indulged in by scientists. The fact is that the *purposive* (or teleological) question cannot be answered by the physical scientist. Causal phenomena for the physicist and chemist are those elemental properties of matter and energy which are changeless and timeless (the doctrine of uniformitarianism). The physical scientist can seek out *functional causes.* He is unable to consider *purposive causes* because the fundamental units with which he deals did not derive their uniqueness as a *consequence of their history.*

The biologist can intelligently ask why a heart pumps blood or a plant manufactures sugars. The answer in both cases is quite obvious: each process is useful to the organism and necessary for its survival. We can delve more deeply and specifically into purpose by extending our "why" to the implication of "what for?" Why is it necessary for plants to manufacture sugars in order to survive? The answer to this is that living things are self-directed energy-using systems. Photosynthesis is a very efficient method of producing stored energy from raw materials that are common almost anywhere on earth.

We can finally extend our "why" to its ultimate implication, "how come?" How and why did the self-directed manufacture of sugars by plants come about? At this point the history of plants becomes germane. A plant is what it is and does what it does *by reason of its developmental history.* It is but a link in a very long evolutionary chain of living things which extends primordially back into the nonliving world.

In this long history any organisms which could improve their chances for survival would be favored by nature. The environment in which organisms live is an exacting master. Food is often not easily obtained. Those living things that could make their own food from raw materials that are common practically everywhere would be selected by nature (that is, not eliminated through starvation). Their numbers could become legion if they incorporated this blueprint for food making into their hereditary material. Succeeding generations of plants would have the purposeful activity of photosynthesis *programmed into them* in a manner analogous to the mathematical prowess programmed into a computer. The purposeful activity of organisms is "built in." More importantly the true nature of purposiveness in biology is unveiled: in the last analysis purpose in living things

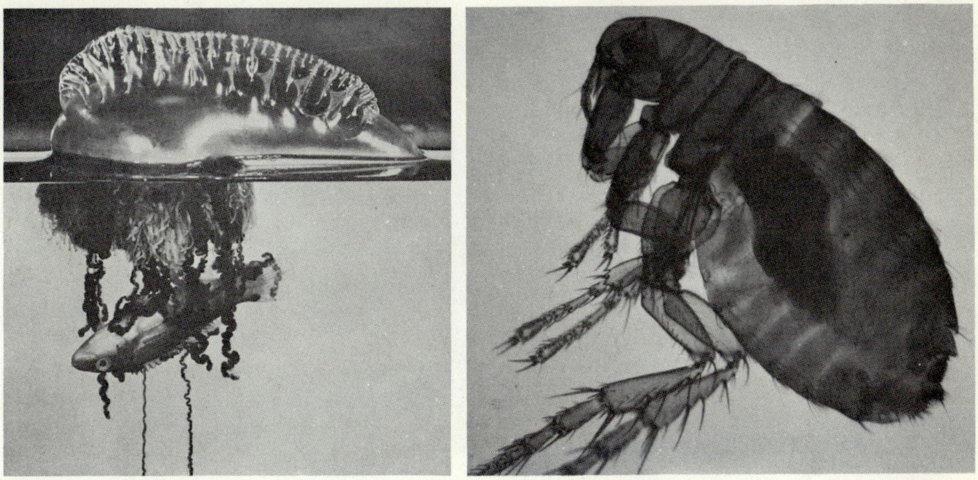

(a) (b)

Figure 1-2 Fitness is reflected in the exquisite adaptations of living things to their mode of life. (a) The Portuguese Man-O-War is a floating animal colony whose stinging tentacles effectively snare its prey. (b) The dog flea is an efficient external parasite. [(a) Douglas P. Wilson, Marine Biological Laboratory; (b) by author]

is *fitness*. This fitness is reflected in the exquisite adaptations of living organisms to their mode of life (Fig. 1-2).

It is apparent from the foregoing discussion that there is much of biology that cannot be explained in terms of physics or chemistry alone. Referring to Figure 1-1, and specifically to the category, "Biology of Populations," we note that a number of biological sciences are listed in which the fundamental issues for the most part are biological in nature. They are concerned with interacting and responding populations of organisms. Here the individual organism is the ultimate indivisible unit. A man with a strong stomach and in-grown toenails may live a long life or a short life. In either case it is lived as a whole man, not as a stomach or a pair of tender feet. Two thousand pounds of metal to which a liberal seasoning of glass and rubber is added is not the recipe for an automobile. Neither is an organism the sum of its chemical components, regardless of the sophistication and complexity of the inventory. An organism is an integrated unit whose complex level of organization, character, and behavior have been impressed on it *as a unit* throughout its history. Although the *functional* explanation of the physical scientist is still useful, it is the *purposive* explanations of the biologist that shed light on the ultimate whys and wherefores of living populations interacting in space and time.

At this point we can profitably turn our attention to an often made assumption relative to the interdependence of explanation and prediction, namely, that an adequate explanation should make reliable predictions possible. Are these two things in fact related or are they independent of one another? It is all too apparent in biology that explanation and prediction are quite independent of one another (the physical scientist, on occasion, has this driven home to him as well). The physiological effect of light on the eye of an octopus can be satisfactorily explained in a number of respects. The reliability of prediction in this instance is

Figure 1-3 Plant succession in which brush (manzanita) is being invaded by pine. (Sol A. Karlin, Los Angeles Pierce College)

very high. That is, the probability of the expected reaction occurring is excellent. Most functional events of this sort have a relatively high predictability. The principles by which plants in a given area begin to invade and establish themselves in a freshly plowed field or lot can be accounted for with equal validity (Fig. 1-3). Different plant species succeed one another in an orderly pattern until a relatively stable "climax" population of plant residents is reached. This, of course, takes a long time. The predictability of this orderly pattern of replacement is, however, low. Many variables are involved, such as insect pests, drought, and rodent populations, which affect the ultimate outcome. Neither an appraisal of their effects nor the vagaries of their occurrence can be reliably predicted. The probability of dependable prediction in this instance is low.

As can be seen in these two examples predictability is largely a matter of probability. Knowing that the probability of rolling a seven in dice is one in six by no means guarantees that we can predict with any certainty the occurrence of one seven in the first six throws of the dice. This in no way undermines the certain knowledge that if we roll dice often enough and long enough, sevens will occur remarkably close to one-sixth of the time. Causal phenomena involving whole organisms are of this nature. The single occurrence of a biological event whose causes are known is unpredictable. This situation is called *indeterminacy*. Let us briefly examine a few reasons for biological indeterminacy.

As we shall see in a later chapter, essentially all of the events regarding the modification, the assortment, and the transmission of hereditary material from parents to offspring involve a large element of chance. A brown-eyed man and woman, if they just happen to be the right brown-eyed man and woman, may have a blue-eyed child. The biological consequences of these variations run the gamut from inconsequential to critical, yet their occurrences are random and cannot be forecast in advance.

Second, no two organisms are ever *exactly* alike. Each individual in a population or species of similar individuals is unique. We may prescribe certain formulations about this *kind* of organism, but the prescription is statistical, not absolute.

Finally, and most significantly, great structural complexity and functional diversity characterize even the simplest organisms. The dynamic architecture of biologic systems defies complete description in any conventional quantitative sense. Numerous biochemical reactions occur with precision and great speed at the right time and in the right place. The repertory for their performance is studded with many alternative standby procedures. As we have observed, an organism is not the sum of its parts nor can its behavior be inferred from the properties of its components. Life is a function of an architectural organization of great sophistication. As matter is integrated at a higher level of organization new and unpredictable qualities emerge which were not latent in the separated components. Sodium is a soft, metallic element that is quite toxic. Chlorine is a poisonous gas. Combined they make sodium chloride, or table salt, which is essential to life. It is this kind of emergent quality that biologic systems manifest to such a remarkable degree.

The Nature of Scientific Inquiry

Up to this point we have confined our discussion largely to the philosophic principles on which scientific inquiry is based and guided. This has included (1) the biologic nature of knowledge, (2) the law of parsimony and uniformitarianism, (3) the relativity of truth, and (4) the nature of explanation and its relation to indeterminacy. There can be no doubt that science, considered as a method of approach to knowledge and understanding, is the embodiment of a particular philosophic position. A scientist is one who is guided by the precepts of this position.

Science has been defined as (1) organized knowledge, (2) a spectrum of disciplines devoted to concise descriptions of the natural world, and (3) a broad system within which natural phenomena can be explained and understood. It is all of these things. For intelligent use it must be organized. For operational efficiency it deals with a vast inventory of all kinds of natural objects and processes which must be accurately and thoroughly distinguished from one another and properly catalogued. Functionally it offers unlimited opportunity to man for satisfying his insatiable appetite to know the unknown and explain the inexplicable. By what processes may this function be realized? What procedures are followed in scientific inquiry?

In a description of any activity involving human beings, the human element is far more important than the mechanical aspects of the activity itself. Scientists are obviously human like anyone else. They are subject to the same frailties, the same highs, the same lows. They like to be right: they are elated with success. They do not like to be wrong: they become depressed with repeated failure. They are subject to bias: the answer based on selected data—the oversimplified answer—is an easy way out of those situations where one's personality has become identified as part of the problem being investigated. Man is a most colorful creature. He is also worthy of the highest dignity when, through the self-disciplined control of his emotional involvements, he brings the full weight of his intellect to bear on a problem. The discipline that scientists impose on themselves is usually referred to as the scientific attitude. This attitude embraces three major considerations.

1. Intellectual Honesty. There should be complete freedom from prejudice, bias, or preconceived notions as to the way something should be or ought to be. A good scientist, after having assumed something may be true, does not set out to prove it is true. His or her objective is always to see *if* it is true. The sole objective of scientific inquiry is truth. This means the investigators of any problem will be willing to accept, record, and consider all the evidence at hand, whether it is favorable or unfavorable to the hypothesis being examined. It also means that they will be as critical of their own work as they are of others.

2. *Accuracy of Observation, Calculation, and Conclusion.* All measurements relevant to a problem under investigation should be as complete and precise as humanly and mechanically possible. The degree of accuracy, of course, should be commensurate with the problem. A weight study on sulfur-bottom whales would not need to be carried to the nearest gram! For a similar study on hummingbirds, however, the nearest gram would not be accurate enough. Those characteristics or factors which are susceptible only to subjective appraisals should be avoided. If subjective evaluations cannot be avoided, they should be clearly identified as such and not treated as if they were objective measurements (any conclusions derived exclusively from such "data" would be considered tenuous indeed).

The mathematical and statistical treatment of collected quantitative data should be accurate and pertinent to the problem at hand. Statistical procedures should reflect the true nature and relevance of all collected data that lend themselves to this sort of analysis. The conclusions drawn from evidence, whether statistically summarized or not, should be logical and totally restricted to that which the evidence unequivocally supports.

3. *Conclusions with a High Degree of Confidence.* An hypothesis may become a theory when it has attained a high degree of confidence. This high degree of confidence can be reached, as we noted, when an hypothesis has a convincing number of consistently confirming observations and has withstood an equally convincing number of attempts to refute it. The "convincing number" in both cases is defined rather precisely. There are accepted statistical procedures (with which we will not burden the reader) that measure the probability of an event occurring by chance alone. These procedures can be applied to the evidence in support of any hypothesis. If such corroborating evidence can be shown statistically to have but one chance in a million of coming out the way it did on the basis of chance alone, it is unlikely that chance is involved. It is therefore likely that the hypothesis is correct, and we may henceforth refer to it as a theory.

Following this code of disciplinary ethics, scientific inquiry may follow a number of paths, depending on the nature of the problem. Considered broadly we may say that a questioning mind poses a problem; the problem fosters inquiry and observation; from inquiry and observation an hypothetical explanation is constructed; an hypothesis requires testing through some research strategy or procedure which the scientist must create and design so as to yield meaningful data pertinent to the hypothesis. Testing and retesting through observation and/or experimentation approve, modify, or reject the hypothesis, and indicate the direction for further study. Let us delve a little more deeply into each of these steps.

1. The problem or objective under consideration should be concise, unambiguous, specific, and in sharp focus. A poorly defined, indefinite, and fuzzy problem is either no problem at all or several unsegregated problems. An objective too broadly stated is actually several problems. We cannot plan on solving more than one problem at a time. Two kinds of

problems may be posed. The descriptive problem asks the question, "What is?" What birds are found on the Galapagos Islands? Here no hypothesis is involved. The dynamic problem asks the question, "Why?" Why is it that these particular birds are found only in the Galapagos Islands? In this case an hypothesis is called for.

2. At some stage of inquiry a search of the literature (library research) is made in order to determine what research has already been done which might bear directly or indirectly on the problem.
3. Based on whatever limited observations have been made, a tentative explanation (hypothesis) is designed to answer the how or why of the problem. At this juncture the law of parsimony is followed religiously. It is also at this point that the scientist joins the artist, the musician, and the poet in a common exercise, that of insight, intuition, imagination, and intellectual creativity. From a bewilderment of seemingly unrelated details he must discern a pattern, a common denominator, which, although unseen, is there.
4. After designing an hypothesis, the next step is equally creative. How can one test this idea? Nature does not answer questions unless they are put to her on her own terms. An organism is responsive only to its own needs. How can we so design a pattern of investigative procedures that some organism, in responding to its own needs, will also answer a meaningful question for us? What species of organism would give us the most straightforward responses? Two brilliant geneticists conducted studies that yielded results of inestimable value to our knowledge of life. A part of their brilliance was due to each man choosing a species that was beautifully fitted for the problem being investigated (Gregor Mendel chose peas, George Beadle a bread mold).
5. An hypothesis will be accepted, modified, or rejected on the basis of a considerable body of evidence and test data. These data must be organized and then compared with any similar information which might have resulted from library research. During this effort new questions almost inevitably emerge which also deserve answers. These questions indicate the direction of future study. The validity of the hypothesis and the nature of further study will hinge on the interpretation of this combined body of evidence. The two methods by which this evidence is secured originally are observation and experimentation.

Both methods may use sophisticated instruments that greatly enlarge the potential field of observation and lend increased precision to measurement. They include a diverse array of apparatus and processes such as the electron microscope and the telemetric recording of free-moving animals. The difference between the two methods is that experimentation manipulates or *controls* the theoretically possible causes or modifiers (called *variables*) of an event, whereas observation does not. A scientific experiment is not at all synonymous with trial-and-error discovery. We may compare these two approaches by referring to an hypothesis cited on page 9, namely, that some fish use the sense of smell to guide them during migration. Destroying the sense of smell surgically *controls* (removes) one of the *variables* (a possible cause, smell). It also adds a variable, surgery. If the fish still orient themselves as if on a migratory run, the sense of smell is eliminated as a cause or means of orientation. If they are disoriented, it is a consequence either of the inability to smell or of surgery. The behavior of certain fish on which similar surgery was performed but whose sensory tracts were left intact would tell the tale.

An observational approach to this same problem might entail placing very small amounts of an organic compound in the water when young fish are being fed; the substance would become associated with feeding. If observations over a period of several months showed that the fish (1) could remember this substance and (2) could detect it at infinitesimally small dilutions, it would imply that smell *could* be utilized in migration. The observational evidence in this case would not be nearly so compelling as the experimental evidence, but nonetheless it would be clearly supportive. Both observation and experimentation are excellent scientific approaches. Each has advantages and each has disadvantages. Experimentation is more restricted in its application (for example, we cannot study extinct organisms in this way); on the other hand, in those many areas in which it can be used it often gives more reliable results because of the increased element of control involved. In many instances experiment is the only means by which fully reliable results can be obtained in the disproof of an hypothesis.

In the interest of stimulating a more intimate appreciation of the scientific approach to problem solving, two actual studies will be described, one observational, the other experimental in nature. Both are concerned with the same problem, animal behavior. In neither case is the procedural approach obscured by the mechanical routines required by elaborate instrumentation or laboratory techniques. We trust the reader will observe, among other things, that there is no single, stereotyped, procedural method used in the solution of scientific problems; methods are as varied and versatile as the imagination of the investigator.

A Case Study in Scientific Observation

A number of investigators, including Adolph Seilacher[1] and Rudolph Richter, have conducted studies of fossil animal burrows found in sedimentary rocks. These rocks were once the unconsolidated silts and sands of ancient ocean bottoms. No actual fossils of the animals that built these trails have been found within the burrows. That these bending, turning mounds and depressions (see Fig. 1-4a-c) had been made by burrowing sediment-feeders (an earthworm is a sediment-feeder) had been established through several observations.

1. The geometric shapes of the trails were often similar to those made by burrowing sediment-feeders today.
2. Burrowing animals disturb the silt or sand immediately adjacent to them; many fossil burrows reflect such a disturbance.
3. Many burrowing forms today have a sticky tonguelike structure to which food and fine sediment adhere; this selected, fine-grained material passes through the animal and is voided in the burrow. The filling in many fossil burrows is similarly different from the material around it.
4. One investigator who was studying living sausage worms found a fossil burrow which was not only constructed identically to that of the modern worm but contained a fossil of the same crab that lives in the sausage worm's burrow today.

[1] Adolph Seilacher, "Fossil Behavior," *Scientific American*, August 1967, 72–80.

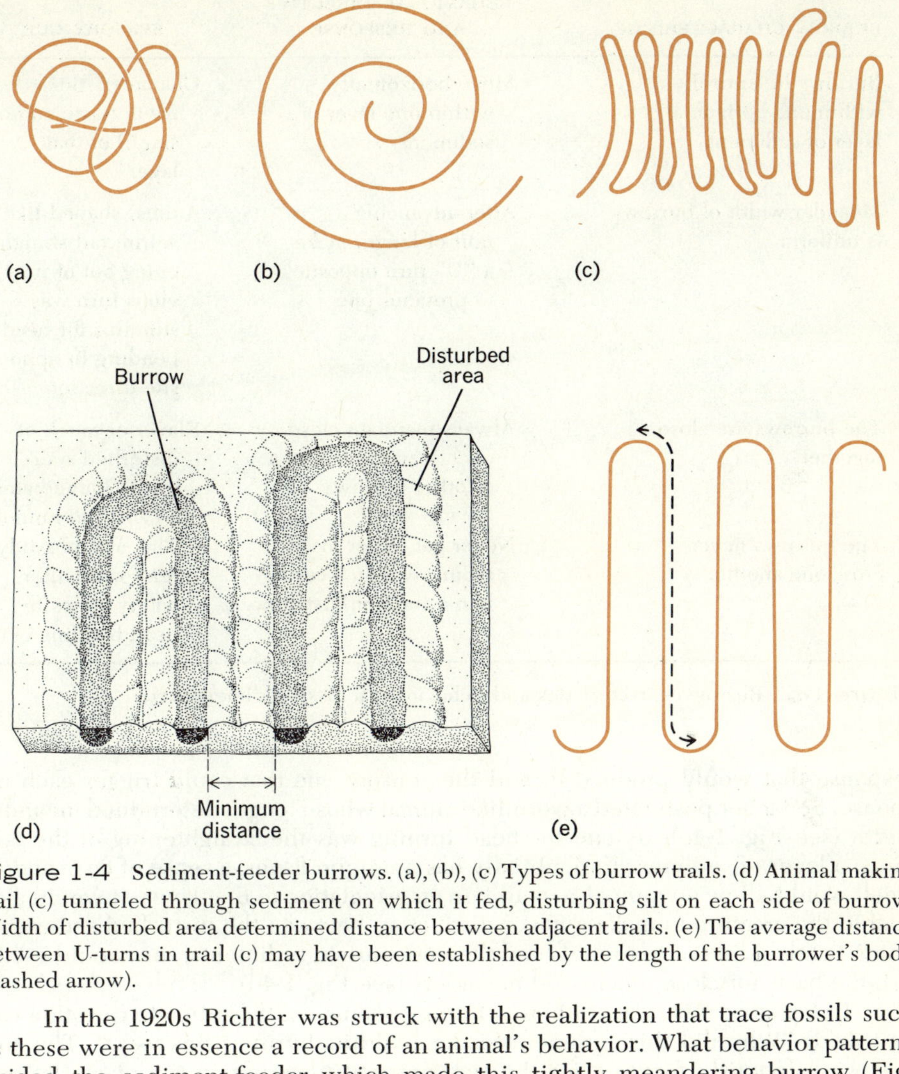

Figure 1-4 Sediment-feeder burrows. (a), (b), (c) Types of burrow trails. (d) Animal making trail (c) tunneled through sediment on which it fed, disturbing silt on each side of burrow. Width of disturbed area determined distance between adjacent trails. (e) The average distance between U-turns in trail (c) may have been established by the length of the burrower's body (dashed arrow).

In the 1920s Richter was struck with the realization that trace fossils such as these were in essence a record of an animal's behavior. What behavior patterns guided the sediment-feeder which made this tightly meandering burrow (Fig. 1-4c)? This is a rather formidable question when one realizes that the animal that made it (1) lived 100,000,000 years ago and (2) has never been seen (at least knowingly). Through careful analysis and measurements of a number of such burrows Richter discovered that they all had four fundamental characteristics in common (see Fig. 1-5). He reasoned that the tunneling organisms responsible for these burrows must therefore have had a repertory of four behavioral directives and responses that corresponded to, and would produce, the four burrowing characteristics. The next logical step was the answer to the question: "What kind of a response could we normally expect an animal to make that would result in each of these?" To this Seilacher added one final step: "What sensory clues could we normally expect the animal to receive that would evoke the proper response?" Figure 1-5 depicts the interrelationships between each tunneling feature, the

BURROW CHARACTERISTIC	BEHAVIORAL DIRECTIVE AND RESPONSE	SENSORY CUE
1. Burrows lie entirely within one horizontal layer of sediment	Move horizontally within one layer of sediment	Characteristic flavor (taste and/or smell) of that layer
2. Meander width of burrows is uniform	After advancing a unit of length make a "U" turn opposite to previous one	Animal shaped like a worm; tail straightening out of previous turn was stimulus for head bending in opposite direction
3. The burrows are close together	Always maintain close contact with your own or other burrows	Flavor or touch of disturbed sediments immediately adjacent to tunnels.
4. The burrows never cross one another	Never get closer to a burrow than a certain minimum distance	This also possibly reinforced alternately opposite head bending in (2)

Figure 1-5 Burrow characteristics and behavior of a fossil sediment-feeder.

response that would produce it, and the sensory cue that could trigger each response. Seilacher postulated a wormlike animal whose length determined meander width (see Fig. 1-4e); its cue for head turning was the straightening of the tail. Since a layer of sediment undoubtedly has a unique flavor, a sense of taste and/or smell would allow an animal to maintain an orientation to that layer. It would also enable the creature to make a flavor distinction between the disturbed sediments immediately adjacent to tunnels and the undisturbed sediments; this would establish the basis for close, uncrossed meanders (see Fig. 1-4d).

In bringing this case study to a fitting conclusion, the ultimate question can be asked: "Why did this sediment-feeder build its burrows this way?" The animal was obviously foraging for the rich nutrient materials in the ocean bottom "ooze." But why the tight meandering trail? An examination of, and comparisons with, other kinds of trails might throw light on this question. One kind of trail described a series of random, overlapping, and open circles (see Fig. 1-4a). This type of foraging pathway is found in sedimentary rocks that are 400 to 600 million years old; in fact, it is the only kind of foraging trail that has ever been found in rocks of this age. A second type of feeding trail described an open, widening spiral (see Fig. 1-4b). This, together with an open, meandering form, were the common types of pathways in rocks about 300 million years old. The circling pathways have never been found in rocks of this, or a more recent age. Tight spiraling and meandering trailways (Fig. 1-4c) are not found in rocks older than 135 million years. It would appear that tunneling behavior underwent an evolutionary change from open, random circles through open meanders and spirals to tight meanders and spirals. Why? The answer to this must be in terms of the reason for foraging in the first place—to obtain food. It is readily apparent, on comparison of the three levels of tunneling techniques, that feeding efficiency is progressively increasing. The random circles leave large areas of unexploited sediment. The open spirals, although yielding more systematic and complete coverage, still leave untouched strips between the spirals. Tight spirals and meanders (some of which even become three dimensional) almost completely exploit the food source. Once again we see the fruitful conquest of life for energy (food). Nature smiles with favor on those sediment-feeders who can come up with a new and better idea.

A Case Study in Scientific Experimentation

The entire group of birds known as gulls are remarkably similar. The external traits that distinguish different but related species are not at all marked. That they are different is attested to by the fact that they do not interbreed, even when different species are nesting along side one another. Under such conditions how is interbreeding prevented? More succinctly, what maintains the biologic isolation of these intermingling species?

Iris color:	Yellow	Yellow	Dark brown	Light medium brown
Eye-ring color:	Orange	Bright yellow	Reddish purple	Reddish purple
	Species (a) (Herring gull)	Species (b) (Glaucous gull)	Species (c) (Thayer's gull)	Species (d) (Kumlien's gull)

(a)

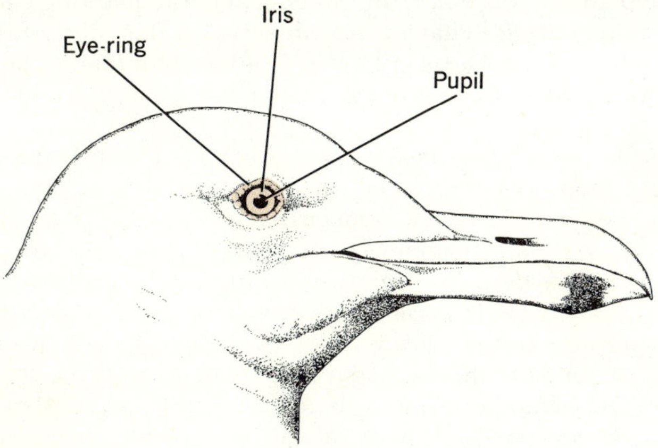

(b)

Figure 1-6 External characteristics of four gulls which nest in the Canadian Arctic. (a) Differences in wing and eye color. (b) Details of eye coloring in species (b) (Glaucous gull).

Neal Griffith Smith investigated this question in a study of four species of gulls that shared nesting grounds in the Canadian Arctic.[2] All are white-bodied, white-headed, and gray-backed. Aside from slight differences in size, they are distinguished by (1) wing tip pattern, (2) intensity of grayness on the back, and (3) the color of the iris of the eye and the eye-ring (Fig. 1-6a). The eye-ring is a narrow, naked strip of flesh immediately adjacent to the eye itself (Fig. 1-6b). Smith adopted an hypothesis suggested earlier by a Danish biologist (Salomonsen, 1950) that the color of the eye-ring was an important feature used by gulls in maintaining their separate breeding autonomies. In a preliminary study he selected a mixed colony of gulls (species a, b, and d) in which mate selection had already occurred. To capture the birds he inserted a capsule containing a drug into pieces of meat. On eating the meat the birds became quickly anesthetized. Once a bird was immobilized its eye-ring color was artificially altered by the judicious application of an oil paint. Colored bands were put on the legs so that individuals could be recognized later.

All females in the colony had copulated with males previous to capture but none had laid any eggs. A small number of these birds (63) were captured. The eye-rings of each of them were changed to resemble that of another species. On release all birds returned to, and were accepted by, their original mates, with these interesting results.

1. Unpainted males that had not been captured no longer mounted any females that had been painted, even though the females were solicitous of this attention; such pairs ultimately broke up and the females left the breeding colony.
2. Five of the above males mated with unpaired females in adjacent areas, with whom they copulated.
3. Unaltered females accepted painted males; copulatory activity was normal.
4. In one pair in which both birds had been painted the behavior was identical to that (1) in which only the female had been painted.

It did seem that an hypothesis was emerging which had merit: eye-ring color in females somehow affects mounting behavior in males. Was it the color of the eye-ring itself or its color value (the darkness or lightness of the color hue)? Could it be the color-value contrast between the entire eye (with eye-ring) and the white head?

To explore these questions the eye-rings of females of each of the four species were painted with light colors (white, yellow, or orange) and dark colors (red or black). It appears from Figure 1-7, which summarizes the results, that color values were being responded to rather than color *per se*, except for species (c). In this species the entire eye, with its dark iris, stands in contrast to the bird's white head, not just the eye-ring alone. At this point Smith took an ingenious step. He painted over the reddish purple eye-ring of a number of species (c) individuals with white to reduce contrast. Outside this he painted the feathers with a "super-eye-ring," using the same combination of colors as before. A significant number of females with light-colored "super-eye-rings" were unable to copulate with males of their own species.

[2] Neal Griffith Smith, "Visual Isolation in Gulls," *Scientific American,* October 1967, 94–102.

SPECIES	COLOR OF IRIS	NORMAL EYE-RING COLOR	EYE-RING PAINTED WHITE, YELLOW, OR ORANGE[a]	EYE-RING PAINTED RED OR BLACK[a]
(a)	Yellow	Orange	Behavior normal	Copulation usually stopped; pairs broken
(b)	Yellow	Bright yellow	Behavior normal except those painted orange	Copulation usually stopped; pairs broken
(d)	Light to medium brown	Reddish purple	Copulation usually stopped; pairs broken (white most effective)	Behavior normal
(c)	Dark brown	Reddish purple	Behavior normal	Behavior normal

[a] Females only.

Figure 1-7 Effect of eye to head color-value contrast on mating behavior.

The study was now ready to begin in earnest. It did appear that mated males selected females with which they would breed on the basis of eye-head contrasts similar to their own. But how did pair formation occur in the first place? Who chose whom in this case? Is eye-head contrast also important here? The preliminary studies had also introduced a number of variables whose possible effects had not been considered: the animals were (1) handled, (2) drugged, and (3) painted. Continued experimentation therefore must include three control groups, respectively: (1) uncaptured gulls whose behavior is observed, (2) gulls that are anesthetized but not painted, and (3) gulls that are anesthetized and painted with their own color.

These final stages of experimentation are briefly summarized in Figures 1-8 and 9. The data in Figure 1-8 show that the female makes the original choice in pair formation, and an important criterion is eye-head contrast like her own. The data in Figure 1-9 corroborate the results of the preliminary studies. Smith's origi-

SPECIES	UNMATED MALES		REJECTION BY FEMALES OF SAME SPECIES	
	Experimental Treatment	*Number*	*Number*	*Percent*
Light-eyed	Painted dark	112	87	77.6
	Painted light	121	15	12.3
	Drugged only	77	5	6.4
Dark-eyed	Painted light	170	124	72.9
	Painted dark	158	23	14.5
	Drugged only	111	9	8.1

Figure 1-8 Pair formation through female choice of proper eye-head contrast in the male.

SPECIES	MATED FEMALES		COPULATION STOPPED; PAIRS BROKEN	
	Experimental Treatment	*Number*	*Number*	*Percent*
Light-eyed	Painted dark	173	132	76.3
	Painted light	163	10	6.1
	Drugged only	71	5	7.0
	Observed, not caught	93	2	2.1
Dark-eyed	Painted light	389	222	57.0
	Painted dark	227	14	6.1
	Drugged only	134	6	4.4
	Observed, not caught	204	12	5.8
	MATED MALES			
Light-eyed	Painted dark	164	4	2.5
Dark-eyed	Painted light	51	3	6.0

Figure 1-9 Copulatory behavior through male choice of proper eye-head contrast in the female.

nal question is answered. Four species of gulls in the Canadian Arctic maintain their separate breeding autonomies in two ways, each of which reinforces the other.

1. Females chose male mates that look like themselves.
2. Mated males will breed only with similar looking females.

An important external feature used in this self-recognition is eye-head contrast. In all scientific inquiry new questions arise as old ones are answered. This one is no exception. For example, how does a gull know what it looks like?

QUESTIONS

1. In what ways is a scientist's concept of truth and reality biological in nature?
2. Why are scientific truths held to be relative rather than absolute?
3. It is sometimes said that a scientist cannot objectively study purposive causes which answer the question "Why?" Discuss the pros and cons of this issue.
4. Knowing the causes for a given natural phenomenon does not necessarily enable one to make a reliable prediction about the occurrence or character of this phenomenon at some specific time in the future. Why is this?
5. Adolph Seilacher[3] has programmed a computer according to the behavior patterns he and Richter postulated for sediment-feeder burrowing. As can be seen in the illustration on the opposite page, the computer-simulated meanders (right) are remarkably similar to the actual burrow (left). Would you consider this computer analysis to be an observational or an experimental approach? Why?

[3] David M. Raup and Adolph Seilacher, "Fossil Foraging Behavior: Computer Simulation," *Science*, **166**, November 21, 1969, 994–995.

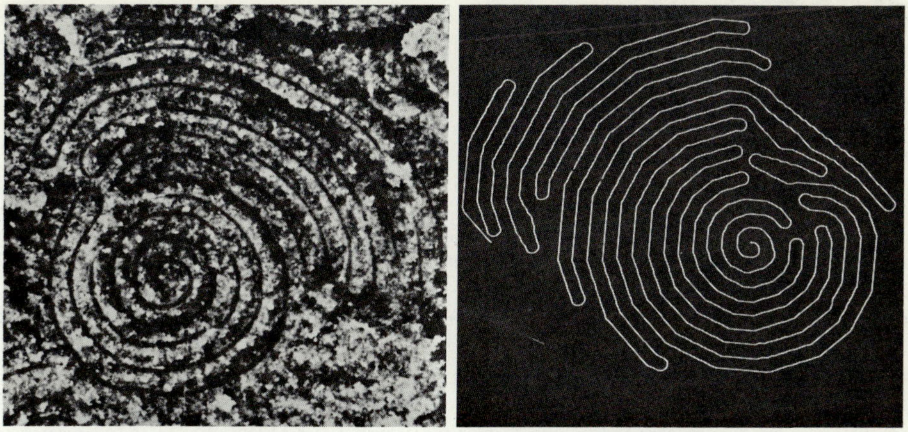

Suggested Readings

Blackburn, R. T. (ed.). *Interrelations: The Biological and Physical Sciences.* Chicago: Scott, Foresman, 1966.

Mayr, E. "Cause and Effect in Biology," *Science,* **134,** November 10, 1961.

Raup, D. M., and A. Seilacher, "Fossil Foraging Behavior: Computer Simulation," *Science,* **166,** November 21, 1969.

Seilacher, A. "Fossil Behavior," *Scientific American,* **217,** August 1967.

Simpson, G. G. "Biology and the Nature of Science," *Science,* **139,** January 11, 1963.

Smith, N. G. "Visual Isolation in Gulls," *Scientific American,* October 1967.

Weaver, W. "Scientific Explanation," *Science,* **143,** March 20, 1964.

CHAPTER 2 Matter, Energy, and Form

Our discussion thus far has attempted to elucidate the nature of science as a system of principles for explaining natural phenomena. The system is logically consistent within itself; its criteria for truth validate the premises on which it stands. Let us now turn our emphasis from the system itself to one of its fundamental premises: the world is real and we can, through observation, continue to learn more and more about it. The scientific system, as we all know, has been turned loose on all manner of things, from plants, animals, and microbes to rocks, air, oceans, and planets. What are these things really? What do they have in common? What is their smallest common denominator? How is this smallest common denominator organized and arranged so that it makes air in one case, a kind of rock in another, and a typhoid bacillus in yet another? *What is matter anyway?*

THE UNITARY ATOM: MATTER IS ENERGY, ORGANIZED

All matter of whatever kind anywhere is composed of combinations of single, unmixed substances called *elements*. Elements cannot be broken down into simpler substances because there are none. There are 92 different kinds which occur naturally. Carbon, oxygen, hydrogen, and nitrogen are familiar ones; they also make up the greater part of living things. Figure 2-1 lists those found in living organisms including their chemical "shorthand" symbols and rough approximations of their relative abundance.

An *atom* is the most minute unit into which an element can be divided. Dividing such a unit further amounts to atomic fission; elemental matter is converted into its component "parts," namely, forms of radiative energy (ionizing radiation). How does energy make matter? Physicists tell us that the three most common forms of energy in the atom often behave as if they were particles. They are consequently known as *elementary particles* ("hot-spots" of energy, as it were, *not* pieces of matter). Two of these particles occupy a central position in the atom known as the nucleus (see Fig. 2-2, carbon). The first of these, the *proton*, has a constant mass or weight, which is given an arbitrary value of 1. It occupies scarcely any space and carries one positive charge. The second, the *neutron*, has the same mass (1) and volume but carries no charge. The third particle is called an *electron*. It may be thought of as spinning at tremendous speeds in an orbit (or shell) around, and at a great relative distance from, the nucleus. The electron has a very small mass (about one two-thousandth that of a proton) but occupies a large volume. It carries a negative charge equal to, but opposite that of, the proton.

The 92 elements differ from one another by the number of protons in the nucleus (from one proton in hydrogen to 92 protons in uranium). Except in the case of hydrogen which has no neutron, the number of neutrons in the nucleus is usually about the same as the number of protons. The sum of neutrons and protons in an atom is its *atomic weight*. There is one electron for every proton. Every element therefore is electrically neutral. Electrons are found in from one to seven

NAME	SYMBOL	RELATIVE ABUNDANCE
Calcium	Ca	+
Carbon	C	++
Cobalt	Co	
Copper	Cu	
Chlorine	Cl	+
Hydrogen	H	++
Iodine	I	
Iron	Fe	
Magnesium	Mg	+
Manganese	Mn	
Nitrogen	N	++
Oxygen	O	++
Potassium	K	+
Phosphorus	P	+
Silicon	Si	
Sodium	Na	+
Sulfur	S	+
Zinc	Zn	

Figure 2-1 Elements found in living organisms.

shells or orbits depending on the element; hydrogen has only one of them, whereas uranium has all seven. The first shell holds a maximum of two electrons; the second holds eight. The third to the sixth electron orbits may hold more than eight electrons; the seventh has no more than two. Figure 2-2 diagrams these relationships for several elements.

One final point should be made. Certain elements have atomic "cousins" called *isotopes*. Carbon-12 and carbon-14 (Fig. 2-2) are good examples. They are chemically identical since they both possess six protons and six electrons. Carbon-14 has two additional neutrons in its nucleus. This makes it heavier. It also makes it *radioactive* since too many extra neutrons tend to decay through the emission of nuclear energy (particles) back to the stable state (in this case, carbon-12). The rate of this decay is quite constant for any given isotope. Because of these characteristics radioactive isotopes have proven very useful in biological studies in two ways: (1) the pathways that certain substances follow in the living organism can be traced by monitoring the energy emissions of radioactive isotopes of that substance and (2) the age of prehistoric organisms up to 40,000 years old can be accurately determined by comparing the proportion of carbon-14 in fossil remains with that found in comparable living types.

In conclusion be assured that our purpose has been only to establish a basis for understanding that matter (substance—the "stuff" of science) is the first level of organization above energy. The memorization of the atomic structure of this or that element is of no consequence to us here except as it contributes to that purpose. An element cannot be understood in terms of "elementary particles" until those particles become arranged in a specific pattern known as atomic structure. Matter is energy, *organized.*

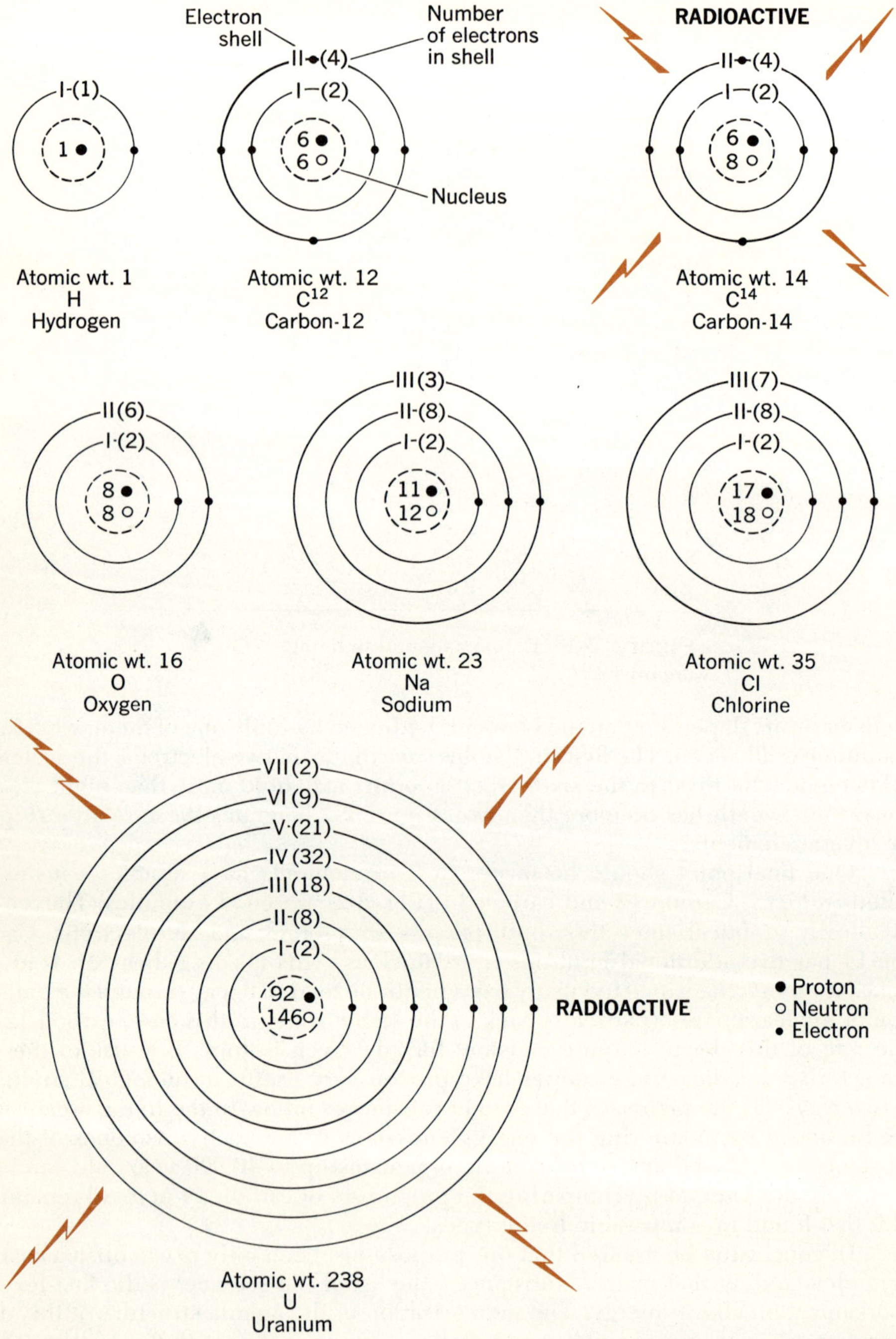

Figure 2-2 Atomic structure of six representative elements. A radioactive isotope of carbon is also shown.

THE MOLECULAR LEVEL OF ORGANIZATION

Elements are most commonly found in nature in combination with one another. The number of different kinds of these chemical unions is exceedingly large. The size and complexity of many such combinations are remarkably sophisticated; the graceful geometry of uncomplicated forms, on the other hand, is artful in its simplicity. For obvious reasons we shall take our examples from the latter. We have already mentioned the combination of sodium and chlorine to form sodium chloride or table salt (page 13). The union of hydrogen and oxygen to make water is another familiar example. The association of two or more atoms in such a fashion forms a new kind of unit, at a higher organizational level, called a *molecule.* The substance so produced is called a *compound.* A molecule is the most minute unit into which it can be divided (with certain exceptions which we will soon detail). A molecule is to a molecular compound as an atom is to an element.

How Molecules Are Held Together: Chemical Bonds

Since a compound is a stable combination of elements, there must be binding forces of some kind which maintain the union. These connective energies are known as *chemical bonds,* of which there are several types. All of them depend directly or indirectly on two important principles: (1) only the outer electron shell of an element determines its chemical reactivity ("combining behavior"); and (2) stable combinations are reached when the outer orbits of the participating atoms contain eight electrons. The lone exceptions to the second principle are those elements (hydrogen and helium) whose outer shell is their only shell (No. 1); in this case stability is achieved with two electrons in the orbit. Helium has two electrons in the atomic state; it is, as a result, chemically inert.

IONIC BONDS AND COMPOUNDS

An *ionic bond* is established when an electron is transferred from one atom (the electron donor) to another (the electron acceptor). Such a case is shown in the compound, sodium chloride (see Fig. 2-3). Sodium (Na) is said to be strongly *electropositive:* it has but one electron in its outer orbit, which it donates readily. Chlorine (Cl), on the other hand, lacks but one electron to complete the eight required: it is strongly *electronegative* and accepts electrons readily. When sodium donates its electron, its stable second shell becomes "exposed." In accepting this electron chlorine's outer orbit (No. 3) becomes complete. In the *electron transfer* from sodium to chlorine, sodium now has a proton with no corresponding electron and bears one positive charge. In this state it is known as a sodium *ion* and is designated Na^+. Similarly, chlorine now has an electron with no corresponding proton. The chloride ion consequently bears a negative charge and is designated Cl^-. The attraction of opposite charges is the ionic bond which maintains the integrity of the combination. Sodium and chloride ions are not the same as atoms because they are electronically unbalanced. We cannot properly speak of molecules, therefore, when referring to these *ionic compounds.* This is particularly evident when ionic compounds are dissolved in water. In such aqueous solutions many ionic bonds are broken, with the result that the ions become *dissociated* as free individual entities:

$$NaCl + \text{water} \longrightarrow Na^+ + Cl^- + \text{water}$$

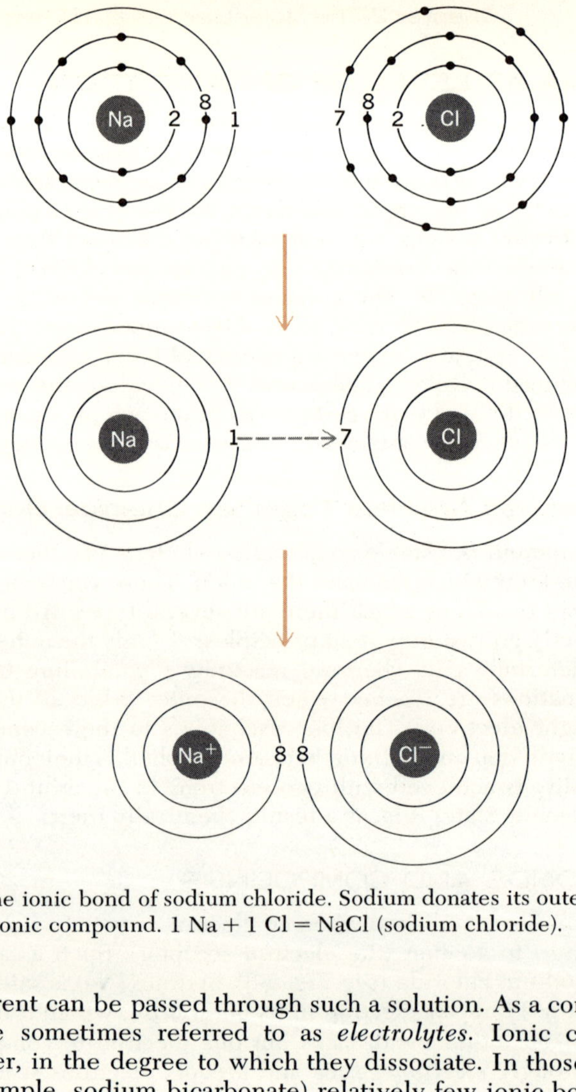

Figure 2-3 The ionic bond of sodium chloride. Sodium donates its outer electron to chlorine, forming an ionic compound. 1 Na + 1 Cl = NaCl (sodium chloride).

An electric current can be passed through such a solution. As a consequence ionic compounds are sometimes referred to as *electrolytes*. Ionic compounds vary greatly, however, in the degree to which they dissociate. In those that dissociate weakly (for example, sodium bicarbonate) relatively few ionic bonds are broken; as a result only a few ions are free at any one time.

In sodium chloride it was noted that sodium was the donor of one electron in its combining action with chloride. This combining value of an atom is called its *valence*. In the case of sodium it is one positive. In the case of chlorine, which accepts sodium's one electron, it is one negative. One sodium atom "satisfies" one chlorine atom. This chemical courtship is expressed by the *formula* NaCl. Two or more elements commonly combine as electron donors or acceptors. Such chemical affiliations are known as *radicals*. As ions they behave as if they were composed of single atoms. Figure 2-4 lists several common radicals, with their valences and formulas. By using valence tables of this kind it is relatively easy to figure out the chemical formulas of ionic compounds. Sodium hydroxide would be NaOH. Ammonium chloride is NH_4Cl. Ammonium carbonate becomes $(NH_4)_2CO_3$. In the latter case the carbonate radical accepts two electrons (see Fig. 2-4); it will therefore accommodate the single donations of *two* ammonium radicals (a chemical polyandry of sorts).

			VALENCE	
			NO. OF ELECTRONS	
NAME	ATOMS COMBINED	FORMULA	DONATED (+)	ACCEPTED (−)
Ammonium	1 nitrogen:4 hydrogen	NH_4^+	1	
Bicarbonate	1 hydrogen:1 carbon:3 oxygen	HCO_3^-		1
Carbonate	1 carbon:3 oxygen	CO_3^{--}		2
Hydroxyl	1 oxygen:1 hydrogen	OH^-		1
Phosphate	1 phosphorus:4 oxygen	PO_4^{---}		3
Nitrate	1 nitrogen:3 oxygen	NO_3^-		1
Sulfate	1 sulfur:4 oxygen	SO_4^{--}		2

Figure 2-4 Common chemical radicals.

Ionic compounds fall into three important groups of substances: *acids*,[1] *bases*, and *salts*. An *acid* is any compound which will yield hydrogen ions as it dissociates. For example,

$$H_2SO_4 + \text{water} \longrightarrow 2H^+ + SO_4^{--} + \text{water}$$

Each of two hydrogen atoms donated their lone electron to the sulfate radical. This freed two hydrogen ions. A hydrogen ion is a hydrogen nucleus—one proton (see Fig. 2-2). A *base* (or *alkali*) is any compound which will yield hydroxyl ions as it dissociates:

$$NaOH + \text{water} \longrightarrow Na^+ + OH^- + \text{water}$$

A *salt* is any compound which will yield positive ions other than those of hydrogen, and negative ions other than hydroxyl as it dissociates. (It is also a compound resulting from the chemical reaction of an acid and base; we will discuss this later.) Ammonium chloride is such a substance:

$$NH_4Cl + \text{water} \longrightarrow NH_4^+ + Cl^- + \text{water}$$

The "strength" or "weakness" of various acids, bases, and salts is dependent on the degree to which they dissociate. A strong acid dissociates almost completely in aqueous solutions. Weak acids dissociate feebly, releasing but small numbers of free hydrogen ions.

MOLECULES AND MOLECULAR COMPOUNDS

Molecular compounds differ from the foregoing in that the atoms from which they are composed are united by *covalent* bonds: electrons are *shared* rather than transferred. These compounds are made of molecules of combined atoms, not latticelike arrays of ions (see Fig. 2-5). One of the water molecules in Figure 2-5 is detailed further in Figure 2-6. Oxygen, with six electrons, requires two more in its outer "ring." Hydrogen, with one, needs one more (page 27). Two hydrogen atoms share their two electrons with oxygen as oxygen shares two of its six electrons with the two hydrogens. Thus two atoms of hydrogen and one of oxygen form a true molecule of water (H_2O):

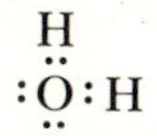

[1] Acids are ionic compounds only when in solution.

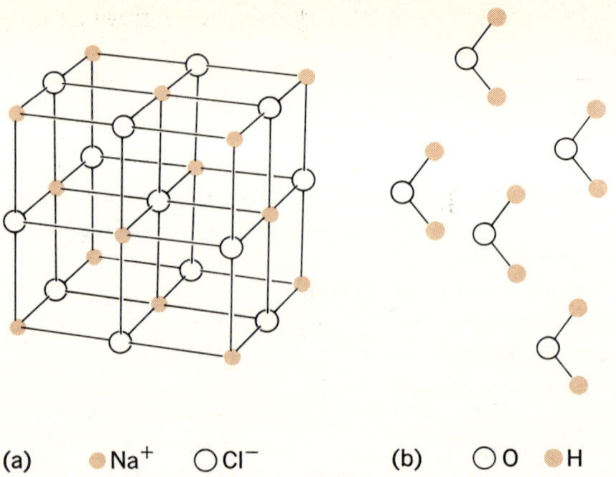

Figure 2-5 Structural difference between ionic and molecular compounds. (a) NaCl—an ionic compound. A sodium ion is not bonded to a particular chloride ion. This accounts for the crystalline structure of ionic compounds. (b) H_2O—a molecular compound. Two hydrogen atoms are bonded to one oxygen atom.

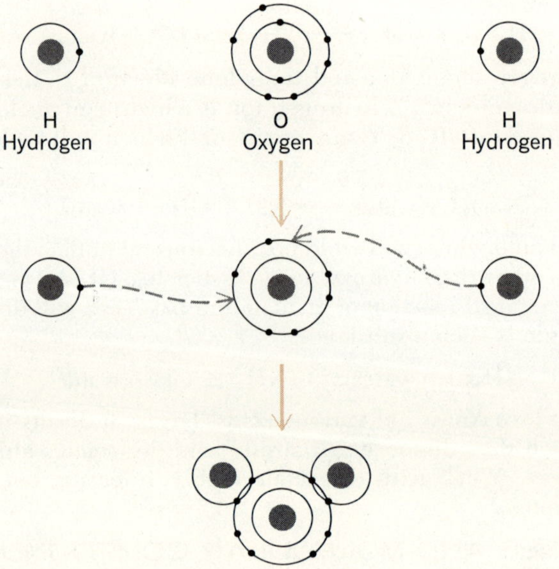

Figure 2-6 Covalent (shared) bonds of the water molecule. Each of the electrons of two hydrogen atoms is shared with one oxygen atom. $2\ H + 1\ O = H_2O$.

In a similar fashion carbon forms covalent bonds in the formation of the gas, methane (CH_4). Carbon, with a valence of four, is neither electronegative nor electropositive. As a result it shares all four of its electrons (Fig. 2-2), in this case with hydrogen:

$$\begin{array}{c} H \\ \cdot\cdot \\ H:\!C\!:H \\ \cdot\cdot \\ H \end{array}$$

It may also share its electrons with electronegative and electropositive atoms as well as other carbon atoms. This is nicely shown in a molecule of acetic acid (vinegar):

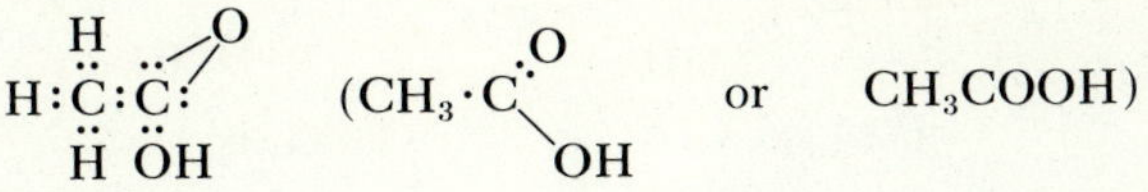

In the methyl group (CH_3^- on the left) carbon bonds to hydrogen and carbon. In the *carboxyl* group (—COOH) carbon bonds to carbon, oxygen, and a hydroxyl group. Any compound containing a carboxyl group will dissociate in water, releasing hydrogen ions:

$$CH_3COOH + H_2O \longrightarrow CH_3COO^- + H^+ + H_2O$$

A third type of bond is of considerable significance to living systems and involves valence only indirectly. It is known as a *hydrogen bond* and serves to link two or more molecules together. It is a weak, "long-distance" bond with about one-twentieth the strength of a covalent bond. It arises when an "end" of one molecule has a strongly electropositive hydrogen atom, and an "end" of another (or the same) molecule has a strongly electronegative oxygen or nitrogen atom. Two such electronegative atoms may be drawn closer together by the intervening proton of hydrogen acting as a bridge. Molecules of water behave in this fashion, as if they were minute bar magnets (Fig. 2-7). The hydrogen "end" is a positive pole,

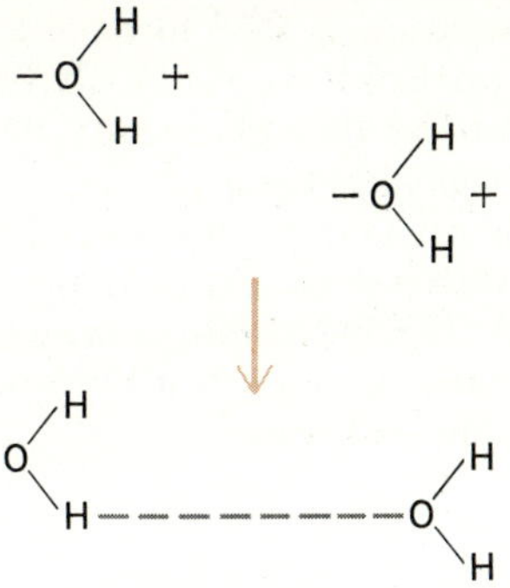

Figure 2-7 Hydrogen bond linking water molecules.

whereas the oxygen "end" is a negative pole. Consequently, water molecules become aligned by hydrogen bonds into molecular complexes. In the same way complex molecules of fatty acids become double molecules by forming hydrogen bonds at their carboxyl ends:

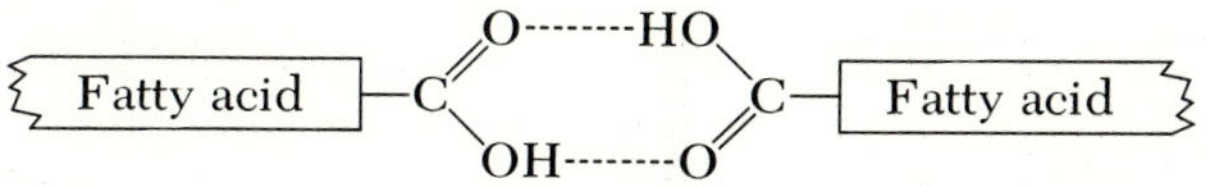

The Interaction of Compounds: Chemical Reactions

The interaction of two or more compounds is called a *chemical reaction*. In an aqueous (water) solution of hydrochloric acid (HCl) and sodium hydroxide

(NaOH) the following reaction takes place:

$$\begin{array}{c} Cl^- \diagup \\ H^+ \diagup \end{array} + \begin{array}{c} \diagdown Na^+ \\ \diagdown OH^- \end{array} \longrightarrow \begin{array}{c} Na^+Cl^- \\ + \\ HOH \end{array}$$

Reactants Products

This is a substitution reaction in which the H^+ ions of the acid and the OH^- ions of the base neutralize one another in the formation of molecules of water. If the water is evaporated away, crystals of the salt, NaCl, will be produced. If the water is not evaporated away, there will be a slight tendency for the reaction to reverse since water dissociates very slightly into H^+ and OH^- ions. It would be more appropriate to show our reaction thus:

$$\begin{array}{c} Cl^- \diagup \\ H^+ \diagup \end{array} + \begin{array}{c} \diagdown Na^+ \\ \diagdown OH^- \end{array} \rightleftarrows \begin{array}{c} Na^+Cl^- \\ + \\ HOH \end{array}$$

The reaction will appear to stop at an equilibrium predicated on a certain ratio of total reactants to total products.

This phenomenon of reaction equilibria is beautifully illustrated in living organisms in *buffering* reactions. Biochemical reactions are very sensitive to changes in hydrogen ion concentration. Most "living processes" are favored by slightly alkaline conditions, yet many of these same processes produce hydrogen ions. Perfect neutrality of H^+ and OH^- ions is found in water where the numbers of each are equal. Such a condition is said to have a hydrogen ion concentration or pH of 7. A solution with a pH of 8 would be slightly basic (slightly more OH^- than H^+ ions); a pH of 14 would be strongly basic (OH^- ions greatly outnumbering H^+ ions. Similarly, a pH of 6 is slightly acidic and a pH of zero strongly acidic. Biochemical processes commonly occur within a pH range of 7.3 to 7.5. Two buffering compounds that are utilized in controlling pH are sodium bicarbonate ($NaHCO_3$) and carbonic acid (H_2CO_3), both of which dissociate weakly. The bicarbonate dissociates a few OH^- ions with a by-product of carbonic acid which decomposes into carbon dioxide and water:

(1) (a) $$\begin{array}{ccc} HCO_3^- + H & \longrightarrow & + H_2CO_3 \\ Na^+ \qquad\qquad OH & & Na^+ + OH^- \end{array}$$

(b) $$\begin{array}{ccc} COO \ (H_2CO_3) & \longrightarrow & + CO_2 \\ H_2 \qquad O & & H_2O \end{array}$$ (as in a carbonated beverage)

It is therefore mildly alkaline in reaction and most of it remains in an undissociated state. If the hydroxyl ions are removed by the presence of a strong acid, the dissociation equilibrium is disturbed and additional bicarbonate dissociates:

(2) $$\begin{array}{ccc} HCO_3^- + H^+ & \longrightarrow & + H_2CO_3 \\ Na^+ \qquad\qquad Cl^- & & Na^+ + Cl^- \end{array}$$

Reactants continue to yield products until the dissociation reaction (1a) is again reached. The sodium bicarbonate acts like a hydrogen ion sponge. In a similar fashion carbonic acid buffers against excessive hydroxyl ions. Carbonic acid commonly decomposes into CO_2 and H_2O (reaction 1b). Small amounts, however, do dissociate into H^+ and HCO_3^- (bicarbonate) ions. If these H^+ ions are removed by a strong base, more of the carbonic acid dissociates:

$$(3) \qquad HCO_3^- + Na^+ \longrightarrow NaHCO_3 +$$
$$H^+ \qquad\qquad OH^- \qquad\qquad H_2O$$

The carbonic acid "soaks up" the OH^- ions. It should also be noted that both of these reactions (2 and 3) reinforce one another. Acid buffering by sodium bicarbonate produces carbonic acid; base buffering by carbonic acid produces bicarbonate.

CHEMICAL ENERGY

The foregoing introduction to the nature of chemical reaction dealt with a class of equations involving chemical substitution. We noted that reactants undergo a reassortment of their chemically active groups to yield new combinations in the products of reaction. Reactions of the sort used to illustrate these principles belong to a group known as *ion-exchange reactions.*

There is another type of chemical interaction in which *electron transfer* takes place. These are called *oxidation-reduction* or *redox reactions.* They are very important to biological systems because they entail the storage and release of chemical energy. This form of energy is, in the last analysis, the only kind of energy on which living things ultimately rely. Green plants can use light energy only when it is transformed by them into the *bond energy* of the organic compounds they manufacture. Bond energy is chemical energy: the mutual forces of electron attraction displayed within ionic, covalent, and even hydrogen bonds. This energy enables plants to send roots into the ground and shoots into the air. It powers muscular contraction in man and rockets into space. Chemical compounds are parcels of stored energy. Some contain a great deal more stored, or potential, energy than others. In some instances the bond-participating electrons are at elevated energy levels or "excited states" and are loosely held by weakly electronegative atoms. Such compounds are relatively unstable and ready suppliers of electrons. At the other end of the spectrum are the relatively stable molecules containing strongly electronegative atoms such as oxygen which are ready acceptors and tenacious possessors of electrons.

Those reactions in which electrons are shifted from a less electronegative to a more electronegative atom are "downhill runs" so to speak: they release energy. Once initiated they are self-sustaining. Such reactions are said to be *exergonic* (*ergon* = energy; *ex* = from). Those reactions, on the other hand, which proceed in the opposite direction from more electronegative to less electronegative atoms are "uphill runs." They require an investment of energy sufficient to move electrons from an electron acceptor to an electron donor. Redox reactions that yield relatively unstable, energy-enriched products are *endergonic* (*ergon* = energy; *end* = in). Endergonic requirements for energy input in the living organism are commonly met by the energy output of exergonic reactions. The green plant, of course, uses the energy of sunlight for this purpose.[2]

A relatively simple example of an exergonic reaction is the oxidation of methane (CH_4). One molecule of methane and two molecules of oxygen yield one molecule of carbon dioxide, two molecules of water, and energy:

$$CH_4 + 2O_2 \longrightarrow CO_2 + 2H_2O + E \text{ (energy)}$$

[2] Is purpose a proper word to use here? How come?

Note first of all that this equation, like any chemical *equation, balances.* That is, there is an equal number of all the atoms represented on each side of the reaction arrow: one carbon, four hydrogen, and four oxygen atoms are on each side, but, of course, in different molecular combinations (an oxygen molecule consists of two atoms of oxygen). The liberated energy (heat and light), as one of the products, existed as potential energy in the methane molecule. As potential (bond) energy in a reactant it is not shown. The bond energy of the methane is released as active (kinetic) energy in the form of heat and light. This occurs because the four hydrogen atoms, each with one electron, move from the less electronegative carbon atom to the more electronegative oxygen atom, thus:

$$\begin{array}{lllll} \quad H_4 + O_2 \longrightarrow & & & + & 2H_2O \\ C \qquad O_2 & & CO_2 & & \end{array}$$

In other words, the four electrons carried by the hydrogen atoms are shifting from an atom with less affinity for electrons (carbon) to one that has a higher affinity for electrons (oxygen). This is the "downhill run" to which we have previously referred. It is as if a newly recycled hourglass had just begun to trickle sand into its empty bottom half. The energy used for lifting the filled half is being liberated as the sand sifts down once more into a stable state (Fig. 2-8). The energy of the reaction (*reaction energy*) is delivered to the environment in both the dropping sand and the heated methane. Carbon, as it loses the electrons furnished by hydrogen and gives them up to oxygen, is said to be *oxidized* by the oxygen. Oxygen, as it receives these same electrons from carbon, is *reduced* by carbon. When an atom loses electrons, it is oxidized; when an atom gains electrons, it is reduced. No atom can gain an electron unless some other atom loses an electron that is available for the gaining. Oxidation and reduction occur simultaneously and hence are called oxidation-reduction or redox reactions.

Carbon is oxidized *and* oxygen is reduced when methane burns. Oxygen acts as an *oxidizing agent* in the oxidation of carbon *at the same time* that carbon acts as a *reducing agent* in the reduction of oxygen. Carbon is a *hydrogen (and electron) donor* as oxygen is a *hydrogen (and electron) acceptor.* When a hydrogen atom passes from donor to acceptor in an exergonic reaction, its electron yields free (kinetic) energy in an amount equal to the difference in energy potential between donor and acceptor.

In clarifying the complementary roles played by oxidation and reduction in both exergonic and endergonic reactions, we can do no better than turn for guidance to the specialists and masters without peer in the "energy-transaction business," namely living organisms. The green plant programs a sophisticated and efficient sequence of redox reactions that begin with carbon dioxide and water. Through the judicious use of sunlight, plus other factors that operate in the dark, organic compounds are produced along with a by-product of oxygen. Although it

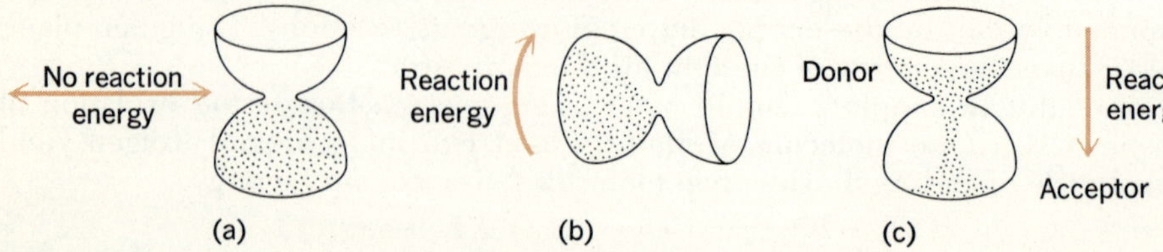

Figure 2-8 Hourglass analogy of energy reactions. (a) Complete stability (maximum entropy); (b) endergonic reaction; (c) exergonic reaction.

is a great oversimplification of what actually happens, we can summarize these events for our present purpose by a chemical equation of sorts:

$$CO_2 + H_2O + E \longrightarrow \text{organic compound} + O_2$$

This reaction does not take place as shown. The reactants to the left of the arrow are the beginning and the products to the right the end of a long chain of redox reactions. A more definitive discussion of these events is presented in a later chapter. In the interim an elucidation of the *kind* of reactions that are occurring here is germane at this point.

Let us suppose that the formula of the organic compound in our equation is CH_2O (it actually is a multiple of these proportions). Our hypothetical reaction, then, reads

$$CO_2 + H_2O + E \longrightarrow \text{“}CH_2O\text{”} + O_2$$

Sunlight (E) supplies the energy which moves the two hydrogen atoms from the highly electronegative oxygen atom (in H_2O) to the electroneutral carbon atom (in CO_2). The two electrons of these hydrogen atoms are being pulled away from an atom that has a strong attraction for them and given to an atom that has a much weaker attraction for them (just the reverse of the methane reaction). In short the oxygen atom in H_2O is being oxidized as the carbon atom of CO_2 is being reduced. This is obviously an endergonic reaction in which sunlight is providing the requisite energy. The bond-participating electrons in the newly synthesized CH_2O molecule are at a higher energy level than they were in the H_2O molecule from which they were transferred.

Returning to the analogy of the hourglass in Figure 2-8, as long as sand is present in the top chamber (sand donor) energy will be expended by sand falling into the bottom chamber (sand acceptor). When all the sand has fallen into the bottom chamber, a state of stability (maximum entropy) is reached. At this point all energy transactions would cease. As long as energy can be mobilized to lift the sand in the bottom chamber (sand acceptor) to the top chamber (sand donor) free energy will be available. The electrons in the newly synthesized CH_2O molecule were lifted from the bottom chamber (electron acceptor) to the top chamber (electron donor) by light energy. The energy differential therefore that latently resides in this molecule is equivalent to the amount of light energy that put it there.

The amount of energy so stored is impressive. This becomes evident when the organic molecule undergoes an exergonic reaction. In this instance we will use an actual compound synthesized by the plant, namely, the simple sugar called glucose. The formula of this compound is $C_6H_{12}O_6$, which is a sextupled multiple of our hypothetical “CH_2O” compound [that is, $(CH_2O)_6$]. Again our equation will be greatly oversimplified, indicating merely the beginning and the ending of a long series of integrated reactions:

$$C_6H_{12}O_6 + 6O_2 \longrightarrow 6CO_2 + 6H_2O + E$$

To appreciate the amount of energy liberated in this reaction we should know how much matter is actually involved in the transaction. How can we determine the actual weights of reactants that provide the proportion of molecules called for in the equation? We can resolve this question by considering the atomic weights of the several atoms in this case (see Fig. 2-2). The combined atomic weights of all of the atoms in a molecule is the *molecular weight* of that molecule. The pro-

	$C_6H_{12}O_6$ +	$6O_2$ ⟶	$6CO_2$ +	$6H_2O$ + E
Reaction weights ⟶	$1 \times 180 = \underline{180}$	$6 \times 32 = \underline{192}$	$6 \times 44 = \underline{264}$	$6 \times 18 = \underline{108}$
	$C_6 = 12 \times 6 = 72$		$C = 12 \times 1 = 12$	$H_2 = 1 \times 2 = 2$
	$H_{12} = 1 \times 12 = 12$		$O_2 = 16 \times 2 = \underline{32}$	$O = 16 \times 1 = \underline{16}$
	$O_6 = 16 \times 6 = \underline{96}$			
Mol. weights ⟶	$\underline{180}$	$O_2 = 16 \times 2 = \underline{32}$	$\underline{44}$	$\underline{18}$

(Atomic weights: carbon = 12; hydrogen = 1; and oxygen = 16)

Figure 2-9 Proportionate weights in glucose oxidation.

portionate reaction weight of each compound in an equation is its molecular weight multiplied by the number of its molecules. This weight provides the same relative numbers of reactant molecules as stipulated in the equation. We can see from Figure 2-9, which shows these calculations, that the molecular weight of glucose is 180 and the reaction weight of six molecules of oxygen is 192. The relative proportion of these two compounds, *by weight*, in this reaction is 180:192. At this point we can simply substitute grams as a standard unit of weight and state that 180 grams of glucose will react completely with 192 grams of oxygen. The molecular weight of a compound expressed in grams is its gram molecular weight; it is called a *mole*. One mole of glucose (180 grams) reacts with six moles of oxygen (192 grams) to yield six moles of carbon dioxide (264 grams), six moles of water (108 grams), and energy. The amount of energy in this case is 380 Kcal (kilocalories).

$$\underset{(180\text{ gm})}{C_6H_{12}O_6} + \underset{(192\text{ gm})}{6O_2} \longrightarrow \underset{(264\text{ gm})}{6CO_2} + \underset{(108\text{ gm})}{6H_2O} + \underset{(380\text{ Kcal})}{E}$$

One hundred eighty grams is about two-fifths of a pound; 380 Kcal is the equivalent of almost 1,172,000 foot-pounds of energy. There is therefore in less than $\frac{1}{2}$ lb of glucose enough energy to lift a 150-pound man over 7800 feet into the air (150 pounds × 7800 feet = 1,170,000 foot-pounds). The utilization of redox reactions by living organisms is indeed impressive. It is especially so when we realize that all of these reactions occur at the low temperatures found in living bodies.

We should not leave redox reactions without brief mention of the fact that electrons may be transferred by means other than the hydrogen atom. They may be transferred directly by themselves. Iron (Fe) is used extensively by living systems in this way. It lends itself nicely to redox reactions because it has two valences: Fe^{++} (ferrous) and Fe^{+++} (ferric). Two electrons can be removed from an iron atom (as in rusting iron), producing the ferrous ion

$$Fe - 2\text{ electrons} \longrightarrow Fe^{++} + 2\text{ electrons}$$

This can be further oxidized by the loss of a third electron:

$$Fe^{++} - 1\text{ electron} \longrightarrow Fe^{+++} + 1\text{ electron}$$

The ferric ion can in turn act as an oxidizing agent and receive an electron from the atom oxidized:

$$Fe^{+++} + 1\text{ electron} \longrightarrow Fe^{++}$$

In all chemical reactions (even those skeletonized above), at least one chemical bond is broken and at least one new chemical bond is established. The energy required to break a bond, or to make a bond, is equal to the energy in that bond. This is a natural consequence of the axiomatic principle that energy can be neither created nor destroyed.[3] Before any chemical reaction can be initiated, therefore, some external source of energy must be available to break the first bond or the first few bonds. This is known as the *activation energy*. It is the means by which, as in the firing of a gun, the reaction is triggered. If the reaction is exergonic, the energy released is sufficient to continue the work of bond breaking and bond making. This situation is well illustrated by a row of dominoes standing on their narrow edges. Once the first domino is tipped (activation energy) the others fall in a cascading sequence. If the reaction is endergonic, however, energy is invested rather than withdrawn. In either event, whether energy is going into the synthesis of an energy-rich compound or is being released from such a compound, the total energy going in or going out is never more or less than the total bond energy of the compound. Once again energy is neither created nor destroyed.

Many factors, aside from the nature of the reactants themselves, affect the speed of chemical reactions. Of these *temperature* and *reactant concentration* are certainly of prime importance. All molecules, atoms, and ions are in ceaseless, random motion, colliding with one another willy-nilly in their constant oscillations. When this action is violent and the particle velocity high, we say the particles are (feel) hot. Temperature is a measure of particle agitation; the higher the temperature the greater the agitation. The greater the agitation the higher the collision rate is between molecules, atoms, and ions. This increased collision rate has two significant effects: (1) it increases contact between reacting particles and (2) it increases the degree of collision impact between reacting particles. The first effect enhances the opportunity for reaction; the second can, in itself, be sufficient to break chemical bonds. It is a generally accepted rule that the rate of chemical reaction is doubled for every 10° C increase in temperature. Biochemical processes are favored by increases in temperature (within tolerated limits, of course). Cold-blooded animals like lizards and snakes furnish excellent examples of these relationships. In warm weather these animals are capable of very rapid movement, which contrasts sharply with their sluggish muscular responses during cold temperatures.

Increasing the concentration of reactants, like increasing the temperature, improves the opportunity for molecular, atomic, and ionic contacts. When the physical contacts between reactants are augmented, there is a corresponding response in the degree of chemical activity.

MACROMOLECULES AND THE CARBON FAMILY OF COMPOUNDS

We have gone to some lengths in elaborating a principle that was stated originally in Chapter 1. Matter is integrated at successive levels of organization. Each level manifests novel quantities which are not predictable from the isolated components of the ordered whole. The atomic energy that makes elemental sodium possesses no "sodium-ness"; the property of being sodium resides in the atomic *arrangement* of that energy. In like fashion the behavior of ions and compounds is not a property that lies incipiently in the atoms from which they came. Sodium

[3] Called the First Law of Thermodynamics or the Conservation of Energy.

chloride is not the sum of elemental sodium and chlorine; it is rather their *ordered arrangement* in a new, ionic form. The current discussion will carry this concept of *emergent qualities* to a third level of organization, which is of the greatest significance to biological systems.

We have thus far dealt, for the most part, with a class of chemical substances known as inorganic compounds. Certain examples cited, however, were organic compounds, such as methane (CH_4), acetic acid (CH_3COOH), the hypothetical CH_2O compound, and glucose ($C_6H_{12}O_6$). It will be noted that all of these compounds contain carbon; all but methane contain hydrogen and oxygen as well. Carbon-containing compounds are characteristically associated with the chemistry of living organisms (Fig. 2-1). In fact, organic chemistry is so dependent on the element carbon that it is defined as the chemistry of the carbon compounds. This is not at all coincidental. The phenomenon of life is predicated on the unique properties of carbon: no other element (not even carbon's closest chemical relative, silicon) stands so precisely electroneutral. It is neither electropositive nor electronegative. Consequently it may share its electrons with electropositive as well as electronegative atoms; it shares its electrons equally well with itself (page 30). Because of its wide-ranging chemical proclivities carbon is found in countless millions of different molecular combinations. Life as we know it was made possible by carbon and would not have evolved without it. Let us therefore turn our attention to this element and its fascinating family of compounds.

Organic versus Inorganic Chemistry

The fundamental principles of chemistry so far described apply equally well to organic compounds. New concepts must be added, however, because of the extreme variety and complexity among carbon compounds. It is quite safe to say that no one knows how many different molecular configurations there are in this area. For example, no two organisms, other than identical twins, represent the same nucleic acid complement (the hereditary material is composed of nucleic acid units). Each species of organism possesses unique combinations of nucleic acid complexes. Species that have been long extinct were quite different from those living today. The diversity among proteins is similarly great. The complexity of molecular types is proportionately remarkable.

Silicon, carbon's closest chemical "cousin," is responsible for the complicated chemistry of soils. Some of the most elaborate silicates, have molecular weights approaching 1500, and these contain heavy elements like lead and uranium. Proteins and nucleic acids have molecular weights that *characteristically* range from about 6000 to 10,000,000 and more. Such molecules are indeed very large and for this reason are called *macromolecules* (*macro* = great). Their intricate geometric shapes make possible varying spatial arrangements of chemically active (bonding) sites. These sites are quite sensitive to chemical changes in the macromolecular environment. Chemical response (reaction) of such macromolecules commonly entails changes in the architectural structure and geometric shape of the molecule itself. Chemical reaction at this level involves a degree of sophistication far beyond that found in inorganic chemistry. Discrete macromolecules may become linked to form fibers visible under the microscope (for example, tropocollagen macromolecules in fibrous connective tissue). Others form linear arrays which actively shorten along their long dimension (the actin and myosin of muscle tissue). One might be tempted to say that such molecules were alive. But, of course, they are not. Some of them have even been synthesized.

It is not our purpose at this time to discuss the biological applications of these great molecules. That will be deferred to Chapter 3. We will consider here, however, some of the chemical characteristics of the carbon compounds and attempt to detail some of the reasons underlying their profusive variety.

Isomerism

Compounds which have the same formula but differ in chemical and physical properties are called *isomers*. Their differences are occasioned by the fact that the component atoms are arranged in dissimilar ways with respect to one another in the molecule. Only *structural formulas* and *structural diagrams* can reveal the differences in such compounds. Figure 2-10a depicts the difference in two compounds having the formula C_2H_6O (single covalent bonds are indicated by one line). These two substances are actually quite different in structure and chemical behavior. Figure 2-10b shows two isomers bearing the formula $C_6H_{12}O_6$. Although these two molecules are quite similar, they are chemically distinct. Exchanging the relative positions of the H and OH of only one carbon atom alters the chemical reactivity of these simple sugars.

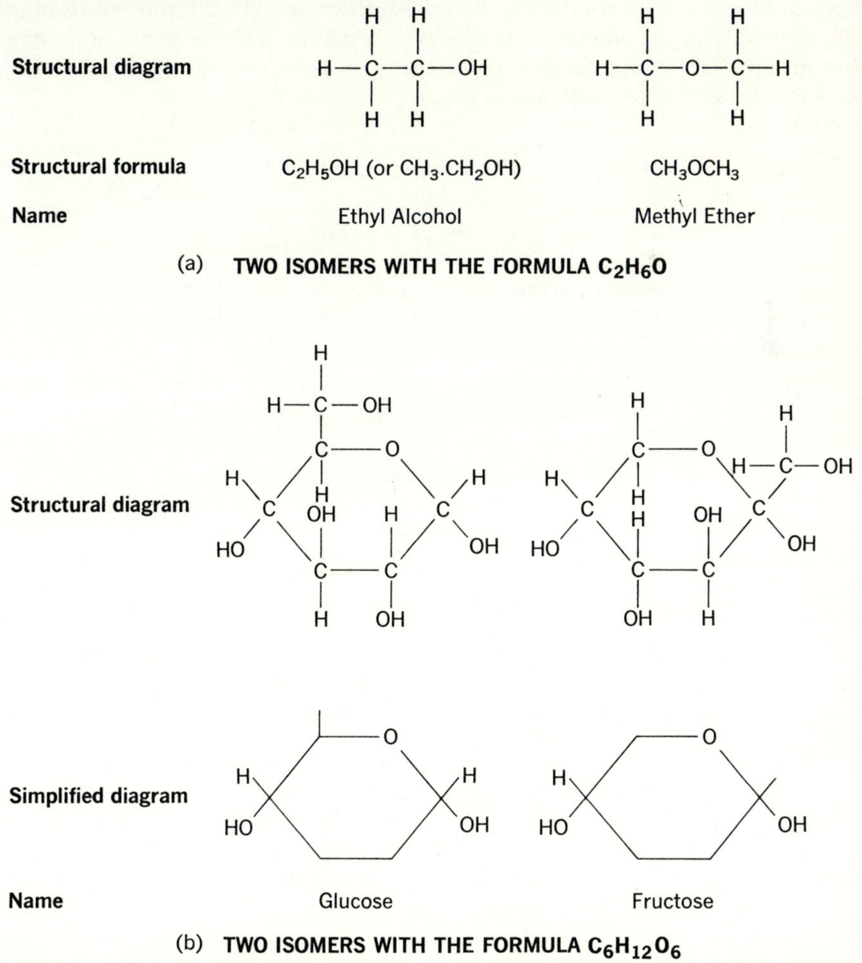

Figure 2-10 Organic isomers (covalent bonds are indicated by single lines).

Carbon-to-Carbon Linkages

The carbon-to-carbon linkages of organic compounds follow two alignment patterns: the *chain* (aliphatic compounds) and the *ring* (aromatic compounds). In both types there may or may not be double (or even triple) bonds between two or more carbon atoms. Those compounds that have one or more multiple bonds are said to be *unsaturated;* if there are many such bonds, the term *polyunsaturated* is used. Those with single bonds are called *saturated.* Figure 2-11 (see page 44) illustrates these various types of carbon linkages; palmitic acid is a saturated chain, linoleic acid is an unsaturated chain (with two double bonds), and the steroid molecule is a saturated ring compound.

Polymerization and Macromolecules

Organic molecules are capable of becoming linked into long chains of repeating units called *polymers.* We have seen how two molecules of fatty acids can become linked at their carboxyl ends by hydrogen bonds. The phenomenon of multiple molecular linkages is particularly well illustrated by a group of substances known as the *amino acids.* Although there are 20 different kinds of amino acids, they all share a common "backbone" segment. This segment consists of an amino group (NH_2^-) on one end and a carboxyl group on the other; the two end groups are separated by one carbon atom.

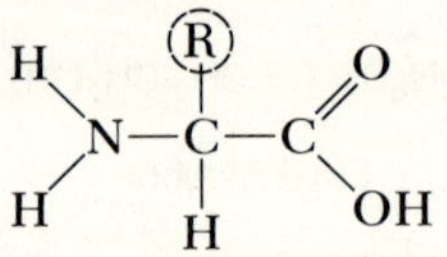

The letter "R" represents various combinations of atoms that may be attached to the "backbone," and distinguish one amino acid from another. Three amino acids, glycine, alanine, and cysteine, differ as follows:

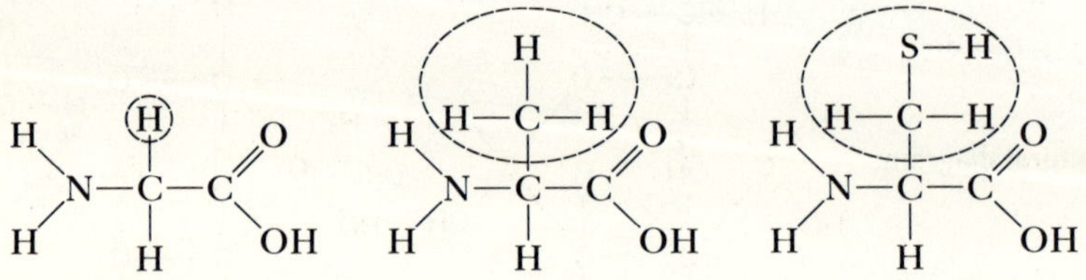

Glycine Alanine Cysteine

By a process known as *condensation* thousands of such amino acids may become linked to one another in a large, chainlike polymer. This is made possible by the removal of an H from the amino group of one molecule and an OH from the carboxyl group of another.

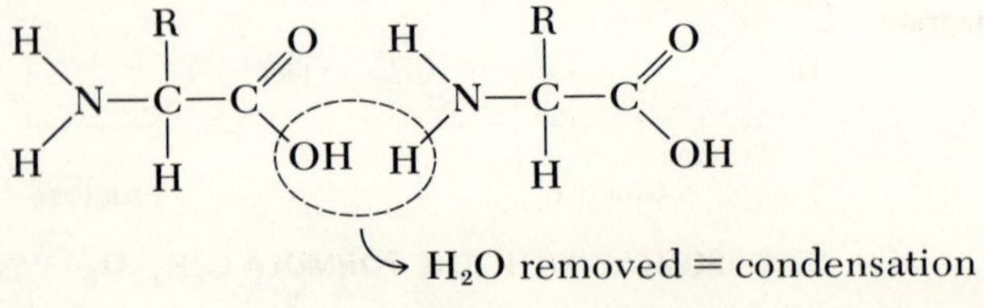

In this fashion many such molecules (*monomers*) are assembled into a polymer; the nitrogens become bonded to the carboxyl carbons:

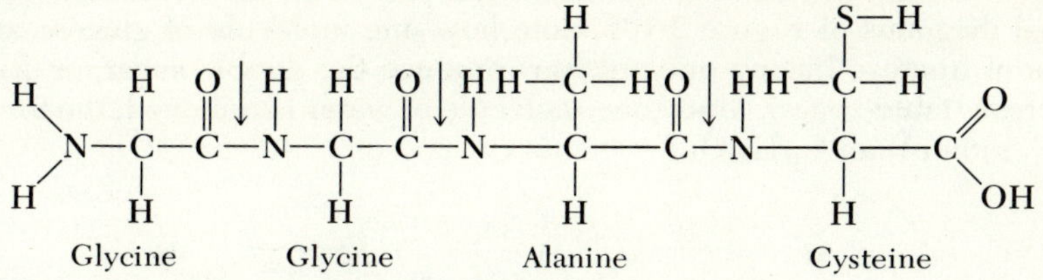

The bonds indicated by the arrows above can be broken by the reverse process of *hydrolysis*. In this event the H and OH are reinserted at the sites from which they were originally removed, thus separating the polymer into its component monomers:

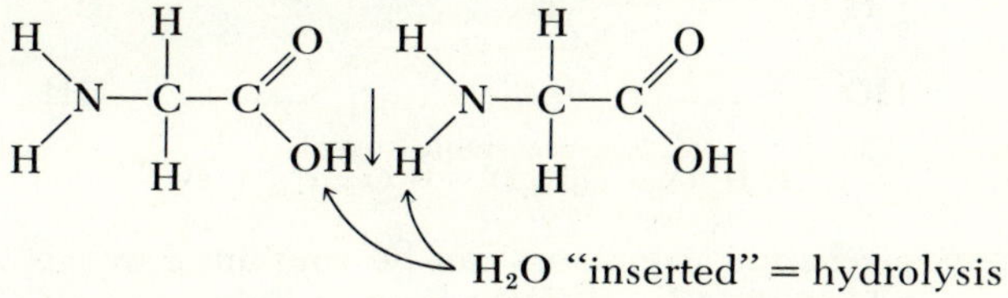

Polymers consisting of hundreds or thousands of amino acid monomers are called *proteins*.

Organic Compounds Important to Life

There are four classes of organic compounds which are of particular significance to living things: *carbohydrates*, *lipids*, *proteins*, and *nucleic acids*. Let us consider each of them in this order.

CARBOHYDRATES

Carbohydrates contain hydrogen and oxygen in the ratio of 2:1, respectively. They are composed of simple units called *monosaccharides* (*mono* = single, *saccharide* = sugar). Figure 2-10b illustrates two such sugars, glucose and fructose. Having the formula $C_6H_{12}O_6$ they are called *hexose* sugars because they contain six carbon atoms. We have discussed the significance of these compounds as an energy source for living organisms. Another group of monosaccharides containing five carbon atoms are called the *pentose* sugars and bear the formula $C_5H_{10}O_5$. Ribose is a good example.

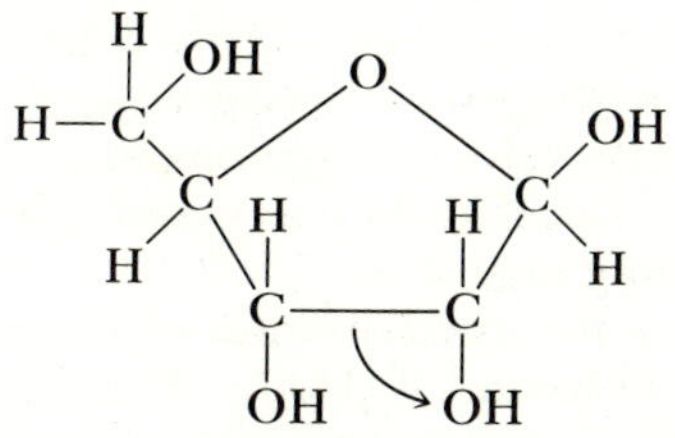

(*Deoxy*ribose is a modification of ribose in which the oxygen atom indicated by the arrow is missing.)

We will return to the pentose sugars when considering the nucleic acids; we will restrict ourselves for the moment to the hexose group.

Monosaccharides undergo condensation just as amino acids do. Using the simplified diagrams of Figure 2-10b, note how one molecule of glucose and one molecule of fructose (facing one another) produce the double sugar, or *disaccharide*, sucrose (table sugar). Since one molecule of water is removed, the formula is $C_{12}H_{22}O_{11}$ rather than $C_{12}H_{24}O_{12}$:

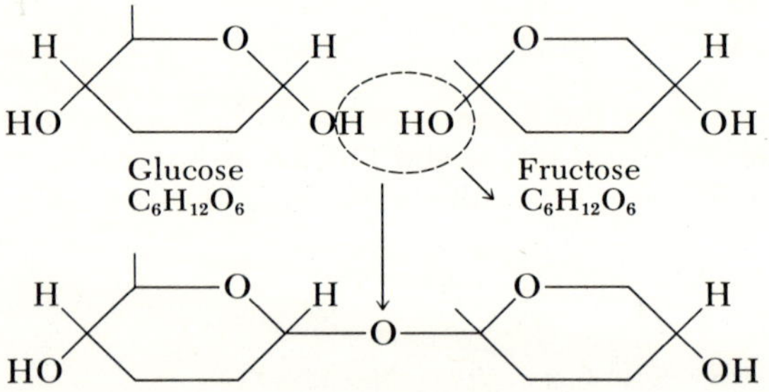

Sucrose (table sugar)

$$C_6H_{12}O_6 + C_6H_{12}O_6 - H_2O = C_{12}H_{22}O_{11}$$

Hundreds of monosaccharide monomers can be combined in this way, yielding complex sugars or *polysaccharides* as in this series of glucose units:

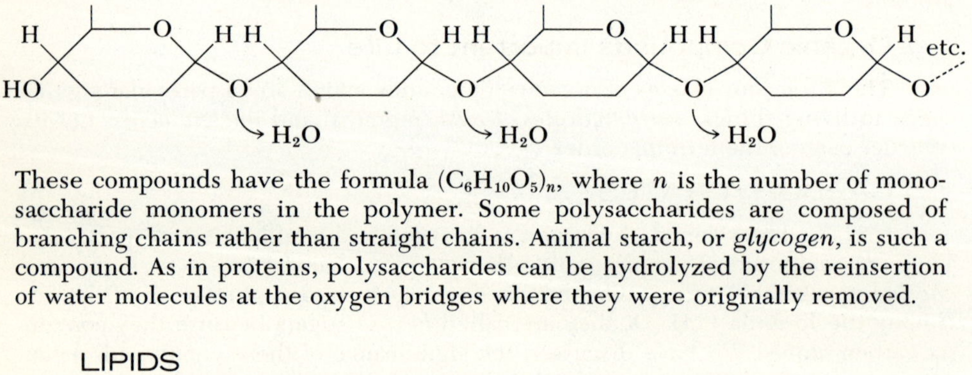

These compounds have the formula $(C_6H_{10}O_5)_n$, where n is the number of monosaccharide monomers in the polymer. Some polysaccharides are composed of branching chains rather than straight chains. Animal starch, or *glycogen*, is such a compound. As in proteins, polysaccharides can be hydrolyzed by the reinsertion of water molecules at the oxygen bridges where they were originally removed.

LIPIDS

Although the lipids contain the same elements found in carbohydrates, their oxygen content is very low. The formula for the fat, palmitin, for example, is $C_{51}H_{98}O_6$. The wide hydrogen-to-oxygen ratio in these substances accounts for their high energy value in biological systems. These relatively unoxidized molecules require an abundance of hydrogen-accepting oxygen atoms for their oxidation. In the process they not only release large amounts of energy but copious quantities of water as well. The fact that they are insoluble in water makes them ideal storage products in the living organism.

Fats or oils are formed by the condensation of three fatty acid molecules (Fig. 2-11) and one molecule of glycerol ($C_3H_8O_3$). The fatty acids may be of the same or different kinds. Palmitin is composed of three palmitic acid molecules. The result is a kind of three-pronged molecular fork (without a handle):

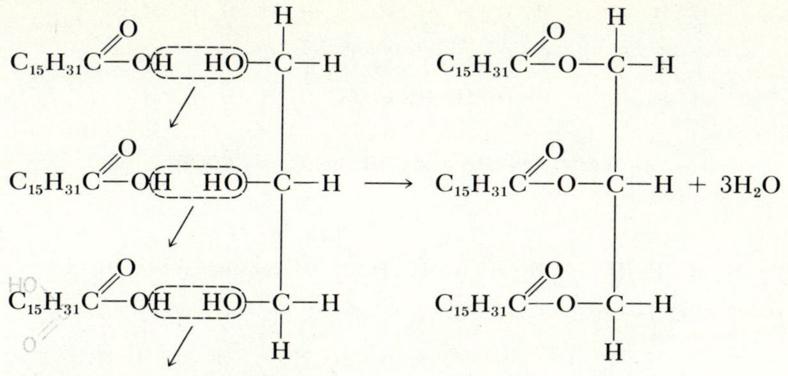

Hydrolysis, as usual, is just the reverse reaction. Our three-pronged molecular fork is reconverted into three palmitic acid tines and their glycerol assembly piece:

$$C_{51}H_{98}O_6 + 3H_2O \longrightarrow 3C_{15}H_{31}COOH + C_3H_8O_3$$

The difference between fats and oils lies in the chain length of the component fatty acids and their relative saturation. Short chains and/or unsaturated fatty acids are characteristic of the oils. Linoleic acid (Fig. 2-11), for example, comes from the oil, linolein (in linseed oil). Longer and/or saturated chains, on the other hand, are found in the solid fats such as palmitin. Plant fats tend more to the unsaturated condition than do fats of animal origin. An unsaturated fat can be converted to the saturated state by exposure to hydrogen (hydrogenation). Plant oils are treated commercially in this way to achieve a smooth, homogeneous texture; creamy (hydrogenated) peanut butter is a good example of this.

The steroids are ring compounds whose structure differs considerably from that which we have just discussed. They are modifications of the basic steroid molecule shown in Figure 2-11. One of the carbons in the terminal hexose ring commonly has an OH attached. *Cholesterol* is such a substance. They too are fat soluble and contain very little oxygen. The steroids are of great significance to animals since they comprise the raw materials out of which many different hormones are made. They are also found widely distributed throughout the plant kingdom.

Before leaving the lipids we should note a fundamental characteristic that pervades both the fats and the steroids. They are polarized compounds. One end is constituted of long carbon chains (or rings) with their attached hydrogen atoms. This end has no affinity with water; it is fat soluble (lipophilic). The other end is either the OH group of the steroid or the glycerol-carboxyl end of the fat. The electronegative oxygen atom(s) confer on this end the capacity to dissolve in water (hydrophilic). Recall (acetic acid, page 31) that the carboxyl group of organic acids dissociates hydrogen ions in water. Acetic acid is, in fact, one of the fatty acids The basic polarity of lipids is utilized by organisms in membrane formation, as we shall see later.

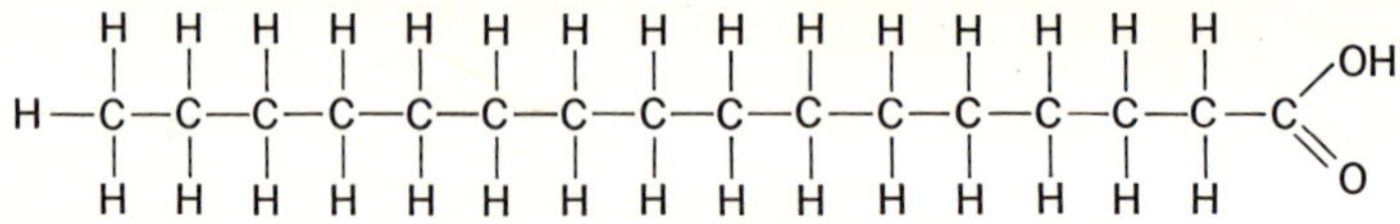

A saturated fatty acid (Palmitic): $C_{15}H_{31}COOH$

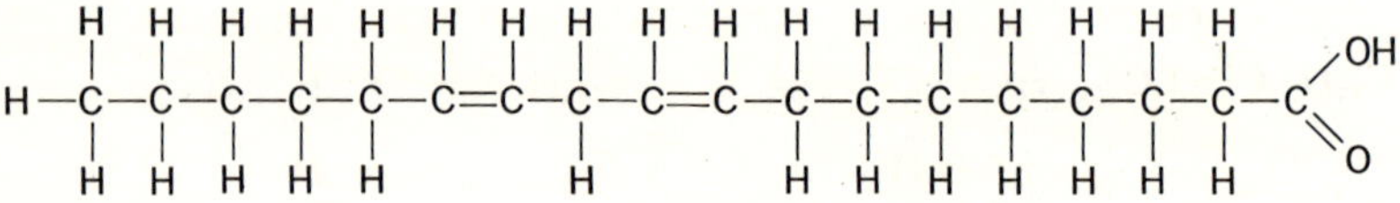

An unsaturated fatty acid (Linoleic): $C_{17}H_{31}COOH$

CHAIN COMPOUNDS

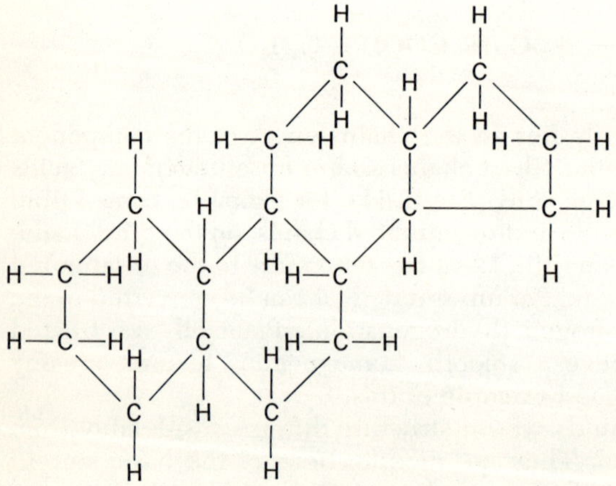

Structural diagram

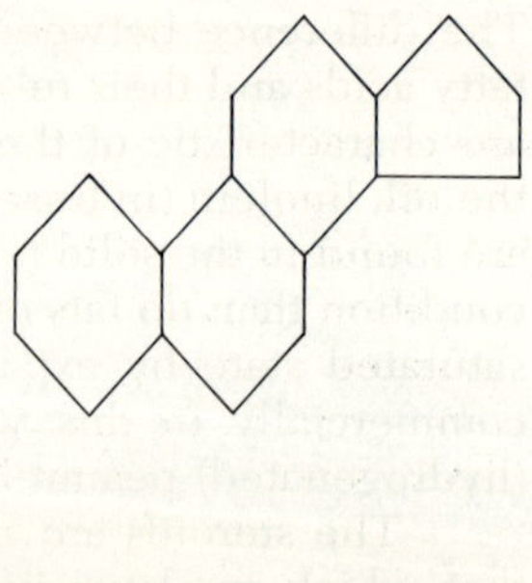

Simplified diagram

A SATURATED RING COMPOUND (THE STEROID MOLECULE): $C_{17}H_{28}$

Figure 2-11 Types of carbon-to-carbon linkages.

PROTEINS

Reference has been made to the fact that proteins are large polymers of amino acids linked at their carboxyl and amino ends. This particular kind of linkage, known as the *peptide bond,* is responsible for the *primary structure* of a protein. Short chains of a few amino acid monomers are called *peptides.* The protein molecule can be thought of as an assembly of several peptides. They are, however, more than just linear chains of amino acid monomers bonded end to end. The chain is commonly twisted on itself, giving rise to a helical spiral (like the red stripe on a barber pole). Thus a *secondary structure,* called the *alpha helix,* is superimposed on the chain. This is brought about by the development of hydrogen bonds between the nitrogen atom of one monomer and the oxygen atom of

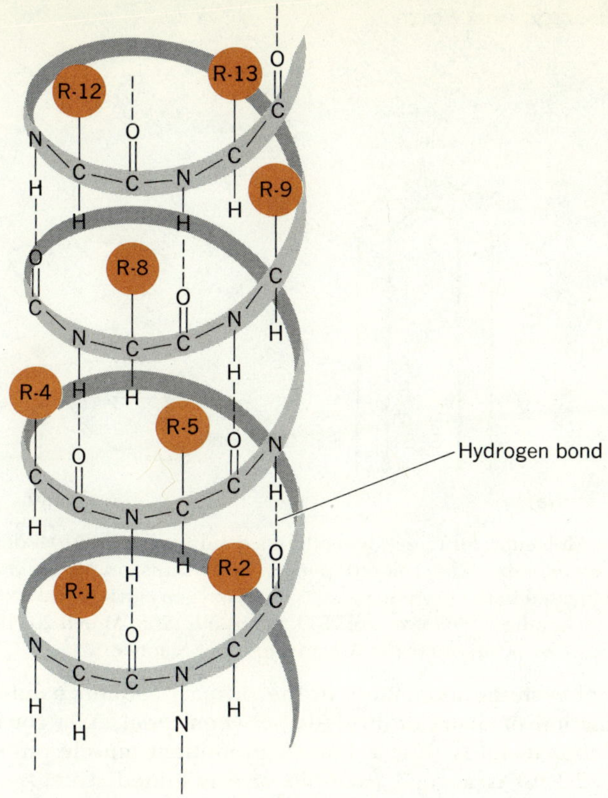

Figure 2-12 The *alpha* helix of a protein molecule. There are 11 amino acids for each three turns of the helix. The amino acids are indicated by their side chains R-1 to R-13 in this 3½ turn segment. Amino acid numbers 3, 6, 7, 10, and 11 on the far side of the helix are not shown for simplicity.

another approximately four peptide linkages removed. Figure 2-12 illustrates this spiral configuration. Each amino acid monomer possesses two hydrogen bonds, one with a monomer behind it and one with a monomer ahead of it.

Further complexity is achieved through folding and looping of the *alpha helix* on itself. This gives rise to the *tertiary* structure of proteins. Folding and looping can occur as a result of what is called the *disulfide bond* or *bridge*. Recall that the amino acid cysteine has an S—H (sulfhydryl) group in its side chain. The oxidation of this group in two cysteine monomers produces a disulfide bond.

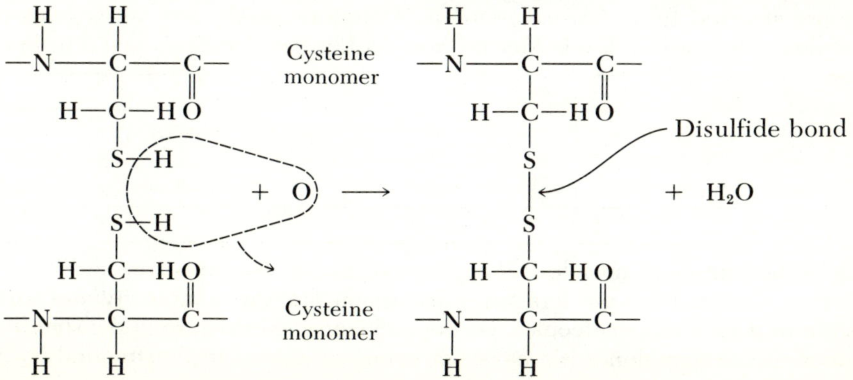

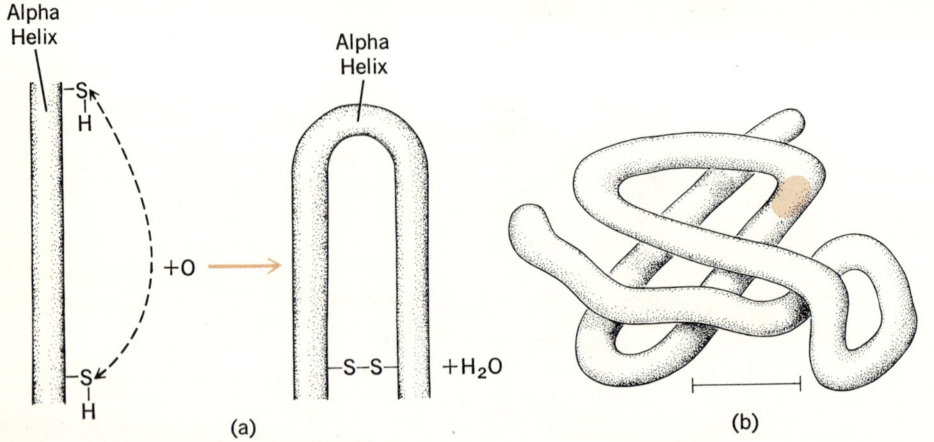

Figure 2-13 (a) Molecular folding as a result of a disulfide bond. (b) Model of the folded, globular protein, myoglobin. The colored portion represents an iron-containing (heme) group which is conjugated to the protein and acts as an oxygen carrier. Scale represents 1 mμ. (Modified from J. C. Kendrew, *Science*, Vol. 139, pp. 1259–1266, March 29, 1963. Copyright 1963 by the American Association for the Advancement of Science)

The occurrence of cysteine monomers in the primary sequence automatically results in the formation of disulfide bridges between them. As a consequence the helix develops loops or folds (Fig. 2-13a). A prominent muscle protein known as myoglobin (Fig. 2-13b) is a good example of this folded structure. Such three-dimensionally folded configurations take the form of a spherical particle rather than a linear chain and are therefore called *globular proteins*. They may undergo a process known as *denaturation* in which molecular unfolding occurs, accompanied by a loss of biochemical activity. This may be induced by a variety of chemical and physical factors. Heat causes an irreversible denaturation (for example, cooked egg white). Denaturation is, on the other hand, often reversible. Plant seeds contain dormant proteins in the denatured state until germination is stimulated; with increased water content these proteins reassume the folded, chemically active state. The globular proteins often become linked to a nonprotein group such as a fat (lipo-proteins), a vitamin (many enzymes), or a pigment molecule (the iron-containing heme of myoglobin or hemoglobin).

Not all proteins share the tertiary, or even the secondary, configurations. These molecules are either in the extended linear chain or are tight, helical spirals. They are referred to as *fibrous proteins*. Disulfide bonds very often cross-link molecules of this sort in fire-ladder fashion, yielding molecules arrayed in line:

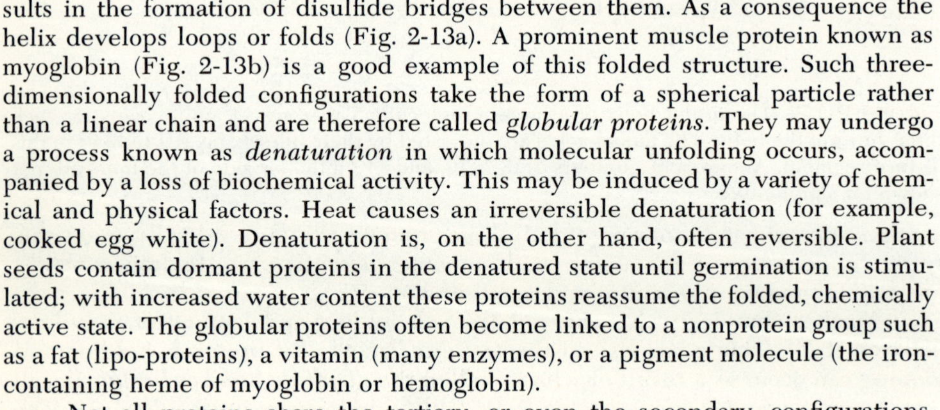

Much of the protein in muscle and connective tissues is of this type.

It is obvious from the foregoing discussion that the complexity and variety possible in these macromolecules are beyond comprehension from the standpoint of geometric design alone. We have not even mentioned triple-stranded coiling.

Nor have we considered the strong possibility that some proteins (one of them in hair and feathers) are very likely superhelices compounded of many alpha helices woven together like the strands in a rope. The degree of diversity possible in primary structure is equally incomprehensible. The primary chain is studded with side groups (R-groups) of the component amino acids. These in themselves are often not simple combinations, and they too can engage in independent molecular unions. The nature of their chemical activity depends not only on their own chemical makeup and that of their neighboring monomers but also on whether they are exposed or buried deep within the folded globule of the molecule. The amino acids arginine, tryptophan, and histidine give some notion of the complexity and potential for chemical reactivity of even these relatively simple groups. Two of them possess unsaturated rings:

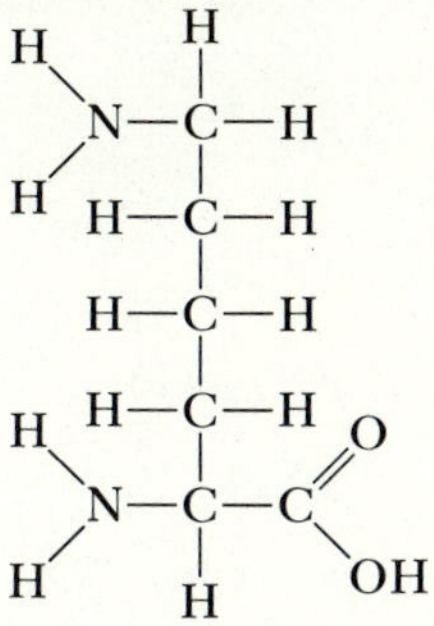
Arginine

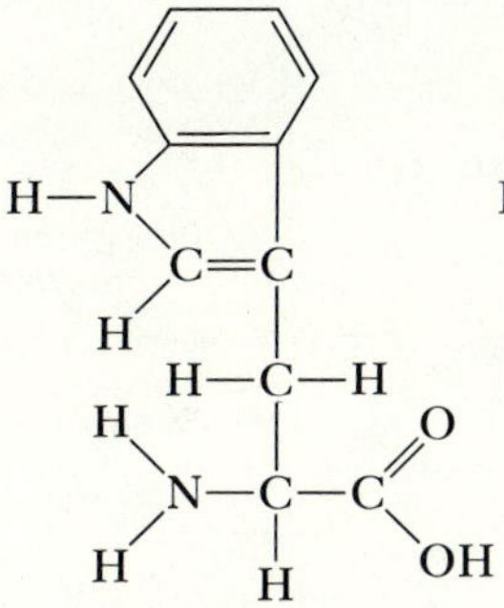
Tryptophan

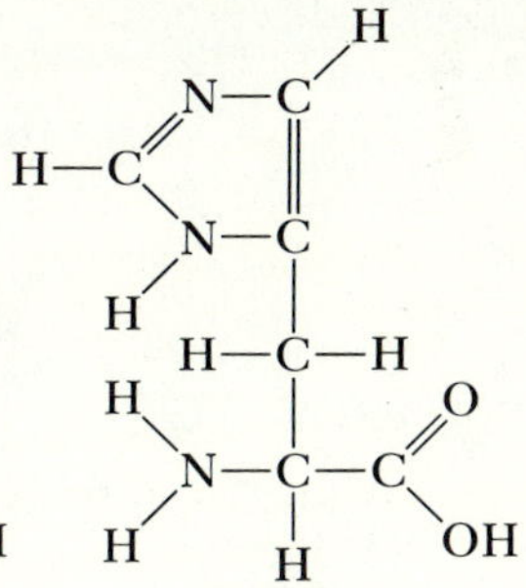
Histidine

In summary, diversity and complexity in the proteins are predicated on three variable factors: (1) the total number of monomers in the primary chain; (2) the relative representation of kinds of amino acids present (out of a total of 20 kinds), including the number of cross-linking cysteine units; and (3) the sequence of amino acids with respect to one another. To gain some idea of the enormous potential for variation implied here let us consider an exceedingly simple peptide hormone consisting of only nine monomers (bradykinin). Five different amino acids are represented; two occur only once, two occur twice, and one occurs three times in the following sequence (amino acids are designated by numbers):

1 - 4 - 4 - 2 - 3 - 5 - 4 - 3 - 1

If we rearrange these numbers in any way (without adding, subtracting, or changing any), a different compound is formed. In how many different ways can they be arranged? Those who are mathematically inclined can discover that these same numbers (amino acids) can be arranged in 15,120 different ways! If this hormone were composed of nine different amino acids, each of which appeared only once, there would be 362,880 different arrangements. What order of magnitude could be expected when rearranging the amino acids in small protein molecules, like the hormone insulin, with 51 monomers, six of which are cysteine? Or how about those proteins with over 50,000 monomers? The obvious reply here, indeed, is how about them!

Amidst the unimaginable diversity of these great molecules there is an element of almost poetic simplicity. The most elegant macromolecular structure is determined entirely by the amino acid sequence of the primary chain. The alpha helix, cross-linking, and folding all occur automatically once the proper sequence of amino acid units is engaged in the peptide bond. The bonding characteristics of the assembled monomers determine macromolecular anatomy.

NUCLEIC ACIDS

The largest of all the macromolecules are the nucleic acids; their molecular weights range from the tens of thousands to the hundreds of millions. In addition to their own great size they can be linked to proteins such as histones, forming huge conjugated *nucleoproteins*. The biological role of the nucleic acids is of a most fundamental nature since it is these substances that are responsible for the synthesis of proteins in the living organism.

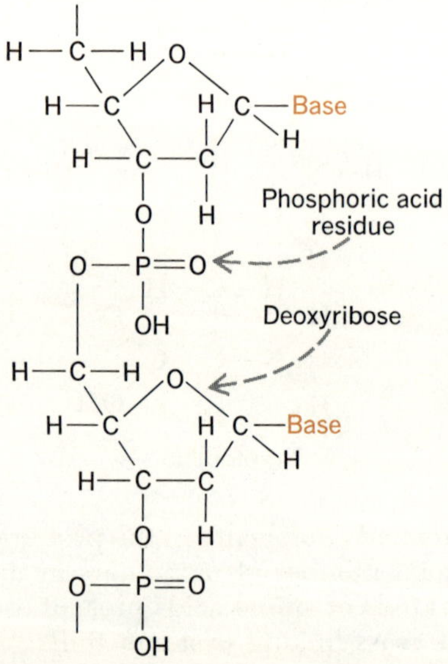

Figure 2-14 Molecular structure of DNA "backbone."

The nucleic acids are, once again, an example of polymerization. In this instance the monomers are *nucleotides*. Nucleotides are fairly complex molecules themselves, and consist of three parts: (1) a phosphoric acid molecule (H_3PO_4) condensed onto a (2) pentose sugar to which a (3) nitrogen-carbon ring compound, or *base*, is linked. The phosphoric acid-sugar condensations comprise the "backbone" of the nucleic acid to which the bases are attached. There are two kinds of nucleic acids depending on which pentose sugar is incorporated into the "backbone." Ribonucleic acid, or RNA, utilizes ribose. Deoxyribonucleic acid, or DNA, contains deoxyribose (Fig. 2-14). We will confine our attention at this time to DNA and return to RNA later. The base which attaches (by condensation) to the sugar is an unsaturated ring compound in which some of the carbon atoms are replaced by nitrogen. There are four different kinds, any one of which may be linked to the sugar. Two of them, *thymine* (T) and *cytosine* (C), are single ring compounds (pyrimidines). The other two, *adenine* (A) and *guanine* (G) are double rings (purines). Figure 2-15 indicates the nature and interrelationships of these four nitrogenous bases. At least two hydrogen bonds can link a purine with a pyrimidine at corresponding points on their rings *if we select the correct pair.* A can bond to T

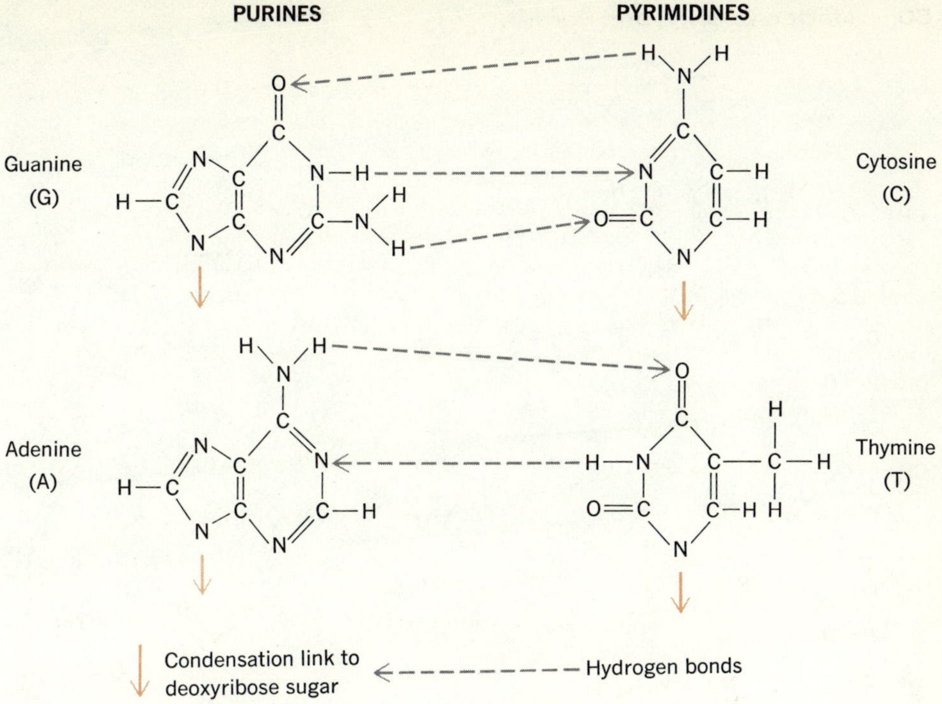

Figure 2-15 Base pairing in the four bases of DNA.

twice; G can bond to C three times. This *base-pairing* of A with T and G with C is of the greatest chemical and biological significance.

The DNA molecule is a side-by-side union of two polymerized nucleotide chains or strands.[4] These strands are held together by the hydrogen bonds connecting their complementary base-pairs: A—T, T—A, G—C, and C—G. These strands, however, are not laid out in parallel lines but are twisted about one another in a double helix (Fig. 2-16). One can imagine a ladder in which the two long supports are the sugar-phosphate linkages and the rungs or steps are the base-pairs. If such a ladder were twisted about itself, a double helix similar to that of DNA would result. DNA contains thousands or even millions of A's with T's and G's with C's. In any one molecule the number of A's and T's are equal, as are the number of G's and C's. The number of A's and G's and of C's and T's, however, is not equal.

As we shall see later, the sequence of bases down one strand of the DNA helix constitutes a chemical language; it is a molecular blueprint, as it were, for protein synthesis. The Morse Code is a communication system founded on two symbols (dots and dashes). The DNA code is a communication system based on four symbols. Each message is a DNA molecule consisting of tens of thousands to millions of paired symbols (A—T and G—C). The possible combinations of the four symbols (bases) in one strand of such a molecule are genuinely astronomical. A small segment of such a molecule is diagrammatically represented on page 51. The double helix has been "unwound" and the sugar-phosphate "backbone" is indicated only by vertical lines.

[4] In a few exceptional cases DNA has been observed in single strands.

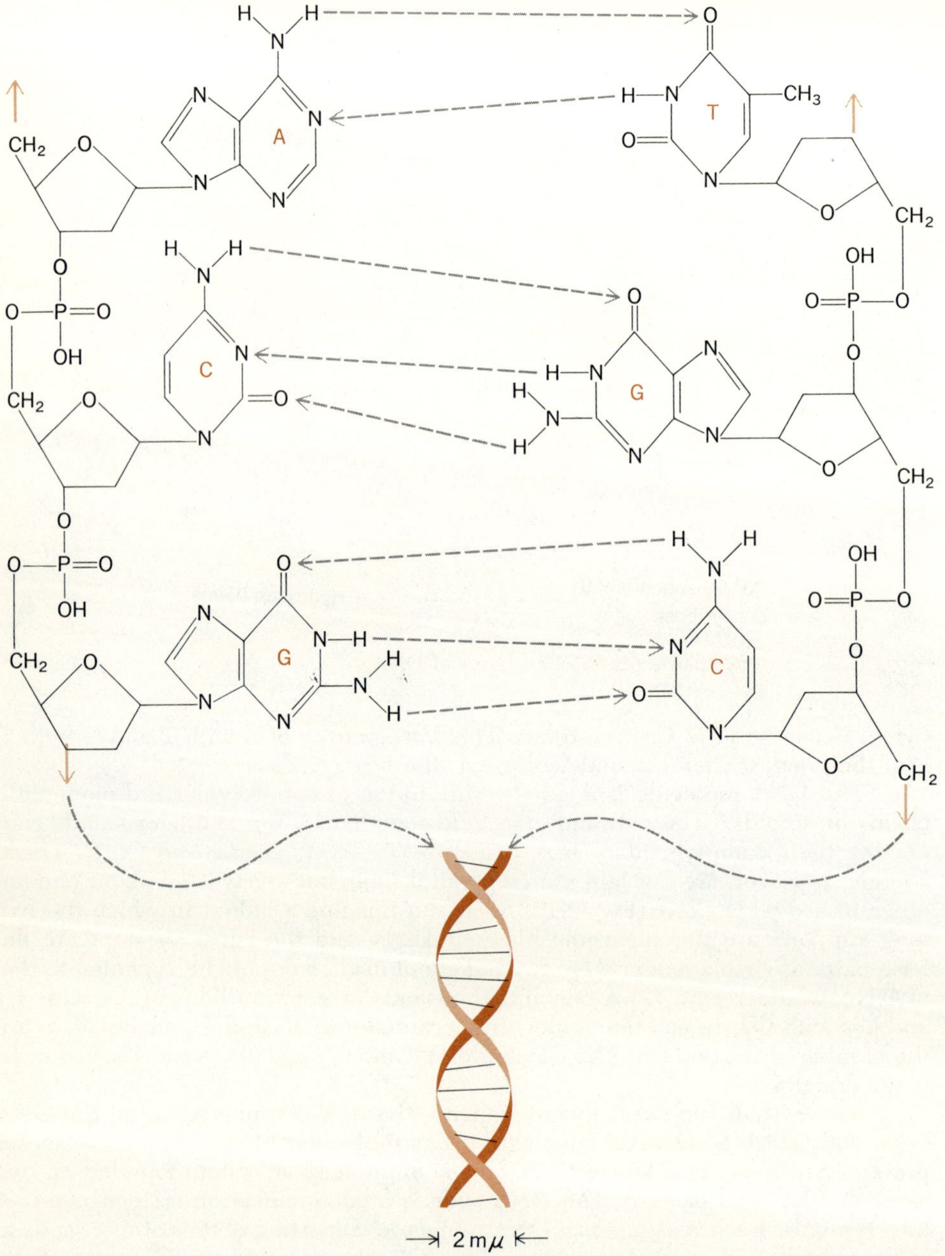

Figure 2-16 The Watson-Crick-Wilkins DNA model. A portion of the molecule is shown below (small figure) as a double helix 2 mμ (millimicrons) in diameter. One complete turn of the spiral includes ten base-pairs (crossbars). The molecule is unwound, flattened, and enlarged above to show the details of molecular structure. Carbon rings are indicated in their simplified form. A = adenine; T = thymine; G = guanine; C = cytosine.

C . . . G
A . . . T
C . . . G
A . . . T
C . . . G
C . . . G
T . . . A
G . . . C
C . . . G
T . . . A
C . . . G
A . . . T
G . . . C
G . . . C
A . . . T

⟵ Of the fifteen base-pairs shown, nine are G—C (three G—C and six C—G) and six are A—T (four A—T and two T—A). The message in the left strand begins, reading downward, C—A—C—A—C—C; that on the right starts with G—T—G—T—G—G. It is especially significant that each strand is a complement of the other, as we shall now see.

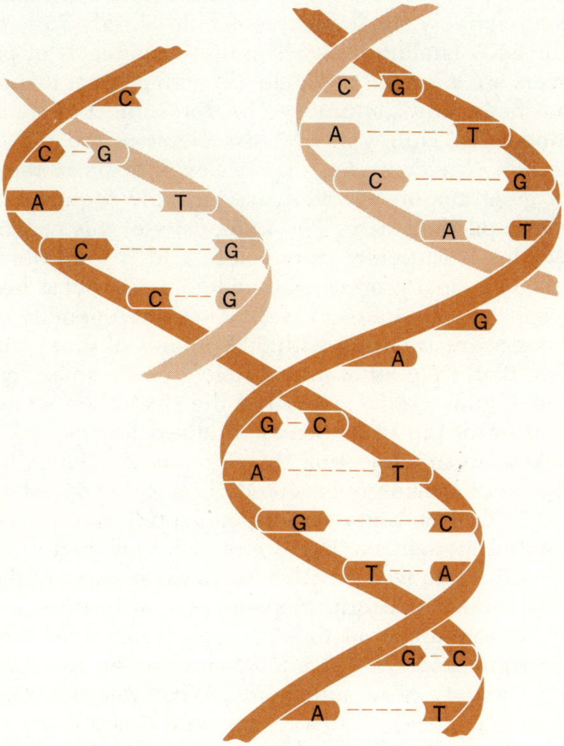

Figure 2-17 DNA replicates a copy of itself by "unzipping" the hydrogen bonds between bases. Each separated strand produces a new strand (light-colored figures) complementary to itself from the free nucleotides that are available in the chemical vicinity. The sequence of bases is the same as that listed to the left.

The DNA molecule is unique in that it can reproduce an exact replica of itself (we will investigate the chemical requirements for this in Chapter 3). This is accomplished by breaking the hydrogen bonds connecting the paired bases. It is done zipper fashion, beginning at the top of the molecule. As the bonds are progressively unzipped through the molecule the strands separate. Simultaneously with the process of separation, each strand begins to synthesize a new strand complementary to itself through the assembly of free nucleotides in the chemical vicinity (Fig. 2-17).

In a comparable fashion DNA can synthesize its chemical cousin, RNA. Following the simile of a DNA molecule as a master blueprint for protein synthesis, RNA can be likened to a work-order copy of this master plan. RNA differs from DNA in a number of particulars. We have already mentioned their differences in backbone sugars. They are further distinguishable in that the base thymine (T) is rarely found in RNA; it is replaced by a closely related pyrimidine called *uracil* (U), which also pairs with adenine. The other three DNA bases are used extensively, but there are a few slightly modified versions of them that occur with some regularity. RNA exists in three different forms, all of which have quite different structures. Some are composed of hundreds or even thousands of nucleotides. One type is a relatively small macromolecule of only 75 to 80 or more monomers. The fact that RNA is almost always single-stranded is of particular interest. How the monomers are arranged in single file with respect to one another in the small RNA form has been, and continues to be, the subject of considerable study.

Robert W. Holley and his associates in the U.S. Department of Agriculture and Cornell University have spent seven years in the analysis of one such molecule. A great amount of effort was initially required in order to isolate the substance in a purified state. The immensity of this task becomes evident when it is realized that a particular ribonucleic acid is found in an organism in association not only with many other ribonucleic acids but a luxuriant variety of proteins, lipids, and carbohydrates as well. The heterogeneity of the original source material is staggering. Starting with 300 pounds of yeast, these researchers extracted a little less than ½ pound of mixed RNA's of the small form. From this mixture they obtained 1 gram (1/450 pound) of the RNA they were seeking, in purified form. This portion of the undertaking required four years. The remaining three years were taken up in subjecting this 1 gram of RNA to hundreds of elaborately designed and painstakingly executed chemical and analytical procedures. The molecule was hydrolyzed into varying nucleotide groups (molecular segments) as well as separated monomers. These had to be purified in turn for accurate identification. Identification was complicated because some of the bases were slightly modified from their usual form. Separation, purification, and identification continued until it became apparent that 77 nucleotides were linked in a single strand that was somehow folded on itself. Seventy-seven monomers can be linked in a bewildering variety of arrangements. What was the sequence of nucleotides with respect to one another? It was reasoned that a comparison of many random segments of the molecule would reveal some segments that possessed nucleotide sequences which overlapped (duplicated) others. Such segments would obviously be connected in the molecule. By persistently matching molecular segments the entire sequence could be determined. This idea ultimately paid off; the nucleotide sequence was established.

Figure 2-18 shows three of the many possible ways in which these nucleotide monomers might be folded, according to the Holley group. By folding the single strand in specific ways it is possible to achieve maximal pairing of bases that are complementary. Those segments containing uncomplementary bases would yield bulges in an otherwise trim figure held tightly in place by a girdle of hydrogen bonds. Of the three configurations shown it appears from an accumulating body of evidence that the clover leaf structure is the correct one.

We are provided here with a good summary of two concepts we considered in the first two chapters.

1. The exercise of human ingenuity and creativity is as essential to scientific inquiry as it is to philosophy, literature, or the arts.

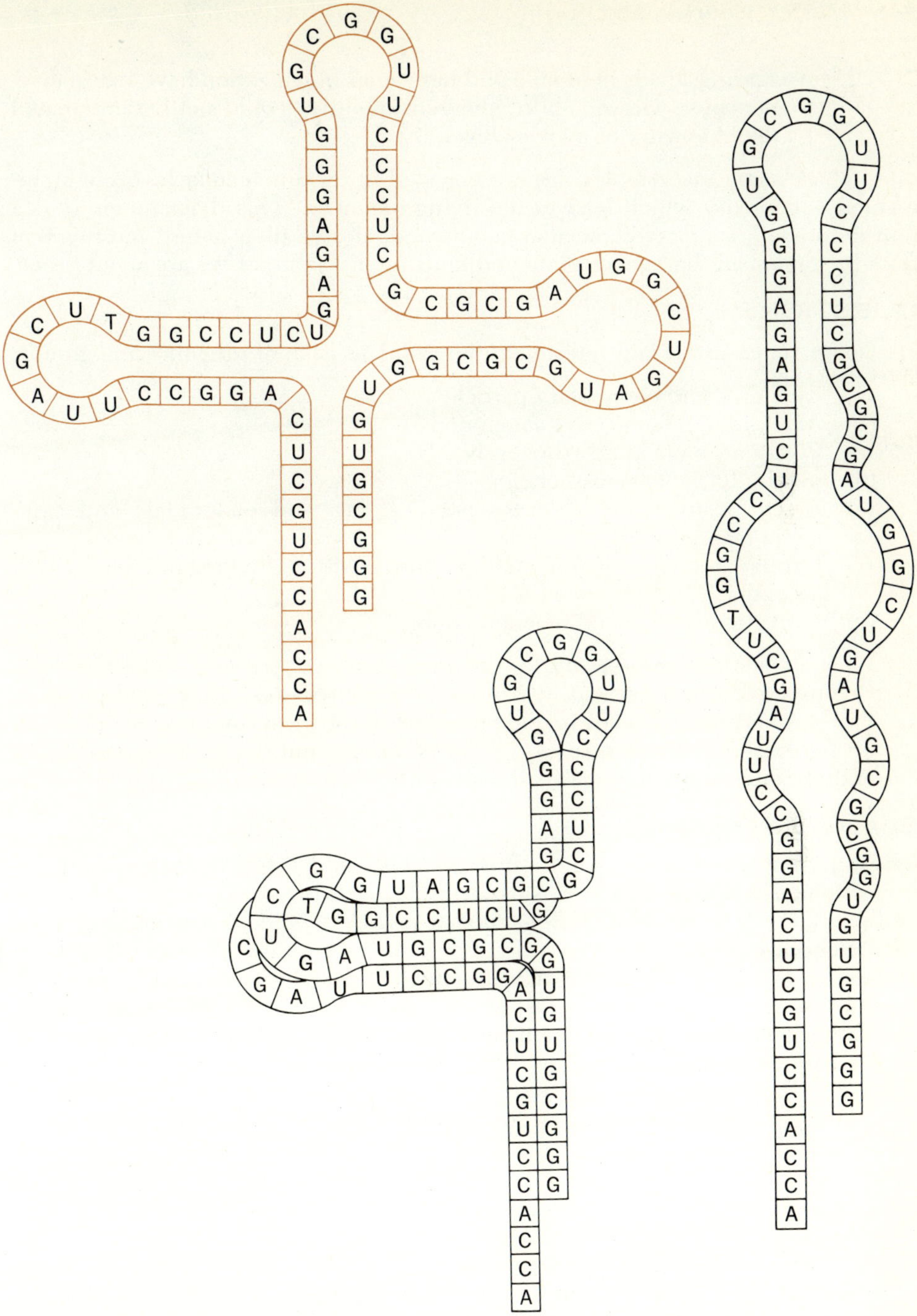

Figure 2-18 Hypothetical models of a small RNA molecule (alanine transfer RNA), showing three of the many ways in which the nucleotide chain might be folded to achieve maximal base-pairing in a single strand. Evidence indicates that the clover leaf structure (color) is correct. Modified bases are not differentiated from their normal forms.

2. Ascending levels of atomic and molecular organization have led to macromolecules whose subtle emergent qualities could not be anticipated by the chemistry of a lower level.

These same macromolecules are engaged in extramolecular levels of architectural complexity which lead to the living organism. This dynamic entity is a unit unto itself. Its macromolecular components derive their nature and function from the organism that makes them and finds them useful, as we are about to see.

QUESTIONS

1. Compare, contrast, and interrelate the terms in each of the following groups:
   (a) Elements and elementary particles
   (b) Atom, ion, element, and compound
   (c) Covalent, ionic and hydrogen bonds
   (d) pH, buffering and dissociation
   (e) Electropositive and electronegative atoms, and endergonic and exergonic reactions
   (f) Oxidizing and reducing agents, and electron (hydrogen) donors and acceptors
   (g) Activation energy and reaction energy
2. What are some of the reasons for the great variety of organic compounds?
3. Compare condensation and hydrolysis in carbohydrates, lipids, and proteins.
4. Proteins and nucleic acids include the largest, the most complex, and the most diverse macromolecules known. What is there about the molecular structure of these substances that makes this possible?

Suggested Readings

Baker, J. J. W., and G. E. Allen. *Matter, Energy and Life: An Introduction for Biology Students*. Reading, Mass.: Addison-Wesley, 1965.

Holley, R. W. "The Nucleotide Sequence of a Nucleic Acid," *Scientific American*, February 1966.

Holley, R. W. *et al.* "Structure of a Ribonucleic Acid," *Science*, **147**, March 19, 1965.

Kendrew, J. C. "The Three-Dimensional Structure of a Protein Molecule," *Scientific American*, December 1961.

White, E. H. *Chemical Background for the Biological Sciences*, Foundations of Modern Biology Series. Englewood, N.J.: Prentice-Hall, 1964.

CHAPTER 3 The Extra-Molecular Level of Life

LIFE AS A SELF-DIRECTED CONQUEST OF ENERGY

The living organism represents a new category of magnitude in architectural complexity. The macromolecules manufactured by living things are the ultimate ordered arrangements of matter at the atomic and molecular levels of design. Organisms, on the other hand, are ordered arrangements of a diverse array of macromolecules disposed in great supramolecular complexes as functioning subunits. To put it proportionately, the simplest organism is to the most complex macromolecule as that macromolecule is to a hydrogen atom. The most complex organisms bear almost the same proportion to the simplest organisms, *quantitatively*. Whether an organism is quantitatively complex or simple in the physicochemical sense, however, bears no relation whatever to its status or efficiency as an organism. Living things succeed or fail, and can be understood and evaluated only as biologic entities, not as hierarchies of physicochemical complexes.

Organisms are obviously highly ordered, nonrandom systems. The same observation can even be made of the macromolecules synthesized by them and incorporated into their structure. The maintenance of this highly organized state requires great expenditures of energy. As a process life is a programmed cycling of endergonic and exergonic reactions whereby the organism is supplied with its requisite needs for free energy. This does not violate the principle of the conservation of energy (the First Law of Thermodynamics). All organisms ultimately rely on sunlight which enables the green plant to synthesize energy-rich organic compounds. This very phenomenon, however, distinguishes living things from the rest of the world. They are *active* participants in the storage and use of the light, heat, and chemical energy available to them. As self-reproducing systems they delay the inevitable consequences of a maxim of nature known as the Second Law of Thermodynamics. This principle states, in effect, that any isolated system tends toward energy degradation and stability. Differences in energy levels are dissipated as hot spots of energy diffuse randomly throughout the system. This stable, randomized, and disordered state in which energy differentials no longer exist is known as maximum *entropy*.

Nonliving things tend to increase their entropy. Rocks roll downhill. They do not possess the potential to roll uphill so that they may roll downhill again. They gradually disintegrate, through physical and chemical erosion, releasing energy as they are decomposed into their simpler components. They do not possess the capacity to maintain their organized status as rocks, to reproduce rocks like themselves, or to reassemble their parts and become rocks again. They are passive subjects to all the natural forces that may be brought to bear against them. Nonliving nature is an hourglass that runs down (Fig. 2-8); it does not possess the ability, within itself, to invert and thereby produce a steady flow of free energy.

Metabolism

Organisms stand in sharp contrast to this state of affairs. They represent a degree of order that is autonomous and self-directed. They are continuously recycled hourglasses. They are, in their essence, internally programmed units which maintain the ordered state and assure a steady flow of free energy. A grass takes in carbon dioxide through its leaves, water and inorganic salts through its roots. Through the photosynthetic process of capturing and utilizing light energy, it converts these raw materials into a great variety of organic compounds and eliminates oxygen. Some of these compounds go into making more plant and some go into structures that will produce a new plant. Still others provide the energy that will send leaves further into the air and push roots deeper into the soil. A cow breathes this oxygen, eats the grass, and drinks some water. It excretes urine, exhales carbon dioxide, secretes milk, eliminates nutrient-rich fecal matter onto the ground, and manages to produce a lot of cow and a few calves. A man eats the cow, or its equivalent, in time, and so on. All of the many and varied physicochemical processes that go into the performance of an organism's total activity are referred to as its *metabolism.*

Integration and Adaptive Behavior

Any living thing is a highly integrated unit. Metabolic activities are efficiently coordinated like the assembly line in a large manufacturing plant. They serve only the organism as a whole. The organism, in turn, serves its interests and those of the species to which it belongs. Its behavior is adaptive and purposeful. It seeks out the best "markets" for obtaining energy. A plant turns its leaves so that they face the sun. A cat stalks and captures a mouse. Conversely, living things behave in ways that serve to prevent their destruction or the impairment of their efficiency. Rodents hibernate through the bleakness of winter. Desert animals are active at night during the summer, resting or sleeping deep in their burrows or beneath rocks during the day. *Integration* and *adaptive behavior* further the cause of metabolism and constitute a second attribute unique to life and totally alien to the nonliving.

Growth

As a plant or an animal successfully exploits the energy about it, it expands its operation. Like a thriving business corporation it grows. Bermuda grass ("devil grass") and ivy extend their area by the proliferation of creeping stems. As the German Shepherd pup grows toward maturity he becomes progressively a more formidable and efficient carnivore. Biologic *growth* is yet another quality that distinguishes life from the nonliving. Although a mineral crystal may increase in size, this kind of growth is simply a matter of similar chemical units in solution added (by accretion) to the parent crystal that acts as a seed or passive template. Organic growth, on the other hand, involves the active transformation of food or nutrient materials into substances unique to the organism. Our dog eats table scraps from the same meal consumed by his master. This may or may not humanize his disposition; it certainly does not alter his biological affinities.

Reproduction and Organic Diversity

The phenomena of growth and reproduction are closely interrelated. Small daughter plants spring up on the spreading stems of the ivy plant, for example, ultimately producing an interconnected network of a number of plants. The two processes are distinguished by the fact that one results in an increase in size of the individual, whereas the other results in the production of one or more new individuals. Most organisms grow old; their biological machinery presumably begins to wear out. As this occurs the incidence of death starts to show a marked rise; the chances of living on become increasingly slim. Long before this condition is reached the plant or animal has made its preparations for the replacement of ageing machinery by the reproduction of a number of new individuals, who in turn will grow and reproduce.

Sometimes only one parent is involved, as in the example of the ivy plant. This is called *asexual* reproduction, and consists of some form of division of a portion, or all, of the parent. Very often reproduction requires two parents and is accomplished by the union of specialized sexual products of some kind from each. This, of course, is *sexual* reproduction. The reproductive products, whether sexually or asexually formed, are living packages of hereditary material. This hereditary material carries chemically coded, macromolecular blueprints (DNA molecules) which not only direct the building (growth) of the new organism but also carry directives for operational procedures (metabolism and integration) as well as the future reproduction of the new organism. This hereditary material has been received by every generation from its parents, who, in turn, received it from their parents, and so on. This applies, quite obviously, to every member of a population of organisms of one kind or species.

There is, particularly among sexually reproducing species, considerable variation, even though members of a species are generally similar to one another. This was pointed out in Chapter 1 in the discussion on indeterminacy. When hereditary material is passed from two parents to one offspring, it is subjected to a reassortment (a card-shuffling kind of thing). It is also modified in other random ways, as we shall see in Chapter 8. No two sexually reproduced organisms, therefore, are ever *exactly* alike. Life is characterized by variety. This susceptibility to change without a necessary loss of biological integrity or effectiveness allows living things to respond, *at the populational level,* in an adaptive way to the environment in which they live. Those individuals which, for one reason or another, are ultimately unable to survive the vicissitudes of their time and place perish or fail to reproduce. Those particular hereditary recipes for animal or plant-making are, as a consequence, "junked." Those populational variants, on the other hand, which effectively survive are selected by nature as fit; their hereditary recipes for animal- or plant-making have been tested and found not to be wanting. In this way the species of organism, *as a heterogeneous population,* is guided and directed by the rigorous selection of the environment in which it lives.

Over periods encompassing millions of years, significant environmental changes have occurred. Under such conditions there is a gradual change in the criteria by which organisms can be considered successes or failures. With changing standards of selection the biologic character of the population changes correspondingly. The new models of parents meet different specifications from the old ones. Given time this ultimately results in a new species of organism. This process is called *organic evolution.* It has produced a million species of animals and almost a half a million species of plants which are alive today. It produced additional millions in the past which have long since become extinct. Such diversity attests to the remarkable adaptive capability of life, as a population of organisms, to seek out sources of energy on this planet wherever they may be. It would be hard to imagine an untapped energy resource on a planet (other than things such as the throat of an active volcano) in which there are one million ways of conducting animal business and almost one half a million ways of carrying on plant business.

Life is a process that is population-oriented. It owes its nature to a long and continuous history of countless individuals distributed over a vast period of time all over the earth. It is a continuum in which living reproductive products containing hereditary material are continuously recycled through one organism and on to the next. These products are a liaison both with the past and the future; as such they actually constitute the continuum and are therefore nonexpendable. Individual organisms are booster pumps that sustain the free flow of energy in the system. They are also testing grounds in which biologic design is kept abreast of current environmental conditions. Although highly essential to the system, organisms are expendable. Individuals die; populations of individuals die; species become extinct. But each in its turn can produce new individuals, a new population, and a new species. That this system is a supremely effective one is attested to by the fossil record of organisms. Life today is much more complex and diversified than it was 500 million years ago and vastly more so than what it must have been at its earliest beginnings more than three billions of years ago. Man, as the current culmination of its progress, has carried raw nature to what is often termed the supra-organic level of human culture. We should entertain no illusions, however, about the biologic foundations of this culture. From the beginning the instrumentality of human culture and technology was designed to assure the biological survival of our species. Disassociating culture from biology is largely responsible for many of our major problems today.

ARCHITECTURAL ORDER AMIDST THE DIVERSITY OF LIFE

The Nature and Extent of Organic Diversity

We noted that organic evolution is responsible for the formation of a species. This is accomplished by a process in which the adaptability of a species to a particular environment is maintained by the selective forces of that environment. If evolution has produced the one and a half million species of animals and plants that are around today, there must be an exceedingly large number of environments. And indeed there are. An organism's environment encompasses all the intricate interrelationships existing between the organism and its surroundings; the surroundings include the physical features of that environment as well as the total population of living things that occupy it. Soil, rainfall, temperature, elevation, geographic location, and so on, are all important. Climate and soil determine the kind and amount of vegetation that will grow. Climate and vegetation affect the kind and numbers of plant-eating animals that will be found, which in turn set the stage for a certain cast of carnivores. All these organisms constitute a food resource for parasites. Death yields carcasses for carrion-feeders and hosts of microscopic organisms that produce decay, and, in turn, die. Unconsumed organic residues return to the soil to be used by soil bacteria and vegetation as the *food chain* is recycled. This complex of "consume-and-be-consumed" is often called the *web-of-life*.

The study of these dynamic processes and interdependencies is *ecology*. It is quite obvious from this brief account that no two areas can be exactly alike. Environments are diverse; mountains, valleys, meadows, deserts, ponds, lakes, rivers, oceans are in themselves extremely varied. The ecologic opportunities for energy exploitation are almost endless. In the example given we could list the following:

1. Tiny microscopic organisms that live on minerals and organic matter in the soil.
2. Organisms that live on these organisms.
3. Large green plants that grow in the soil and carry on photosynthesis.
4. Herbivores who eat these plants.
5. Carnivores who eat these herbivores.
6. Animal, plant, and bacterial parasites who live on these plants and animals.
7. Parasites who live on these parasites (and who sometimes have their own parasites!).
8. Carrion feeders like vultures and flies that live on dead carcasses as well as the bacteria which, through their feeding, bring about decomposition and decay.
9. Plants, like toadstools, and animals, like earthworms, that live on decayed organic matter in the soil.

In letting not the slightest ecologic opportunity go unchallenged, life, through organic evolution, has produced an impressive staff of highly competent specialists and experts to which the pictures on these pages bear witness. It is obvious that some of these organisms belong to the plant kingdom and some belong to the animal kingdom. Others are not so easily classified, particularly some of the small, microscopic forms and their larger relatives. Certain of these behave like an animal at night by eating other organisms; during the day they carry on photosynthesis like a plant! Some forms of algae manufacture their own food like typical plants, but are capable of rapid, wormlike movements. This relatively simple, but highly heterogeneous group of organisms include the protozoa, algae (except one group that is classed with the bacteria), fungi, and molds. They are usually included in a separate kingdom called the *Protista*.[1] (For a more complete, illustrated classification of organisms, the reader should consult the Appendix at the end of this book.) The bacteria and viruses claim both the smallest and the simplest of living things. In fact, viruses are not even considered to be true living organisms. Bacteria are definitely alive but differ in important ways from protists as well as plants and animals. The smallest bacteria range from about 0.1 to 1μ[2] (even the latter, however, contains billions of molecules). The smallest viruses range down to 30 mμ. None of the small bacteria (or any of the viruses) can be accurately photographed except by the electron microscope, in which magnifications of 20,000 to 200,000 and more are possible. Bacteria and viruses are placed in a fourth kingdom called the *Monera*.[3]

It should not be assumed that the distinction between plants and animals is necessarily obvious. It is ordinarily taken for granted that all animals can move about, yet sponges and corals, as well as many other animals, are firmly attached. It is also customary to think of plants as being able to manufacture their own food,

[1] In our estimates of species numbers in the text, the protists were more or less arbitrarily apportioned between plants and animals for simplicity's sake.

[2] The unit of measurement for microscopic organisms is the *micron* (μ), which is 0.001 millimeter (mm); the *millimicron* (mμ) is 0.001 μ, or 0.000001 mm. An *angstrom* (Å) is 0.1 mμ. A thumbnail is approximately 0.5 mm thick.

[3] Estimates of species numbers in the text assumed that bacteria are plants.

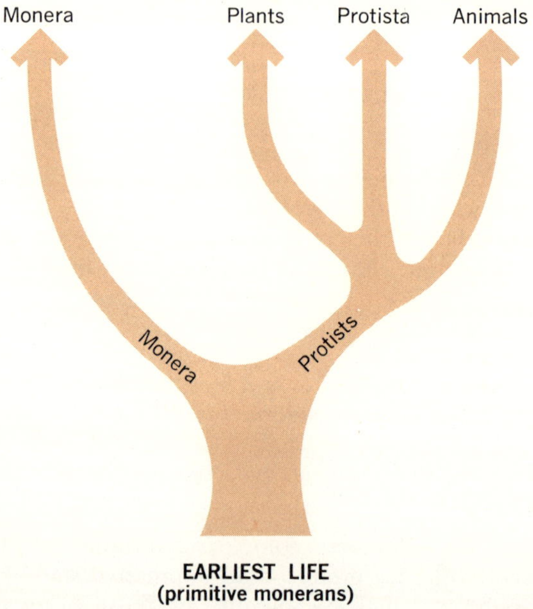

Figure 3-1 Probable evolutionary lines of descent of the four kingdoms.

whereas animals cannot. There are plant species, however, that are quite incapable of doing so; they derive their nourishment from decayed organic matter or as parasites on some living organism. It is nonetheless in the area of nutrition that *most* plants differ from *most* animals. The exceptions to this general rule are easily classified on the basis of their structure and form. Those plants that cannot manufacture their own food resemble plants much more closely than animals.

The amount of overlapping between diverse forms of life should not come as any great surprise to us. Organic evolution has been producing new kinds of organisms for more than three billion years. Most biologists believe that the first living things were not greatly dissimilar from modern monerans. Over millions of years many new species began to appear, each one adapted to a different environment. This kind of evolution is said to be *divergent*.[4] Over eons of time some of the newer forms had become so changed and altered that they were no longer monerans. This was presumably how the early protists came into being. As additional hundreds of millions of years became history, the protists, through divergent evolution, gave rise to plants and animals. This theoretical course of evolutionary events is represented schematically in Figure 3-1.

The divergence that led to the formation of four kingdoms was a response to the highly varied ecologic opportunities for energy exploitation that exist in the world. Each kingdom, in turn, boasts of great diversity within itself for the same reason. So varied are the members in each kingdom that they overlap extensively in the kinds of food sources and environments they utilize. Living on dead organic matter are monerans (soil bacteria), protists (toadstools), plants (snow plant), and animals (earthworms). Living as parasites are disease-producing bacteria, "athlete's

[4] Divergent evolution is a widely occurring phenomenon. Parasitic bacteria today, for example, are known to adapt to certain bacteria-killing (bactericidal) agents through the reproduction of resistant strains.

foot" fungus, the dodder plant, and tapeworms. As a consequence an environment includes diverse organisms which have converged upon it through their dependence on similar food resources.

Referring to our example of the web of life, we can, in summary, note two reasons for the variety of life in a given environment:

1. The consume-and-be-consumed dependencies that are part of the web itself.
2. The convergence of those diverse organisms who are dependent on similar food sources.

Levels of Organization in Organic Structure

It is a singular fact that living things, despite their great diversity, are constructed along remarkably similar lines. They are all composed of one or more usually microscopic units called *cells.*[5] The cell is the smallest structural unit capable of carrying on all the living functions (metabolism, responsiveness, growth,

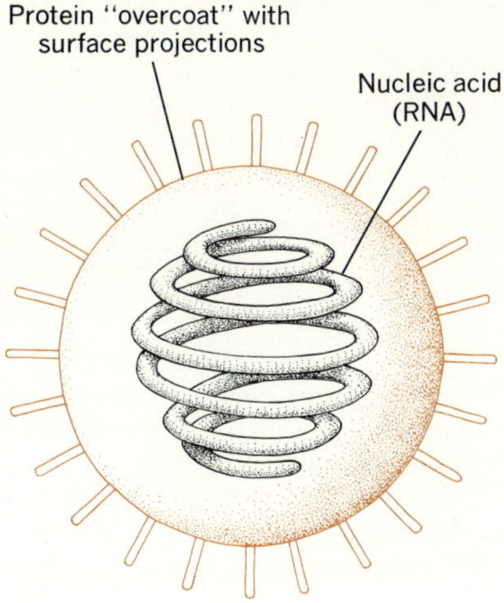

Figure 3-2 A schematic illustration of the influenza (flu) virus magnified 700,000 times. The virus is about 100 mμ in diameter.

and reproduction).[6] Some types of organisms, like a bacillus (Plate 2), *Ameba* (Plate 5), or *Paramecium* (Plate 7), consist of only one cell and are said to be *unicellular.* Sometimes unicellular organisms exist in cell aggregates known as *colonies;* the cells may be aligned end to end, forming filaments (*Spirogyra,* Plate 8) or chains (blue-green alga, Plate 4). Colonies may assume various shapes. In *Volvox* (Plate 9) the cell aggregate is spherical. A few large cells toward the rear of the sphere are specialized for reproduction only. The numerous smaller cells are capable of all vital functions except reproduction. This state of cellular specialization and interdependency is characteristic of the many-celled or *multicellular* organisms. We cannot properly speak of *Volvox* as a colony of independent, uni-

[5] This generalization is referred to as the *cell theory.*

[6] It is in this context that biologists hesitate to consider viruses as living organisms. They generally consist of a nucleic acid "core" imbedded in a protein "overcoat" (Fig. 3-2). They will reproduce and carry on metabolic activity only as parasites within living cells. Outside of cells they are biologically inactive and are often clumped together in crystalline aggregations.

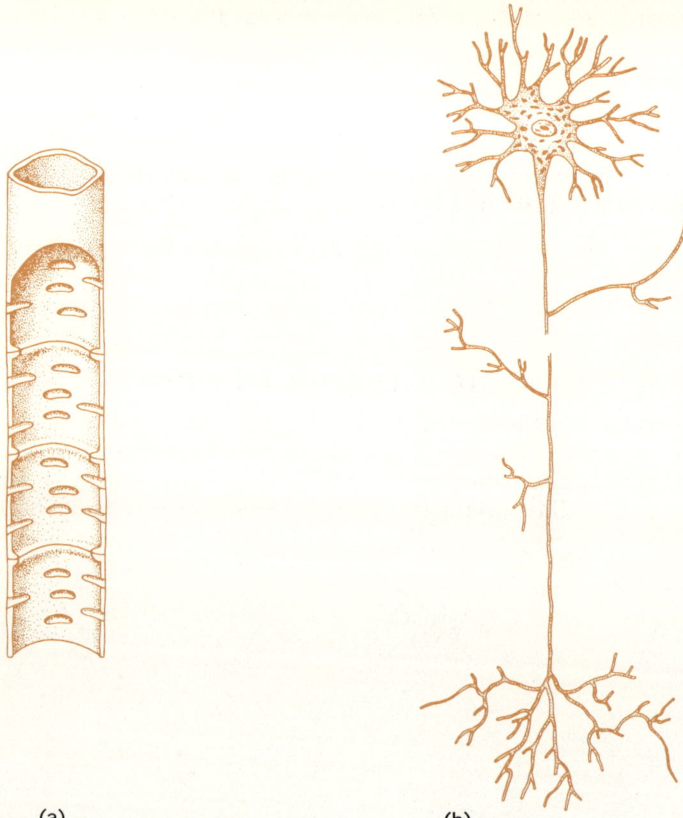

Figure 3-3 (a) A vessel from plant vascular tissue. It is a tube formed by hollowed-out cells lined up end-to-end. (b) A nerve cell from an animal. Although these cellular structures are microscopic in every other dimension, both may be several feet long.

cellular organisms; it behaves as a many-celled unit. This relatively simple level of multicellular structure is found in the larger protists such as toadstools (Plate 11) and sea weeds (Plate 12).

The division of labor whereby certain cells become specialists in the performance of a particular function is best demonstrated in plants and animals. Specialized cells which are grouped together and are usually similar in both structure and function are called *tissues*. The cells that compose tissues are often highly modified for the performance of their particular function. The nerve cells of animal nervous tissues possess long processes that conduct nerve impulses throughout the animal body. The vessels of plant vascular tissues are hollowed-out cells that are lined up end-to-end, forming a tube; they conduct water and minerals from the roots to the leaves (Fig. 3-3). The sponge (Plate 24) is a good example of an animal at the tissue level of organization.

In more complex organisms we find that several different tissues take part in the formation of an *organ*. An organ is a very complex structure that performs an equally complicated function. The leaf of a plant and the stomach of an animal are organs. The stomach contains muscle tissue for its churning action, nerves to stimulate the muscles, a glandular lining to secrete digestive juices, connective tissue to hold it together, and so on.

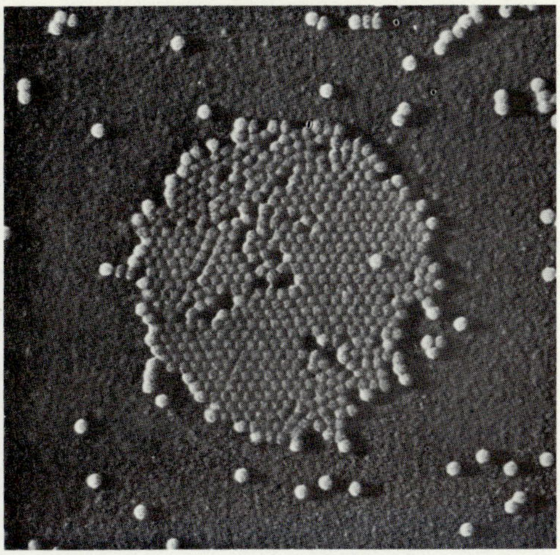

Plate 1 Polio virus (Dr. R. C. Williams) ×50,000

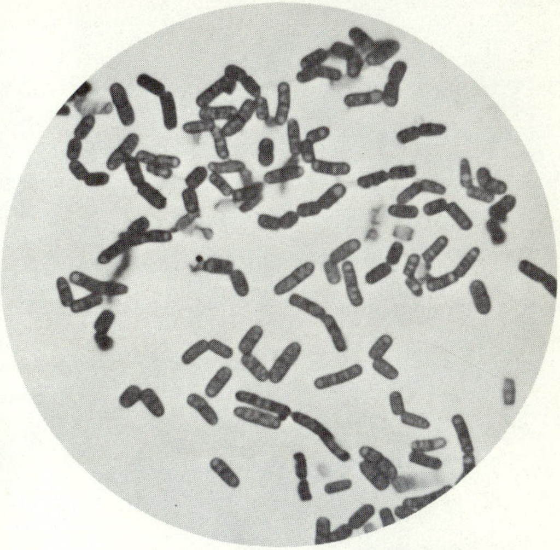

Plate 2 Bacillus (General Biological Supply House) ×2000

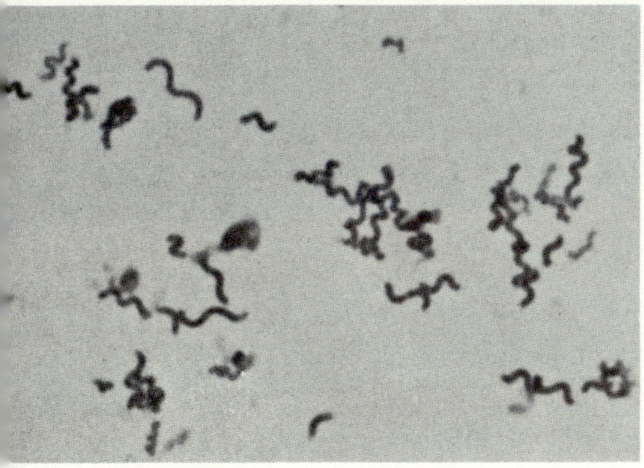

Plate 3 Spirillum (photograph by author) ×1000

Plate 4 Blue-green alga (Monkmeyer) ×320

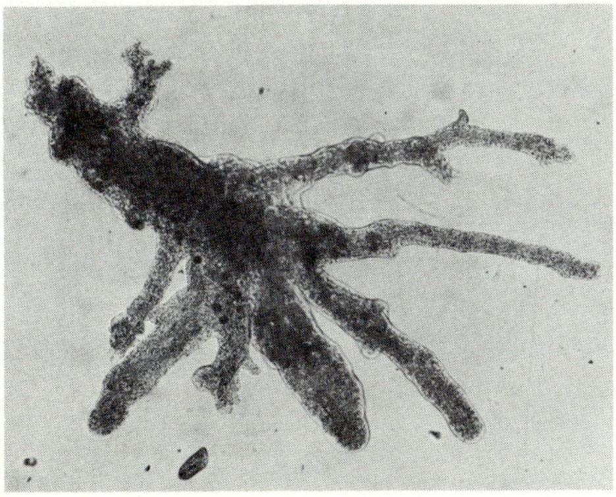

Plate 5 Ameba (Eric Grave, Photo Researchers) ×90

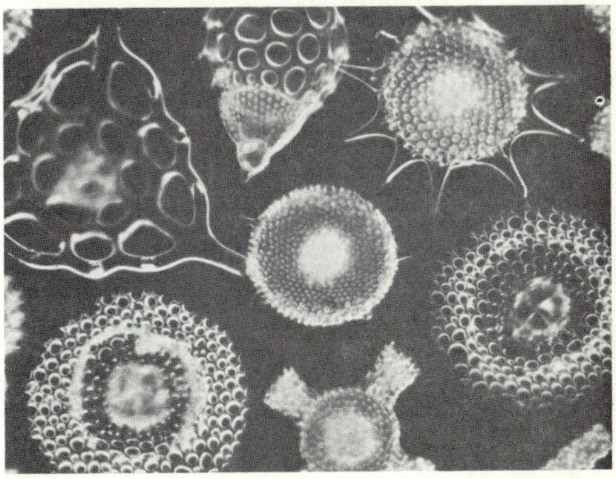

Plate 6 Radiolaria (Carl Struwe, Monkmeyer) ×150

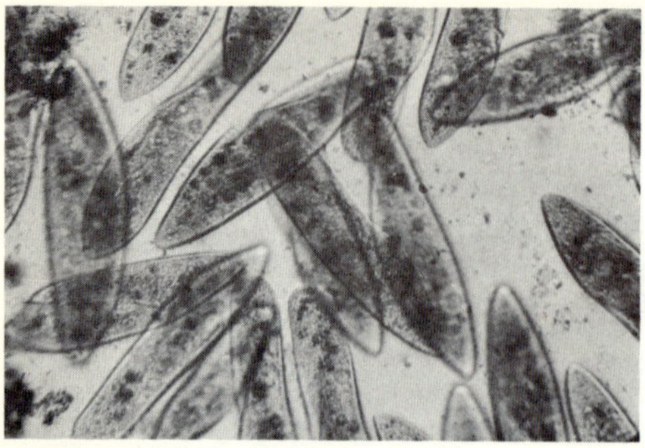

Plate 7 Paramecium (Hugh Spencer, Monkmeyer) ×300

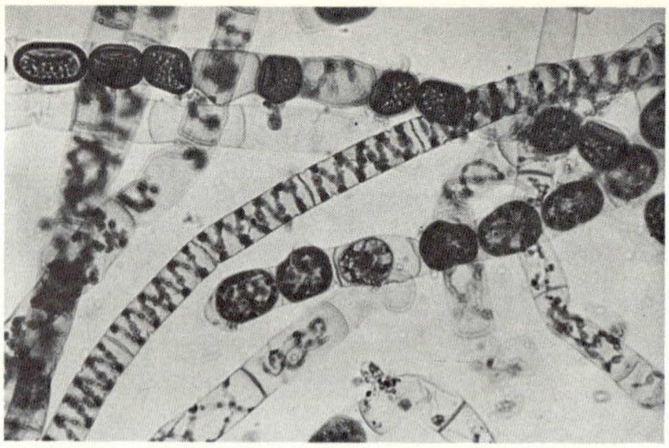

Plate 8 Spirogyra (Russ Kinne, Photo Researchers) ×300

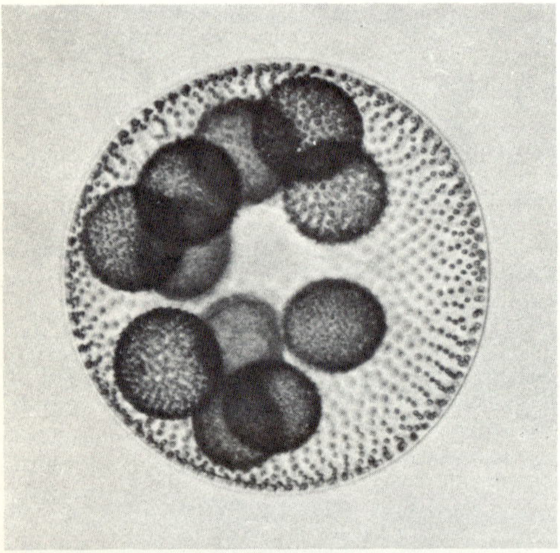

Plate 9 Volvox (Eric Grave, Photo Researchers) ×70

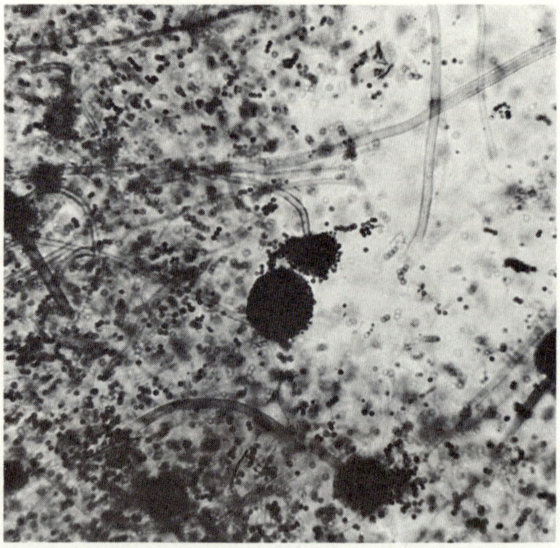

Plate 10 Bread mold (Russ Kinne, Photo Researchers) ×350

Plate 11 Toadstools (Elizabeth Hibbs, Monkmeyer)

Plate 12 Sea weed (kelp) (Phil Palmer, Monkmeyer)

Plate 13 Club moss (Dr. Wrice)

Plate 14 Horsetails (Dr. E. R. Degginger)

Plate 15 Sphagnum moss (American Museum of Natural History)

Plate 16 Ferns (Hugh Spencer, Monkmeyer)

Plate 17 Cycad (T. H. Everett)

Plate 18 Pine (U.S.D.A.)

Plate 19 Oak trees (T. H. Everett)

Plate 20 Coconut palm trees (C. Dolfinger, Photo Researchers)

Plate 21 Snow plant (Dr. E. R. Degginger)

Plate 22 Dodder (Hugh Spencer, Monkmeyer)

Plate 23 Grass (John Gerard, Monkmeyer)

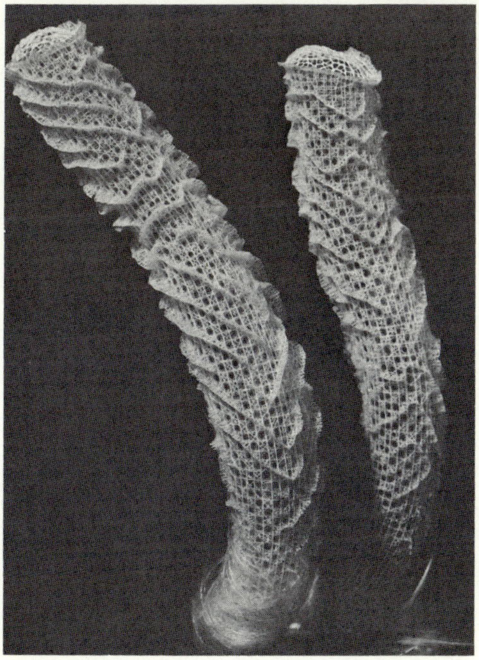

Plate 24 Glass sponges (American Museum of Natural History)

Plate 25 Jellyfish (Ron Church)

Plate 26 Coral (Russ Kinne, Photo Researchers)

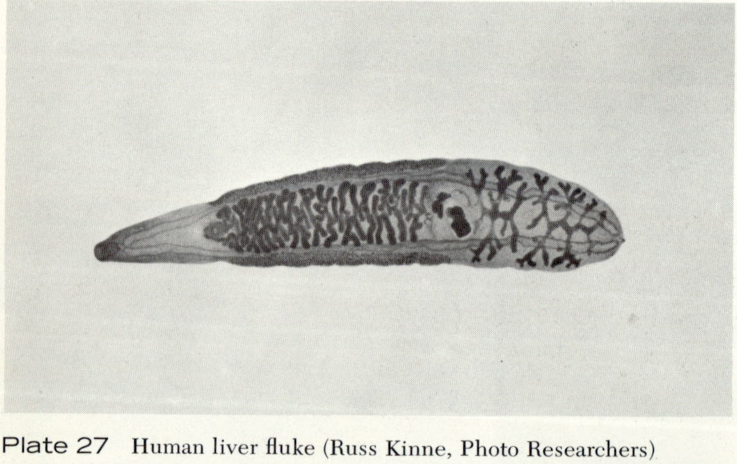

Plate 27 Human liver fluke (Russ Kinne, Photo Researchers)

Plate 28 Earthworm (John H. Gerard, Monkmeyer)

Plate 29 Snail (Laurence Pringle, Photo Researchers)

Plate 30 Octopus (Jim Annan)

Plate 31 Starfish (Pro Pix, Monkmeyer)

Plate 32 Lobster

Plate 33 Grasshopper (Rue, Monkmeyer)

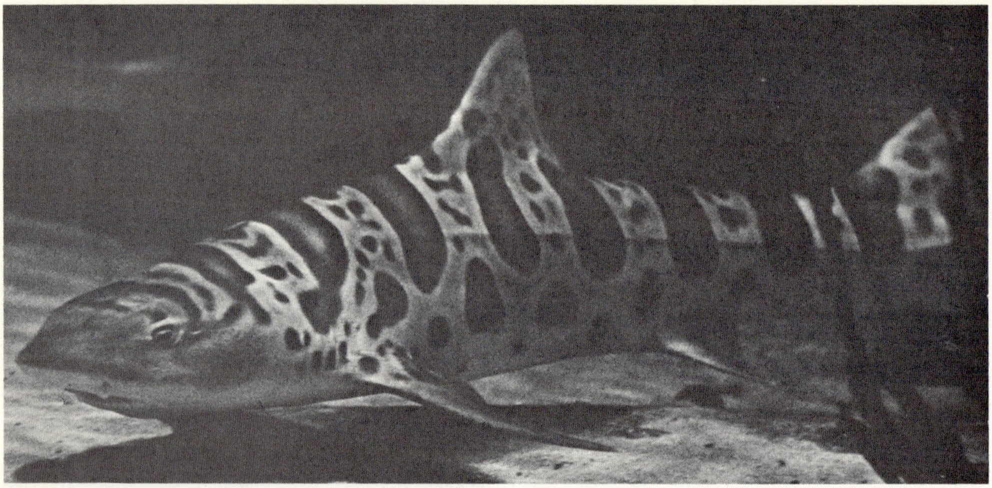

Plate 34 Leopard shark (Russ Kinne, Photo Researchers)

Plate 35 Tree frog (Russ Kinne, Photo Researchers)

Plate 36 Box turtle (Karl Maslowski, Photo Researchers)

Plate 37 White lipped snake (John Brownlie, Photo Researchers)

Plate 38 Condor (Russ Kinne, Photo Researchers)

Plate 39 Hummingbird (Russ Kinne, Photo Researchers)

Plate 40 Gopher (Verna R. Johnston, Photo Researchers)

Plate 41 Blue whale (Rijkmuseum van Natuurlijke Historie, Leiden)

Organs in turn are arranged into *organ systems* of varying complexity. A plant consists of a *shoot system* and a *root system.* The shoot system is composed of its several organs; stem, leaves, and reproductive organs (for example, flowers). The higher animals are even more elaborately structured, consisting of many coordinated organ systems (*skeletal, muscular, digestive, reproductive, nervous,* and so on). An organ system, particularly in animals, is usually associated with all of the activities related to some one vital function. The digestive system in man, for example, accommodates the function of nutrition, and entails the collaboration of many organs acting in concert: mouth, throat (pharynx), gullet (esophagus), stomach, gut (intestines), rectum, anus, and the associated glands and their secretions.

Whether the organism is simple unicellular, colonial, simple multicellular, or at the tissue, organ, or organ system level of organization, it is beautifully adapted to its mode of life and capable of the entire gamut of living processes. The phenomenon of life is found only in cells acting singly or in concert. We would therefore be remiss by not devoting the remainder of this chapter to them.

THE LIVING CELL

Cell Shape and Size

We have seen that living cells are found in a variety of shapes which are adaptive to their function (Plates 2 to 7 and Fig. 3-3). They also span a broad range of sizes. Some are relatively large, such as chicken eggs (less shell, shell membrane, and albumen). These egg cells are large because of the great bulk of nutrient material stored in them for the nourishment of the chick prior to hatching. They are exceptional. Excluding them (and others like them), the largest cells appear as tiny specks to the naked eye. We could place 1000 of the smallest bacteria end to end along a line representing the diameter of a human hair (100 μ). We could align 13 human red blood cells along such a line or three plant cells taken from a growing root tip. Why are most cells so small?

The answer to this lies in the surface-to-volume ratios of small versus large bodies. A cell takes in nutrients and gives off wastes through its surface. The total surface area is critical because it affects a cell's ability to exchange materials with its environment. The amount of nutrients needed for metabolism depends on cell

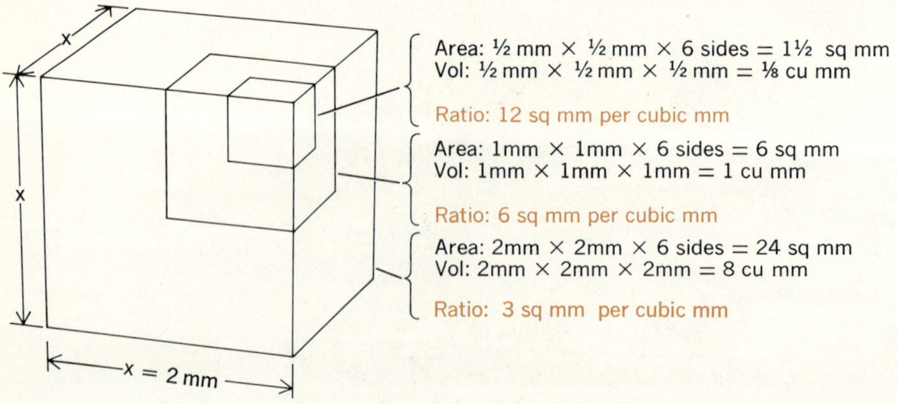

Figure 3-4 The effect of size on surface area. The surface, per unit of volume enclosed, increases as the volume is diminished.

volume. Small bodies have more surface per unit of volume than large bodies, as can be seen from Figure 3-4. No cell can have a volume whose metabolic requirements exceed the exchange capability of its surface. Sizes that provide the necessary surface for chemically active cells, as a consequence, fall within the microscopic range. The most minute bacteria, on the other hand, may have reached the minimum volume necessary to accommodate the molecular machinery of life. The P.P.L.O.'s (*p*leuro*p*neumonia-*l*ike *o*rganisms), at 100 to 150 mμ, may represent this limit.

Cell Structure as Seen through the Light Microscope

Under the light microscope a cell appears to consist of a thin, outer *cell*, or *plasma membrane*, within which a grayish, viscous fluid called *protoplasm* is contained (Fig. 3-5). Outside the cell membrane, plant cells have a rigid, nonliving *cell wall* of cellulose (plant fiber is composed of cellulose). Some protists are similarly enclosed in a nonliving, flexible *pellicle* (*Paramecium*), or a rigid *test* (shell). Animal cells, however, are limited only by their plasma membranes. The protoplasm usually includes a roundish body, the *nucleus* (plural = *nuclei*). With the exception of the blue-green algae and bacteria this nucleus is enclosed within a membrane. (Membrane-enclosed structures commonly found in protist, plant and animal cells are characteristically lacking in bacteria and blue-green algae, as is noted later in the chapter.) All cells presumably have nuclei, although these structures may, in some cases, be lost later in the life of the cell (for example, human red blood cells).

In cells having membrane-enclosed nuclei that have been stained with dyes to accentuate structural contrasts, the nucleus has the appearance of an irregular cobweb that is out of focus. This web, with its tiny dark splotches and granules, is known as the *chromatin network*. The chromatin material is actually a number of highly dispersed, long threadlike structures known as *chromosomes*, each of which is composed of complex nucleoproteins (page 48). The nucleus also contains one or more spherical, dense bodies, called *nucleoli* (singular = *nucleolus*, a "little nucleus"). They are exceedingly rich in RNA, which is produced by DNA in the nucleolus itself. The DNA molecules in the chromosomes of the nucleus have two preeminent functions which we elaborate upon later in this chapter: (1) the regulation of cellular activity and (2) the transmission of hereditary information. It is

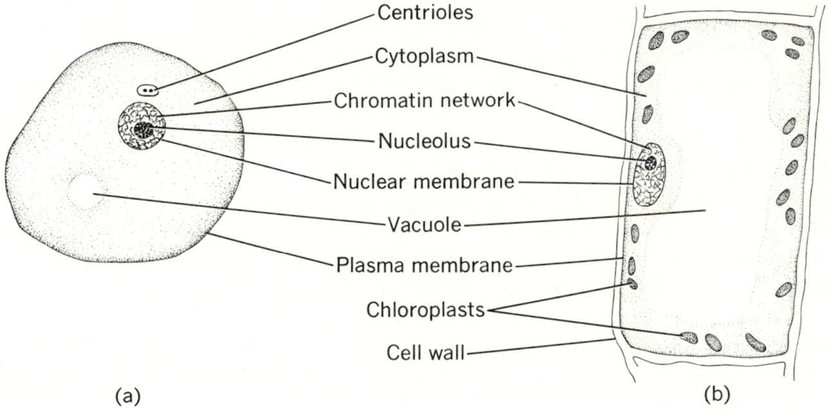

Figure 3-5 A typical animal cell (a) and plant cell (b).

consequently with good reason that the nucleus is sometimes called the head office of the cell. The protoplasm in the nucleus is referred to as *nucleoplasm*,[7] and is contained within the *nuclear membrane*.

The remaining protoplasm outside the nucleus is called *cytoplasm*. If the nucleus can be likened to the head office, the cytoplasm can certainly be compared to the operations and manufacturing division. As in any thriving corporation the nucleus (management) is as dependent on the cytoplasm (operations) as the cytoplasm is dependent on the nucleus. Within the cytoplasm a number of different structures appear which are encased in membranes. These structures are called *organelles*. The *vacuole* is a relatively simple organelle containing nutrients (for example, an organism that an ameba has just captured), water, or a metabolic product (such as waste material). The vacuoles of mature plant cells are usually much larger than those of animal cells. *Plastids* are considerably more complex organelles found in the cells of plants and some protists. *Chloroplasts* are a common form of plastid containing the green pigment, *chlorophyll*, by means of which the green plant carries on photosynthesis.

In animals and some protists a pair of dotlike objects, the *centrioles*, can be seen under the proper conditions lying near the nucleus. They serve an important function in cell division, as we shall see later. The remainder of the cytoplasm usually appears as a more or less homogeneous, translucent fluid, unless special staining techniques or optical systems (for example, phase optics) are used. That cytoplasm, and indeed protoplasm in general, is a fluid mixture is witnessed by the fact that water makes up 75 to 90 percent of the content of most cells. Protoplasm very likely contains in the neighborhood of 20,000 water molecules for every protein molecule in its structure. Let us turn our attention to the aqueous character of the living stuff.

Protoplasm as an Aqueous Mixture

Water is a unique substance in many ways. Were it not for its distinctive properties, life would not be possible. In fact, no single ingredient of protoplasm is regulated as meticulously as its water content. In the first place water is unusual in that it will absorb or give up large amounts of heat before changing appreciably in temperature itself. The *gram calorie* is defined as that amount of heat that must be added (or subtracted) in order to raise (or lower) the temperature of 1 gram of water 1° C. Almost all solvents have a much lower *specific heat* than this; that is, much less than 1 gram calorie of heat is required to change the temperature of 1 gram 1° C. Aqueous systems such as protoplasm therefore tend to change in temperature very slowly. They are less affected by external changes in temperature or by internal (exergonic) reactions that produce heat. In addition, water is an excellent conductor of heat, having about 25 times the heat conductivity of air. As a consequence of these thermal properties, water is not only an efficient buffer against radical temperature changes but is also an effective medium for heat conservation, transport, and removal.

A second unusual characteristic of water is its unparalleled solvent properties. It can almost be called a universal solvent. The fact that it is relatively inert chemically enhances its role in this regard. With very few exceptions water is the exclusive solvent in cellular mixtures. The variety of substances that may be involved in such mixtures is exceedingly large, as we noted in Chapter 2. A number

[7] The term *nuclear sap* refers to the nuclear fluid in which the chromosomes and nucleoli are bathed. Nucleoplasm includes the entire nuclear content.

of different inorganic compounds make up about 1½ percent of protoplasm. Lipids and carbohydrates add another 3 to 4 percent. The large macromolecules of proteins and nucleic acids contribute between 7 to 10 percent. The size of the molecule has a great deal to do with the nature of the mixture.

In a *solution* the solvent (water) is mixed with one or more substances (solutes) whose ions or molecules are quite small (0.1 to 1.0 $m\mu$). They are too small to be discernible even with the electron microscope. The solute particles do not differ in size greatly from the small hydrogen-bonded complexes of water molecules. A solution is homogeneous: the solute and solvent are completely indistinguishable. The solute will not settle out nor can it be separated by filtering the solution. Protoplasm is a highly complex solution in which a great number of ionic and molecular solutes are dissolved in water. Some are gases (for example, oxygen and carbon dioxide). Others consist of inorganic salts such as phosphates, bicarbonates, and chlorides of calcium, magnesium, sodium, and potassium.[8] The smaller organic molecules are involved also, including the solid monosaccharides and the liquid fatty acids, glycerol, and amino acids.

At the other extreme of the size scale there are substances whose particle dimensions are much greater than those just described, ranging around 500 μ and larger. When solid particles of this size, for example, sand grains, are dispersed in water, a *suspension* is formed. It is only by constantly agitating such a mixture that the sand grains are kept from settling to the bottom. Mixtures of this sort are heterogeneous; the dispersed substance is separate and distinct from the water. The chloroplasts in many plant cells and protists form a suspension; they are kept suspended by cytoplasmic "streaming" (flowing), cell movement, or other devices.

A third kind of mixture is produced when the dispersed particles are too large for a solution but too small for a suspension. The particles may be macromolecules or clusters of molecules. Such mixtures are called *colloids*. The dispersed particles are known as the *dispersed phase;* the dispersion medium, or solvent, is called the *continuous phase* (in protoplasm this, of course, is water). Smoke is a colloid in which the dispersed phase is carbon particles and the continuous phase is air. In homogenized milk the dispersed phase is a liquid (fat droplets) suspended in a liquid (water). Boiled corn starch is a colloid of solid particles dispersed in water.

The colloidal state is unique in that it involves the physics and chemistry of reactive surfaces more than it does the physics and chemistry of compounds themselves. The amount of surface that the dispersed phase presents to the continuous phase is immense. To illustrate this let us consider a cube that is 10 mm wide, high, and deep. The area of such a cube is 600 sq mm (10 mm $\times$ 10 mm $\times$ 6 sides), or approximately 1 sq in. If we were to divide this cube into colloidal particles 1 $m\mu$ wide, high, and deep, the combined area of the particles would be nearly 1½ acres! The number of particles comprising this area would be expressed by writing 21 zeros after the number 1 (10^{21}).

Colloids, unlike suspensions, do not settle out. There are a number of reasons for this. First, colloidal particles are small enough to show thermal agitation (the kinetic energy of moving molecules, page 37). Not only do they collide with one another, but they are in turn bombarded by similarly active water molecules. This kind of Brownian[9] movement is seen in the vibrating fat droplets of milk under a

[8] Plus small amounts of several other elements. See Figure 2-1.

[9] Named after Robert Brown who first described the phenomenon in 1827, based on observations of pollen grains under the microscope.

microscope. In the second place colloidal particles are large enough to *adsorb* ions on their surfaces (a thin film of water adheres to your finger by adsorption). Colloids consequently become either positively or negatively charged. Since like charges repel, this tends to keep the similarly charged particles dispersed. Some colloids, for example, smoke, will settle out if these charges are neutralized. Sometimes the dispersed phase has a high adsorptive capacity for water, which further increases dispersion. Each particle attracts a concentration, or halo, of water molecules around it, often with great tenacity, forming a stable colloid. The protein colloids of protoplasm are of this nature. Relatively dry protein (for example, in seeds) imbibes (adsorbs) large amounts of water in this way. When gelatin has adsorbed four times its own weight in water, it will continue to take in more.

Colloids differ from solutions in a number of ways. In the first place colloidal particles are large enough to be photographed in the electron microscope. They will scatter light, so that we can see shafts of light containing colloidal dust (for example, light filtering through trees in a forest). They are too large to pass through the pores of a cellophane or collodion membrane; solutes (crystalloids) in solution have no trouble in doing so. Biological membranes similarly restrict the passage of large colloidal particles. Another difference has great significance to protoplasmic structure; colloids have the capability of forming mixtures with such mechanical characteristics as firmness, elasticity, and tensile strength. This property relates in particular to the protein colloids we have mentioned which have an attraction for water. Let us examine this feature in more detail.

The protein gelatin, when mixed with hot water, forms a semifluid colloid. When this is placed in a refrigerator, it becomes a familiar edible jelly. If the mixture is again warmed, it assumes the original semifluid state. The semifluid state is said to be a *sol;* the semisolid state is called a *gel.* In this example the process of becoming a sol (*solation*) or a gel (*gelation*) depends on temperature. Transformations back and forth between gel and sol may occur readily. This colloid is consequently said to be *reversible.* Gelation and solation may be brought about by a number of factors other than temperature. In the living cell the more important of these are (1) the relative abundance of such ions as those of sodium, potassium, calcium; (2) the concentration of the dispersed phase; and (3) hydrogen ion concentration. The pH can affect the electric charge carried by a colloid.

The biological importance of the gel state lies in its high viscosity; the colloidal particles resist flowing by or over one another. The mixture is relatively firm and elastic, as if it were made of rubber. Most protoplasmic gels are formed by long polymers of fibrous protein (page 46). These are disposed in intertwining, three-dimensional arrays which have been likened to a brush heap (Fig. 3–6a). The brush-heaped chains are linked by bonds of varying strengths (such as covalent and hydrogen bonds). If the polymers are interlocking, branched chains linked by many covalent bonds, the gel will be very stable and *irreversible.* If, on the other hand, the chains are of varying length and secured in place by a relatively few hydrogen bonds, the gel will be less stable and readily converted to its sol. It is interesting to observe in Figure 3–6a that a very stable, irreversible gel could easily contain 80–90 percent water and still retain the firmness, elasticity, and strength required of the structural matrix out of which an organelle could be built.

This observation becomes all the more convincing in the appearance of *coacervates. Simple coacervates* are formed when several colloidal particles join or coalesce into one larger particle or element (Fig. 3-6b). *Complex coacervates* result when several colloidal particles from two different colloidal systems bearing

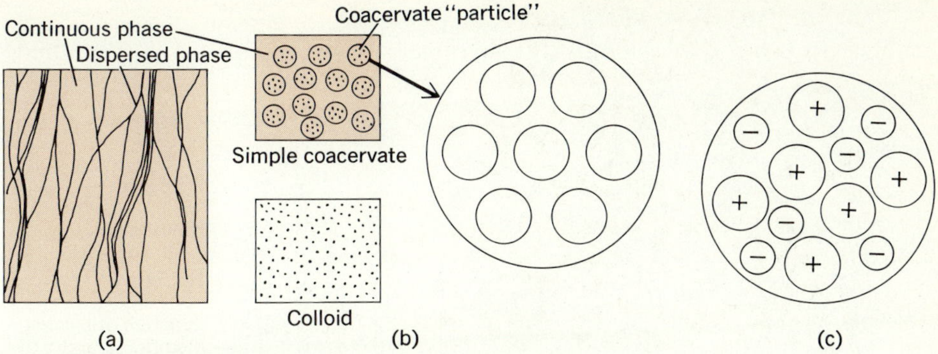

Figure 3-6 Colloids and coacervates. (a) Brush-heap gel of fibrous protein; (b) colloid and simple coacervate compared; (c) complex coacervate "particle."

opposite charges coalesce into a larger particle (Fig. 3–6c). These are frequently found in protoplasm. The nucleoproteins, which are an important part of chromosome structure, are complex coacervates incorporating basic proteins (histones) and nucleic acids.

In summary, protoplasm is indeed a complex mixture. Even in a small part of one cell it will contain several different proteins (kinds of dispersed phases). Some of these will be fibrous proteins forming structural gels. Others will be metabolically active globular proteins in the sol state. The same water which serves as the continuous phase for all of these proteins is also the common solvent for many different, chemically active solutes. In this solution-colloid complex additional substances or bodies form suspensions. The stable gels form the structural framework (or matrix) of diverse organelles. This framework is encased, and in some cases partitioned or layered, by membranes which, as we will see later, are complex (but remarkably uniform) coacervates themselves. This structured, colloidal gel is bathed in chemically active solutes and protein sols that can alter the physical structure of the gel. This in turn regulates the varied biologic activity of the organelle. Other less stable gels form temporary or transitory organelles. When their purpose has been served, they undergo solation so that the cell is not encumbered with inactive structures. Still other colloidal reactions occur in relatively homogeneous cytoplasm. Chloroplasts move around the plant cell in a flowing sol. An ameba moves by flowing out fingerlike projections (pseudopods) in the direction of travel. These then undergo gelation as the remainder of the organism becomes a sol and flows into them.[10]

With this introduction to the principles of subcellular architecture behind us, we can now profitably direct our discussion to these ultrasmall structures themselves.

The Ultramicroscopic Structure of Protoplasm

The invention of the light microscope late in the sixteenth century led to the discovery of the cell and man's first look at the "living mixture," which came to be known as protoplasm. The development of the electron microscope in the mid-1930s has now extended man's horizons into the ultramicroscopic world of molecular complexes. At magnifications of 700,000 and more protoplasm looks a great deal

[10] This *ameboid movement* is characteristic of many protists. It also occurs in certain white blood cells of man and other vertebrates.

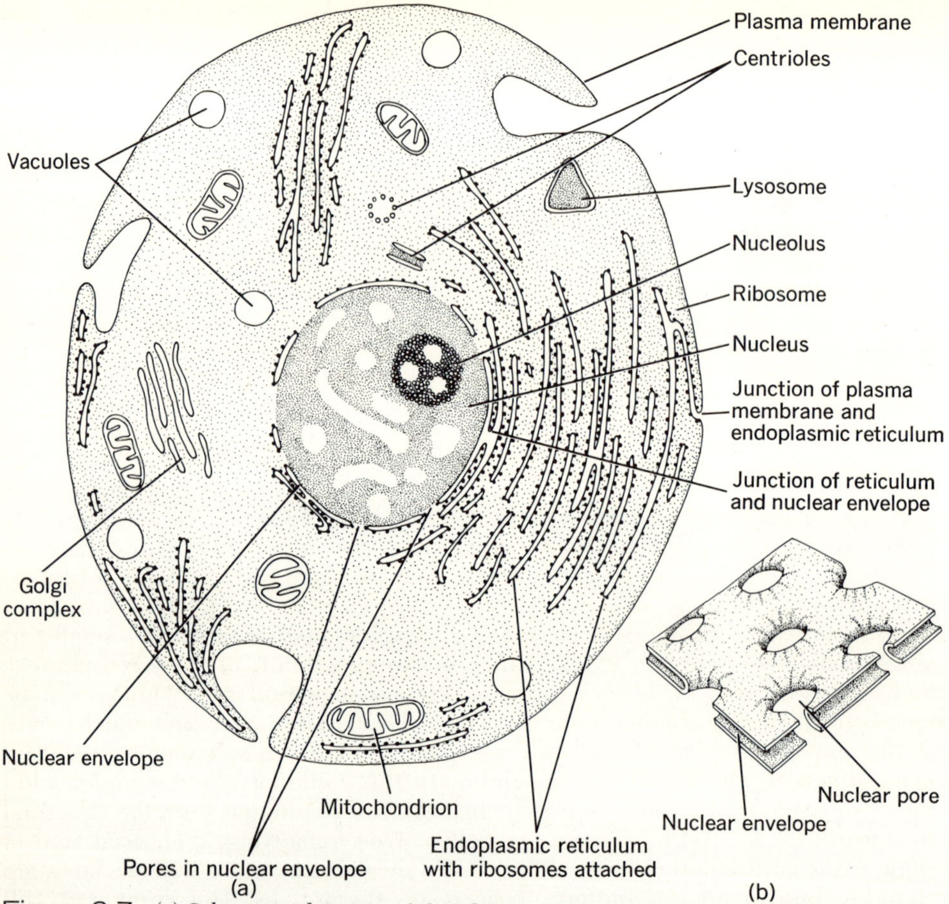

Figure 3-7 (a) Schematic drawing of the ultramicroscopic structures of a typical cell; (b) detailed drawing of portion of nuclear envelope.

different from that at magnifications of 2000. The vague impression of a complex fluid is replaced by an architectural design of elaborate detail (Fig. 3-7). A variety of organelles, with dimensions of 6.5 to 700 mμ or more, appear as sausagelike vesicles, channels, stacked plates, flattened sacs, or bundled filaments. A term as nebulous as protoplasm no longer carries any real meaning for this highly ordered, colloidal machinery. A brief survey of these ultramicroscopic structures will verify this.

MITOCHONDRIA

The mitochondria (singular = *mitochondrion*) (Figs. 3-7 and 3-8a and b) are organelles of singular importance since they are the primary supplier of cellular energy. Their shape and number vary in different kinds of cells. They have not been found in monerans, but are very common organelles in all protists, plants, and animals. Shapes vary from spheres to ovals, rods, or long filaments. Their diameter usually falls within a range of about 0.2 to 1.5 μ. Length, of course, varies with shape. Some of the filamentous forms are up to 400 μ long. Mitochondria are quite numerous in unicellular organisms, particularly the larger species. One of the giant amebas may contain 500,000. In multicellular organisms they are most numerous in metabolically active cells. Even a small liver cell (a cube approximately 12 μ on a side) will have several hundred to well over a thousand. Gland, muscle, and nerve cells contain similarly large populations.

In some highly specialized cells (for example, muscle tissue) the mitochondria occur in fixed positions, whereas in other cells they appear to be capable of movement; they are by no means rigid structures. Some are swept along by cytoplasmic streaming. They expand or shrink in size depending on the metabolic activity within them. Within any given cell they tend to congregate where the action is. Within a dividing cell, for example, they will be concentrated around the division plane. They are capable of independent division themselves. As a cell divides their numbers will be more or less evenly apportioned between the two resulting daughter cells. In fact, they have been shown to contain DNA, and are capable of synthesizing some of their own protein by means of mitochondrial DNA-RNA system.

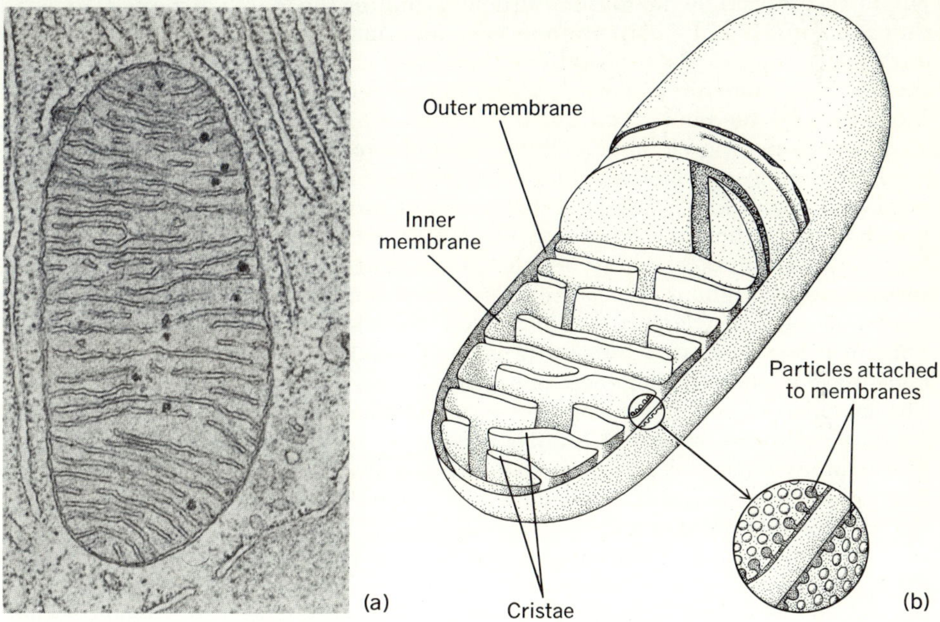

Figure 3-8 The mitochondrion. (a) Electron micrograph of a mitochondrion, magnification 30,000×; (b) schematic drawing. [(a) Dr. K. R. Porter, Harvard University]

Structurally a mitochondrion is composed of an outer and an inner membrane. The inner membrane is folded into incomplete partitions called *cristae*. These may be oriented in a crosswise (transverse) fashion or in a lengthwise (longitudinal) manner. In some protists the cristae are tubular rather than flat projections of the inner membrane. In any case the cristae greatly increase the surface of the inner membrane. Attached to this surface are large numbers of molecules which are thought to be distributed in orderly arrangements. These molecules are globular proteins called *enzymes*, and are characterized by great metabolic activity. We will have more to say about them later on in this chapter. Enzymes are also found between the two membranes and in the inner space of the mitochondrion itself. In the monerans enzymes similar to those found in mitochondria appear along the inner surface of the plasma membrane (except, of course, in the viruses).

Mitochondria have been aptly called the powerhouses of the cell. Practically all of the energy requirements for cellular activity are met by redox reactions which take place in these organelles. By an ingenious mechanism, which we will investigate later, mitochondrial energy production is regulated by the rate of cellular energy consumption.

PLASTIDS

Plastids (Figs. 3-5b and 3-9a and b) are organelles that are found only in plants and those protists which are capable of photosynthesis. Of the several types of plastids the *chloroplasts* are the most important as well as the most common. Wherever chlorophyll is found it is always contained, with the exception of blue-green algae and bacteria, in these organelles. The chloroplasts of protists are of various shapes, usually rather large, and few in number. In *Spirogyra* (Plate 8) the chloroplast is a single, long, spiral structure running the length of the cell. In plants they are considerably more uniform in shape, smaller, and more numerous. They are usually shaped like a magnifying lens with a diameter between 4 to 8 μ. Although 50 or more are commonly found in one cell, their number will vary in the life of the cell, the kind of cell, and the species of plant from which the cell is taken.

The bounding membrane of the plant chloroplast is similar to the mitochondrial membranes. In fact, all cellular membranes are so remarkably alike that the term *unit membrane* is used. (A discussion of this structure occurs later.) Within the chloroplast there are 40 or more structures, each of which bears a resemblance to a stack of hotcakes. These are the *grana* (singular = *granum*). The grana are distributed throughout the matrix or *stroma* of the organelle. The stroma consists mostly of proteins (largely enzymes) and granules of RNA in a sol state. Each granum is made up of a number of layers (hotcakes) called *lamellae* (singular = *lamella*). A lamella is a flattened, membranous sac, like a collapsed rubber balloon. The lamellae of one granum appear to interconnect with the lamellae of adjacent grana, possibly by means of tubular extensions similar to the neck of a balloon. These interconnections pass through the stroma. Each lamella contains chlorophyll. The connecting channels between the lamellae of two or more grana may also contain this green pigment.

It is by means of chlorophyll that the plant captures the radiant energy of sunlight. The chloroplast utilizes this energy, in a series of endergonic reactions, to produce energy-rich organic compounds, such as glucose. This photosynthetic process is a very efficient one. Its efficiency is predicated on the spatial arrangement of chlorophyll and the enzymes of the stroma. The distribution of chlorophyll in stacks of flattened, membranous sacs is an excellent device for the absorption of solar energy. This proliferation of reacting membrane surface also makes possible a much greater opportunity for contact between reactant molecules. This is further enhanced by the immersion of the grana and their connecting channels in the enzyme-rich sol of the stroma. Chemical reactions among the countless macromolecules spaced along these expansive surfaces occur with great speed and precision. Chlorophyll and its associated enzymes in photosynthesis manufacture organic compounds by reason of their ordered distribution within the architectural framework of the chloroplast. In performing this function chloroplasts not only serve the photosynthetic organisms of which they are a part, but the entire living world as well. All living things ultimately depend on the energy-rich organic compounds synthesized by them.

Plants and plantlike protists also contain two additional kinds of plastids. *Leucoplasts* lack chlorophyll. Some synthesize starch. A large part of a potato, for

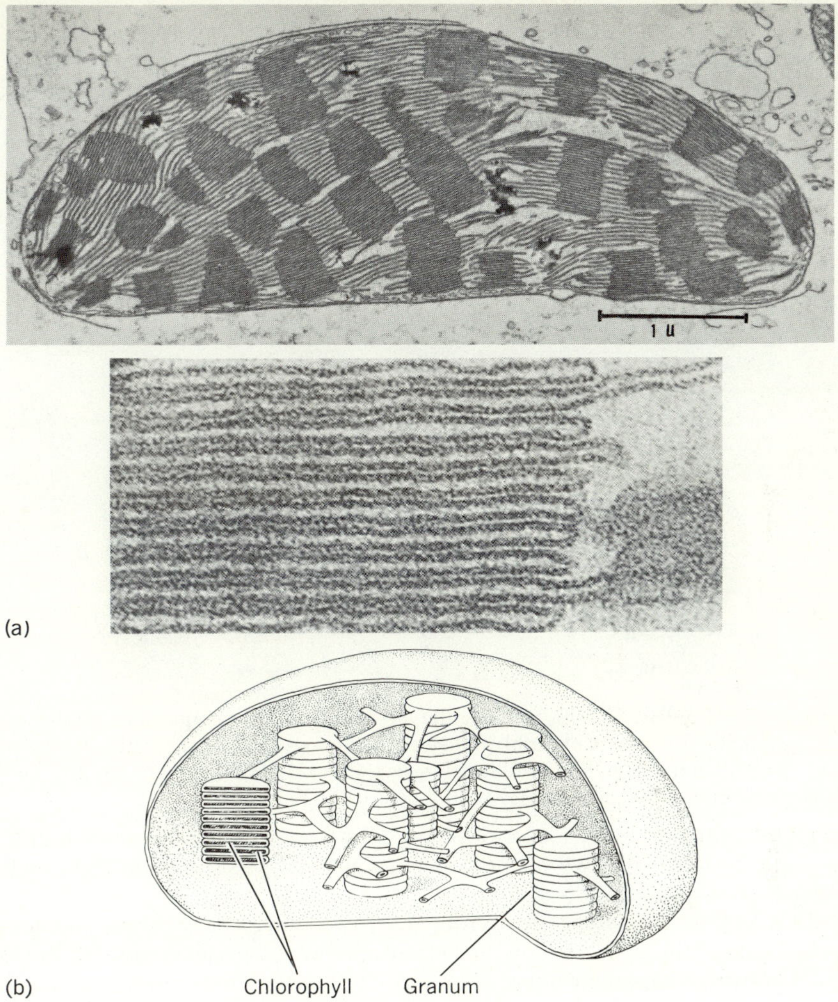

Figure 3-9 The chloroplast. (a) Electron micrograph of chloroplast of corn leaf cell, ×38,500. Inset shows the detail of a position of an individual granum, ×250,000; (b) schematic drawing. [(a) Dr. Jaime Rosado-Alberio, University of Puerto Rico; inset Dr. L. K. Shumway, Washington State University]

example, consists of leucoplasts and the starch stored in them. Some synthesize fat; others manufacture protein. *Chromoplasts* represent formerly active chloroplasts; they are responsible for the colors (other than green) of fruits, flowers, and autumn leaves.

Like mitochondria, plastids are also capable of self-reproduction and possess a DNA-RNA system which enables them to synthesize some of their own protein.

CENTRIOLES, CILIA, AND FLAGELLA
(Figs. 3-7 and 3-10a, b, and c)

Centrioles are found in animal cells, many protists, and the reproductive cells of some plants. They are interesting structures in a number of ways. Unlike the mitochondria and plastids, they do not appear to possess a bounding membrane. They are formed instead of 27 filaments which are grouped into 9 triplets. The triplets are arranged in the form of a short tube or cylinder about 150 mμ in di-

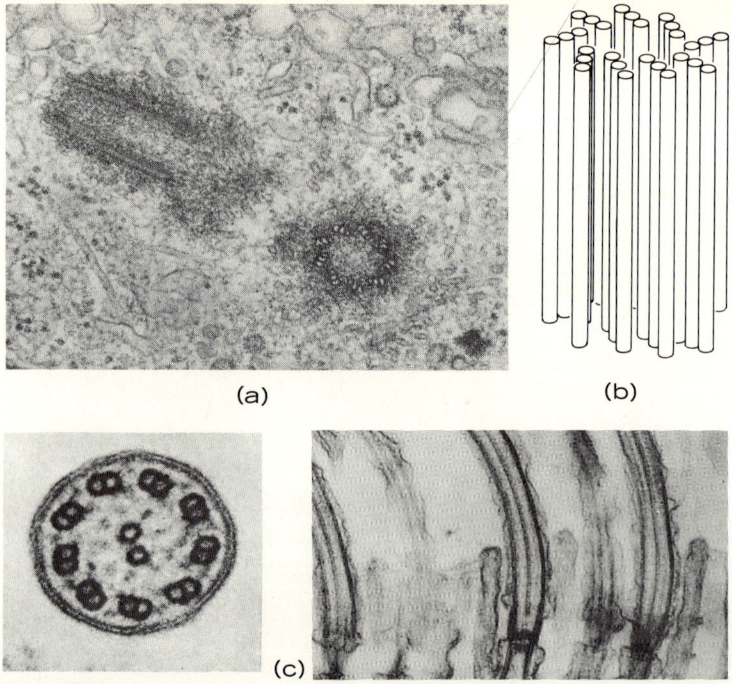

Figure 3-10 Centrioles, cilia, and flagella. (a) Electron micrograph of centrioles in bone cell; (b) schematic drawing of centriole; (c) electron micrographs of cilia (right, in longitudinal section, ×34,000) and a flagellum (left, in cross section ×127,500). [(a) D. F. Bainton and M. G. Farquhar, University of California; (c) I. Gibbons, University of Hawaii]

ameter and 300 mμ long. Centrioles are found near the nucleus of a cell, and there are almost always two of them. One is usually positioned at right angles to the other.

As a cell is preparing to divide, the two centrioles separate and move to opposite ends of the cell. As they do so they appear to be responsible for the formation of a spindle-shaped structure (the *mitotic figure*). This spindle mechanism is of great significance to cell division because it makes possible the equal distribution of hereditary material between the two daughter cells. We will discuss this process at considerable length in later chapters. It is interesting to note at this point, however, that each daughter cell receives one of the two centrioles present in the original cell before it divided. The single centriole in each daughter cell then divides (reproduces itself), so that each cell has its normal complement of two centrioles.

Another rather singular feature of these structures is their similarity to, and close association with, *cilia* and *flagella.* These hairlike projections of the plasma membrane are used by protists, through a whipping action, as a means of locomotion (*Paramecium,* Plate 7 and *Volvox,* Plate 9). Although cilia are much shorter and usually more numerous than flagella, they are otherwise identical in structure. They are by no means confined to the protists. A number of plants produce reproductive cells that are motile because they possess flagella. The sperm of an animal moves by reason of its tail, which is a flagellum. The surfaces of the air passages to our own lungs are cleansed by currents of mucus which are swept along by the beating cilia of lining cells. The widespread use of these structures in the living world is made all the more significant by the fact that their structure is so remarkably consistent regardless of where they are found. Only bacterial flagella are exceptional.

Each cilium (or flagellum) originates from a basal body just within the plasma membrane. The structure of the basal body bears a striking resemblance to a centriole. It too is capable of duplicating itself (and producing another cilium). If not a centriole itself, it is probably at least derived from one. The cilium is similar in basic design to a centriole. It is, of course, a much longer body that projects outside the cell. It is also invested in a membrane which is actually an extension of the cell membrane. But the arrangement of fibrils within the axis of a cilium or flagellum is but a slight modification of the pattern seen in the centriole: nine pairs of fibrils in a ring, with two additional filaments in the center.

A seemingly unusual application of these structures to biological function is found in the light-sensitive cells of the eye. The outer segments of these cells (the *rods* and *cones*) are modified cilia which develop from centrioles.

THE ENDOPLASMIC RETICULUM AND RIBOSOMES

Under the electron microscope much of the cytoplasm of plant and animal cells is shown to be occupied and partitioned by an expansive complex of tiny membrane-bounded cavities which commonly run parallel to one another. It is often less well developed in protists and absent in monerans. This membrane system is known as the *endoplasmic reticulum* (Figs. 3-7 and 3-11a and b). It is actually a system of tiny spaces varying in shape from tubular canals to roundish vesicles and broad, shallow caverns. These in turn are variously joined by interconnecting channels. Their bounding membranes are also continuous with the plasma membrane externally and the double membrane of the nucleus internally. As a result the minuscule channels and sacs of the reticulum are continuous with the exterior of the cell as well as the cavities within the double membrane of the nucleus. Both the plasma and nuclear membranes are a part of this extensive membrane system.

A number of functions have been ascribed to the endoplasmic reticulum. It is thought that the reticulum serves as a transport system for the movement of substances throughout the cytoplasm. Channels are available whereby materials outside the cell have access all the way in to the nuclear membrane. These same channels, of course, also allow for the movement of substances synthesized within the cell to the exterior. The folded membranes of the reticulum also take part in the synthesis as well as the transportation of cellular products. Just as in the case of the mitochondria, enzyme molecules are attached to these highly folded membranes also. The extensive surface area of the membranes is an ideal design for the synthesis of molecules. For example, metabolically active enzyme molecules are known to be instrumental in the manufacture of lipids in certain parts of the structure.

The endoplasmic reticulum also plays an important accessory role in the synthesis of certain proteins. Protein synthesis is carried on in the cytoplasm through the assembly of amino acids by innumerable spherical bodies called *ribosomes* (Figs. 3-7 and 3-11a and b). They are about 15 $m\mu$ in diameter and are composed of RNA and proteins. Ribosomal RNA is of the same type as that found in the nucleolus, and is supplied to the ribosomes by that nuclear structure. Cells that are actively engaged in protein synthesis contain many ribosomes, which are distributed throughout the cytoplasm in varying arrangements, depending primarily on the kind of cell and its activity. Some of them are attached to the cyto-

(a)

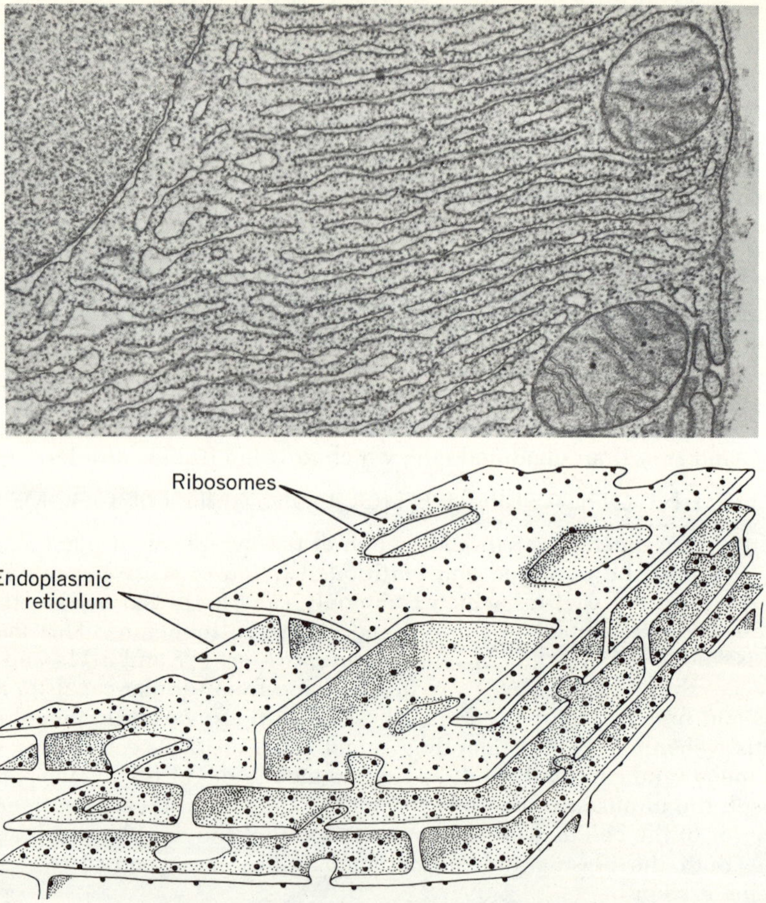

(b)

Figure 3-11 The endoplasmic reticulum and ribosomes. (a) Electron micrograph of endoplasmic reticulum of a pancreas cell, ×35,000; (b) schematic drawing. [(a) Dr. G. E. Palade]

plasmic edge of the reticular membrane. These portions of the endoplasmic reticulum to which ribosomes are attached are said to possess rough membranes. Reticular membranes that lack attached ribosomes are called smooth; those referred to previously as synthesizing lipids are of this type. The distributional pattern of ribosomes on rough membranes apparently has a great deal to do with the kind of protein being manufactured. The difference in function between the attached and unattached ribosomes is still not thoroughly understood. It may be that the unattached ribosomes in the cytoplasm are primarily concerned with the synthesis of structural proteins, whereas those comprising the rough membrane system are involved with chemically active proteins (enzymes) and others that serve some purpose outside the cell. The evidence for this, however, is inconclusive.

THE GOLGI COMPLEX

The *Golgi complex* (Figs. 3-7 and 3-12a and b) has been a source of controversy among biologists ever since its discovery in 1898. There is still much to be learned about the function of these organelles. They seem to be particularly common and well developed in animal cells which produce secretions, such as gland cells and nerve cells. They are not restricted to these kinds of animal cells, however, nor are they limited to animals. The Golgi complex occurs widely in the cells of both protists and plants.

These organelles commonly take the form of stacks of about four to eight or more membrane-bounded sacs. In the central region the piled sac membranes are close to, and parallel with, one another. At the margin the sacs are usually expanded into bulbs. Closely associated with the margin of the Golgi complex are separate, spherical vesicles and vacuoles which resemble the Golgi material. It appears as if these vesicles arise by a pinching off of the expanded margins of the sacs.

Recent studies indicate that the Golgi complex plays an important role in the synthesis of carbohydrate macromolecules. These substances are used in the production of important secretions, such as the mucus of animal cells and the cell wall materials (cellulose and pectin) of plant cells. The synthesized molecules are then packaged in the Golgi vesicles and carried to the cell surface, where they are discharged.

It is also believed that the Golgi apparatus may give rise to lysosomes.

(a)

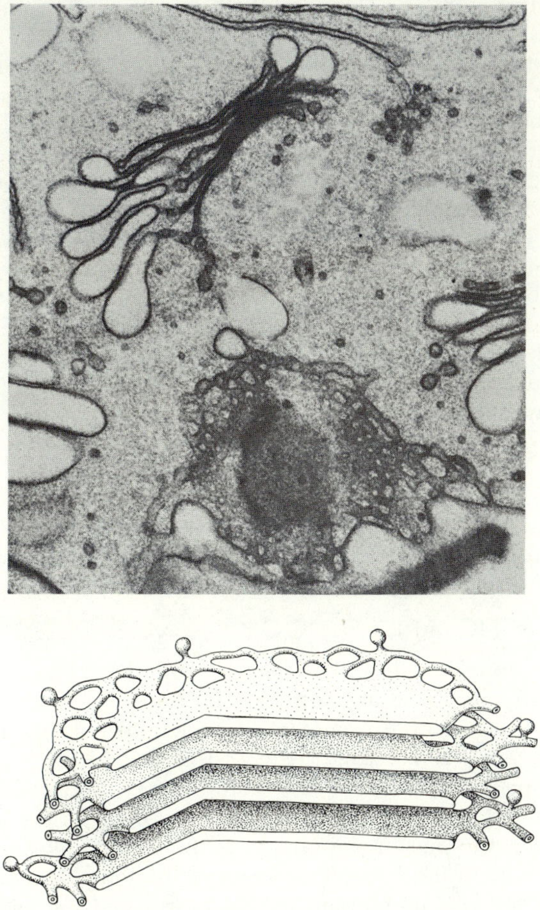

(b)

Figure 3-12 The Golgi complex. (a) Electron micrograph of Golgi complex of corn root cell in surface (bottom) and transverse (top) views; (b) schematic drawing of Golgi complex and vesicles. [(a) Dr. Hilton H. Mollenhauer]

LYSOSOMES

Lysosomes (Fig. 3-7) are spherical, membrane-bounded vacuoles which contain hydrolytic enzymes. These enzyme molecules, in other words, are capable of bringing about the hydrolysis of organic compounds. As a matter of fact, lysosomes carry a sufficient variety of enzymes to hydrolyze, or digest, any of the organic macromolecules: carbohydrates, lipids, proteins, or nucleic acids. Although these organelles vary in size, they are usually about as large as average-sized mitochondria. Lysosomes have not as yet been observed in plants, but they are commonly seen in the cells of animals, animal-like protists that capture their food (for example, *Paramecium* and *Ameba*), and some of the plantlike protists, notably the fungi and yeasts.

The lysosomal enzymes are very likely synthesized in the rough membranes of the reticular system. From here they are presumably transported, via the reticular channels, to the Golgi complex where they are packaged into the baglike lysosomes. It is essential that these enzymes be separated at all times by a membrane from the rest of the cellular contents, otherwise the cell could be destroyed by them.

The destruction of the cell, strangely enough, seems to be one of the functions of lysosomes. For this reason they have been called "suicide bags." There are many instances in which the removal of cells is desirable, for example, dead or injured cells or cells that no longer serve a purpose. As a tadpole changes, or metamorphoses, into a frog it gradually loses its tail. This loss of a no longer useful structure is accomplished in part by lysosomes in the cells of the tail. As we grow from infants to adults our bones change in shape as well as size; these changes are made possible by the lysosomal destruction of bone tissue here and there as well as by the addition of new bone. In cases of this kind the membranes of the lysosomes are either ruptured, dissolved away, or become no longer capable of restraining the contents and leak them out. Lysosomes are also useful in the removal of wornout organelles. Partially digested mitochondria have been photographed within lysosomes. In this way aging cellular "machinery" or organelles with strictly temporary functions are prevented from cluttering up the cell. It is small wonder indeed that lysosomes have a highly varied internal structure; their contents can run the gamut of a considerable miscellany. In protists like *Paramecium* and *Ameba* lysosomes assume a digestive function. They empty their contents into food vacuoles containing organisms the protist has captured. Lysosomes in our own white blood cells perform a similar service on bacteria the cells have "eaten."

PHAGOCYTIC AND PINOCYTIC VESICLES

The food vacuoles just described can be formed in two rather different ways, depending primarily on the nature of the food. First, if the food is fairly good-sized, solid material, such as another organism, the vacuole is formed by *phagocytosis*, which literally means "cell eating" (Fig. 3-13a and b). It is commonly accomplished by flowing extensions of the cytoplasm called *pseudopodia*. Liplike pseudopodia gradually surround and enclose the prey. As the pseudopodia meet they fuse. What was the plasma membrane of the facing edges of the pseudopodia is now the membrane of the vacuole containing the food. The vacuole membrane is detached from the cell membrane at this point, and the vacuole is free to be moved about the cell. Lysosomal enzymes discharged into the vacuole now proceed to digest the contents. The smaller molecules formed by hydrolysis (monosaccharides, fatty acids, amino acids, and so on) have no trouble passing through the

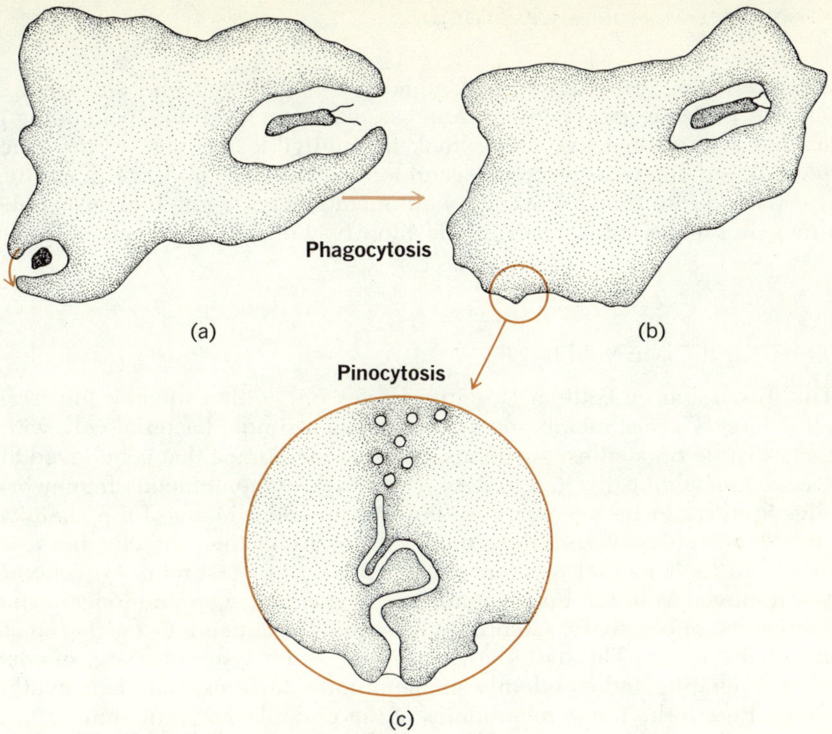

Figure 3-13 Phagocytosis and pinocytosis in an ameba. (a) Fingerlike processes called pseudopodia flow around a small protist. An old vacuole is also shown discharging its undigested contents to the outside. (b) The small protist has been completely surrounded and enclosed in a vacuole. The inset (c) illustrates a pinocytic channel and several pinocytic vesicles.

membrane of the vacuole into the surrounding cytoplasm. Any indigestible residues that remain are voided by a process that is the reverse of vacuole formation. At this time the depleted food vesicle is much reduced in size. It is moved to the plasma membrane. The membranes once again fuse as the vesicle disappears, releasing its contents to the exterior.

A second method of vacuole, or vesicle, formation involves substances that are very small, but not small enough to pass through the cell membrane. Macromolecules, for example, are too big to pass (diffuse) through the submicroscopic, molecular "pore spaces" of the plasma membrane. They are obviously much too small to be taken in by phagocytosis. Yet macromolecules, such as proteins, are taken into cells. The process by which this is done is called *pinocytosis* or "cell drinking" (Fig. 3-13c). It occurs widely in the cells of animals, many protists, and at least some plants. Pinocytosis entails a response in which the substance to be taken in becomes bound to the plasma membrane. Thereupon small indentations begin to appear in the cell membrane. These may take the form of cups up to 50 μ across. More often they are narrow, cleftlike channels, sometimes no more than 1 to 2 μ in diameter, that may extend deep into the cytoplasm along an irregular path. The membrane-bound molecules or particles pass down these pinocytic channels. Near the bottom, but above a group of molecules or particles, the channel pinches off, forming a small pinocytic vesicle, which then moves through the cytoplasm. The course of its progress at this point is similar to that of the phagocytic vacuole or vesicle.

An interesting experiment with amebas showed that one substance can stimulate the pinocytosis of another. Amebas placed in a solution containing glucose labeled with radioactive atoms took in negligible amounts of the labeled molecules. When a certain protein was added to the solution, large quantities of the sugar passed into the amebas and were metabolized. Some biologists believe that certain hormones may act in this fashion by helping other molecules enter cells.

THE UNIT MEMBRANE

The discussion on cellular structure leaves one with a singular impression. The cell is largely a vast membrane system. Even a simple bacterial cell, with its lack of any visible organelles, possesses a plasma membrane that is believed to be highly active metabolically in a variety of ways. The membranous framework of organelles appears to be necessary to biologic function. Many of the chemically active macromolecules of cells lose much, if not all, of their effectiveness when separated from their natural membranes and placed in a test tube. Mitochondrial enzymes, removed from the inner membrane of the cristae, are no longer capable of normal metabolic activity. Chlorophyll is equally dependent on the lamellar structure of the grana. The hydrolytic enzymes of the lysosomes are, of course, adapted to operating independently of membrane surfaces, but their synthesis seems to be tied to the rough membranes of the endoplasmic reticulum.

There appear to be at least two reasons why metabolically active macromolecules are so directly dependent on cellular membranes. First, they may be attached to, or a part of, the membrane structure. In this event the membrane molecules become an essential part of the biochemically active machinery themselves. Second, an orderly proliferation of membrane surfaces allows for particular *spatial arrangements* of molecules. Enzymes may be arrayed along these surfaces in highly ordered complexes, such as the design of an assembly line in automobile manufacture. It is certain that the layering of chlorophyll in the lamellae of grana is eminently more efficient in trapping the energy of sunlight than a solution of chlorophyll in a test tube would be. We are confronted once again with the fact that the phenomenon of life is based on order and organization.

At the level of the cell, structural order takes the form of organized membrane systems. Two of these are formed directly from (and out of) the plasma membrane, namely phagocytic and pinocytic vesicles. Three additional structures are continuous with the cell membrane and comprise one vast system of folded membranes and channels: (1) the endoplasmic reticulum, (2) the Golgi complex, and (3) the nuclear membrane or envelope. Another organelle, the lysosome, develops by a pinching off of the margin of a Golgi sac. Only two membrane-bounded organelles remain which are not continuous with, or derived from, the plasma membrane: mitochondria and plastids. Photographs of these organelles and the cell membrane at magnifications approaching 300,000 or more reveal very interesting comparisons, however. All the membranes appear to have the same structure. Pictures of other organelles at comparable magnifications show membranes of similar design. Chemical analyses bear out a common, although not identical, macromolecular composition. Such considerations have led to the hypothesis that all internal membranes either are, or originally were, derived by an infolding of the plasma membrane. This kind of membrane has therefore been aptly called the *unit membrane* (Fig. 3-14).

The unit membrane varies from about 7.0 to 10 mμ in thickness. It is a

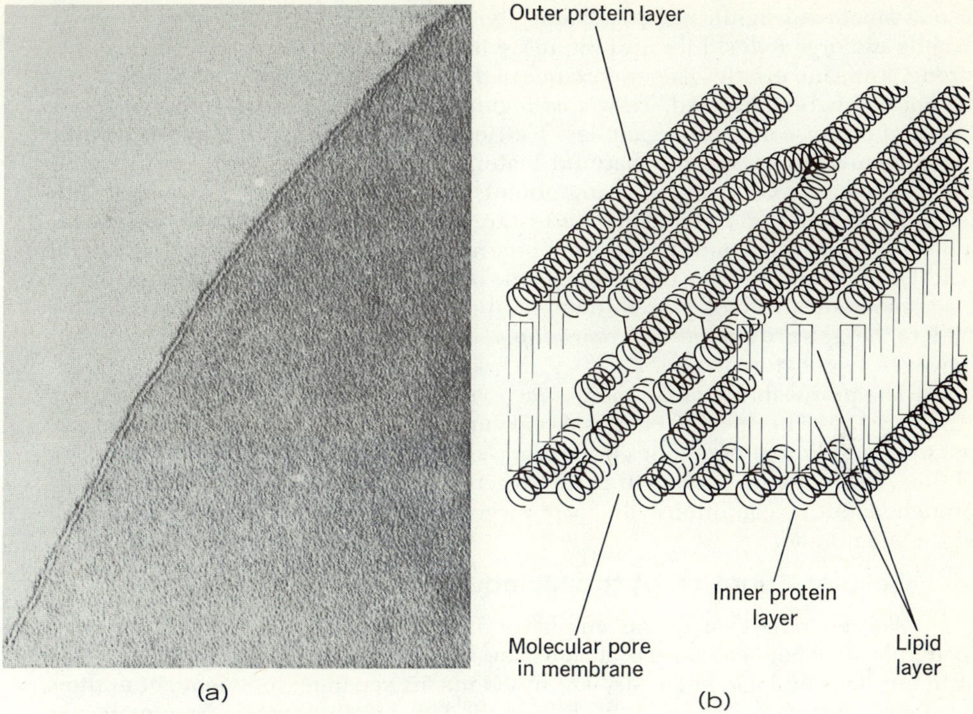

Figure 3-14 The unit membrane. (a) Electron micrograph of cell membrane of red blood cell, ×360,000. (b) Schematic drawing. Fibrous, helix-shaped protein molecules are thought to make up the outer and inner layers, whereas lipid molecules comprise the middle layer. The lipid molecules are arranged at right angles to the surface of the membrane. The protein molecules run parallel with the surface. It may be that the outer and inner protein fibers run at different angles, as in the wood veneers of plywood. They may follow random courses. These details of membrane structure are not yet known. [(a) J. David Robertson]

complex coacervate consisting of lipids[11] and fibrous proteins. A dark outer layer and inner layer "sandwich" a lighter area between them (Fig. 3-14a). It is thought that each dark layer is a single thickness of protein molecules, and the lighter area is two layers of lipid molecules. Pores through the membrane are apparently lined by protein molecules. X-ray diffraction studies indicate that the molecules of the light area (lipids) are arranged at right angles to the membrane surface, whereas the outer and inner molecules (proteins) are parallel to the surface. The polarized lipids (page 43) are believed to have their fat-soluble carbon chains or rings facing one another in the interior of the membrane, whereas their water-soluble ends lie next to the protein layers at the surface. Thus the water-soluble ends of one lipid layer face the watery contents of the cell (or organelle); the others face the watery exterior of the cell (or organelle). The result is a precisely organized coacervate gel in the form of a protein-lipid-protein sandwich, the *interior* of which is water-*in*soluble.

As we have previously stressed, the unit membrane is intimately and directly involved in the dynamic metabolic activities which occur along its surfaces. It is

[11] Largely phosphorus-containing lipids known as phospholipids and/or steroids.

also of functional significance mechanically, imparting structural form and strength to cells and organelles. Fibrous proteins, with their helical secondary structure, are strong, firm, and elastic. The importance of the unit membrane in yet a third respect cannot be overemphasized. This is its regulatory role in the transport of materials into and out of cells and organelles. Nutrients and necessary raw materials must enter. Many unnecessary or harmful materials are excluded (by no means all of them, however). Critically vital ingredients must be retained while many manufactured products and metabolic wastes are allowed to leave. Water, many small molecules, and many ions pass through (permeate) the pores of the unit membrane readily. Sodium ions, on the other hand, have difficulty passing through. Large macromolecules cannot pass through at all. The unit membrane consequently is said to be *selectively* or *differentially permeable.* Its behavior in this regard, however, is not stereotyped.

The permeability of the membrane to a given substance is subject to change; it is variable. The specific chemical makeup of the membrane is believed to vary not only in different organelles but in the same segment of membrane over a period of time. The chemical state within the structure is one of continual flux and change, brought about by conditions which are varying in different directions on either side of the membrane.

Cellular Function at the Molecular Level

We can now change our emphasis on cellular function from its structural to its physicochemical aspects. It must be stressed again, however, that biologic structure and biologic chemistry are by no means separate, independent entities, as already noted. We are merely shifting the focus in this section from molecular structures to the behavior of molecules themselves. In doing so we shall restrict ourselves to three fundamental issues: (1) the transport of materials; (2) the manner in which macromolecules perform biologic services; and (3) the mechanisms by which macromolecules regulate cellular activity.

THE TRANSPORT OF MATERIALS

We have just observed that the movement of materials into and out of cells and organelles occurs through or across unit membranes which are variably and selectively permeable. To understand how materials are thus transported it will be necessary to reexamine some of the factors responsible for molecules moving from one place to another. We should draw particular attention again to the phenomenon of thermal agitation.

DIFFUSION

Molecules, molecular-sized particles, and ions are in constant motion, colliding with one another. If molecules of one kind are concentrated in one place, they will spread or disperse into surrounding areas where their concentration is not so great. This dispersal continues until ultimately the entire available space is occupied by randomly and uniformly distributed molecules. The airborne molecules of onions cooking on the stove pervade the kitchen, then the entire house. The use of deodorants (as well as the failure to do so) is evident to other people nearby. A drop of colored antiseptic solution added to a glass of water will ultimately spread uniformly throughout the glass without stirring. These are all examples of molecular movement by *diffusion.* Any substance diffuses from regions where its concentration is greater toward regions where its concentration is less. There are two reasons for this.

First, there are more possible molecular pathways away from the center of concentration than there are toward the center of concentration. The chances are, then, that more molecules will be moving outward than inward. Second, those molecules that happen to move outward have less chance of colliding with other molecules. This is particularly true at the periphery of the region of molecular concentration. Here molecules which randomly move outward will have less probability of being bounced back by a molecular collision. They will therefore be more likely to enter the surrounding space.

OSMOSIS

The movement of water into or out of a cell or organelle is a rather special case. Here diffusion is taking place through a selectively permeable membrane. This membrane acts as a physical barrier to a great variety of solutes; they cannot pass through it, but the movement of water through the membrane is unrestricted. Just exactly why this is so is not known for certain. It is thought that the unit membrane possesses tiny pores which are too small to be photographed by the electron microscope with present techniques. These pores are probably spaced far apart and also very likely unstable in number as well as cross-sectional shape, constantly opening and closing as minute, irregular spaces between the filaments of protein molecules. In any event the pores are presumably important because they allow free passage of water molecules through the membrane but restrict the movement of other molecules and most ions. The diffusion of water through such a differentially permeable membrane is called *osmosis.* This process is of great significance to the cell because it concerns the transport of life's most precious, single ingredient.

To illustrate osmosis let us imagine a selectively permeable membrane which separates two liquids (Fig. 3-15). The liquid below the membrane is pure water. The liquid above the membrane is a solution of sugar in water. The membrane is permeable to water but impermeable to sugar. Pure water is 100 percent water molecules; it is highly concentrated water. The sugar solution, on the other hand, has fewer water molecules because a part of its volume is occupied by sugar molecules. The greater the number of sugar molecules the fewer the number of water molecules. A thick, syruplike solution would have many less water molecules per unit of volume than a thin, dilute solution. Inasmuch as osmosis is the diffusion of water, there will be a net movement of water molecules upward through the membrane, from their region of higher concentration to their region of lesser concentration. The fluid level in the tube of the apparatus gradually rises as water molecules continue to diffuse through the membrane. The height the column of solution finally reaches in the tube depends on the difference in water concentration on either side of the membrane. The stronger the sugar solution is the higher the column will rise. The final height the column reaches reflects the *osmotic pressure* of the sugar solution. At this point the weight of the column of sugar solution is forcing, or *filtering,* as many water molecules back out through the membrane as are diffusing upward. The osmotic pressure of a given solution, then, is that minimum pressure which will prevent osmosis between the solution and pure water.

More familiar examples of osmosis are found in the kitchen. Soaking dried fruits, such as prunes, in water causes them to swell. The sugar solution within the prune has a high osmotic pressure. The osmotic pressure of the surrounding water is essentially nil (if it was distilled water, it would be zero). There is a greater

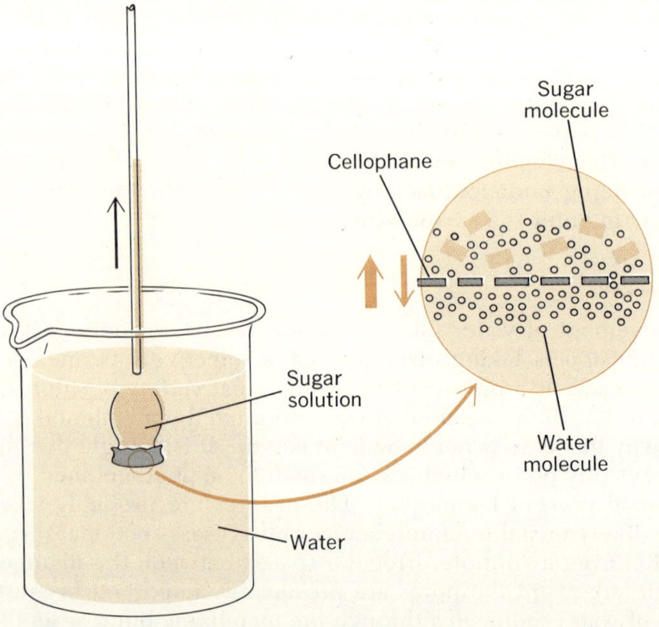

Figure 3-15 Osmosis demonstration. Cellophane covering the mouth of funnel (thistle tube) acts as a selectively permeable membrane. Osmosis occurs between water in the beaker and the sugar solution in the funnel. Larger arrow in inset indicates more water molecules diffuse upwards into the funnel than downwards into the beaker because sugar solution has fewer water molecules at the membrane surface.

concentration of water outside the prune than inside. The surrounding water is therefore said to be *hypotonic* to the prune's insides, and water diffuses into the prune. A jar of fruit jam needs no preservatives. Its osmotic pressure is high enough that most microorganisms which chance to fall into it can do no worse than water down the jam to an infinitesimal degree as they pay the price of dehydration and death. There is a lower concentration of water in the fruit than in the microorganisms. The surrounding medium is consequently said to be *hypertonic* to the microorganisms and water diffuses *out* of them.

In living cells we find many cases (the tissues of our own bodies, for instance) where the osmotic pressure of the cytoplasm and the fluids that surround the cells is the same. Such fluids are said to be *isotonic* to the cellular contents. The rates at which water diffuses out of the cell and into the cell are equal. An interesting point should be made at this time relative to osmosis involving living cells. Cells derive a great deal of their osmotic pressure from colloids. The macromolecules that make up cellular colloids loom large in the total solid bulk of the cell. They are relatively few in number, per unit of volume, compared with small molecules and ions. Osmosis involves differences in the relative numbers of molecules, not the weight or size of molecules. As a result we find that a sodium chloride solution which is 99.1 percent water by weight is isotonic with human red blood cells which are about 72 percent water by weight. The isotonic saline solution has the same number of water molecules per unit of volume as the contents of the red blood cells. Similarly, sea water, which is about 97 percent water by weight, is isotonic with many marine protists (for example, Radiolaria, Plate 6) whose cytoplasm is approximately 85 percent water by weight.

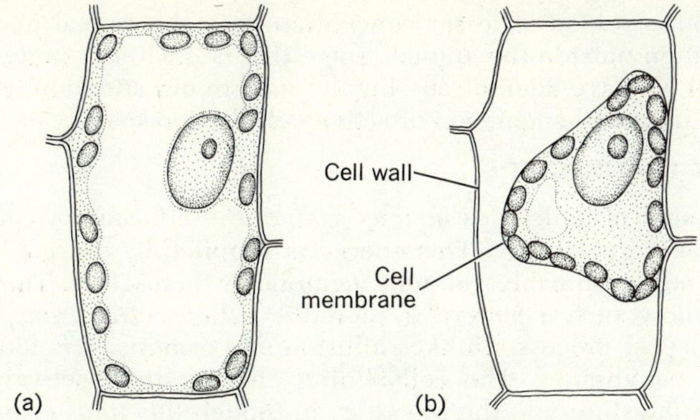

Figure 3-16 Osmosis in plant cells. (a) Plant cell in fresh water. Water moves into the cell, pressing the plasma membrane tightly against the rigid cell wall. (b) The same cell in a hypertonic salt solution loses water to the external medium and shrinks away from the cell wall.

The cells of living organisms are of course, not always immersed in isotonic solutions. When they are not, the cells must have some mechanism for water removal (if the surroundings are hypotonic) or water conservation (if the surroundings are hypertonic). *Ameba* and *Paramecium* live in fresh water which is hypotonic to them. The continual influx of water could ultimately burst their cell membranes, just as it does the skins of prunes that are soaked too long. These protists remove the water by collecting it in one or more vacuoles. When the vacuoles fill, their contents are discharged to the exterior by contraction of the wall of the vacuoles. They are consequently called *contractile* vacuoles. Contrary to the osmotic inflow of water which is passive and automatic so far as the organism is concerned, its removal requires an expenditure of energy on the part of the cell.

The cells of plants which are immersed in fresh water (for example, aquarium plants) solve the problem in a different way (Fig. 3-16a). The rigid cell wall prevents rupturing of the plasma membrane. As water continues to diffuse into such a cell the pressure within begins to build. When this pressure, which is known as *turgor pressure*, equals the osmotic pressure, further osmosis ceases. As much water is filtering out through the cell membrane because of turgor pressure as is entering because of diffusion. Such cells have a definite firmness and rigidity. If we were to place these plants in sea water, however, turgor pressure would drop to zero (Fig. 3-16b). Water would now leave the cells and they would shrink away from the embracing cell walls. This is responsible for the wilting of plants. Leaves and stems droop because of the loss in turgor pressure.

Organisms living in hypertonic environments require some way of preventing water loss. In marine vertebrates this is some specific structure (such as gills, kidneys, or glands) whose cells conserve water and/or eliminate salt. In sharks a pair of glands empty into the end of the intestine. The fluid secreted by these glands is saltier than the sea water the animal drinks and about twice as salty as the animal's blood. Oceanic birds, such as the "gooney birds" or albatrosses, have similar glands at their other ends, which open in front of their eyes. The cells of these glands have the ability to desalt the water the animal uses to meet its water needs. As in the ameba and its contractile vacuole, this requires an expenditure of energy on the part of the cells. Here, metabolic energy is being used to remove sodium and chloride ions rather than water. In both cases, however, materials are being moved from a region of lesser concentration in the animal to a region of higher concentration outside the animal. Since this is obviously quite a different process from that we have been discussing, let us turn our attention from passive diffusion to this unique phenomenon of active cellular transport.

ACTIVE TRANSPORT

The movement of molecules or ions against the diffusion or concentration gradient always requires energy. This energy is supplied by exergonic reactions within the cell and, presumably, the unit membranes themselves. The transfer of materials that follow such a pattern is therefore called *active transport,* and is obviously a biological process. Unlike diffusion and osmosis, it is found only in living cells and membranes. The cells lining our intestines actively transport glucose from the intestine into the blood, even though this transfer may be, and often is, against the concentration gradient. Certain glandular cells in the wall of our stomach produce (secrete) hydrochloric acid. In doing so they increase the concentration of hydrogen ions one million times or more above what it is in the blood that supplies them.

It is particularly significant that cells are able to move ions against an electrochemical gradient. As observed in Chapter 2, like charges repel and opposite charges attract one another. Despite this fact active transport brings about a concentration of negative ions along the inner surface of the plasma membrane and a concentration of positive ions along its outer surface. As a consequence an electric potential (difference in charge) is created between the membrane surfaces. This is a very common and important method by which cells increase the amount of free energy in a living system. It appears that the cellular transport of sodium and potassium ions has a great deal to do with this. Most cells contain much more potassium than the fluids around them. Conversely, most cells seem to pump sodium out; the concentration of sodium ions is much greater outside than inside. This sodium pump is believed to be responsible for the net negative charge on the inside of the cell membrane. The sodium ions on the outside do not diffuse back into the cell because of the relative impermeability of the membrane to them. Any

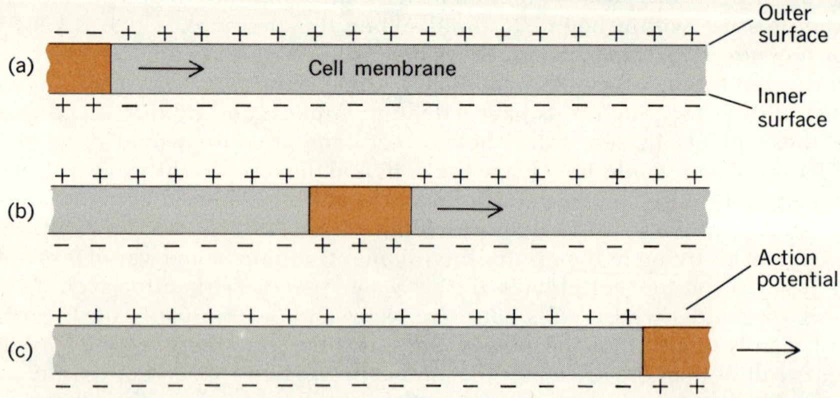

Figure 3-17 Action potential. The action potential is a reversal in membrane polarity caused by the inrushing positive ions of sodium along a segment of membrane made temporarily permeable to them (color). This permeable segment moves along like the lighted fuse of a firecracker at speeds ranging from about ½ inch per second, in some algae, to 200 miles per hour along the membranes of nerve cell processes. As it passes a given point the normal polarity is restored (see text).

form of energy that would momentarily make the membrane permeable at some point to sodium ions would allow them to rush into the cell, not only because of the diffusion gradient but also because of the electrochemical attraction of opposite charges. Such energy would be activation energy (page 37) that releases the free energy across the membrane as kinetic energy in the form of an *action potential* (Fig. 3-17). The action potential is the energy produced as the polarity of the membrane is suddenly reversed.

Inrushing sodium ions momentarily leave the outside of the membrane with a negative charge and establish a positive charge at its inner surface. The action potential usually continues to move along the membrane to the ends of the cell. As it passes a given point, the impermeability of the membrane is restored and the sodium pump reestablishes the membrane potential.

The action potential appears to be a universal form of kinetic energy in living things. Activation energy (stimulus) for its release is varied. Heat, light, contact, a variety of chemicals, and electricity are all known to evoke the initial change in membrane permeability. The energy so released may perform a great variety of functions. Nerve impulses are action potentials. The action potentials of contracting heart muscle are singularly impressive; they can be picked up by electrodes placed on the chest and fed into an electrocardiograph to record heart beat. The predatory electric eel utilizes specialized muscle fibers in the form of a series of plates in an

(a) (b)

Figure 3-18 Action potentials precede movement in plants. (a) Sensitive plant (*Mimosa*) before and after stimulation. (b) Venus flytrap. Inset shows captured fly. (T. H. Everett)

electric organ that is capable of delivering a shock of several hundred volts. In fact, Alessandro Volta (1745–1827) used such an organ as a model when he designed the first voltaic cell (battery). Action potentials also occur widely throughout the protist and plant kingdoms. A fresh water alga (*Nitella*) produces an action potential of about 0.1 volt. It travels along the cell membrane similar to a nerve impulse. Action potentials are known to precede movement in the sensitive plant (*Mimosa*) and the carnivorous Venus flytrap (Fig. 3-18a and b). Here we see once again, in these few examples, the masterful and oftentimes unusual ways in which living things put energy to work in their self-interest.

The mechanism by which cells move materials against diffusion and electrochemical gradients is not known. Bacterial studies indicate that enzymes are required. It does seem certain, with the possible exception of a few small ions, that active transport moves materials into and across the unit membrane rather than through its minuscule pores. To do this the transported material would have to be in a fat-soluble form in order to pass through the lipid (water-insoluble) center of the membrane (Fig. 3-14). To accomplish this, fat-soluble *carrier molecules* have been postulated (Fig. 3-19); these molecules lie within the membrane. There are presumably different carrier molecules for different substances. Under the influence of enzyme molecules (page 81) which are equally specific, the carrier molecules combine with the molecules to be transported. This may occur at either the inner or outer surfaces of the membrane, depending on whether the material is leaving or entering the cell. Any given carrier molecule then ferries its load across the membrane. At the opposite surface the molecules are separated from their carriers by different unloading enzymes. Inasmuch as the loaded carriers would be

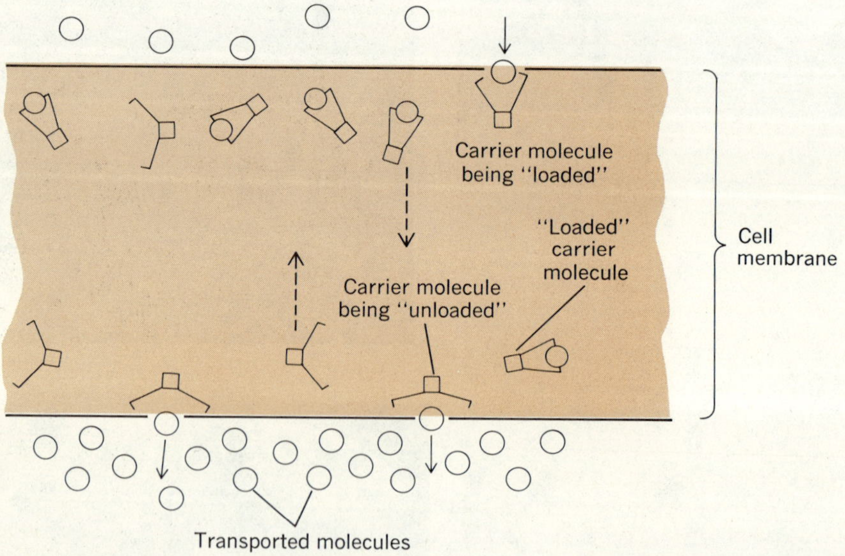

Figure 3-19 Active transport. Active transport has been hypothetically explained by carrier molecules which combine with the molecules or ions to be transported across the membrane. These carrier molecules are fat-soluble so they can pass through the lipid center of the membrane. A special enzyme molecule "loads" the carrier at one surface. A different enzyme "unloads" the carrier at the opposite surface. Loaded and unloaded carriers are ferried across the membrane along diffusion gradients (dashed arrows) in opposite directions (see text).

at a higher concentration at the loading surface, they would follow the diffusion gradient to the unloading surface. The unloaded carriers would have a higher concentration at the unloading surface and would consequently follow a diffusion gradient in the opposite direction.

Although the model of active transport just described is hypothetical, it is supported by some indirect evidence and is based on types of biochemical processes which are common in living systems. The hemoglobin of red blood cells, for example, is a well-known carrier molecule of oxygen. As for enzymes it is exceedingly doubtful if any strictly biological process takes place without them. These protein molecules are the workhorses of the cell. Every exergonic and endergonic reaction that has been studied in a living system has been found to be dependent on them. Metabolic reactions are commonly long sequences of single reactions, each one prerequisite to the next. A separate enzyme appears to be necessary for the successful completion of each step in all such sequences. We should therefore devote some time to these highly essential expediters of biologic function.

ENZYMES

If the activation energy of metabolic reactions were supplied by heat, the cell would be destroyed by the irreversible denaturation (page 46) of proteins alone. Cellular chemistry, even in the warm-blooded animals, must be a relatively cold chemistry. It is essential to any living system that energy, particularly in the form of heat, be used and released at temperatures below that which would damage organic structures. For this reason life is dependent on molecules and ions that act as *catalysts*.

A *catalyst* is a substance that increases the rate at which a chemical reaction takes place. It does this by reducing the amount of activation energy needed for reaction. The catalyst usually combines very briefly with one or more reactants. When the reaction is completed with the formation of one or more products, the catalyst becomes disengaged from its temporary union and is free to combine with additional reactant(s). It is a singular substance in that it remains unchanged by the chemical activity it has stimulated. Having evoked reaction, it remains unaltered and unaffected, and is free to promote further reaction. The ions of many metals, such as iron, manganese, and magnesium, are known to act as catalysts. A standard example used to demonstrate catalytic activity is the rate at which a solution of hydrogen peroxide (H_2O_2) releases oxygen ($2H_2O_2 \rightarrow 2H_2O + O_2$). An uncapped bottle of this substance will gradually produce a few bubbles of oxygen, particularly if exposed to light. Adding iron (ferric) ions to such a solution accelerates the liberation of oxygen. If a little manganese dioxide (MnO_2) is added instead of the ferric ions, oxygen bubbles are formed quite rapidly. These catalysts are not forcing something to happen that cannot or does not happen anyway. They are merely increasing reaction speed by lowering the activation energy threshold for reaction.

The living cell employs catalysts of a special kind called *enzymes*. They are protein molecules, with molecular weights ranging from about 10,000 up into the millions. The smallest enzyme is made up of about 100 amino acid monomers. Most enzymes appear to be globular proteins, although certain fibrous proteins (for example, the myosin of muscle tissue) are known to display some enzyme activity. It is distinctly possible that all cellular proteins[12] may play an enzymatic role under certain conditions.

[12] The fibrous proteins making up unit membranes may act as enzymes in active transport.

The action of enzymes differs from that of inorganic catalysts in two important respects. In the first place enzymes are much more efficient catalysts. An enzyme (catalase)[13] displays ten billion times more catalytic activity in the decomposition of hydrogen peroxide than the ferric ions we mentioned. One molecule of enzyme will catalyze more than 33,000 molecules of peroxide per second in freezing water! It will moreover continue to break down millions of molecules without being broken down itself. Another enzyme brings about the hydrolysis of large protein molecules at a rate of 4000 molecules per second per enzyme molecule.

A second unique feature of enzymes is their high *specificity*. A given enzyme usually affects but one reaction involving very specific reactants. It commonly stimulates a reaction in which one substance, called the *substrate*, is broken down, such as in hydrolysis. The substrate may, on the other hand, be altered (for example, reduced or oxidized). More than one substrate may be involved, as in the condensation of two monosaccharide molecules to form a disaccharide molecule. Whatever the particular case, an enzyme apparently forms an intimate union with the substrate molecule. This union is referred to as the *enzyme-substrate complex* and is quite specific. The enzyme that hydrolyzes starch will not hydrolyze a closely related polysaccharide, glycogen. Some enzymes will not even catalyze closely related isomers (page 39) of their substrate. They are equally particular in the kind of chemical job they perform. The complete hydrolysis of a protein molecule requires several steps in disassembly. One enzyme specializes in peptide bonds (page 44) between certain amino acids, whereas another enzyme specializes in others, and so on. Varying environmental conditions provide additional opportunity for enzyme specificity. A protein-splitting enzyme in our stomachs (pepsin) is active only in an acid medium. A protein-splitting enzyme in our intestines (trypsin), however, is active only in an alkaline medium.

The high efficiency and specificity of enzymes are most essential to cellular function. As many as 1000 different kinds of chemical reactions may be going on in one cell. One enzyme molecule can go a long way toward meeting the metabolic schedule of any one of them. High specificity provides the precision necessary to guide each of these reactions along its separate path. In long chains of exergonic reactions, large amounts of energy can be released a little at a time in a stepwise fashion. In this way energy can be packaged, distributed, and efficiently managed with minimal changes in temperature. The energy conquest of life is thoroughly dependent on enzymatic activity.

Considerable research has been directed toward the nature of the enzyme-substrate complex. Experimental evidence over the last several years indicates that enzyme and substrate mutually affect one another. A substrate induces its own fit in or on an enzyme molecule (Fig. 3-20). Like feet change the shape of a pair of socks, a substrate induces a change in the structure of the enzyme molecule. One portion of a protein-splitting enzyme (carboxypeptidase) is known to shift one of its dimensions by 1.4 mμ (about one-fourth of the width of the entire molecule) in order to engage a part of its substrate. Figure 3-20 schematically illustrates the manner in which the enzyme-substrate complex is believed to form. Ordinarily the substrate is much smaller than the enzyme, as shown in the illustration. When the substrate is larger, the enzyme attaches to a small portion of the substrate. In either event the enzyme has a relatively small specific surface (shown as colored spheres) on or into which the substrate fits, known as the *active site*.

[13] The names of most enzymes carry the suffix *ase*. The remainder of the name either indicates the substance or the kind of reaction that is catalyzed.

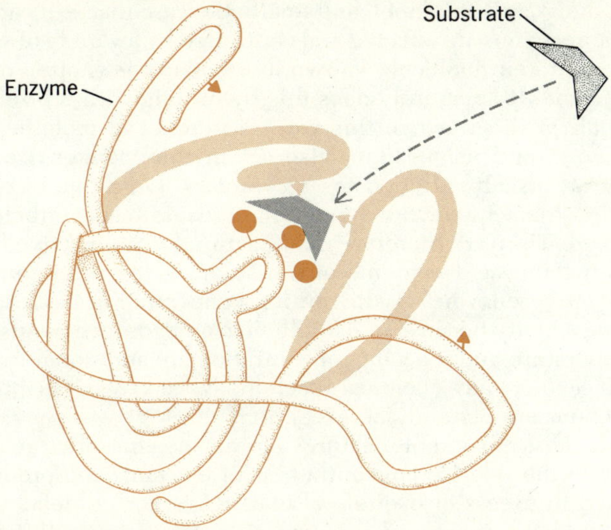

Figure 3-20 The enzyme-substrate complex. The enzyme molecule is schematically shown as a folded globular protein. Three amino acid monomers (spheres) comprise the active site. The two triangles represent amino acid monomers which serve as catalytic groups. The substrate becomes attached at the active site and induces a change in the configuration of the enzyme. This orients the catalytic groups with respect to the sites on the substrate molecule which are to be catalyzed.

As the enzyme and substrate combine, the *catalytic groups* on the enzyme (shown as colored triangles) become oriented with respect to the site that will be catalyzed on the substrate molecule. Both the spheres and triangles represent those specific amino acids that comprise the active and catalytic sites, respectively. Other amino acid monomers (not shown) operate indirectly, but nonetheless importantly in that they maintain the correct three-dimensional shape of the enzyme molecule. The formation of the flexible enzyme-substrate complex is obviously dependent on molecular geometries that mutually fit one into the other. Disulfide bonding between certain cysteine monomers would be of great architectural significance in this regard. Many other monomers serve no purpose either directly or indirectly. Their removal from the enzyme molecule appears to have little if any effect on the biological activity of the enzyme. In fact, large regions of the molecule seem to fall into this category.

The dynamic geometrical fit of the substrate with the flexible active site of the enzyme helps to account for the high specificity and efficiency of enzymes. The precise placing of catalytic groups at the surface of the substrate molecule will bring about exceedingly rapid catalysis. The velocity of these reactions would be vastly greater than those dependent on the chance meeting and random orientations of separate substrate and catalyst molecules in solution. This becomes particularly convincing when one recalls that cellular enzymes operate in tiny spaces of only a few millimicrons. They are usually attached to, if not a part of, the unit membrane, along which they are probably arranged in highly ordered patterns (mitochondria, page 81; grana, page 82; endoplasmic reticulum, page 85). Complicated reaction sequences can take place with unbelievable rapidity under such conditions as reactant molecules are passed from one enzyme to the next in line. Substrates are literally trapped, like so many cans on a conveyor belt awaiting their charge of beer, their cap, their label, and finally the case that will carry them away.

Enzymatic reactions are affected and controlled by a variety of substances and conditions. A nonprotein molecule known as a *coenzyme* is often needed to activate a given enzyme. It apparently does this by uniting with it briefly. Many vitamins (perhaps all of them) act in this way. A number of metallic ions (such as copper, manganese, and magnesium) also act in this manner. Because of their simplicity they are usually alluded to as *cofactors*. Digestive enzymes, such as the pepsin in the stomach which we mentioned, could damage the cells in which they are produced. They are therefore manufactured in an inactive form (pepsinogen). On reaching the acid contents of the stomach the inactive form becomes converted into active enzyme. Hydrogen ion concentration has a very important effect on enzyme activity because of its influence on hydrogen bonds in the protein molecule. Temperature and reactant concentration are additional variables of considerable significance to any chemical reaction, as we saw in Chapter 2 (page 37). Enzymatic reactions are certainly no exception.

One of the most remarkable features of enzyme-controlled reactions is their high sensitivity to the relative concentrations of reactants and products. Products are not produced in excess of overall cellular needs nor do deficiencies develop in the face of suddenly increased requirements. Changing cellular demands for some metabolic product are met by corresponding changes in enzymatic activity. Although much has yet to be learned about this phenomenon, two mechanisms are known which serve this regulatory function.

The cellular synthesis of a metabolic product requires many separate reactions, each one dependent on the one preceding it. Each step in such a reaction sequence is made possible by a different enzyme. The starting material is commonly the source material for more than one reaction sequence. These additional sequences are critically important to cellular function in their own right. Figure 3-21 diagrammatically represents such a case. Substance *E* is converted to product *I* in four steps, requiring four different enzymes. Substance *E* can also be converted

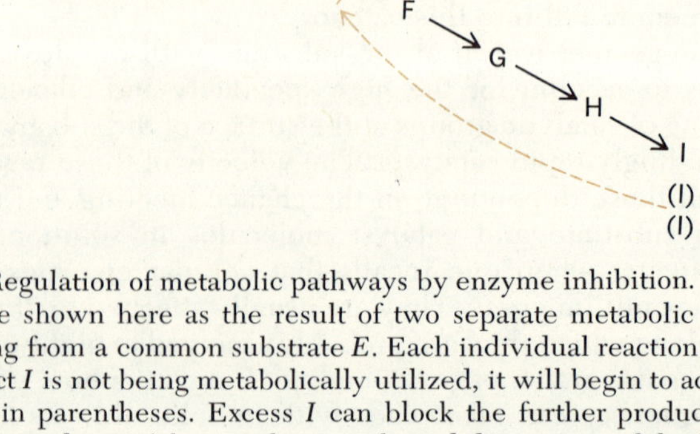

Figure 3-21 Regulation of metabolic pathways by enzyme inhibition. Two different products (*Z* and *I*) are shown here as the result of two separate metabolic pathways (reaction sequences) leading from a common substrate *E*. Each individual reaction requires a different enzyme. If product *I* is not being metabolically utilized, it will begin to accumulate as shown by additional *I*'s in parentheses. Excess *I* can block the further production of *F* (and consequently its own production) by attaching to the *inhibitory site* of the $E \rightarrow F$ enzyme (see Fig. 3-22 and text).

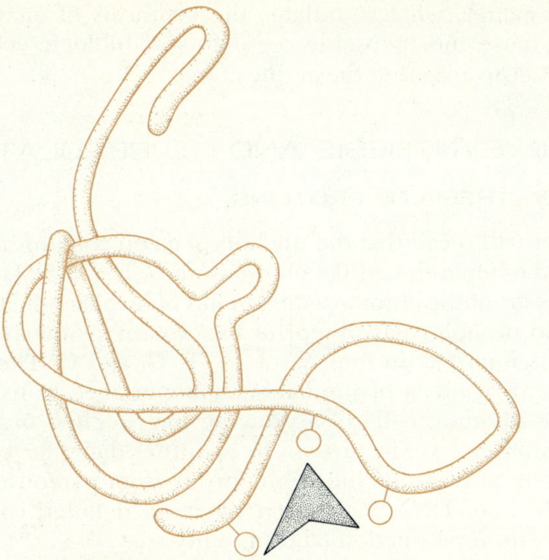

Figure 3-22 Enzyme inhibition. The inhibitory site of the enzyme illustrated in Figure 3-20 is represented here by three amino acid monomers (circles). A substance (angular figure) which combines with the enzyme at this site prevents the formation of enzyme-substrate complex (as shown in Fig. 3-20).

to product Z along a different series of reactions, involving different enzymes. If substance *I* is being used for some metabolic purpose (and consequently being removed), more *I* is produced by the enzyme that converts *H* to *I*. This same kind of thing happens, as we saw in Chapter 2, when the dissociation of sodium bicarbonate is affected by the removal of hydroxyl ions. If, on the other hand, *I* is not being used, it will begin to accumulate.

At this point a phenomenon known as *enzyme inhibition* commonly takes place. The enzyme which converts *E* to *F* has an *inhibitory site* in addition to an active site. The active site, of course, is specific for substrate *E*. The inhibitory site, however, is specific for substance *I*. Figure 3-22 illustrates this, using the same enzyme molecule as shown in Figure 3-20. As *I* increases in amount it begins to attach to the inhibitory site on the $E \rightarrow F$ enzyme. In so doing it prevents (inhibits) the enzyme from engaging its normal substrate *E* in an enzyme-substrate complex. As a result the entire reaction sequence from *E* to *I* is "turned off"; no more *I* is produced. The metabolic reaction sequence, or pathway, from *E* to Z is undisturbed. *E* can now be converted to Z. Note the highly ordered efficiency of this system. Were *I* to inhibit any other enzyme, energy would be wasted or the supply of products impaired. If it inhibited the $G \rightarrow H$ enzyme, there would be needless production of *F* and *G*. If the $D \rightarrow E$ enzyme were blocked, the supply of a different, but nonetheless essential, product Z would be cut off.

The second method by which metabolic reactions are regulated is through the manufacture of enzyme molecules themselves. When cellular needs demand the production of certain products, the cell becomes actively engaged in the synthesis of those particular enzymes required for the job. When, on the other hand, there is no longer any need for a given product, the cell stops producing those enzymes responsible for the synthesis of that product. The production stoppage of enzymes which are no longer in demand is called *enzyme repression*. Just how does this occur? For that matter, what stimulates the synthesis of enzymes? More importantly, how are these indispensable regulators of biologic activity designed and manufactured? Let us examine these questions.

PROTEIN SYNTHESIS AND ITS REGULATION

THE SYNTHESIS OF PROTEINS

The reader will recall that the nucleus of a cell contains a chromatin network that is composed of long, threadlike chromosomes (Fig. 3-5). DNA molecules make up a significant part of the chromosomes. It has been observed that DNA molecules constitute a kind of molecular blueprint for protein synthesis. The chemical language of DNA is founded on four bases, A, T, G, and C. The fact that DNA can replicate copies of itself is of fundamental importance to living things because this enables the dividing cells of a growing many-celled organism to pass their blueprints for protein manufacture on to resulting daughter cells. It also enables reproductive cells to transmit these blueprints from parent to offspring. We will defer these aspects of DNA replication for more detailed consideration in later chapters on growth, reproduction, and heredity.

Of equal significance, however, is the fact that DNA can transcribe copies of itself in the form of RNA. It is in this role that DNA instigates the manufacture of enzymes, carrier molecules, and all the other protein molecules on which metabolic reaction and cellular function are so dependent.

As pointed out, DNA produces three distinct forms of RNA. The first is *ribosomal RNA*. Ribosomes, which are the active sites where protein synthesis occurs, are composed of ribosomal RNA and proteins. The DNA of the nucleoli is the prime source of this RNA. Ribosomal RNA, as well as the other two types, gain access to the cytoplasm and the membrane system of the endoplasmic reticulum through the pores of the nuclear membrane or envelope (Fig. 3-7).

The second kind of RNA is called *messenger RNA* (abbreviated mRNA). The term "messenger" is applied to it because of its commonly accepted role in carrying a message from DNA to ribosomes. A long series of bases on the DNA molecule spells out in chemical language the exact sequence of amino acids that make up a particular protein molecule. That portion of a DNA molecule which encodes the entire primary structure (page 44) of a specific protein molecule is called a *structural gene*. An mRNA molecule is a work-order copy of this structural gene and is transcribed from one of the two DNA strands. Under the influence of appropriate enzymes the single-stranded mRNA molecule is reproduced from the master code of the DNA strand in a fashion similar to that shown in Figure 3-23. Bear in mind, however, that the T's of DNA are replaced in RNA by U's. The sequence of bases in each strand constitutes a message. Like any message it is read in one direction (arrows), starting from a specific point on the nucleic acid. The message carried by the mRNA in this instance reads U A G A U G G C U, and so on.

The successful translation of this gene language has resulted in a dictionary called the *genetic code*. The accomplishment of this task in the last decade has been a great scientific achievement, crowned with many beautifully designed and executed experiments performed by many investigators. Although much is yet to be learned about the language of the gene, one remarkable fact appears to be established. The language is essentially the same between organisms as different as bacteria and guinea pigs. A species of bacteria (*Escherichia coli*, the colon bacillus) uses a somewhat different "dialect" from guinea pigs, but most of the "words" are

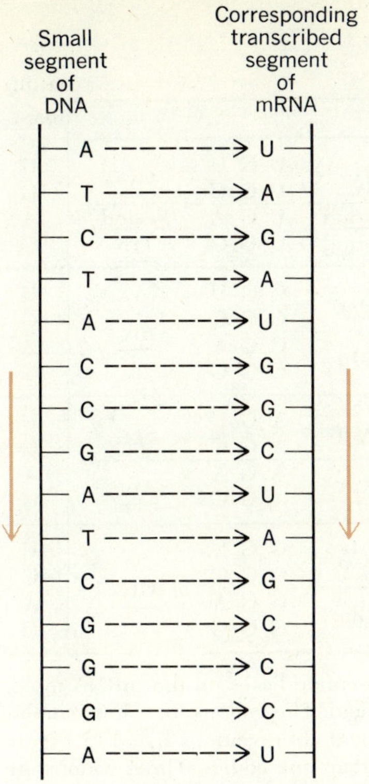

Figure 3-23 The synthesis (transcription) of mRNA. Under the influence of appropriate enzymes one strand of DNA reproduces (transcribes) a complementary mRNA copy of itself. The T's of DNA are replaced in RNA by U's. The sequence of bases in each strand constitutes a message. Like any message it is read in one direction (arrow), starting from a specific point on the nucleic acid segment. Message carried by mRNA segment reads: U A G A U G G C U, and so on.

the same. The gene language of an African toad (*Xenopus laevis*) is even closer to the guinea pig language. The genetic code for this language is depicted in Figure 3-24 in terms of mRNA base sequences. Figure 3-25 lists the abbreviations of the 20 different amino acids which are found in proteins and are coded for. Note from Figure 3-24 that three bases constitute a code "word," or *codon*, for some amino acid. All amino acids, except methionine (met. No. 13) and tryptophan (trp. No. 18), have more than one codon. In other words, there are usually synonyms in this language.

Returning to the mRNA base sequence in Figure 3-23, we may now recognize, and consequently translate, the following codons:

U A G	A U G	G C U	A G C	C C U
Period— start new message	Met (13) Methionine	Ala (1) Alanine	Ser (16) Serine	Pro (15) Proline

The first codon does not code for any amino acid. Two other such "nonsense words" are in the dictionary: U A A and U G A. Any one of these codons says, in effect, "terminate previous message and begin a new one"; in the grammar of the genetic code they serve the purpose of periods. In our illustration the mRNA segment represents the beginning portion of an mRNA molecule. We have arbitrarily placed a period (U A G) in front. The coded message would therefore read: Previous directions for protein synthesis complete; begin synthesis of a new protein whose primary structure (amino acid sequence) is methionine, alanine, serine, proline, and so on.

The actual synthesis of a protein molecule requires still a third kind of RNA.

FIRST BASE	SECOND BASE: U		SECOND BASE: C		SECOND BASE: A		SECOND BASE: G		THIRD BASE
U	U U U	Phe	U C U	Ser	U A U	Tyr	U G U	Cys	U
	U U C	Phe	U C C	Ser	U A C	Tyr	U G C	Cys	C
	U U A	Leu	U C A	Ser	U A A	Period	U G A	Period	A
	U U G	Leu	U C G	Ser	U A G	Period	U G G	Trp	G
C	C U U	Leu	C C U	Pro	C A U	His	C G U	Arg	U
	C U C	Leu	C C C	Pro	C A C	His	C G C	Arg	C
	C U A	Leu	C C A	Pro	C A A	Gln	C G A	Arg	A
	C U G	Leu	C C G	Pro	C A G	Gln	C G G	Arg	G
A	A U U	Ile	A C U	Thr	A A U	Asn	A G U	Ser	U
	A U C	Ile	A C C	Thr	A A C	Asn	A G C	Ser	C
	A U A	Ile	A C A	Thr	A A A	Lys	A G A	Arg	A
	A U G	Met	A C G	Thr	A A G	Lys	A G G	Arg	G
G	G U U	Val	G C U	Ala	G A U	Asp	G G U	Gly	U
	G U C	Val	G C C	Ala	G A C	Asp	G G C	Gly	C
	G U A	Val	G C A	Ala	G A A	Glu	G G A	Gly	A
	G U G	Val	G C G	Ala	G A G	Glu	G G G	Gly	G

Figure 3-24 The genetic code of mRNA. Three consecutive bases on the mRNA molecule constitute a code "word" or codon for some amino acid. The codons for the 64 combinations of three bases is shown above, using the amino acid abbreviations listed in Figure 3-25. All but two amino acids (Met and Trp) have more than one codon. Three codons are "nonsense" triplets (U A A, U A G, and U G A) which do not translate to any amino acid; they serve the purpose of "periods" or message terminators.

It is known as *transfer RNA* (abbreviated tRNA) and is illustrated in Figure 2-18. This relatively small macromolecule serves to recognize and transfer a given amino acid to its proper site in protein synthesis. There is at least one kind of tRNA for each of the 20 amino acids. Each kind of tRNA will bind to (combine with) but one kind of amino acid. Every tRNA molecule has two binding (combining) sites. Figure 3-26 illustrates this, using a simplified version of the clover leaf model depicted in Figure 2-18. One site bears a triplet which is the complement of the codon for the amino acid specified in mRNA, in this case alanine. The C G G of tRNA is a complementary triplet to the codon G C C for alanine. The tRNA triplet

NO.	NAME	ABBREVIATION	NO.	NAME	ABBREVIATION
1	Alanine	Ala	11	Leucine	Leu
2	Arginine	Arg	12	Lysine	Lys
3	Asparagine	Asn	13	Methionine	Met
4	Aspartic acid	Asp	14	Phenylalanine	Phe
5	Cysteine	Cys	15	Proline	Pro
6	Glutamic acid	Glu	16	Serine	Ser
7	Glutamine	Gln	17	Threonine	Thr
8	Glycine	Gly	18	Tryptophan	Trp
9	Histidine	His	19	Tyrosine	Tyr
10	Isoleucine	Ile	20	Valine	Val

Figure 3-25 Names and standard abbreviations for the 20 amino acids.

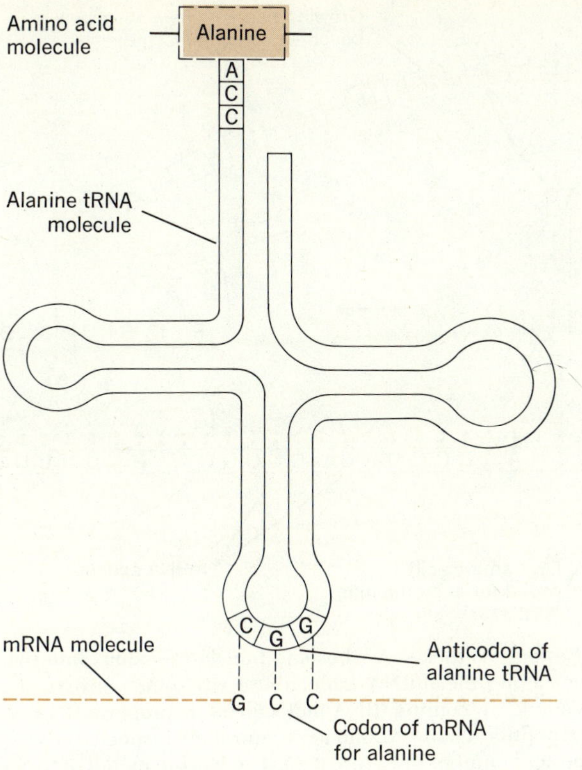

Figure 3-26 The relationship of tRNA to mRNA. The tRNA molecule represented is a simplified version of the clover leaf model of Figure 2-18. The middle loop bears a triplet of bases complementary to the mRNA codon (in this case for alanine). It is therefore called an anticodon. The open end opposite this loop always ends in the triplet C C A and binds to the amino acid for which the molecule is specific.

is consequently called an *anticodon* and will bind to its codon on mRNA. The other site always ends in the triplet C C A. It is to this end that the amino acid is bound. The molecular conformation and primary structure of each kind of tRNA presumably determine which particular amino acid will be bound.

We are now in a position to see how protein synthesis is believed to take place in the ribosomes of the cell (Fig. 3-27). Each ribosome is composed of two parts. One part positions each in-coming tRNA molecule with respect to its mRNA codon. The other part holds the growing polypeptide chain as amino acid monomers are added to it from their tRNA carriers. As a ribosome moves along on mRNA the peptide chain gradually grows to completion and a protein molecule is synthesized. In fact, several ribosomes move successively along the same mRNA molecule (forming a *polyribosome*), each one progressively completing the synthesis of a protein molecule from the same mRNA template.

Every reaction involving the transcription of molecules, the union of molecules, or the separation of bonded molecules requires the assistance of enzymes. These same enzymes are produced by the very process they assist. The same can be said for the many additional enzymes whose reactions and products supply the energy needed for protein synthesis and a multitude of other cellular processes.

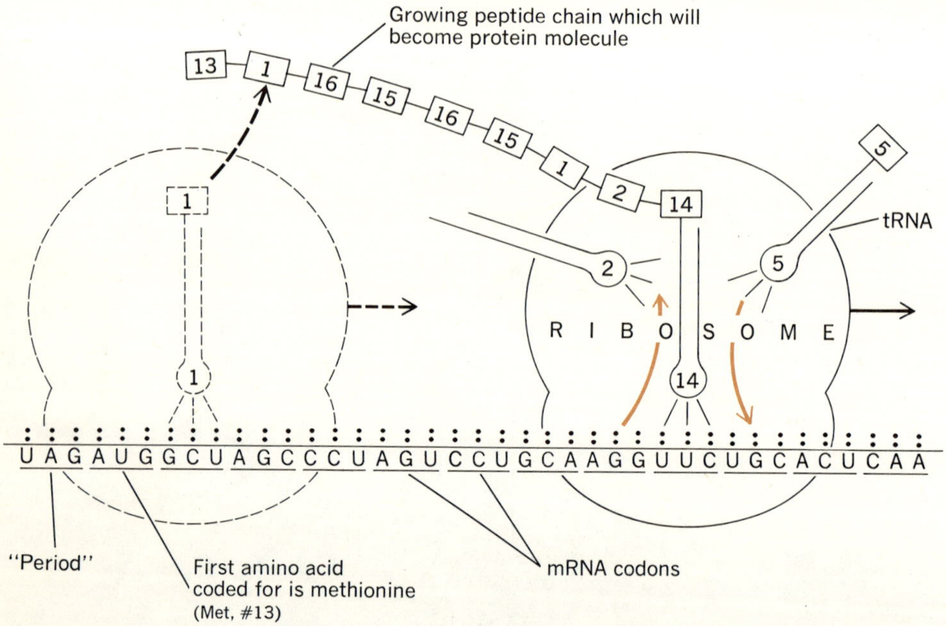

Figure 3-27 Protein synthesis. A ribosome translates codons into the peptide chain of a protein as it moves along the mRNA molecule. A ribosome consists of two parts: one part positions each "loaded," in-coming tRNA molecule at its proper mRNA codon; the other part holds the growing peptide chain. Amino acids and their respective tRNA carriers are designated by the amino acid number in Figure 3-25. The hairpin model of tRNA is used to simplify the illustration.

THE REGULATION OF ENZYME SYNTHESIS

The question as to how enzyme synthesis is regulated by the cell has puzzled biologists for a long time. Considerable research has gone into the problem, but much of what is known today is still speculative. In 1961 Francois Jacob and Jacques Monod proposed an hypothesis to explain the regulation of enzyme synthesis based on their experimental studies with bacteria. A decade of extensive experimentation by many investigators (studying monerans principally but not exclusively) has failed to refute their hypothesis, modify it significantly, or yield a better one.[14] It will not be a waste of time, therefore, to consider the Jacob-Monod model. In elucidating this hypothesis we will use the hypothetical reaction sequence illustrated in Figure 3-21 as a frame of reference.

A specific metabolic pathway, such as the sequence $E \rightarrow F \rightarrow G \rightarrow H \rightarrow I$ in Figure 3-21, behaves as a unit. Product I is either synthesized or the reaction sequence is stopped at substrate E. It is believed that the structural genes specifying the four enzymes which convert E to I comprise a continuous linear sequence along the DNA molecule (Fig. 3-28). Their spatial arrangement along the DNA helix corresponds to the metabolic pathway catalyzed by the enzymes they code for. All four of these structural genes act as a unit; they either transcribe each of their respective kinds of mRNA molecules or they transcribe none. In this way the

[14] Both French biologists were awarded the Nobel Prize for the work which led to the formulation of their hypothesis.

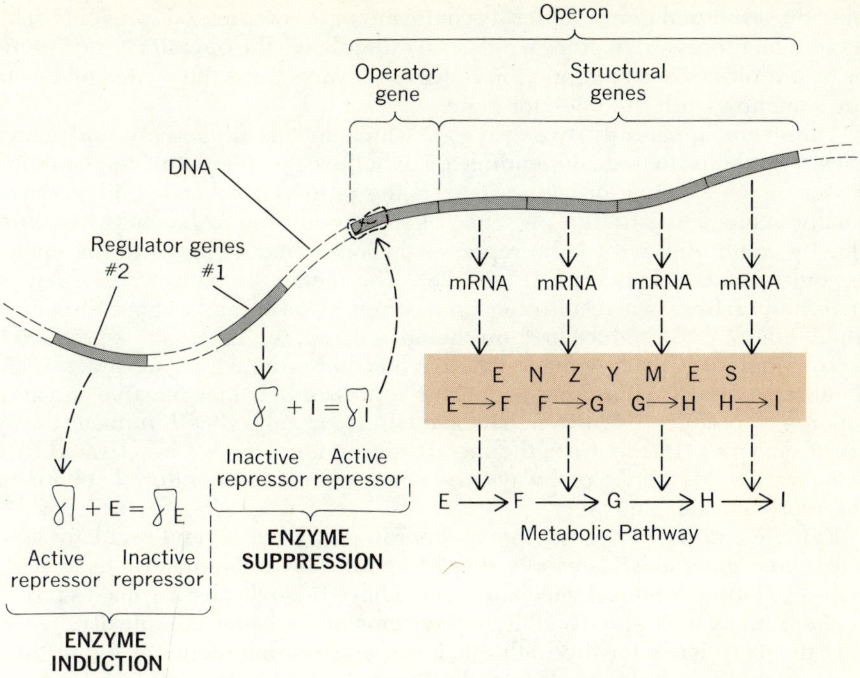

Figure 3-28 The regulation of enzyme synthesis (Jacob-Monod model). The structural genes which specify the enzymes in a metabolic pathway occur together and operate as a unit under the control of an operator gene. This operating unit, or operon, is controlled by independent regulator genes through the repressor substances they produce. Repressors produced in an inactive state are activated by combining with an end product (*I*). Thus activated they combine with the operator and prevent mRNA transcription (enzyme suppression). Repressors produced in an active state are made inactive by combining with available substrate (*E*). Thus unrepressed, the operon transcribes mRNA (enzyme induction).

entire metabolic pathway is either fully operative (supplied with enzymes) or inoperative (deprived of enzyme supply). The mechanism by which a related sequence of structural genes could be so regulated is thought to be dependent on two "controller" genes. One, the *operator gene,* is a DNA segment immediately adjacent to the first structural gene in the series. Under the influence of this operator gene the four structural genes transcribe their mRNA as a unit.

The operator gene, in other words, controls the activity of all four structural genes. Since the operator and its structural genes behave as a unit, the entire DNA segment incorporating them is referred to as an *operon.* The second controller gene may be located some distance from the operon; that is, it is a segment of the DNA helix which is not necessarily adjacent to the operon. This gene, known as the *regulator gene,* is responsible for the production of a kind of metabolic monitoring agent called the *repressor* (or *repressor substance*). The repressor is the liaison or communications contact between DNA transcription in the nucleus and metabolic activity in the cytoplasm. For any given metabolic pathway, a repressor reacts, through a sort of sensing mechanism, to the accumulation of substrate or end-product molecules. The outcome of this reaction is that the operator gene is either allowed to activate the structural genes under its command or it is repressed (prevented) from doing so. The repressor, in other words, can either leave the operator gene "switch" on or turn it off. An active repressor, however, always turns the switch off by combining somehow with the operator gene.

There are apparently two ways in which metabolic activity and enzyme synthesis are coordinated, depending on whether the regulator gene produces repressor in the inactive or active state. If the repressor substance is inactive, it is usually made active by the presence of some end-product, such as I in our reaction. By combining with I the repressor becomes active, binds to the operator gene, and turns the switch off (Fig. 3-28). The metabolic pathway $E \rightarrow I$ would become inoperative as a result of *enzyme suppression* brought about through the accumulation of end-product I. If, on the other hand, the repressor is produced in the active state, it is usually made inactive by combining with a substrate, such as E in our reaction. By combining with E the repressor becomes inactive and allows the operon to produce its mRNA. The availability of substance E induces the synthesis of enzymes. In this case, the metabolic sequence $E \rightarrow I$ is activated by *enzyme induction.* Metabolic pathways may consequently be monitored at both ends by the repressor mechanism.[15]

Evidence to date suggests that an operon may have several regulator genes. One regulator gene may also serve several operons. The fine control exercised by the self-regulating, feedback mechanisms we have been discussing makes possible the extraordinary complexity of living systems. This same complexity, we will find, is the very basis for the high efficiency and self-autonomy of living things. One is tempted to ask the obvious and ultimate question. How did such a beautifully integrated system get its first start? How did the design and directions for life, coded in DNA molecules, ever get to the design stage in the first place? How did life originate?

THE ORIGIN OF LIFE

All organisms today spring from the organisms that produced them. All cells come from previously existing cells. Only life can beget life.[16] Until a century ago, however, it was believed that organisms could develop spontaneously from nonliving material. Such *spontaneous generation* was thought to account for the appearance of microorganisms in spoiling broth. It was even suggested that mice could be produced from a sweaty shirt if it is kept in the dark and properly seasoned with several pinches of wheat! It was Louis Pasteur who ultimately scuttled these notions. He boiled a broth in flasks with long S-shaped necks which excluded microorganisms. After boiling, the flasks were allowed to stand for indefinite periods. Only those flasks whose necks he broke ever developed microscopic life. He thus established, indeed, that even microbes come from other microbes, at least under the conditions prevailing on earth now. But where did the first microorganisms come from? Was our planet always the same as it is now? Could primordial living things have developed eons ago from nonliving matter on a young earth quite different from the one we know?

The earliest evidence to date of life preserved in the rocks comes to us from South Africa. One geologic formation there contains algalike fossils which may be

[15] The applicability of the Jacob-Monod hypothesis to protists and multicellular organisms has not as yet been established.

[16] Referred to as the theory of biogenesis.

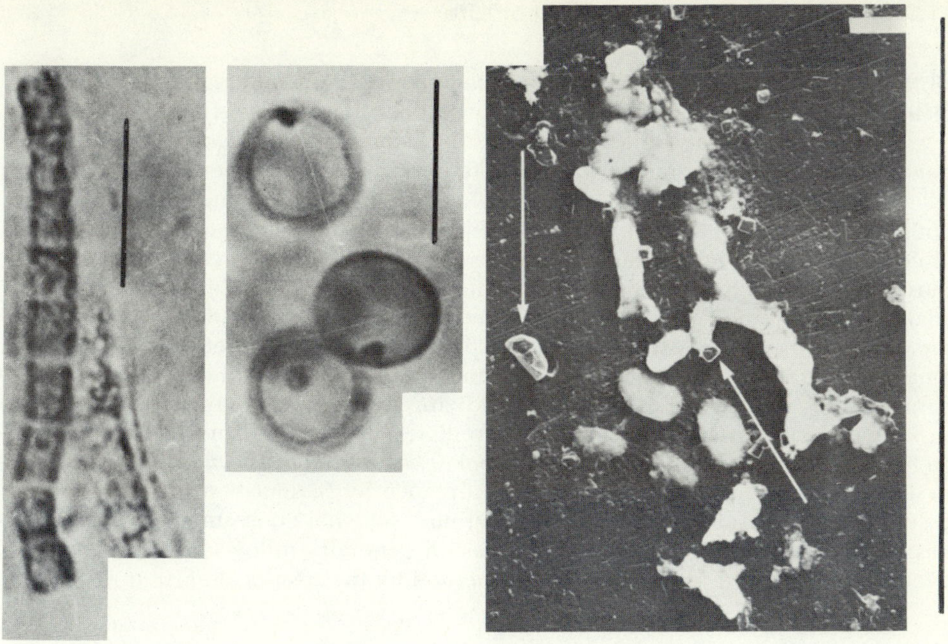

Figure 3-29 Ancient fossils of microorganisms. Bacteria on the right are 2 billion years old. Black vertical lines represent 10 μ. (E. S. Barghoorn, Harvard University)

as much as 3.5 billion years old. A younger formation known as the Fig Tree Chert[17] has been dated by radioactive isotope techniques (see page 25 and Chapter 9) at 3.1 billion years old. These deposits contain traces of 19 different amino acids and the fossils of organisms which appear to be monerans. In the Soudan Shale of Minnesota (2.7 billion years old) are the fossil remains of what are believed to be blue-green algae. E. S. Barghoorn, through elegant chemical and microscopic techniques, has taken beautiful pictures of these and other ancient monerans with the electron microscope (Fig. 3-29).

All of these fossil microorganisms seem to show a remarkable similarity to the monerans in Yellowstone National Park that live and grow in hot springs where the temperature of liquid water is essentially at the boiling point. Since the earth is believed to be about 4.5 billion years old, the South African fossils tell us that primitive life had made its appearance about one billion years after the earth was formed. What actually happened in those one billion years that led to the phenomenon of life?

We can safely say, without reservation, that an incomprehensible number of *kinds* of events transpired for the simple reason that one billion years is an order of magnitude of unimagineable immensity. We can, however, at least make an attempt to enhance our appreciation of such a vast period. Let us suppose that the distance from Los Angeles to San Francisco, about 400 miles, represents 1 billion years. At this scale the length of time since the first Egyptian Dynasty (about 3300 B.C.) would be 11.1 feet, the Christian Era 4.2 feet, and a human lifetime of 75 years 1.9 inches. One cannot possibly have experienced any fraction of the changing scenic vistas of a 400 mile journey as he finishes the first 2 inches of the trip at its start on his bed! In similar fashion, events which would be considered impos-

[17] Chert is a rock formed by water with a very high silica content.

sible in a lifetime, and virtually impossible in 5 millenia, would be certainties in a billion years—just as the physicist finds it impossible to predict the motions of a few molecules but is absolutely certain of the behavior of a billion. It is with this perspective that the events we are about to describe must be viewed.

Astrophysicists tell us that the early earth, unlike older views, was probably not a molten, fiery ball at all. Having been formed by the condensation of a cloud of cosmic dust and gas, its temperatures were very likely not greatly dissimilar to those experienced today. Its atmosphere, however, was quite different. Based upon analyses of the atmospheres of other planets and other evidence, it is believed that the early earth was invested in an atmosphere consisting of methane (CH_4), ammonia (NH_3), molecular hydrogen (H_2), and water. It was a reducing atmosphere dominated by hydrogen rather than an oxidizing atmosphere dominated by oxygen. A. I. Oparin, a Russian biochemist, had predicted this in 1924, and also developed an hypothesis for the origin of life. His book, *The Origin of Life on Earth,* appeared in 1936. As a scientific hypothesis its acceptance by biologists over the world is essentially universal. It is a cornerstone around which a great amount of research has and is being undertaken. Our discussion generally follows the sequence of events as postulated by Oparin, supplemented by the results of current research.

Nonbiologic (Abiotic) Synthesis of Organic Compounds

ABIOTIC SYNTHESIS OF ORGANIC MONOMERS

As water vapor condensed and fell on the young, prebiological earth, pools of water were formed. This water contained some of the H_2, CH_4, and NH_3 carried by the rain from the atmosphere. These molecules reacted under the influence of available sources of energy: heat, electricity (lightning), and ultraviolet radiation. The products of these reactions gradually accumulated in considerable variety such as amino acids, monosaccharides, fatty acids, purines, and pyrimidines. The pools of water became a dilute broth or soup of organic monomers that continued to accumulate slowly because there were no organisms to consume them nor was there an oxidizing atmosphere to decompose them.

In 1953 Stanley Miller was the first to demonstrate that such a series of events *could* transpire by simulating the conditions described. He subjected a mixture of methane, ammonia, hydrogen, and water vapor, confined in a sterile glass apparatus, to a continuous electric spark for a period of one week. At the end of the period an appreciable quantity of several amino and fatty acids had been formed as well as other organic compounds. Since this historic work, similar studies by a number of investigators, using heat and ultraviolet light as well as electricity as an energy source, have increased the yield to include at least 14 amino acids, ribose, deoxyribose, purines, pyrimidines, and several macromolecules (peptide chains). One study, in fact, showed that a large number of amino acids can be produced at 1000° C.

ABIOTIC SYNTHESIS OF MACROMOLECULES

The dilute broth yielded macromolecules through the polymerization of its monomers. Thus proteins and nucleic acids came to be formed. The catalytic activity of metallic ions (of iron, magnesium, manganese, and so on), phosphoric acid, and the particle-surfaces of the colloidal clays aided the energy sources in these reactions.

Laboratory confirmation that such events could have taken place has been most impressive. The most successful syntheses have occurred when the reactants were just slightly moistened rather than thoroughly immersed and dissolved in water. Under simulated, prebiological conditions all 20 amino acids have been polymerized into peptide chains, some of which had a molecular weight of 300,000. Investigators have used different energy sources, each of which has proven adequate for synthesis; for example, ultraviolet light, heat at 180° C without phosphoric acid, and heat at 70° C with phosphoric acid. Some syntheses have even been reported utilizing only the catalytic activity of acid clays. The synthetic proteins produced by these techniques appear to be normally bonded molecules because bacteria can hydrolyze and use them for their protein nutrition.

All five nucleotides have been produced, utilizing a mixture of purines, pyrimidines, ribose, and inorganic phosphate. Either ultraviolet light or heat will energize the reactions, although a preliminary study suggests that the job can also be done through the catalytic activity of clays. From these nucleotides nucleic acids have been synthesized abiotically consisting of 60 to 200 monomers. Some of these man-made nucleic acids have been used by biochemists as templates from which peptide chains were synthesized. In fact, this was an important technique utilized in breaking the genetic code. Biochemists found, for example, that a peptide chain consisting only of phenylalanine monomers was translated from a man-made nucleic acid made up only of U's. Note from Figure 3-24 that U U U is a codon for phenylalanine.

In 1967 Kornberg and Goulian of Stanford University, jointly with Sinsheimer of the California Institute of Technology, mixed the four DNA nucleotides with two enzymes[18] and the DNA strands of a virus that parasitizes the colon bacillus (*E. coli*). The DNA of this virus is unusual in that it is normally single stranded. The mixture yielded a synthetic DNA strand complementary to the natural viral DNA. It contained 5500 nucleotides (about 5 to 6 genes). Since the original nucleotides had been radioactively labeled, it was possible to separate the synthetic from the natural DNA. The synthetic complementary strands were now mixed with nucleotides and the two enzymes to produce synthetic replicas of the natural DNA. The replicated synthetic DNA was then separated out and placed in dishes containing *E. coli*. This DNA proceeded to invade the bacterial cells and produce large numbers of viral particles indistinguishable from the natural virus in both form and effect. The bacterial cells were killed.

From the standpoint of current research and evidence it would appear that it is not so much a question about whether organic compounds and macromolecules could have been produced on the prebiological earth as it is a question of what could have possibly prevented their formation.

The Appearance of Supra-Molecular Aggregates and Complex Coacervates

Macromolecules began to accumulate in number and variety and occurred on the moist surfaces of rocks and soil particles along the edges and in the shallows of ponds and the primordial seas. The bathing broth to which they were intermittently exposed was becoming gradually richer in simple organic compounds and a variety of inorganic salts. Coacervate forms (tiny discs, spheres, and other shapes) of increasing complexity (Fig. 3-6) began to appear when different macromolecules became enjoined. These complex coacervates showed widely varying degrees of

[18] DNA polymerase and DNA ligase. DNA ligase is a ring-forming enzyme.

stability. Some were not only quite stable but had a highly ordered internal structure that was bounded by a membrane of considerable selectivity. The orderly arrangement of molecules favored catalytic activity. Such systems were able to adsorb substances in solution and incorporate them into the coacervate complex. They grew in size, and on reaching a certain optimum mass would divide or fragment.

None of the characteristics described are at all foreign to known nonliving coacervate complexes. S. W. Fox has synthesized proteins by subjecting amino acids in a semimoist state to heat. The polymerizing amino acids had a definite influence on their own order of sequence in the peptide chains. When hot water was added to these synthetic proteins tiny spheres about 2 μ in diameter were formed (Fig. 3-30). These microspheres were bounded by a double membrane, consisting of an outer member of oriented water molecules and an inner member of surface-active coacervate molecules. The microspheres shrank when placed in a sodium chloride solution hypertonic to that in which they were produced. They were very stable structures and capable of both catalytic activity and division. Other complex and multiple coacervates have been formed, including some by Oparin, that adsorb and incorporate surrounding substances (sometimes in astonishing quantity), undergo an orderly internal organization, and develop increasingly selective membranes. Some have been capable of a kind of self-fragmentation in which the basic character of the coacervate remains unaltered (they have "reproduced" themselves). Some have even been mistakenly identified by competent biologists as a new species of bacteria.

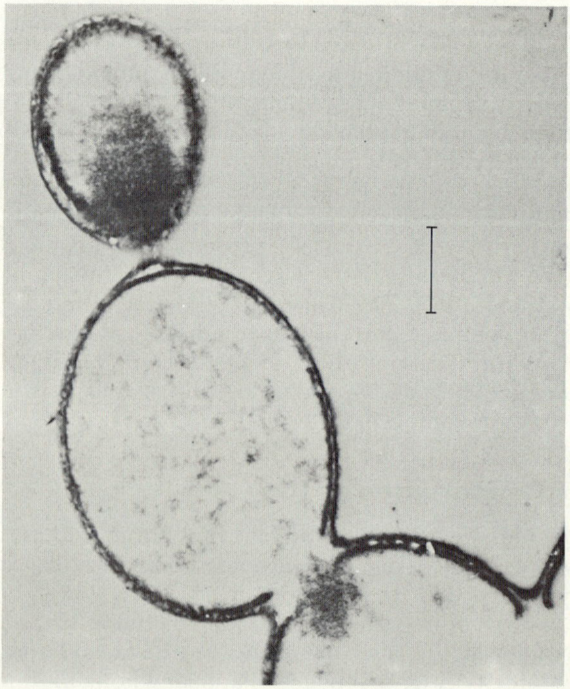

Figure 3-30 Coacervate microspheres of "synthetic protein." Note double-layered membranes. The vertical line represents 1 μ. (Dr. S. W. Fox)

The Appearance of Living Things

NATURAL SELECTION AMONG NONLIVING SYSTEMS

The very stability of the more stable coacervates was a selective advantage in their favor. Stable kinds of molecular combinations would tend to increase in number, whereas unstable combinations would remain few in number because of their inability to persist. A very natural kind of selection would favor the stable coacervates and restrict the unstable. This *natural selection* would similarly favor well-ordered and integrated systems incorporating a variety of inorganic salts, organic substrates, proteins, and nucleic acids. Harmonious spatial arrangements between reactants and catalysts would greatly enhance the potential for both endergonic and exergonic reactions. Such fortuitous circumstances were not at all unlikely.

A few metabolic pathways essentially identical to their counterparts in living cells have already been reproduced in the laboratory under simulated prebiological conditions. Natural selection would be particularly effective on those well-structured coacervate systems which could, on division (or fragmentation), reproduce themselves with such fidelity that their basic character would remain unchanged. This may well have been accomplished in the beginning without recourse to nucleic acid replication and transcription. Whether such systems would be called living is an arbitrary point. The phenomenon of life came about gradually. It would be impossible to point with any accuracy to the precise state in which the nonliving became the living—even if the entire panorama of actual events were laid before our eyes. Certainly the gradual incorporation of an information system predicated on coded nucleic acids[19] would begin to yield systems we might more readily recognize as alive. Such systems would be highly favored by selection, both because of increased precision in the performance of endergonic and exergonic reactions and increased fidelity with which the entire machinery would be replaced when the system reproduced itself. Even so the first living organisms were without a doubt vastly less efficient and much less complex than the most primitive bacteria or blue-green algae around today.

EARLY BIOLOGIC EVOLUTION

The earliest organisms probably lived on the abiotically synthesized organic compounds in the dilute broth of ancient pools and seas. Such organisms, living on organic substances they themselves do not synthesize, are called *heterotrophs.* The early heterotrophs undoubtedly increased rapidly. The rate at which organic substrates were consumed greatly exceeded the rate at which these substrates were being manufactured abiotically. As a result some of the most commonly used substances (the favored, high-energy kind) gradually came to be in shorter and shorter supply. The handwriting was clearly on the wall. Those heterotrophs that could convert some of the more abundant second-choice substrates into first-choice compounds would have an advantage. The favored heterotrophs, in other words, would possess the necessary additional enzymes to convert second-choice into first-choice compounds. This would be successful until the second-choice substances began to dwindle. In this way the stage was set for the evolution of more and more elaborate and efficient metabolic pathways as third-choice, then fourth-choice and so on, compounds came successively into short supply.

[19] Indeed, a few biologists feel that the "nucleic acid informational complex" came first. (Reminds one of chickens and eggs.)

There was not, however, an unlimited supply of choices, and life had become prolific. The ancient heterotrophs furthermore lived in an oxygen-free atmosphere; their metabolic business was conducted on a fermentative basis like the yeasts. In one such reaction, one mole (Fig. 2-9) of glucose yields CO_2, alcohol, and 28 Kcal of energy.

$$C_6H_{12}O_6 \longrightarrow 2CO_2 + C_2H_5OH + 28 \text{ Kcal}$$

Living things were being challenged on two fronts. First, all abiotically synthesized organic substrates of any use to life were being seriously depleted or overly exploited. Second, CO_2 was accumulating in the atmosphere as a result of biologic activity. Not only is CO_2 a poor substitute for methane in abiotic synthesis but it also tends to filter out ultraviolet light, an important energy source for abiotic synthesis. These new challenges, like those before, were once again successfully met by life's inexorable conquest of energy.

Heterotroph evolution to this point had been characterized by a proliferation of metabolic pathways. Some of these organisms had gone so far in this regard that they could synthesize first-choice substrates out of CO_2 and H_2O using light energy. They were no longer heterotrophs but *autotrophs.* These, the first photosynthetic organisms and their later descendants, guaranteed the indefinite continuance of life. Through a sophisticated system of metabolic pathways, living systems could now synthesize energy-rich compounds from raw materials which were abundant and an energy source that was dependable according to the following scheme:

$$6CO_2 + 6H_2O + \text{light} \xrightarrow[\text{intermediate steps}]{\text{Many enzymes and}} C_6H_{12}O_6 + 6O_2$$

As a result of photosynthetic activity free molecular oxygen began to enter the atmosphere for the first time. With the increasing availability of oxygen living things were free to extend their metabolic pathways leading to energy release. As a consequence organisms gradually evolved that exceeded the energy-yielding capacity of heterotrophs like the yeasts by more than 13 times. One mole of glucose could now release 380 Kcal of energy:

$$C_6H_{12}O_6 + 6O_2 \xrightarrow[\text{intermediate steps}]{\text{Many enzymes and}} 6CO_2 + 6H_2O + 380 \text{ Kcal}$$

As hundreds of millions of years passed, the earth's atmosphere gradually changed through biologic activity from a reducing atmosphere dominated by hydrogen to an oxidizing atmosphere dominated by oxygen. Probably sometime between 600 million and 1.2 billion years ago the stage had been set for the evolution of oxygen-loving multicellular life. The new regime, however, very likely ushered out the old. An oxidizing atmosphere and a world teeming with life would decompose and/or consume any organic substrate or coacervate produced the old-fashioned abiotic way. Spontaneous generation did occur, we believe, a long time ago when the earth was quite different. As for now it would appear that Pasteur was right.

Life on Other Worlds

Since life has arisen on this planet it is most natural and logical to inquire if it could not have arisen elsewhere in the universe as well. The consensus among scientists competent to judge is that life is probably not an uncommon phenomenon at all. Astronomers tell us that there are at least 10 billion earthlike planets among the stars of the heavens. The stars (suns) of these planets are the same size and type as ours. These planets are all earth-sized and at about the same distance from their sun as we are from ours. They follow an orbit similar to ours, and so on. With this many earthlike planets the probability is good that some of them have intelligent life. This does not mean that such beings would resemble us, as we will see in Chapter 9. The nearest possible earthlike planet, outside our own solar system, would belong in the planetary system of Tau Ceti, a star about 10 light years[20] away. That is a long trip, even at 186,000 miles per second with no stops.

QUESTIONS

1. In what ways are living things distinguished from the nonliving?
2. Stated in broad terms, how does organic evolution take place?
3. What accounts for the great diversity of life?
4. Despite the great diversity of life there is a fundamental similarity among living things. In what ways are diverse species similar architecturally?
5. What are some of the characteristics of water that make it indispensable to life?
6. Discuss the various kinds of aqueous mixtures in cells and how these mixtures serve critical biological functions.
7. Describe the unit membrane and indicate where it is found in cells. What is its significance to living processes?
8. Distinguish between the following:

   (a) Osmosis and active transport
   (b) The action of an inorganic catalyst versus that of an enzyme
   (c) The active, catalytic, and inhibitory sites of an enzyme
   (d) Ribosome and polyribosome
   (e) Codon and anticodon
   (f) tRNA and mRNA
   (g) Operator, structural, and regulator genes

9. How does protein synthesis take place in the cell?
10. Explain the Jacob-Monod hypothesis for the regulation of enzyme synthesis in bacteria.
11. What are the principal elements in the Oparin hypothesis for the origin of life? Does scientific evidence support or refute this hypothesis? Cite some of this evidence.

Suggested Readings

Allison, A. "Lysosomes and Disease," *Scientific American*, November 1967.

Buchsbaum, R. *Animals Without Backbones*. Chicago: University of Chicago Press, 1948.

Clark, B. F. C., and K. A. Marcker. "How Proteins Start," *Scientific American*, January 1968.

Coult, D. A. *Molecules and Cells*. The Principles of Modern Biology Series. Boston: Houghton Mifflin, 1966.

Giese, A. C. *Cell Physiology*. Philadelphia: Saunders, 1962.

[20] A light year is the distance light travels in a year. Light travels at a speed of 186,300 miles per second.

Gottlieb, J. E. *Plants: Adaptation Through Evolution.* Selected Topics in Modern Biology. New York: Reinhold, 1968.

Green, D. E. "The Mitochondrion," *Scientific American,* January 1964.

Grobstein, C. *The Strategy of Life.* San Francisco: Freeman, 1965.

Horne, R. W. "The Structure of Viruses," *Scientific American,* January 1963.

Hurry, S. W. *The Microstructure of Cells.* Introductory Studies in Biology. Boston: Houghton Mifflin, 1966.

Keosian, J. *The Origin of Life.* Selected Topics in Modern Biology. New York: Reinhold, 1964.

Kornberg, A. "The Synthesis of DNA," *Scientific American,* January 1968.

Koshland, D. E. "Correlation of Structure and Function in Enzyme Action," *Science,* **142,** December 20, 1963.

Loewy, A. G., and P. Siekevitz. *Cell Structure and Function.* Modern Biology Series. New York: Holt, Rinehart and Winston, 1969.

McCoy, C. J., Jr. *Vertebrates.* Selected Topics in Modern Biology. New York: Reinhold, 1968.

Nass, M. M. K. "Mitochondrial DNA: Advances, Problems and Goals," *Science,* **165,** July 4, 1969.

Neutra, M., and C. P. Leblond, "The Golgi Apparatus," *Scientific American,* February 1969.

Nomura, M. "Ribosomes," *Scientific American,* October 1969.

Oparin, A. I. *The Origin of Life.* New York: Dover, 1953.

Rustad, R. C. "Pinocytosis," *Scientific American,* April 1961.

Sistrom, W. R. *Microbial Life.* Modern Biology Series. New York: Holt, Rinehart and Winston, 1969.

Vickerman, K., and E. G. Cox Francis. *The Protozoa.* Introductory Studies in Biology. Boston: Houghton Mifflin, 1967.

SECTION II The Self-Directed Organism: unit of life

PROLOGUE The Indivisibility of the Organism as a Living Entity

In Chapter 3 we learned that the phenomenon of life is based on an elaborate and elegantly organized system called an organism. The spectrum of life's affairs, which we call the properties of life, cannot be found at levels of organization below that of the entire living being. Those characteristics or properties which distinguish the living from the nonliving were identified as (1) metabolism, (2) integration and adaptive behavior, (3) growth, and (4) reproduction and organic diversity.

In this section we will delve more deeply into each of these four "properties." As we focus our attention on one "property" and then another, it should constantly be borne in mind that none of these "properties" exists without the others. Metabolism does not take place without integration and the structures provided by growth. Growth in turn requires both integration and metabolism. Reproduction depends on growth, metabolism, and integration, but without reproduction there would be nothing to grow, metabolize, or integrate. The living organism is an indivisible, self-directed unit.

Mushrooms growing in horse dung. (Photograph by author)

CHAPTER 4 The Self-Directed Exploitation of Energy: metabolism

In our discussion of the origin of life, we noted that one of the major achievements of ancient heterotrophs was the proliferation of metabolic pathways. Complicated chains of chemical reactions evolved in order to utilize efficiently a dwindling supply of organic substrates. Autotrophs finally arrived on the scene, capable of capturing light energy and subsisting on CO_2, H_2O, and very small amounts of a few inorganic salts. The precise nature of these ancient metabolic pathways may never be known. It is a singular fact, however, that all living organisms from bacteria to man are quite similar in the manner in which the basic synthesis of carbohydrates, lipids, proteins, and nucleic acids is carried out, as well as the means by which the flow of free energy is maintained. On the basis of this rather remarkable state of affairs, it would seem reasonable to presume that the primitive organisms of ancient seas performed their principal metabolic processes in ways similar to those found in organisms today. In any event it seems that all living organisms share a common heritage in the performance of their metabolic activities. This heritage is undoubtedly very ancient and is the focus of our attention in this chapter.

THE GENERAL NATURE OF METABOLIC PROCESSES

As we learned in Chapter 3 metabolism is a term referring to all the chemical processes involved in the performance of an organism's total activity. These processes may be logically divided into three categories: nutrition, anabolism, and catabolism.

Nutrition

Nutrition is concerned with the manner in which living things secure the raw materials they need from their environment. It also includes whatever preparation may be required of these materials in order to make them ready and available for use by the organism. Such use may involve the synthesis of cellular structures and products (anabolism), the release of chemical energy (catabolism), or both. Metabolism starts with the raw materials provided by the environment in which the organism lives. Heterotrophic organisms, like animals, satisfy their needs for building materials and energy by the consumption of energy-rich organic compounds called foods. These compounds must in turn be readied for use by being broken down into simpler components. These preparatory reactions are hydrolytic and are referred to as *digestion*. Carbohydrate, lipid, and protein foods are consequently hydrolyzed (digested) into their constituent monosaccharides, fatty acids, and amino acids through the action of various carbobydrate-, lipid- and protein-splitting enzymes. The heterotroph is now able to use these simple organic substances directly in meeting its need for synthesis and/or energy.

Figure 4-1 Vegetative sprouts from a potato.

Were it not for autotrophic organisms, however, there would be no energy-rich organic foodstuffs for the heterotrophs to consume. It is through autotrophic nutrition that these complex organic foods are synthesized from common inorganic substances in the environment. The green plants and protists (and blue-green algae) not only satisfy their own needs for raw materials and energy but manufacture practically all[1] of the foodstuffs used by heterotrophs as well. Like the heterotrophs the autotrophs also find it necessary to hydrolyze the molecules which they manufacture. The potato plant, for example, synthesizes starch and protein which it stores in the packages we know as potatoes. When a potato sprouts (Fig. 4-1), the young potato shoots rely on the nutrients stored in the potato. These nutrients (polysaccharides and proteins) must first be hydrolyzed into their constituent monosaccharide and amino acid monomers. The amino acids thus formed may then undergo anabolism and become synthesized into the protein structures which will go into making the cells out of which the shoots are constructed. The energy required for this synthesis comes from the oxidation (catabolism) of the monosaccharides. The same sort of thing occurs in any young plant sprouting from a seed. It lives as a heterotroph, subsisting on the stored food in the seed, until such time that it develops its own photosynthetic organs (leaves and/or stems).

Anabolism

We have referred to the biosynthesis of materials within the cells of all organisms as anabolism. Anabolic processes include the most complex and highly ordered reactions that occur in living cells. (The discussion in Chapter 3 on protein synthesis and its regulation is an excellent case in point.) Anabolism begins with the products of nutrition such as monosaccharides, fatty acids, glycerol, amino acids, nucleotides. From these building block, or precursor molecules, polysaccharides, lipids, proteins, and nucleic acids are manufactured that are characteristic of the particular cell and organism in which they are produced. Some of the lipids and proteins go into the repair, replacement, and new construction of membrane systems, and into the building of new organelles. Other proteins are enzymes or any of a wide variety of cellular secretions. Many macromolecules are "earmarked" for future use and held on reserve inside the cell as storage products (fat and animal starch, or glycogen, in animals; starches, fats and proteins in the seeds of plants).

[1] Autotrophic iron, sulfur, and nitrifying bacteria synthesize organic compounds by using the energy released in the oxidation of iron compounds, H_2S, and NH_3, respectively. Their role, however, is quite minor in total organic synthesis.

All these activities contribute not only to the maintenance and operation of cells (and organisms) but to their growth and reproduction as well. As a consequence of biosynthesis cells grow in size. On reaching optimum size they divide and the resulting daughter cells proceed to grow in turn. If the organism is unicellular, cell division amounts to reproduction. If the organism is multicellular, cell division may result in growth; reproduction occurs either through the formation of special reproductive cells or by some kind of asexual division of the organism, as in the ivy plant (page 56).

The high degree of organization which we find in all organisms is maintained only through the continual synthesis of cellular components. The metabolic machinery operates unceasingly and at a level of efficiency unmatched by any man-made contrivance. The membranes of the endoplasmic reticulum and mitochondria, for instance, together with their associated enzyme systems, are in constant need of repair. Chemical components altered through incessant use must be replaced by newly synthesized macromolecules. The biosynthesis of these components is an endergonic process. Activation energy (supplied by catabolism) is required in order to assemble simple organic precursors into macromolecules. Special enzymes are also needed in order to direct the synthesis of the specific compounds required. In meeting these demands anabolic pathways are not only efficient but adaptable as well. In the biosynthesis of nucleic acids, proteins, lipids, and carbohydrates cells have at their disposal many alternative or optional metabolic routes. People can become fat by eating too many starchy foods because the individual is transforming glucose into fatty acids, which in turn are being converted into fat. Similarly, amino acids can be changed into either fatty acids or nucleic acids. The autotroph can even produce amino acids from fatty acids. We will investigate some of these alternate pathways later in this chapter. Suffice it to say at this point that such transformations between groups of organic compounds are of great adaptive value to the organism under those circumstances in which some particular group of foodstuffs happens to be in short supply.

As stated anabolism requires energy—in fact, a great deal of energy. Rapidly growing and dividing cells may divert 80 to 90 percent of their total energy output into protein synthesis alone. The cells of secreting digestive glands, during the digestion of a meal, need as much energy as muscle cells during arduous exercise. The anabolic energy requirements for the synthesis of a given product are by no means restricted to the metabolic pathway leading directly to that product. Numerous allied and accessory reactions must also be energized. All of the enzymes required for the biosynthesis must be produced in adequate amounts. Ribosomes may be in need of repair or replacement. Ribonucleic acids must be synthesized. Each one of these reactions has its own batteries of allied and accessory reactions. Numerous additional enzymes must be produced whose primary function is to supply energy to the entire system. So interwoven are the processes of anabolism and catabolism that it becomes virtually impossible to determine where one leaves off and the other begins. Let us therefore direct our attention to this other side of the coin.

Catabolism

Nothing so clearly distinguishes the living from the nonliving as the endless transformation and flow of free energy in living systems. The dynamism of life as well as its elegant state of organization is made possible only by its relentless intake of energy. The autotroph captures light energy and transforms it, through the endergonic process of biosynthesis, into the potential chemical energy of organic compounds. From these energy-enriched compounds autotrophs and heterotrophs alike release free, kinetic energy via the exergonic process of cellular respiration, or catabolism. Autotrophs utilize the energy they themselves have stored. Heterotrophs who live on autotrophs (such as plant-eating animals or *herbivores*) also use this same energy. Indeed, even heterotrophs who live on heterotrophs (for example, meat-eating animals or *carnivores*) are merely using the potential energy of autotrophic synthesis second-hand.

In all these conversions of potential to kinetic energy a variety of energy transformations may take place. The energy stored in glucose molecules is released as active chemical energy when the glucose is oxidized. Some of this chemical energy may be transformed as in the firefly into light. In the electric eel some of it is transformed into electricity. In contracting muscle chemical energy is converted into both mechanical energy (body motion) and heat. In all these transformations energy is being neither used up nor made. It is merely being changed from one form of energy into another. In muscle cells, for instance, the total amount of mechanical and heat energy that is released, plus the untapped chemical energy of metabolic by-products, is exactly equal to the chemical energy from which all of it was originally derived.

In the organism as in a machine, energy transformations occur for a purpose: they carry out some necessary function. In serving such useful ends energy conversions are said to perform work. Igniting gas in the combustion chamber of an automobile engine produces heat which expands the gas and drives the piston downward. Changing chemical energy into electric energy in a nerve cell results in nerve impulses that cause muscle fibers to contract and bones to move. In the course of such energy conversions useful energy is never passed along with 100 percent efficiency. Every transformation from one form of energy to another is always accompanied by a certain loss of *useful* energy.[2] This is a natural consequence of the Second Law of Thermodynamics (page 55). When chemical energy is released from its concentrated, highly ordered, and unstable state in energy-rich compounds, it disperses randomly into the environment. In doing so a certain amount of the original potential energy is invariably converted into some form of useless energy that usually turns out to be heat. Our automobile engine loses a great deal of the potential chemical energy in gasoline through the wasted dissipation of heat into the walls of the combustion chamber as well as the frictional heat produced by moving parts.

Since every energy conversion is accompanied by a decrease in useful energy, it is apparent that the more often energy is transformed in the performance of a given function, the greater will be the amount of energy that cannot be put to work. As we will see cells convert 55 percent or more of the potential energy in glucose into useful chemical energy. When this energy is transformed by muscle cells into body movement, net efficiency drops to about 40 percent in some well-trained athletes and to about 25 percent in the rest of us. Even so we, together with other organisms, rate equally well as mechanical machines with man-made engines. The highest rated engine (the diesel) has a maximum efficiency of about 40 percent. The gasoline engine converts only about 25 percent of its fuel energy into work.

An organism, however, is a great deal more than a mechanical machine. Living things are chemical engines. Most of their work is chemical work involving the flow of free chemical energy. In performing this work the rate of conversion of potential energy into useful energy runs in excess of 50 percent. Macromolecules are

[2] "Useless" energy is that which is unavailable for performing work.

synthesized. The electric energy of action potentials is produced through the chemical energies of active transport. The organism as a chemical machine is an unusual engine; it repairs and/or replaces its worn-out components continuously. Long before the whole machine deteriorates from the accumulated ravages of age, use, and abuse it has reproduced new models of itself to carry on the work. It provides itself with fuel. It drives and governs itself by restricting its fuel consumption to its need for energy. And it performs these astonishing feats at temperatures ranging from lukewarm to cold.

The control exercised by the "cool chemistry" of cellular catabolism is made possible through a universal medium for energy exchange. All living organisms store, package, and distribute energy in a standardized "parcel." This "parcel" is known as ATP (adenosine triphosphate); it is invariably involved in the performance of any and all kinds of biological work.

ATP

The ATP molecule is essentially an adenosine nucleotide (see Fig. 2-16) which possesses three phosphate groups rather than just one (Fig. 4-2). The bonds that join the two additional phosphatic groups to the molecule are of especial signif-

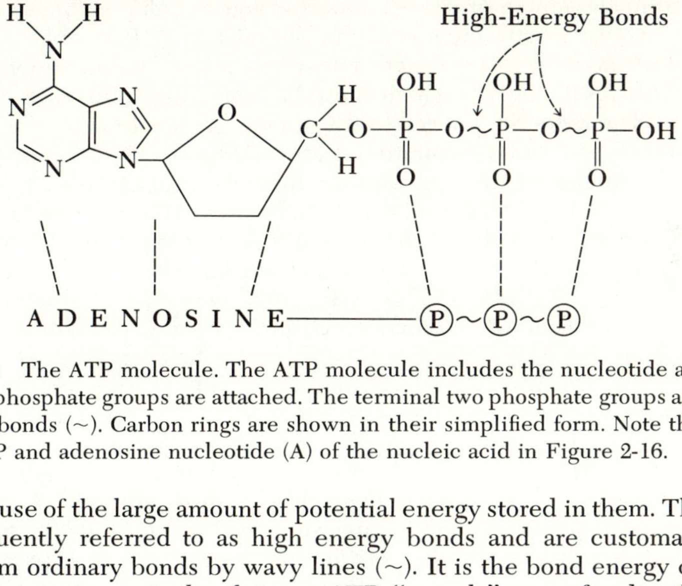

Figure 4-2 The ATP molecule. The ATP molecule includes the nucleotide adenosine to which three phosphate groups are attached. The terminal two phosphate groups are linked by high-energy bonds (~). Carbon rings are shown in their simplified form. Note the similarity between ATP and adenosine nucleotide (A) of the nucleic acid in Figure 2-16.

icance because of the large amount of potential energy stored in them. These bonds are consequently referred to as high energy bonds and are customarily distinguished from ordinary bonds by wavy lines (~). It is the bond energy of the third phosphate group in particular that our ATP "parcels" carry for distribution and ready release. When energy is to be transferred, an enzyme "opens the parcel" by breaking the bond of the terminal phosphate group. This liberates the energy of the phosphate bond and converts ATP into ADP (adenosine diphosphate) plus a free (inorganic) phosphate group.

$$\underset{\text{(ATP)}}{A-\textcircled{P}\sim\textcircled{P}\sim\textcircled{P}} \xrightarrow[\text{(enz.)}]{} \underset{\text{(ADP)}}{A-\textcircled{P}\sim\textcircled{P}} + \textcircled{P} + E$$

The units of energy so delivered to sites of metabolic activity are small enough that they can be released in quantities adequate to meet cellular needs, with a minimum waste of unused energy. These units have been likened to a kind of energy currency in which energy transactions can be fully paid in exact change to the nearest penny. A metabolic function that actually costs 23.4 cents worth of energy will require an investment of 24 cents to get the job done. The overpayment of 0.6 cents is minimal. If the unit of exchange were dollar bills, energy overpayments would be truly exorbitant.

This energy currency, of course, must not only be kept in circulation but it must be added to continuously because, as we have seen, it cannot be used with 100 percent efficiency. There must consequently be a ceaseless packaging of energy into new ATP "parcels" as long as the cellular demands for energy are not fully met. This is accomplished by the "refilling" of already opened "parcels" (ADP). Available free energy (plus an enzyme) makes possible the resynthesis of ATP from inorganic phosphate and ADP.

$$A{-}\textcircled{P}{\sim}\textcircled{P} + \textcircled{P} + E \xrightarrow[\text{(enz.)}]{} A{-}\textcircled{P}{\sim}\textcircled{P}{\sim}\textcircled{P}$$

This process by which a phosphate group, $\textcircled{P}$[3] is incorporated into a molecule is known as *phosphorylation.* ATP, in other words, is synthesized by the phosphorylation of ADP.

The ultimate source of energy mobilization for ATP synthesis is the autotroph. The photosynthetic organism phosphorylates ADP through the chemical energy of electrons raised to "excited states" by captured light energy. The ATP so synthesized carries the energy which is used in the endergonic reactions of glucose biosynthesis. The potential energy in the glucose molecule is released by the exergonic reactions of cellular respiration. In the autotroph, the heterotroph who eats the autotroph, and the heterotroph who eats the heterotroph, the energy released by cellular respiration is packaged into ATP molecules. All these reactions involving energy flow and utilization occur in long sequences of many small steps. Chemical energy is released a little at each step as electrons flow down energy gradients from their elevated state in an electron (or hydrogen) donor (such as glucose) to the next lower energy level in an intermediate electron (hydrogen) acceptor or carrier. (The reader may benefit from a review of pages 33–36 at this point.)

In this chain of reactions the accumulated bits of released energy (mostly chemical plus some heat) equal the energy that could have been obtained if we had actually burned the original energy-rich compound in the laboratory. The chemical energy of living systems, however, has several advantages over the heat energy of physical combustion: (1) it does not damage or destroy cellular components, (2) it can be managed and controlled more precisely and thriftily, and (3) it lends itself to more efficient distribution within the cell and organism. Lest it be assumed that long metabolic sequences might be clumsy and slow in the release of energy, it should be remembered that all these reactions are mediated by enzymes. Complicated reaction sequences take place with astonishing rapidity. A nerve cell, for example, may produce several hundred nerve impulses (action potentials) per second.

The most remarkable feature of these metabolic pathways is their self-direction and the precision with which production is adjusted to cellular needs. The rate at which the cell releases free energy, for instance, is regulated by the rate of energy utilization. This is accomplished by linking a number of separate steps in a long

[3] 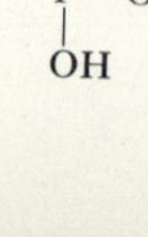 (See Figure 4-2.)

series of catabolic sequences to the presence of inorganic phosphate Ⓟ and ADP. If either inorganic phosphate or ADP is lacking, cellular respiration conveniently ceases. Certain critical steps in cellular respiration are consequently blocked in the absence of these two substances and retarded when they are in short supply. If the energy requirements of the cell have been met (or nearly met), all (or most) of the ADP and inorganic phosphate will have been converted to ATP and no (or little) ATP will be reverting to ADP. This coupling of ADP phosphorylation to the oxidative processes of cellular respiration constitutes a control system by which the rate of energy utilization governs the rate of energy production.

Let us direct our attention to a somewhat more detailed study of these processes. We may then better comprehend the extraordinary capacity of life for managing its metabolic affairs in furthering its own interests.

METABOLISM AT THE CELLULAR AND SUBCELLULAR LEVELS

Cellular Respiration

The exergonic reactions in which free energy is released and transferred are divided into three phases: (1) glycolysis, (2) the cytochrome system, and (3) the Krebs or citric acid cycle. Let us consider each one of these briefly.

GLYCOLYSIS

We have indicated that glucose is the principal energy source in living systems. It is not surprising therefore that we should consider the oxidation of glucose first because cells certainly do. It may come as a surprise, however, to learn that the early stages in glucose oxidation can (and usually do) begin in the absence of free oxygen.[4] This beginning series of oxygen-independent reactions takes place in the cytoplasm and is called *glycolysis.*

Like most exergonic reactions, glycolysis requires activation energy. This is accomplished by a reaction in which ATP donates its terminal phosphate group to glucose (symbolized below as $\boxed{C6}$):

(1) $$\boxed{C6} + ATP \xrightarrow[\text{(enz.)}]{} \textcircled{P}-\boxed{C6} + ADP$$

In the three reactions that follow this, glucose is (1) changed to a related isomer, (2) phosphorylated once more by an additional molecule of ATP and (3) broken into two molecules of a 3-carbon compound, each of which bears a phosphate group. These reactions may be summarized as follows:

(2) $$\textcircled{P}-\boxed{C6} + ATP \xrightarrow[\text{(enz.)}]{} 2\textcircled{P}-\boxed{C3} + ADP$$

Each 3-carbon molecule is now simultaneously oxidized and phosphorylated. Two hydrogen atoms are removed from each molecule to a nucleotide coenzyme which is widely used in living systems as a hydrogen acceptor and carrier (symbolically represented as an open box, $\sqcup$). The energy released by the removal of the hydrogen atoms attaches a second phosphate group to the 3-carbon molecule, this time with a high-energy bond:

(3) $$2\textcircled{P}-\boxed{C3} + 2\textcircled{P} + 2\,\sqcup \xrightarrow[\text{(enz.)}]{} 2\textcircled{P}-\boxed{C3}\sim\textcircled{P} + 2\,\underset{\sqcup}{H_2}$$

[4] Conditions in which there is no free or molecular oxygen are said to be *anaerobic.*

In the next reaction, the high-energy phosphate group of the 3-carbon compound is transferred to ADP, resulting in the synthesis of ATP:

(4) $$2\textcircled{P}\!-\!\boxed{C3}\!\sim\!\textcircled{P} + 2ADP \xrightarrow[(enz.)]{} 2\textcircled{P}\!-\!\boxed{C3} + 2ATP$$

Note at this point that the two molecules of ATP which have just been produced in reaction 4 replace the two ATP molecules which were used in reactions 1 and 2 for activation energy. In the two reactions which follow and complete the sequence of glycolysis, two more molecules of ATP are synthesized, both of which represent a net gain in ATP energy storage. In the first of these reactions, the 3-carbon products of reaction 4 are oxidized by the loss of hydrogen atoms (in HOH), and the released energy converts the remaining phosphate bond into the high-energy variety:

(5) $$2\textcircled{P}\!-\!\boxed{C3} - 2H_2O \xrightarrow[(enz.)]{} 2\textcircled{P}\!\sim\!\boxed{C3} + 2H_2O$$

The final reaction transfers the energy-rich bond just formed to ADP:

(6) $$2\textcircled{P}\!\sim\!\boxed{C3} + 2ADP \xrightarrow[(enz.)]{} 2\boxed{C3} + 2ATP$$

The symbol $\boxed{C3}$ in reaction 6 represents *pyruvic acid.* It is the end product in the glycolytic oxidation of glucose. At this point cellular respiration can follow two different metabolic pathways, depending on whether oxygen is or is not present. In the continued absence of oxygen, the pyruvic acid is reduced by receiving the two hydrogen atoms carried by $\underline{\;H_2\;}$ in reaction 3. Several different enzymes will bring about this reduction. One of them is commonly found in muscle cells and converts pyruvic acid into lactic acid. (Note that the reaction is reversible.)

$$\text{Pyruvic acid} + \underline{\;H_2\;} \xrightleftharpoons[(enz.)]{} \text{lactic acid} + \underline{\qquad}$$

Another is used by some microorganisms (for example, yeast), reducing pyruvic acid to ethyl alcohol and CO_2.

$$\text{Pyruvic acid} + \underline{\;H_2\;} \xrightarrow[(enz.)]{} \text{alcohol} + CO_2 + \underline{\qquad}$$

This reaction has a certain friendly and familiar ring to many of us. The term *fermentation* which we apply to it, however, is equally applicable to any reduction of pyruvic acid by $\underline{\;H_2\;}$.

It is tempting to speculate that the oxidative processes of glycolysis and fermentation were metabolic processes developed by ancient heterotrophs before the appearance of free oxygen on this planet. If such were the case, these primordial organisms, like their counterparts today, were getting two ATP "parcels" of energy for every molecule of glucose they consumed. That this is a rather meager return on the energy investment market is attested to when we consider the energy performance possibilities in an oxygen-rich environment. Let us therefore examine some of the metabolic pathways open to pyruvic acid when oxygen is available.

THE CYTOCHROME SYSTEM

When oxygen is present, the 2 $\underline{\;H_2\;}$ molecules of glycolytic reaction 3 are no longer required for the reduction of pyruvic acid. They are now free to pass

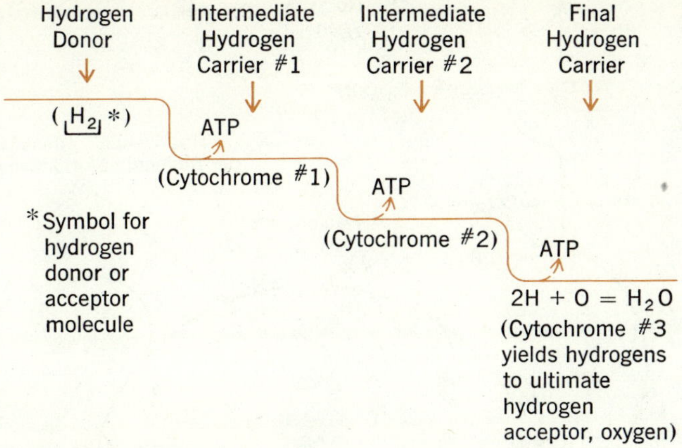

Figure 4-3 The cytochrome enzyme system. Cytochrome carrier enzymes transfer hydrogen atoms (or electrons) to successively lower energy levels. The energy released by two electrons at each drop in level is packaged into one ATP molecule. The last carrier enzyme releases its two hydrogen atoms (or electrons) to oxygen, forming water.

their four hydrogens down a chain of hydrogen (electron) carrier molecules known as *cytochromes*. The cytochromes are thought to be aligned along the cristae of the mitochondria in highly ordered arrays; they may even comprise some of the roundish particles that are attached to the mitochondrial membranes (Fig. 3-8). In any event the four hydrogen atoms (and their electrons) cascade down a gradient of steadily decreasing energy levels in somewhat the same fashion that a stream drops over a series of small waterfalls (Fig. 4-3). Electrons are passed from one intermediate carrier molecule (cytochrome) to the next lower in line. Most of the energy so released is captured at each step. For every two electrons (or hydrogen atoms) which "shoot the rapids" enough energy is released to package three ATP molecules. Since four electrons make this trip, six molecules of ATP are synthesized. It is at this point that the cascading sequence ends as oxygen, playing its role of ultimate hydrogen acceptor, is reduced by the hydrogens to form water.

The cytochrome carrier system of mitochondria therefore adds six more molecules of ATP to the two supplied by cytoplasmic glycolysis, per molecule of glucose started with.

The mitochondria have yet another enzyme system, however. This system is designed to utilize most of the energy still stored in the two pyruvic acid molecules of glycolytic reaction 6. It is known as the Krebs citric acid cycle.

THE KREBS CITRIC ACID CYCLE

A preparatory series of reactions takes place in which each of the two molecules of pyruvic acid is broken down into a 2-carbon compound plus CO_2. The 2-carbon compound is immediately activated by becoming bonded to a coenzyme known as coenzyme A (abbreviated CoA). In the process each pyruvic acid molecule gives up two hydrogen atoms. We can simplify these reactions as follows:

$$2\boxed{C3} + 2CoA + 2\ \underline{\qquad} \xrightarrow[\text{(enz.)}]{} 2\boxed{C2}\text{—}\text{(CoA)} + 2CO_2 + 2\ \underline{H_2}$$

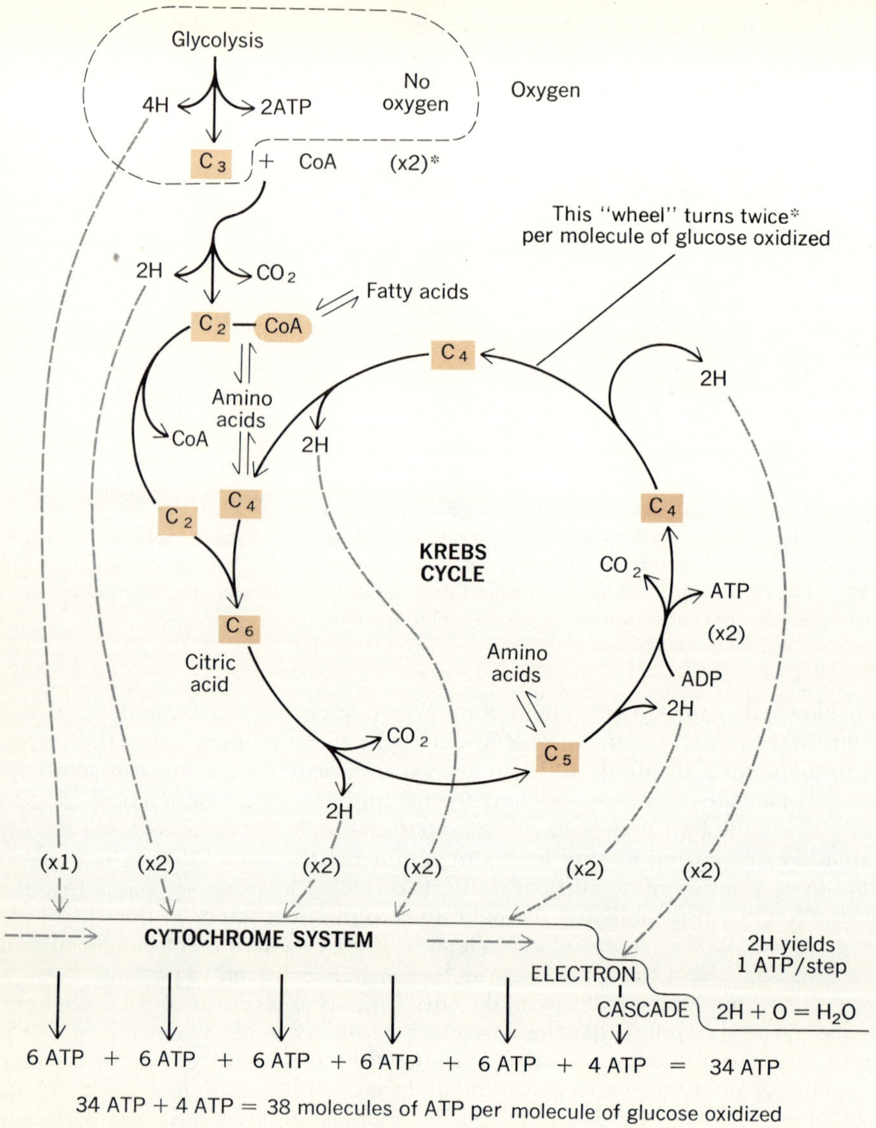

Figure 4-4 A simplified model of cellular respiration. Two Krebs cycles are required for each molecule of glucose oxidized. Hydrogen atoms contributed by glycolysis and the Krebs cycle are passed along to the electron cascade of the cytochrome system. For every two hydrogen atoms contributed to this system, one molecule of ATP is synthesized at each of three successive steps in the electron cascade. Four hydrogen atoms passed along all three steps yield six molecules of ATP.

The four hydrogens of two H_2 are transferred to the cytochrome system (Fig. 4-4). The C2 molecule next becomes freed from its bond with CoA. It now joins a 4-carbon molecule already present in the Krebs cycle to form a 6-carbon compound known as *citric acid* (C6 in Fig. 4-4). The citric acid is progressively broken down first into a 5- and then a 4-carbon compound by the loss of two molecules of CO_2. It is at the same time oxidized through a series of reactions that remove a total of eight hydrogen atoms and synthesize one molecule of ATP (between

[C5] and [C4]). All these hydrogen atoms are donated to the cytochrome system. Two of them, however, do not pass through the entire chain and consequently yield two molecules of ATP rather than the usual three. The citric acid molecule has by now been converted into the [C4] compound with which our beginning [C2] molecule united to form citric acid. This [C4] compound has actually served as a carrier molecule by which [C2] was conveyed through the citric acid cycle and completely oxidized into CO_2 and H_2O. Having carried one load through the "processing plant" it is ready, so to speak, for the next.

THE SIGNIFICANCE OF CELLULAR RESPIRATION

Cellular respiration is of singular significance because of its efficiency and its versatility. Its efficiency is attested to by inspecting Figure 4-4. The respiratory chain of the cytochrome system converts the four hydrogens contributed by glycolysis into six molecules of ATP. It also packages six ATP molecules from the four hydrogens released in the oxidation of the two pyruvic acid molecules. And it finally synthesizes an additional 22 molecules of ATP from the 16 hydrogens removed from the two [C2] molecules which are processed in the Krebs cycle. To this we add two molecules of ATP supplied directly by glycolysis and two more supplied directly from the citric acid cycle. The grand total of ATP molecules that are packaged from one molecule of glucose is 38. A mole, or gram molecule of ATP, releases about 10,000 gram calories (10 Kcal) when it is converted to ADP. A mole of glucose yields 690,000 gram calories (690 Kcal) when completely oxidized. The efficiency of cellular respiration is therefore 380,000/690,000, or about 55 percent. As we have seen this is exceptionally high and far above that of any man-made machine.

The versatility of cellular respiration lies in its adaptability and plasticity. As pointed out, many alternative, metabolic routes are open to living systems. In glycolysis, for example, the (P)—[C3] compound of reaction 2 can be derived from glycerol as well as glucose. Pyruvic acid can be produced from certain amino acids.[5] The entire glycolytic pathway is reversible, so that glucose can be resynthesized from pyruvic acid. In Figure 4-4 note that the [C2]—(CoA) molecule can originate from either fatty acid or amino acid sources. It can also be changed into fatty acids or certain amino acids. Two of the compounds in the citric acid cycle ([C5] and the [C4] carrier molecule) can be either converted into certain amino acids or derived from them. These few examples not only indicate the remarkable versatility of metabolic pathways but also stress the intimate relationships between cellular respiration and biosynthesis. Reaction pathways are commonly reversible. They are also intersected by other pathways. Because of this, portions of a metabolic route that yield energy very often overlap others that lead to biosynthesis.

Let us now consider some reaction sequences in which the emphasis is placed on biosynthesis rather than oxidation.

Photosynthesis

The biosynthesis of glucose and other organic compounds by the green plant is a very complicated metabolic pathway. Although the energy for this process

[5] Reactions of this kind require the removal of amino (NH_2^-) groups by a process called *deamination.*

comes from captured light energy, the actual synthesis itself occurs through an involved series of reactions that do not require light at all. They have consequently been referred to as the *dark reactions*. They occur in that part of the chloroplast where there is no chlorophyll, namely the stroma (page 82).

THE "DARK REACTIONS"

Carbon dioxide which has entered the chloroplast from the atmosphere is reduced to glucose in the reaction series shown in Figure 4-5. This is a highly simplified version of the carbon cycle, as it is oftentimes called, insofar as it is understood today. A number of phosphorylated carbon compounds are involved in a cycle of reactions. For simplicity the phosphate groups of these compounds are not shown. The cycle begins with three molecules of a 5-carbon compound (C5) resembling the sugar *ribose* (page 41). Each of these molecules receives a molecule of CO_2 and, through energy supplied by ATP, is converted into two 3-carbon compounds (C3). The resulting six molecules of C3 (two for each combination of C5 and CO_2) are in turn changed by being reduced to a different C3 compound. Each of these reactions requires energy from ATP and hydrogen atoms from our reducing agent (hydrogen carrier) H_2. Five of the reduced C3 molecules are finally reconverted into the three molecules of C5 that we started with. In completing the cycle, however, an additional C3 molecule has been synthesized by the reduction of three molecules of CO_2. This has been accomplished through the agency of a carrier molecule (C5) that is comparable to the part played by the C4 carrier in the Krebs cycle. One more "turn" of the carbon cycle "wheel" produces another C3 molecule.

This 3-carbon end product of the dark reactions is actually the same 3-carbon compound which is produced in glycolytic reaction 2 (Ⓟ–C3). It therefore has a profusion of metabolic pathways open to it, as previously suggested. It may go "up" the glycolytic pathway to glucose; or it may pass on "down" into the Krebs cycle and out alternative pathways that lead to amino acid and fatty acid synthesis. And, of course, many of these simple monomers may follow metabolic routes which ultimately lead to the synthesis of polysaccharides, lipids, proteins, and nucleic acids.

Since all of these biosynthetic reactions are endergonic, energy must be available not only to activate them but also to carry them through to completion. An examination of Figure 4-5 reveals that nine molecules of ATP and six molecules of H_2 are required for every "turn" of the "carbon wheel." Since two turns of the wheel are needed for every glucose molecule, this amounts to 18 molecules of ATP and 12 molecules of H_2. The autotroph, unlike the heterotroph, does not require glucose as a source of this energy. Here lies the basic distinction between heterotrophs and autotrophs. They both carry on biosynthesis in much the same way.[6] But autotrophs secure the energy for biosynthesis by transforming light energy into chemical energy. They thereby avoid the necessity of obtaining ready-made organic compounds as an energy source and are consequently the only organisms that can *accumulate supplies* of the energy-rich molecules on which the rest of life is so totally dependent. They have accomplished this by carrying life's quest for energy to its ultimate source—the sun. Let us examine how this has been done.

[6] The reduction of CO_2 to glucose is basically glycolysis in reverse and occurs in many animals, including ourselves.

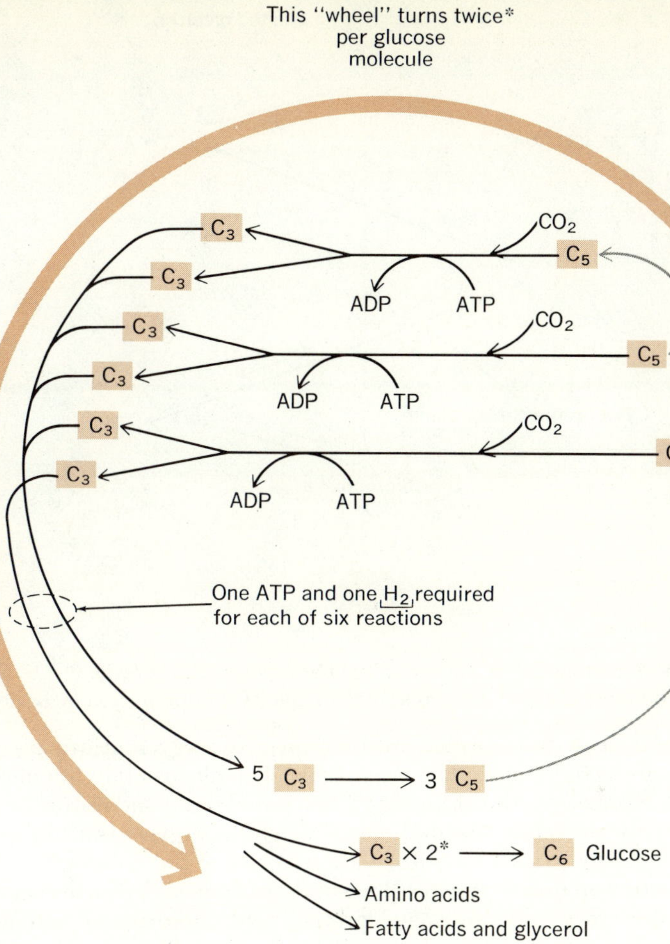

Figure 4-5 Simplified model of the dark reactions. Two complete cycles of the reaction sequences shown result in the synthesis of one molecule of glucose. Six molecules of CO_2, 12 molecules of H_2, and 18 molecules of ATP are required for this.

THE "LIGHT REACTIONS"

It has been known for a long time that photosynthesis was restricted to those parts of plants, protists, and blue-green algae that contain the green pigment chlorophyll. In plants and protists this pigment is confined to the chloroplasts; it is particularly concentrated in the lamellae of the grana. The chlorophyll molecule plays a critical role in photosynthesis that entails reaction sequences whose fundamental purpose is to provide energy for the endergonic processes of biosynthesis we have just investigated. This role involves the transformation of light energy into chemical energy. It is therefore only natural that we should refer to these transformations as the "light reactions."

When chlorophyll is illuminated by white light (light containing all the colors of the rainbow), it absorbs certain colors (wavelengths of light energy) and allows others to pass through it or to be reflected from it. The colors it absorbs are the reds and the blues. The remaining colors (collectively adding up to greenish) are reflected or transmitted and give the pigment its characteristic color. We say the reds and the blues have been absorbed because these particular wavelengths of

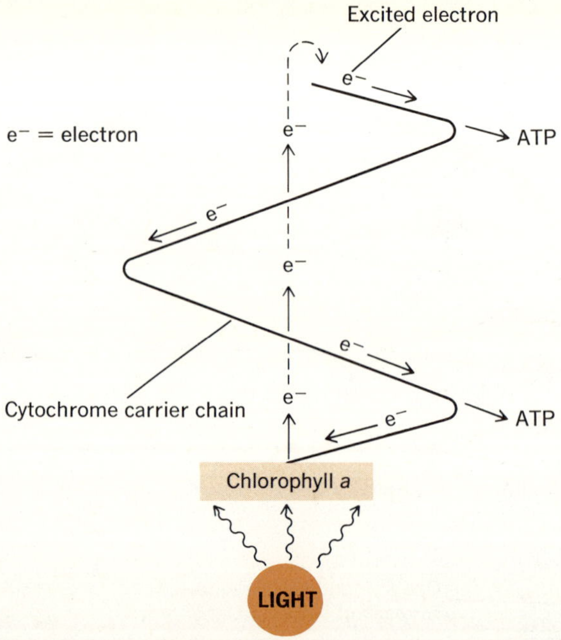

Figure 4-6 Simplified model of the cyclic light reactions. The specific electron acceptor molecule which delivers excited electrons (e^-) to the cytochrome system is not shown.

light energy have been transformed into chemical energy and are therefore no longer visible to us as light. What happens is this. The absorbed light energy somehow removes electrons from the chlorophyll macromolecule and raises their energy level to a relatively unstable "excited state." Light energy is transformed to the chemical energy of energized electrons. Variant forms of the chlorophyll molecule perform this function in different ways. Two of them, chlorophyll *a* and chlorophyll *b*, operate together in plants and many protists and comprise an integrated energizing system. The *a* form takes part in what we may call *cyclic light reactions.* The *b* form is involved in *noncyclic light reactions.*

CYCLIC LIGHT REACTIONS

In the cyclic reactions, illuminated chlorophyll *a* releases energized electrons. These electrons, now elevated to an excited state, are received by a type of electron acceptor molecule that delivers them to a cytochrome system (Fig. 4-6). As these electrons pass down the electron cascade of the cytochrome carrier system, energy is released and packaged into ATP molecules just as it is in cellular respiration. On reaching the bottom of the cascade the deenergized electrons are returned to the chlorophyll molecules from which they were originally expelled. With continuous illumination the same electrons may be recycled over and over again along this metabolic route. Under such conditions this elegant mechanism for phosphorylating ADP directly from light energy very likely produces enough ATP to accommodate all of the organism's energy needs for biosynthesis.

NONCYCLIC LIGHT REACTIONS

If we refer back to Figure 4-5 we note that the reduction of CO_2 to glucose requires the presence of both hydrogen carriers ($\lfloor H_2 \rfloor$) and ATP. These hydrogen carrier molecules, as well as additional ATP molecules, are derived from noncyclic reactions which have their origin in chlorophyll *b* (Fig. 4-7). When light is

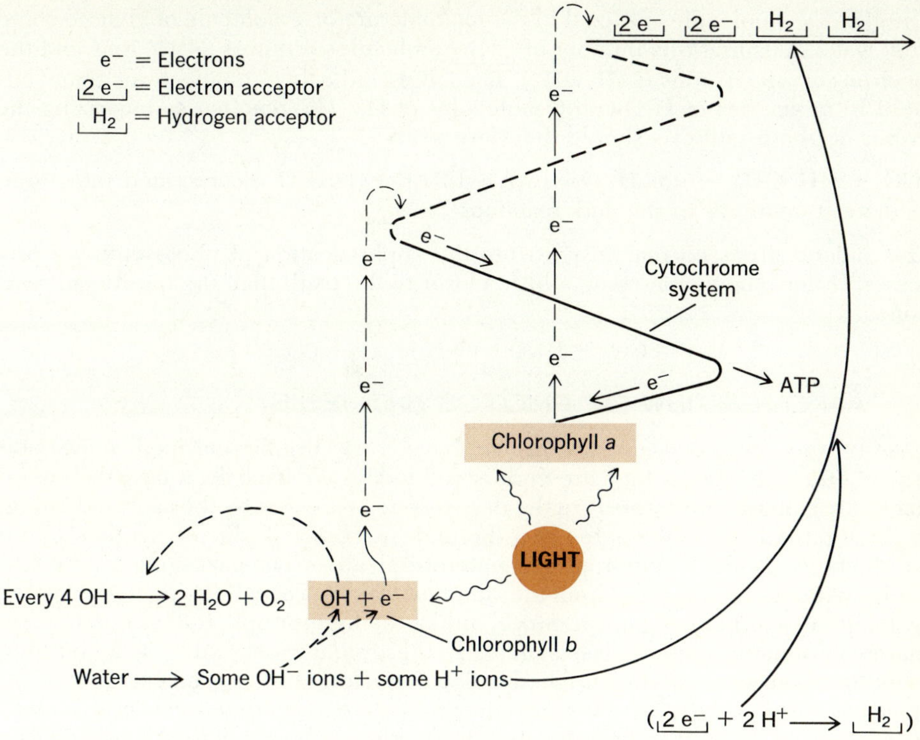

Figure 4-7 Simplified model of the noncyclic light reactions.

absorbed by this pigment, the loss of electrons leaves it with a positive charge. At this point the H^+ and OH^- ions which are always present in water become intimately involved in the process. The positively charged pigment, acting as an oxidizing agent, removes electrons from the OH^- ions. For every 4 OH^- ions so oxidized two molecules of water and one molecule of oxygen are formed:

$$4OH \longrightarrow 2H_2O + O_2$$

These captured electrons may now in turn be energized to the excited state. The path followed by electrons from chlorophyll *b* is not completely known. They do pass down a cytochrome carrier system, however, which terminates in chlorophyll *a*. Figure 4-7 assumes that this carrier chain is a portion of the same system utilized by chlorophyll *a*. In passing through this cytochrome system electrons from chlorophyll *b* release energy which is packaged into additional molecules of ATP before they finally come to rest in chlorophyll *a*. Here they are once again elevated to a high-energy state by absorbed light. Because these electrons exceed the normal complement of electrons held by chlorophyll *a* they are "picked off" by electron carrier molecules ($\lfloor 2e^- \rfloor$) and kept at their high-energy level. The high affinity these electron carriers have for the H^+ ions we left behind (in the water) converts them into hydrogen carriers ($\lfloor 2e^- \rfloor + 2H^+ \rightarrow \lfloor H_2 \rfloor$).[7]

In this manner the noncyclic reactions not only contribute to ATP synthesis but also supply the $\lfloor H_2 \rfloor$ molecules needed for the reduction of CO_2 to glucose.

[7] Actually two different acceptor molecules are involved here, a fact which need not concern us in this short survey.

The 12 $\lfloor H_2 \rfloor$ molecules utilized in the manufacture of a molecule of glucose come from water. Loading this number of $\lfloor H_2 \rfloor$ molecules requires 24 H^+ ions and the electrons of 24OH^- ions ($24H^+OH^-$). The 24OH radicals (with electrons removed) yield 12 molecules of H_2O and 6 molecules of O_2. The reaction summarizing the events of photosynthesis should therefore read:

$6CO_2 + 24H^+OH^- \longrightarrow C_6H_{12}O_6 + 6O_2 + 12H_2O$ ($+6H_2O$ incorporated into intermediate compounds in the dark reactions).

This indeed gives no real insight into the sophistication of photosynthetic processes. It does, however, come a little closer to the truth than the traditional reaction:

$$6CO_2 + 6H_2O \longrightarrow C_6H_{12}O_6 + 6O_2$$

AN OVERVIEW OF PHOTOSYNTHESIS

In the foregoing discussion we have very briefly outlined some photosynthetic pathways as they are understood today. We have seen how the energy needs for glucose biosynthesis in the dark reactions are met by the captured energy of the light reactions. We noted that the dark reactions of glucose synthesis were substantially similar in autotroph and heterotroph alike. The packaging of ATP molecules with energy captured from the sun, however, is a capability possessed solely by the photosynthetic organism. Since, unlike the heterotroph, it does not need the energy in organic compounds to synthesize these same compounds, it can build up great storehouses of energy-enriched molecules. *The solution to an energy problem is clearly a dual solution, for it takes care of the matter of organic synthesis as well.* Again we must note that the intimate interrelationships which exist between free-energy flow and biosynthesis do not permit us to make hard-and-fast distinctions between them. A case in point are the light reactions. One integrated metabolic system not only mobilizes reaction energy into ATP parcels but also yields the $\lfloor H_2 \rfloor$ molecules which will give up their hydrogens to convert CO_2 into glucose.

The efficiency of this system under optimum conditions appears to be slightly less than 50 percent. Not quite one-half of the light absorbed by chlorophyll is retained as chemical energy stored in glucose molecules. When one considers the complexity of photosynthetic pathways, we may assume that the preliminary light reaction must be taking place with even greater economy. At this point we scarcely need reiterate the relevance of chloroplast structure to metabolic efficiency. The disposition of chlorophyll molecules, cytrochromes, and so on, in convenient spatial arrangements along the lamellae of the grana not only allows for maximum light absorption but greatly facilitates the transfer of electrons and the speed and precision of chemical reaction. It is no coincidence that chlorophyll in a test tube solution does not perform as it does in the living cell. In the chloroplast, as in the mitochondrion, architecture, organization, and function are inseparable partners in the efficient management of energy transactions.

In the multicellular organism, with its multitude of cells, successively higher levels of organization become superimposed on those at the cellular level. In this way cellular activities are channeled and coordinated to serve the organism as a whole. This allows for the luxury of specialized cells with a high proficiency in certain functions. As observed in Figure 3-3 such cell specialists are grouped into various types of tissues. Tissues in turn may be arranged to form organs. And organs, especially in animals, can become interconnected and assembled into complex organ systems.

Let us direct our discussion of metabolism to these higher levels of organization that are exhibited in multicellular organisms.

METABOLISM AT THE ORGANISMIC LEVEL

Metabolism in Plants

In examining metabolism in the multicellular plants we will focus our attention on those flowering land plants which are familiar to all of us: the oak trees, rose bushes, strawberry plants, and so on, which grow in our yards, gardens, parks, and farms. These plants are not only remarkably similar in their fundamental characteristics but share many significant features with a variety of other plants as well.

NUTRITION

Most of the photosynthesis carried on by the majority of plants takes place in the leaves, which are admirably adapted to this function. Their large flat surfaces favor the absorption of light; their thinness brings cells close to the air around the

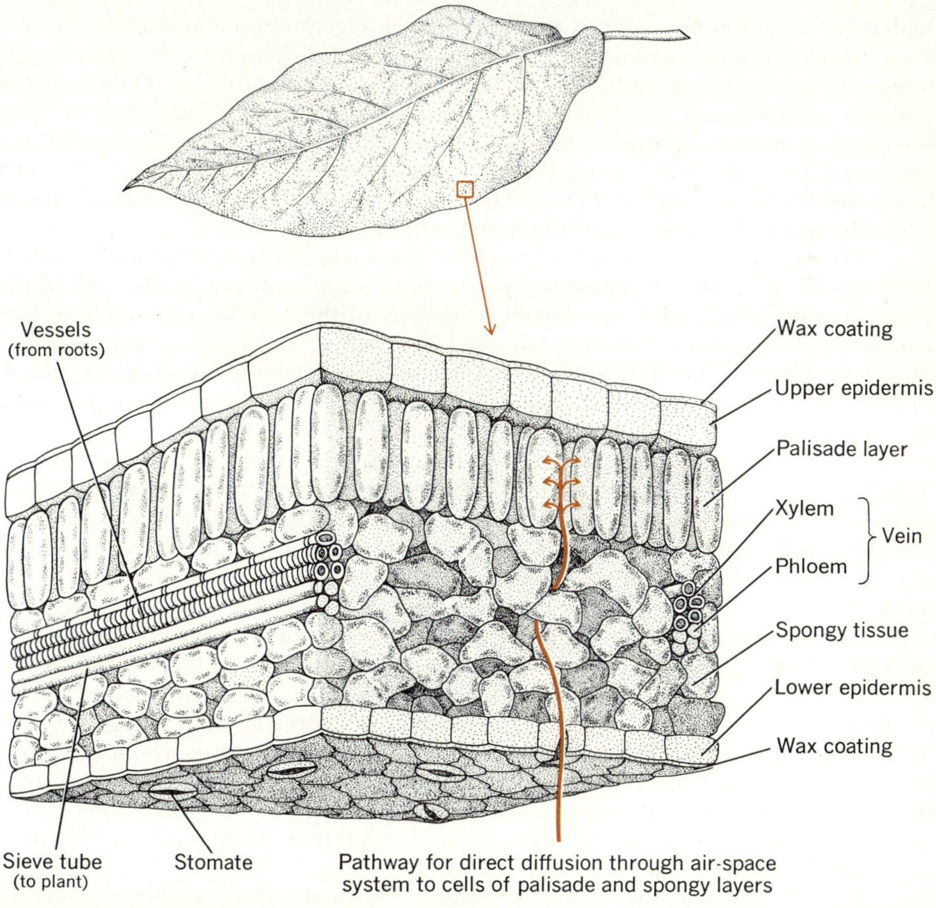

Figure 4-8 The microscopic structure of a leaf.

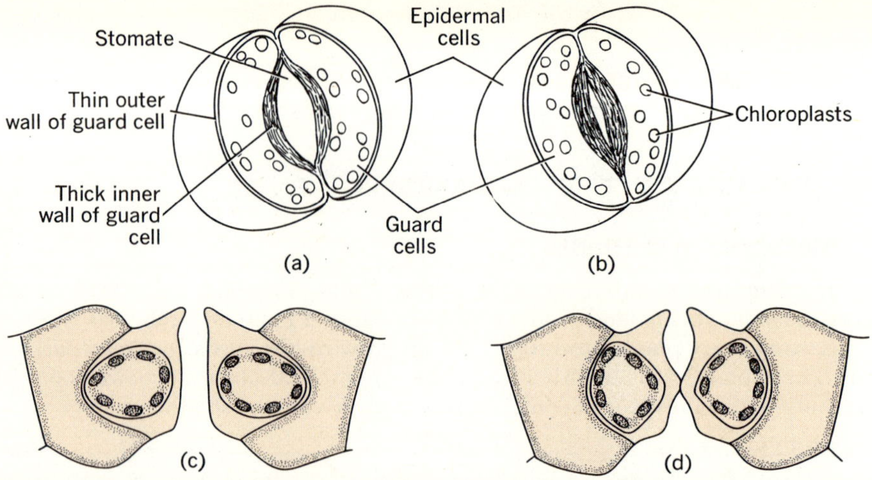

Figure 4-9 Stomates. Drawing of a stomate in the lower epidermis of a leaf (a) when open and (b) when closed. Diagrams of a stomate in cross section show the shape of the guard cells when the stomate is open (c) and closed (d).

leaf. If we cut through a leaf and examine the cut edge microscopically, we observe a labyrinth of interconnecting air spaces inside (Fig. 4-8). These spaces are more extensive in the lower *spongy tissue* than in the *palisade tissue* above. The elongated palisade cells contain more (and larger) chloroplasts than the cells below. Their long axis is oriented vertically to maximize the efficiency of light absorption and minimize the scattering of light. Both the upper and the lower surfaces of the leaf are covered with a transparent layer of cells (the *epidermis*) which bears a coating of wax. The epidermis keeps water loss at a minimum.

Passing through the spongy tissue are *veins* which bring water and minerals from the roots below and carry the products of photosynthesis to the rest of the plant. Usually confined to the lower epidermis of the leaf are thousands of tiny openings, the *stomates*, each one bounded by a pair of sausage-shaped *guard cells* (Fig. 4-9). The guard cells regulate the opening and closing of the stomates, which in turn control the exchange of gases (CO_2, water vapor, and oxygen) between the air surrounding the leaf and the intercellular air spaces within.

The stomates open in the morning when the leaf is illuminated and close at nightfall. This responsive and adaptive phenomenon seems to be caused by turgor (Fig. 3-16) in the guard cells. When the two guard cells bordering a stomate become turgid, the thinner wall of each cell bulges outwards, drawing the thick inner wall with it. This opens the stomate. When the cells lose their turgor, the inner walls are allowed to return to their original, unstressed state as thick-walled "doors" that close the stomate. Since an increase or decrease in turgor is associated with corresponding changes in osmotic pressure, guard cells must be capable of changing the concentration of solute molecules in their cytoplasm. This they can bring about by photosynthesis since they possess chloroplasts. In the presence of light, guard cells manufacture glucose that increases their osmotic pressure and consequently their turgor. They also reconvert some of the (insoluble) starch which they synthesized the night before back into glucose. The enzyme action which hydrolyzes starch in this case is promoted by the lowered CO_2 concentration brought about by active photosynthesis in the leaf.

Light therefore opens stomates because it initiates photosynthesis. Nightfall brings photosynthesis to an end and closes the stomatal doors.

GAS EXCHANGE

During the day the chlorophyll-bearing cells of the leaf are consuming CO_2 for photosynthesis. The CO_2 concentration in the intercellular air spaces is thereby lowered below that in the surrounding atmospheric air. As a consequence the CO_2 of the outside air diffuses through the stomates into the air spaces. This inward diffusion of CO_2 continues as long as photosynthesis takes place. The movement of oxygen is in the reverse direction. The release of O_2 in the noncyclic light reactions creates a concentration gradient which results in the diffusion of O_2 outward through the stomates.

The concentration of water vapor within the leaf is such that the intercellular air is saturated with moisture. This is necessary to prevent the loss of life-giving water from the cell by evaporation (a problem which all terrestrial organisms must successfully face). Since the air surrounding the leaf is seldom saturated with moisture, there usually is a diffusion of water vapor out of the leaf as long as the stomates are open. This process is known as *transpiration;* its rate depends on how dry or relatively unsaturated the outside air is. The amount of water lost in transpiration must be replaced by water coming into the leaves through the veins from the roots. On hot, dry days the root system may be incapable of doing this and the plant reacts effectively in its own self-interest. The decreased water supply to the cells of each leaf causes them to lose turgor and the leaves wilt (Fig. 4-10). In the guard cells loss of turgor results in closure of the stomates. Transpiration is drastically reduced (along with photosynthesis) and moisture is conserved. Here is a fine example of the finesse with which life designs its working machinery to serve purposeful and adaptive ends. The same mechanism which is triggered by light to supply the CO_2 for photosynthesis also conserves moisture when water transport becomes critical.

The stems and roots of plants also possess facilities for gas exchange in order to accommodate the needs of cellular respiration. These needs are modest compared to that of animals, since the metabolic release of energy is relatively low. Structures known as *lenticels* (Figs. 4-11 and 4-12) occur in both stems and roots. They are made up of loosely arranged cells with rather conspicuous intercellular spaces through which O_2 and CO_2 may diffuse to and from the deeper tissues.

Figure 4-10 Temporary wilting in the leaves of the squash plant. *Left:* In the late afternoon of a hot day leaves wilt and stomates close. *Right:* Early the next morning. The plants have absorbed water and recovered from wilting during the night. The stomates are open. (C. Wilson and W. Loomis, *Botany,* 4th ed. Holt, Rinehart and Winston, 1967)

Figure 4-11 *Right:* Lenticels of alder. This stem was 4 inches in diameter. (C. Wilson and W. Loomis, *Botany*, 4th ed. Holt, Rinehart and Winston, 1967)

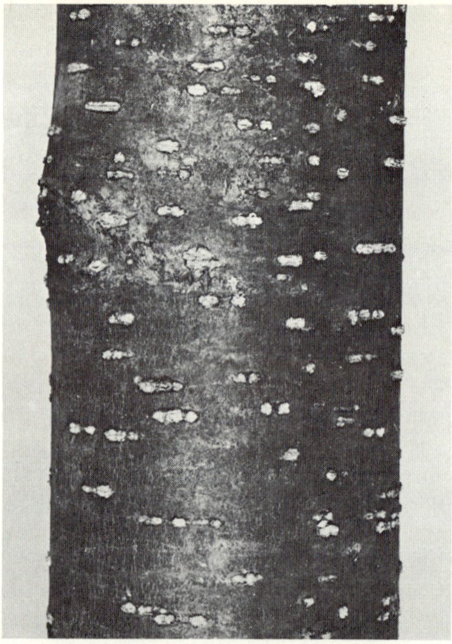

Figure 4-12 *Below:* Lenticel of black cherry tree, as seen in cross section of a young stem. (Dr. L. H. MacDaniels)

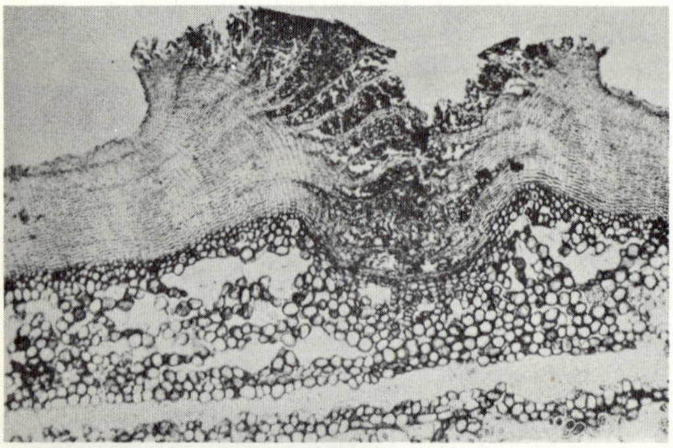

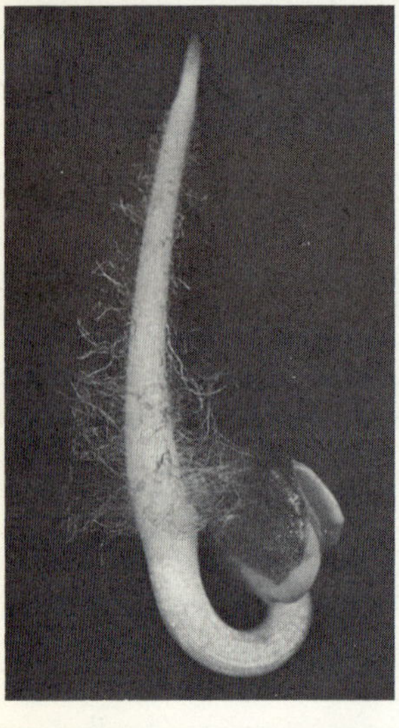

Figure 4-13 Root hairs of young radish plant. (Ward's Natural Science Establishment)

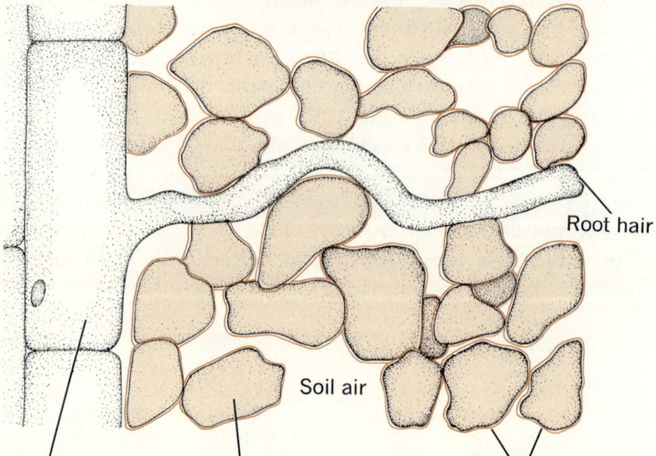

Figure 4-14 The relationship of root hairs to soil particles, soil water, and soil air.

The threadlike terminal portions of the root system (rootlets) are equipped with cellular extensions called *root hairs* (Figs. 4-13 and 4-14). These tiny filaments grow out into the spaces between soil particles. The air trapped between these particles and the moisture adhering to their surfaces is readily accessible for rapid absorption. O_2 and CO_2 may diffuse into and out of the root hairs as well as through the intercellular spaces between the epidermal cells of the rootlet. The moisture, with its dissolved solutes, may enter the plant through these same pathways by osmosis, diffusion, and active transport.

THE TRANSPORT OF MATERIALS

Referring back to Figure 4-8 note that the veins of the leaf possess two kinds of elongated cylindrical structures, the *xylem vessels* and the *phloem sieve tubes.* Both vessels and sieve tubes constitute a continuous transport system extending from the leaf through the stem to the roots. The vessels are slender nonliving pipes of microscopic bore whose inner walls are reinforced by ringlike or spiral thickenings (Fig. 4-15). They are formed by long columns of cells aligned end-to-end. The cell walls between adjacent cells gradually disappear, and the cells die leaving an inert microscopic conduit of considerable length (and strength). Vessels are confined to that part of the stem and root known as the *xylem*[8] (Fig. 4-17). The sieve-tube elements are elongated cells without nuclei, which are also aligned into long columns. Each cell, however, remains alive (Fig. 4-16). Some of the cytoplasm of one cell may flow into adjacent cells through perforated sieve plates that are actually modified cell walls. The sieve-tubes are found in that part of the stem and root called the *phloem* (Fig. 4-17).

As a plant grows in girth new phloem and xylem tissue is added to by the proliferation of cells from a thin generative layer called the *cambium.* As shown in Figure 4-17, growth results in pushing the phloem peripherally as more xylem is added. The contrast between slow summer growth and the rapid growth in spring is reflected in the xylem as "growth rings."

[8] There are other types of conducting structures in the xylem which need not concern us here.

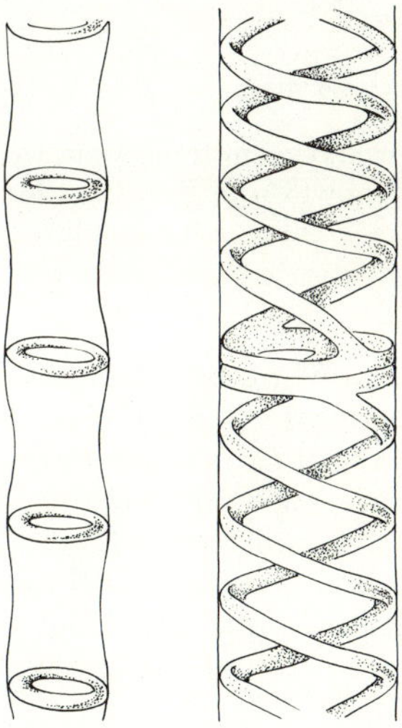

Figure 4-15 Vessels in plants.

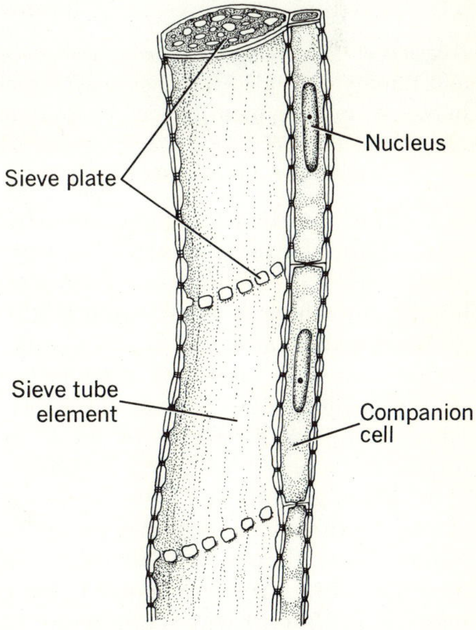

Figure 4-16 The sieve-tube element of plants. The nonnucleated sieve-tube elements probably remain alive because of their close association with nucleated companion cells.

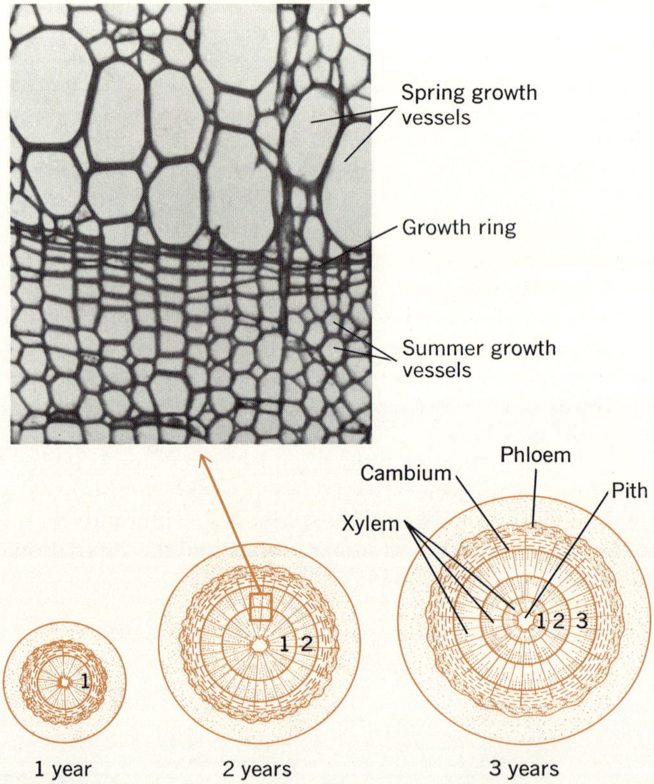

Figure 4-17 Diagrammatic cross sections of growing woody stems (one year, two years, and three years old) The inset is a photomicrograph of a growth ring showing the contrast between spring growth (wood) and summer growth (wood). (Photomicrograph by author)

WATER TRANSPORT

The vessels of the xylem constitute a mechanism whereby water and its dissolved solutes can be transported from the soil to the leaves and other parts of the plant (Fig. 4-18). When considering the volume of water that is transpired and the heights to which it is commonly lifted, this is an unusual phenomenon. In one growing season an acre of corn plants transpires enough moisture to flood the field under almost 15 inches of water if it were applied all at once in liquid form. The stately sequoias of California lift water from their roots to their topmost leaves, a distance commonly in excess of 300 ft. How is water lifted so high? A gravity or vacuum pump can lift water no higher than 32 feet (Why?).

Many hypotheses have been advanced. Only one has currently achieved the status of a theory. A large and accumulating body of evidence supports the *cohesion theory*. This theory capitalizes on the well-known fact that water is relatively incompressible: it is resistant to changes in volume. This same property manifests itself in a different way when water is confined in a closed tube with a very small diameter (such as a xylem vessel). In this case water also resists being stretched: it is relatively inextensible. The force of adhesion holds the water firmly to the walls of the vessel just as a film of water "sticks" to your hand. The column of water therefore exerts no pressure downward because of its weight.

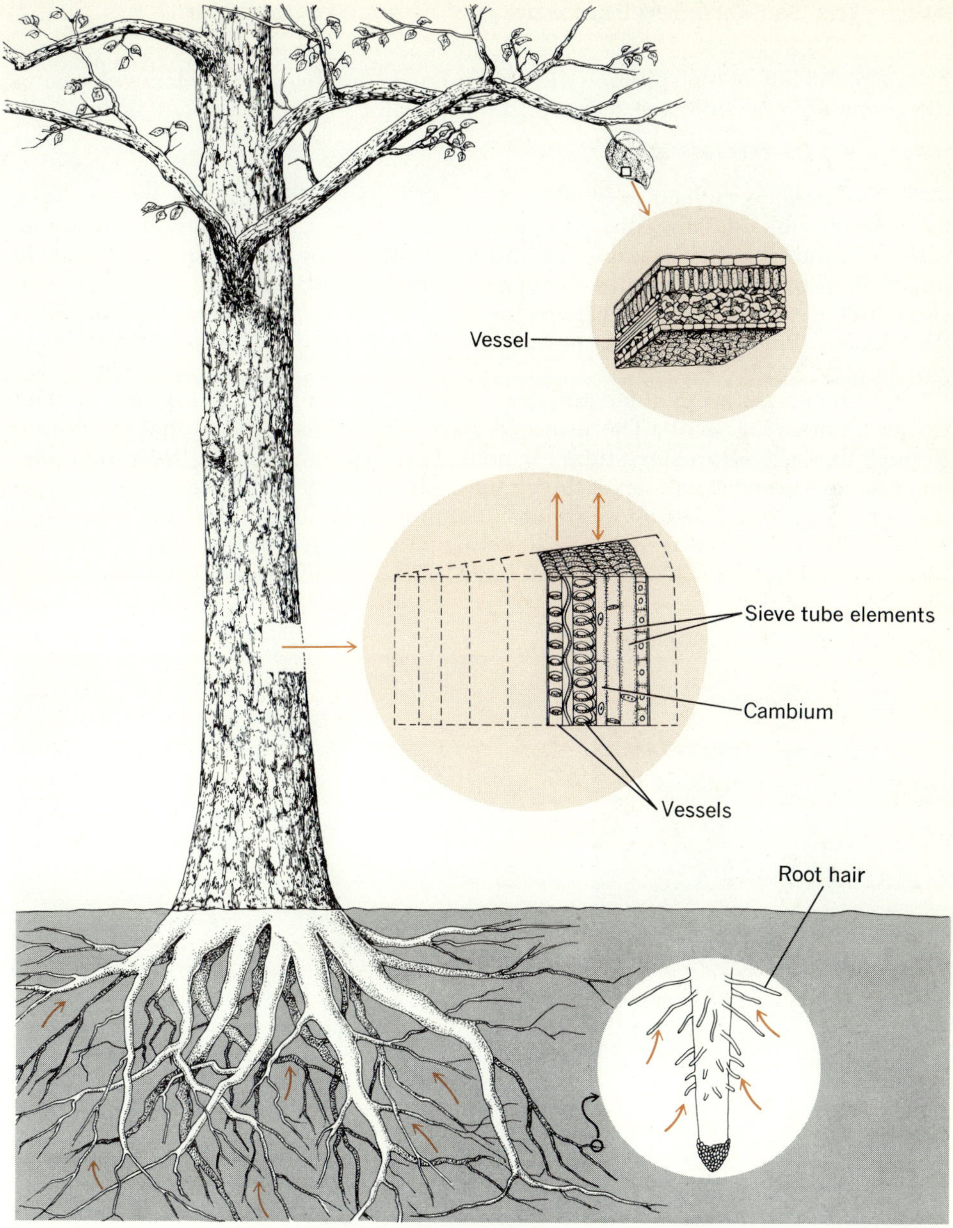

Figure 4-18 Transport system in a tree.

The *cohesive force* of intermolecular attraction between water molecules is so great that the tiny filaments of water within the vessel can be pulled upon as if they were steel wires. The pull is provided by transpiration. The loss of water from a vessel in a leaf puts the entire column of water in that vessel in tension. This tensile pull is reflected down into the root where the xylem vessel begins, and water is literally pulled from the soil by the root. The leaf, in short, pulls water from the ground. This was beautifully demonstrated by a study in which the stem of a tall climbing vine was cut near its base and inserted into a sealed container of water.

Although the rate of transpiration did not diminish, the vacuum which developed in the container was sufficient to bring the remaining water to a spontaneous boil.

FOOD TRANSPORT

The organic molecules synthesized by plants are transported by the sieve tubes in the phloem. How this takes place is not clear. Interfering with metabolic activity in the plant (for example, by chilling or curtailing oxygen supply) seems to impair food transport. It has been suggested that cytoplasmic streaming between sieve-tube elements may be responsible for movement of materials up and down the phloem. Phloem transport, however, is usually more rapid than cytoplasmic streaming.

An ingenious method for studying food transport utilizes the services of plant lice or aphids (Fig. 4-19). These insects have a long beak (stylet) that penetrates through the bark into a sieve-tube element. Turgor pressure in the sieve-tube element forces the sugar-rich sap of the phloem into (and out of) the insect. By anesthetizing the insect and cutting its stylet, phloem flow can be studied and measured. With the help of insects we may yet solve the problem. In the meantime the mechanisms of food transport remain in that large inventory of unsolved problems which

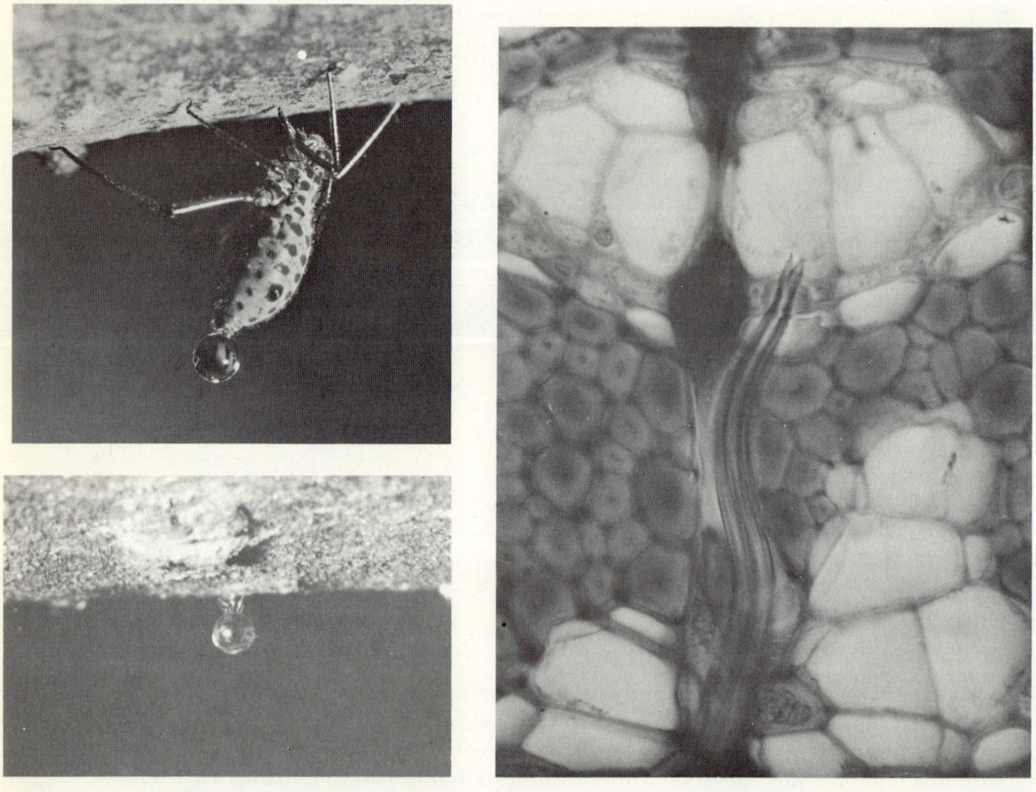

Figure 4-19 Aphid feeding on branch. *Left:* (Top) Photograph of aphid feeding on lower side of tree branch. (Bottom) Photograph of sap flowing through the aphid stylet after the rest of aphid has been removed. *Right:* Photomicrograph of aphid stylet penetrating sieve-tube element (cross-sectional view). (Martin H. Zimmerman)

continue to challenge and enchant those curious souls who ultimately come up with some answers.

Metabolism in Animals

The discussion on plant metabolism was directed specifically to the level of tissue type and function as it relates to organs (leaves, stems, and roots) and the organism as a whole. The following investigation of animal metabolism changes the emphasis to higher levels of organization. We now lay stress on organs and organ systems as they relate to whole organisms. In doing so, we direct our attention to some of the so-called higher animals, with special reference to three familiar subjects: the earthworm, the mosquito, and the frog.

NUTRITION

In their exploitation of energy-enriched organic foodstuffs animals capitalize on every type of available resource. Carnivores and herbivores in inconceivable numbers and kinds have evolved for the consumption of an equally impressive list of the preyed upon. *Omnivores* living on both plants and animals add many more kinds to the vast assemblage. All of these, in turn, are fed on by a host of animal parasites, such as tapeworms and fleas. In this consume-or-be-consumed order of things not one source of organic nutriment is left untapped. Even the dead remains of animals and plants and the organic residues of their decomposed bodies and waste products are utilized.

There are diverse techniques by which various kinds of food may be secured. The aphid of Figure 4-19 is a *fluid-feeder* on plant sap. The adult mosquito is a fluid-feeder on both plant fluids (usually nectarlike liquids from flowers) and the blood of animals (Fig. 4-20). The aquatic larva which will become a mosquito, however, is a *filter-feeder* (Fig. 4-21). Two small food brushes around the mouth are copiously endowed with hairlike structures. When feeding, the brushes move rap-

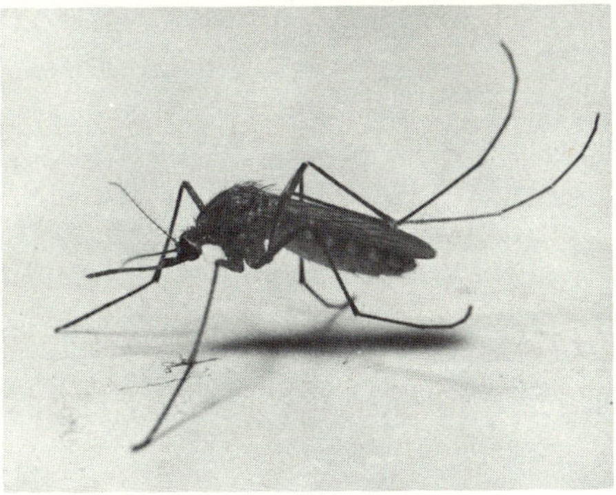

Figure 4-20 Like the aphid the adult mosquito is also a fluid feeder. (Russ Kinne, Photo Researchers)

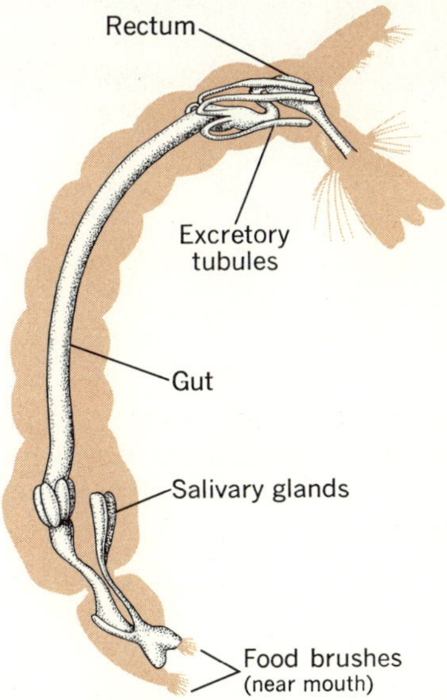

Figure 4-21 Mosquito larvae are filter feeders.

idly in a scooping action that entraps bacteria and other microscopic organisms in the water, which are then passed into the mouth. Filter-feeding is a widely employed and efficient means of obtaining food. The largest animals that have ever lived (the blue whale and related species) feed in this fashion.

Sediment-feeders, such as the earthworm (Plate 28), were discussed in Chapter 1. They engulf soil or sediments, living on decayed organic matter as well as any living microscopic organisms contained therein. A relatively large amount of bulk is consumed because of the low nutritive value of the food. The food of carnivores is much more nutritious. Relatively small amounts of bulk meet the dietary needs of the organism, but the food must be captured and held. Devices such as beaks and teeth serve to hold the prey. In mammals (animals with hair who suckle their young) teeth are also commonly used to tear and chew the food and actually begin the process of (mechanical) digestion. In frogs and toads the tongue is attached in front and can snare insects (Fig. 4-22); teeth in the upper jaw securely hold the victim. Many of the tadpole larvae of these animals, however, are herbivores. Their diet consists of the foliage of aquatic plants and the algae growing on submerged objects. Although such food need not be captured, it is not as nutritious as animal flesh because it contains much more water and considerable amounts of

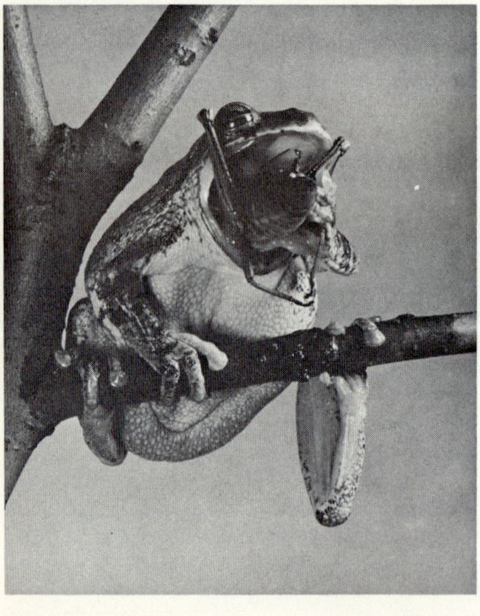

Figure 4-22 Frog which has captured a grasshopper with its tongue. Front leg is used to push the insect down. (Jim Annan, Photography)

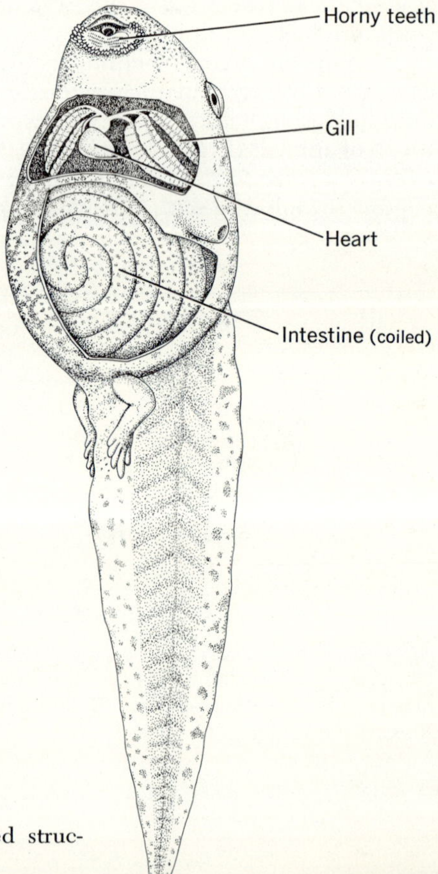

Figure 4-23 Digestive organs and associated structures of the tadpole.

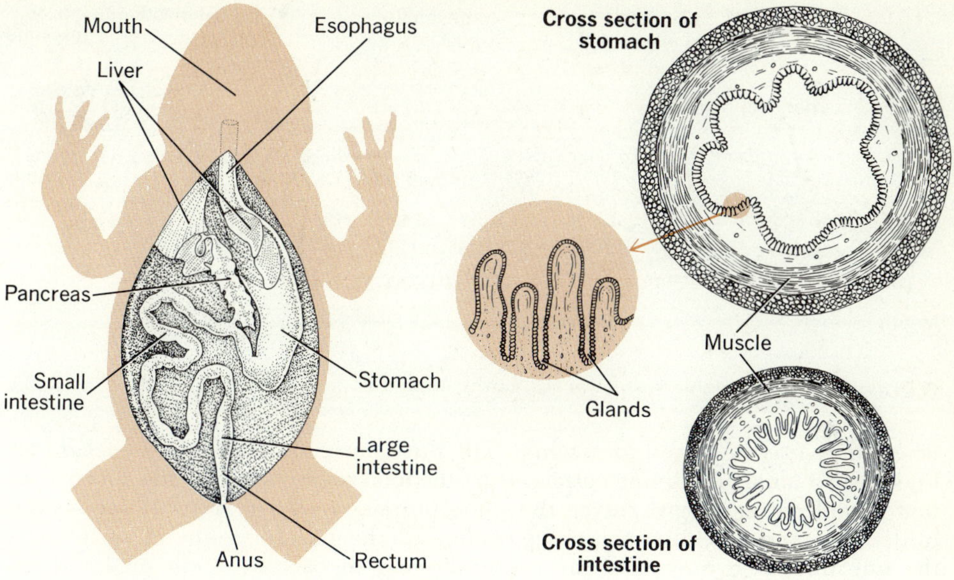

Figure 4-24 Digestive system of the frog.

relatively tough and indigestible plant fiber (from the cellulose of plant cell walls). The digestive process is more complex. Consequently, we find the tadpole's mouth well equipped with horny teeth and its gut (intestines) extended into a long spiral coil (Fig. 4-23).

THE DIGESTIVE TRACTS OF ANIMALS

The adult frog has a simpler digestive tract because its food is more nutritious, easier to digest, and need not be consumed in large amounts (Fig. 4-24). From the mouth food passes through a tubelike *esophagus* into the *stomach*. The churning action of the stomach's muscular walls begins the mechanical breakdown of the food into smaller pieces. This facilitates the action of hydrolytic enzymes, secreted by glands in the stomach lining. At the exit from the stomach a doughnut-shaped ring of muscles controls a narrow constriction (valve) which leads into the *small intestine*. Here additional enzymes are contributed by the pancreas and gland cells in the intestinal lining (epithelium). Bile from the liver aids in the digestion of fats. The intestinal contents, now highly liquefied, are available for absorption into the tiny blood vessels (capillaries) of the intestinal wall. Simple organic molecules pass by active transport through the cells of the intestinal lining into the bloodstream. Prominent longitudinal folds in this lining greatly increase the surface through which absorption can take place. The remaining material passes into the *large intestine* where the absorption of water occurs. It is also in this organ where large numbers of bacteria and other microorganisms are responsible for the microbiological digestion of the more resistant food residues. The undigested material which remains (feces) is temporarily stored in the *rectum* and periodically eliminated through the *anus*.

The digestive system of an earthworm is also an open-ended tube (Fig. 4-25). Soil is drawn through the mouth by the sucking action of powerful muscles in the wall of the *pharynx*. The contents are passed through the esophagus into the *crop*

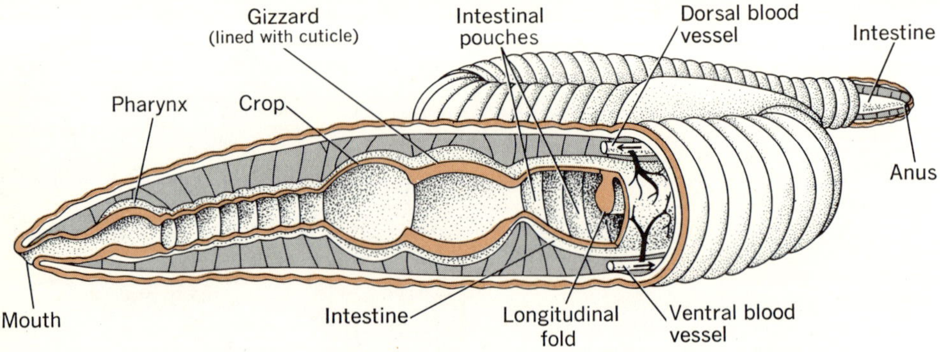

Figure 4-25 Digestive system of the earthworm.

where they may be stored for a while. The thick muscular walls of the *gizzard* are lined with a tough, nonliving cuticle. Here the food is ground up by the abrasive action of soil particles (grit serves the same purpose in the gizzard of seed-eating birds). The mechanically fragmented contents of the gizzard are then transferred to the intestine. The bulges in the intestinal wall in each segment of the worm, together with a prominent longitudinal fold, greatly increase the surface for digestion and absorption of food. A rich supply of blood vessels in the intestinal wall receives the absorbed nutrients in much the same way as it occurs in the frog. Undigested substances and soil particles leave the worm through the anus.

It might well be asked at this point why digestive enzymes produced by these animals are prevented from digesting the frogs and earthworms that make them. The answer to this lies in the fact that digestive enzymes are secreted in an inactive form. When they reach the normally acid contents of the stomach (or the basic contents of the intestine in the case of intestinal enzymes), they are converted to the active form. The lining epithelium of digestive organs is also protected with a layer of mucus which is secreted over them by epithelial mucus glands.

GAS EXCHANGE

Respiratory organs and organ systems are designed to meet the needs of metabolically active tissues for oxygen supply and CO_2 removal. A transport mechanism of interconnecting blood vessels (or channels) is usually, although not always, necessary to facilitate the process of gas exchange, just as with digestion. In large animals the problem of transport is a formidable one because of the great distance between active tissues and the delicate respiratory structures (membranes) through which the organism exchanges gases with its environment. If the animal is terrestrial, it is beset with the problem of excessive water loss through exposed respiratory membranes. If the animal is aquatic it must face the fact that water contains only about one-fourtieth to one-twentieth as much oxygen as air does. (Air contains almost 21 percent oxygen.) Let us examine some of the ways animals have resolved these problems.

THE SKIN-BREATHING EARTHWORMS

The long, narrow shape of the earthworm exposes a large skin surface relative to the weight of the worm. This skin is a thin layer of epidermal cells overlain by a delicate, protective cuticle (Fig. 4-26). The epidermal cells are highly supplied with blood vessels. Metabolizing tissue has a large skin area available through which the

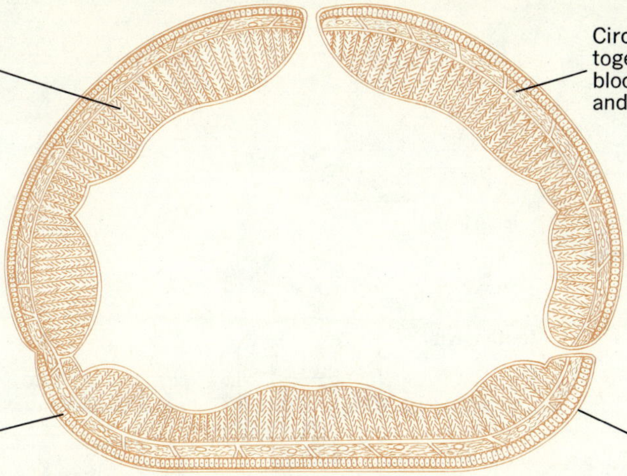

Figure 4-26 Cross section through an earthworm, showing skin, body wall musculature, and associated blood vessels.

diffusion of gases can take place. No part of the worm is very far from this skin. Since cellular respiration consumes oxygen and releases CO_2, these gases will diffuse along opposite gradients. The bloodstream will consequently receive oxygen and eliminate CO_2 through the respiratory epidermal covering and its cuticle. Since the air respired by the worm is the moist air between soil particles there is no problem of water loss.

GILL BREATHERS

Skin breathing is admirably adapted to the earthworm and its mode of life. For active, aquatic organisms, however, it possesses certain disadvantages. Except for minute organisms the skin usually does not present enough surface for rapid diffusion to take place. Second, a skin that is permeable to O_2 and CO_2 is also permeable to other small molecules such as those of water. This poses a problem to freshwater animals whose body fluids are hypertonic (page 94) to the water they live in; they have to be equipped with continuously operating bilge-pumps! As a consequence many active, aquatic animals (as well as some that are not so active) have relatively impervious skins and restrict their respiratory functions largely to organs known as *gills*. These include varied types such as the clam, octopus, crab, shrimp, fish, and the tadpole larva.

The gills of a tadpole (Fig. 4-23) and a fish (Fig. 4-27) consist essentially of flat plates composed of numerous *gill filaments*. Water is sucked into the mouth cavity as the floor of the mouth is depressed and the gill cover is pulled outward. The valve of the gill cover is closed by this action. When the floor of the mouth is elevated and the mouth closed, water passes over the gill filaments and out through the valve in the gill cover. As shown in Figure 4-27 the gill filaments are abundantly supplied with blood pumped directly from the heart. Gas exchange occurs between the water that bathes the filaments and the blood in microscopic capillaries just under the gill surface.

LUNG BREATHERS

A respectable number of land-dwelling animals breathe by means of lungs, such as the amphibious frogs and toads, and the essentially terrestrial reptiles,

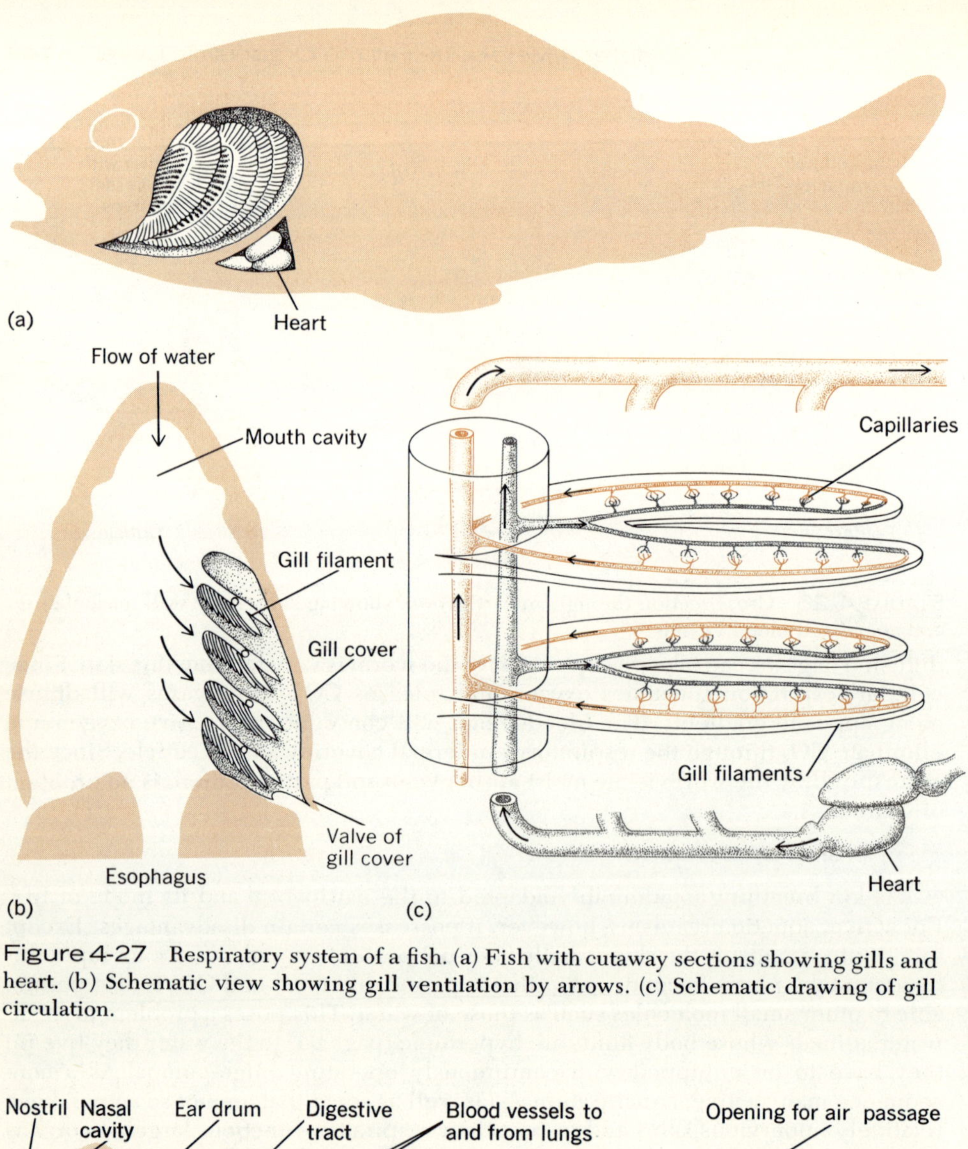

Figure 4-27 Respiratory system of a fish. (a) Fish with cutaway sections showing gills and heart. (b) Schematic view showing gill ventilation by arrows. (c) Schematic drawing of gill circulation.

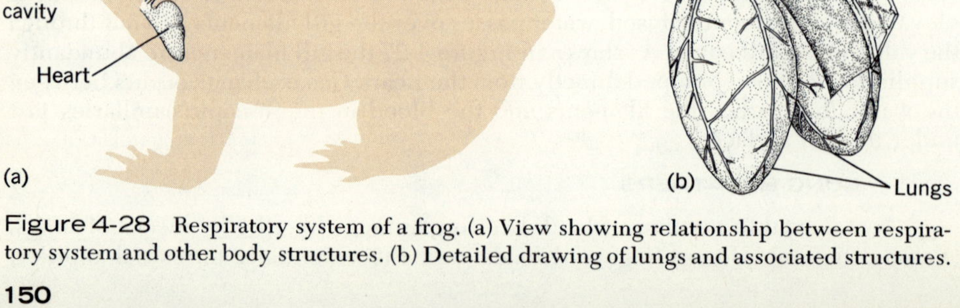

Figure 4-28 Respiratory system of a frog. (a) View showing relationship between respiratory system and other body structures. (b) Detailed drawing of lungs and associated structures.

birds, and mammals. Even a few fishes (the lung-fishes of Australia, Africa, and South America) utilize lungs and can therefore withstand those seasonal periods when their ponds and streams run dry.

Lungs are paired saclike organs connected to a single tube that opens into the digestive tract just behind the mouth cavity. The epithelial lining of the lungs is structured in such a way that it presents an extensive surface for gas exchange. In the frog the lining bears many shallow folds which divide the surface into innumerable small chambers (Fig. 4-28). An extensive network of blood vessels allows for the ready exchange of gases. Ventilation is accomplished by swallowing the air taken into the mouth through the nostrils. Air is exhaled by the contraction of muscles in the body wall. Unlike reptiles, birds, and mammals, frogs are capable of a limited amount of skin breathing.

The respiratory systems of birds and mammals are exceedingly efficient, with expansive surfaces for gas exchange (the respiratory surface in human beings is about 100 times greater than that of the total skin surface). This is necessary to maintain the high metabolic rate of these warm-blooded creatures. Warm bloodedness is a desirable trait because it allows animals to carry on more or less constant metabolic activity regardless of tolerable variations in outside temperatures. In animals that cannot control their temperature, metabolic activity is approximately halved for every 10° C drop in temperature.

TRACHEA BREATHERS

Insects and a few related animals have a respiratory system that does not depend on blood vessels for transport; it provides its own transport facilities. Air diffuses through the body in an elaborate system of open, branching tubes called *tracheae* (singular = *trachea*). These are shown in the mosquito larvae in Figure 4-29. The two tracheal trunks open externally through a pair of apertures (*spiracles*), each of which can be opened and closed by a valve. Small tubules branch off from the tracheae. Each of these, in turn, divides repeatedly, spreading into every part of the body. The smallest tubules may be no more than a micron in diameter. Oxygen and CO_2 move along these tiny, open passageways by diffusion. In the large tubes air movement is facilitated by body contractions.

Since insects are very active animals, there are maximum limits in body size beyond which the diffusion of oxygen through microscopic tubules would be too slow to accommodate metabolic needs. This is probably one reason why insects are not large animals.

TRANSPORT OF MATERIALS

The importance of internal transport, or the movement of circulatory fluids throughout the animal body, has by now doubtless been impressed on the reader. Absorbed nutrients and oxygen must be distributed to, and metabolic waste products must be eliminated from, the entire organism. Particularly in large, active animals this function can be a critical one.

The circulatory fluid (blood) that acts as the vehicle for this transport is a complex salt solution that is isotonic with the tissues it supplies. It also carries metabolic regulators (hormones) supplied to it by glands, many different proteins, and phagocytic blood cells that engulf foreign substances. It commonly contains *respiratory pigments*, such as *hemoglobin*,[9] which serve as highly efficient oxygen carriers.

[9] Hemoglobin may or may not be confined to blood cells. In the earthworm it is not; in the frog it is contained within red blood cells. Most insects have no need for respiratory pigments at all.

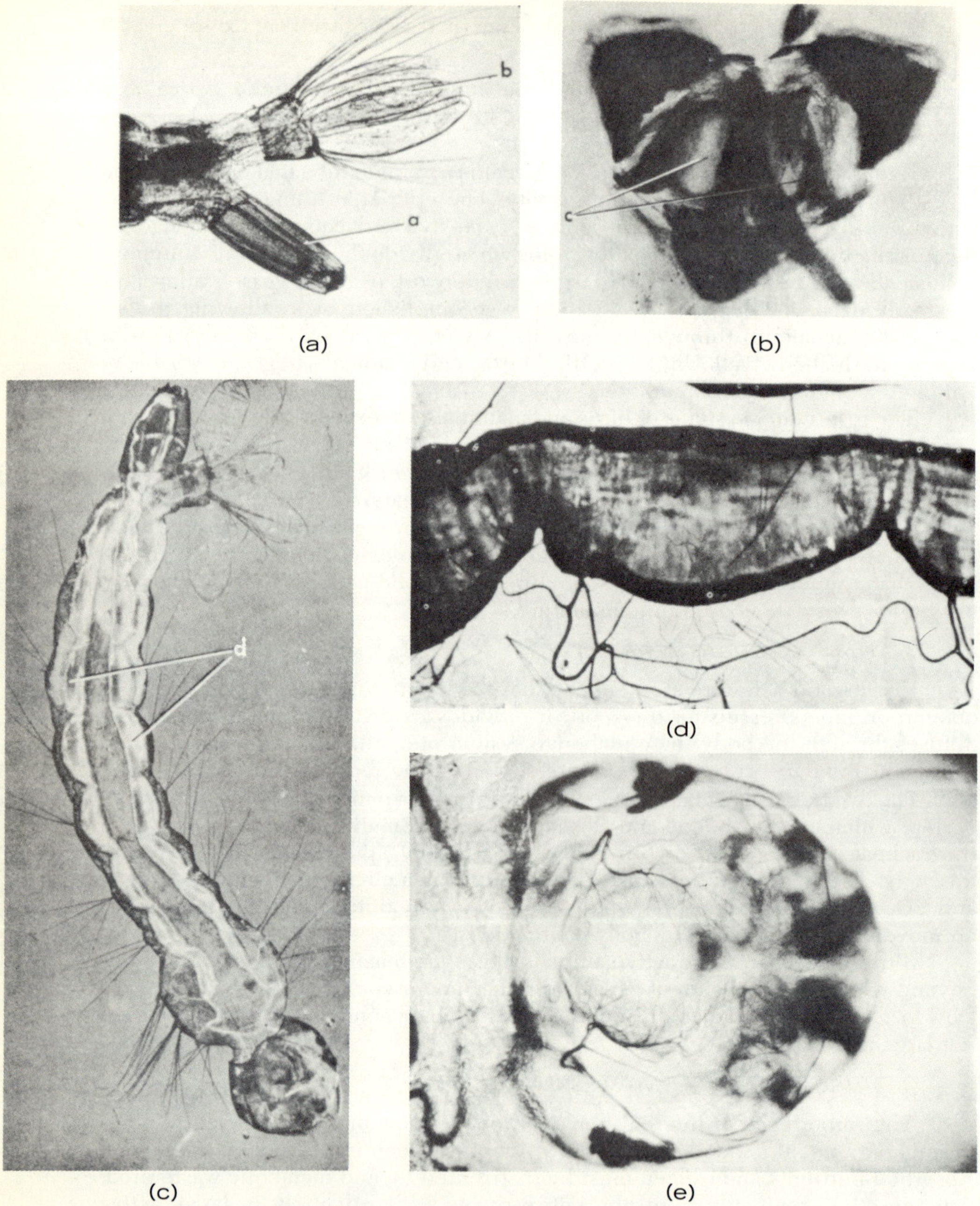

Figure 4-29 The respiratory system of a mosquito larva. In the adult mosquito there are several pairs of spiracles located along the sides of the body, an arrangement typical of most insects. For orientation see also Figure 4-21. (a) Photograph of terminal portion of larva. The letter a indicates process bearing spiracles; the letter b, anal papillae. (b) Magnified end view of process, showing two spiracles (letter c points to the spiracles). (c) Photograph of larva with two air-filled tracheal trunks indicated by the letter d. (d) Photomicrograph of portion of tracheal trunk with small tracheae branching from it. (e) Photomicrograph of head of larva with tiny tracheal branches. (A. Burnett and T. Eisner, *Animal Adaptation*. Holt, Rinehart and Winston, 1964)

These unique macromolecules release their oxygen to metabolically active sites where oxygen is being consumed and oxygen pressures lowered. When they are transported back to a respiratory surface, such as a gill, they are reloaded because of the higher oxygen pressures prevailing there.

Blood is transported through the animal body in two different ways, known as the *open system* and the *closed system*. Many animals actually utilize both. Let us examine each of these transport systems.

THE OPEN SYSTEM OF THE INSECT

Insects have a remarkably simple yet surprisingly efficient circulatory system. Since the tracheal system of respiration does not depend on a separate circulatory system, these animals have largely dispensed with the necessity for an elaborate distributive system of blood vessels. Consequently they do not require a powerful pump (heart) to push blood through these structures.

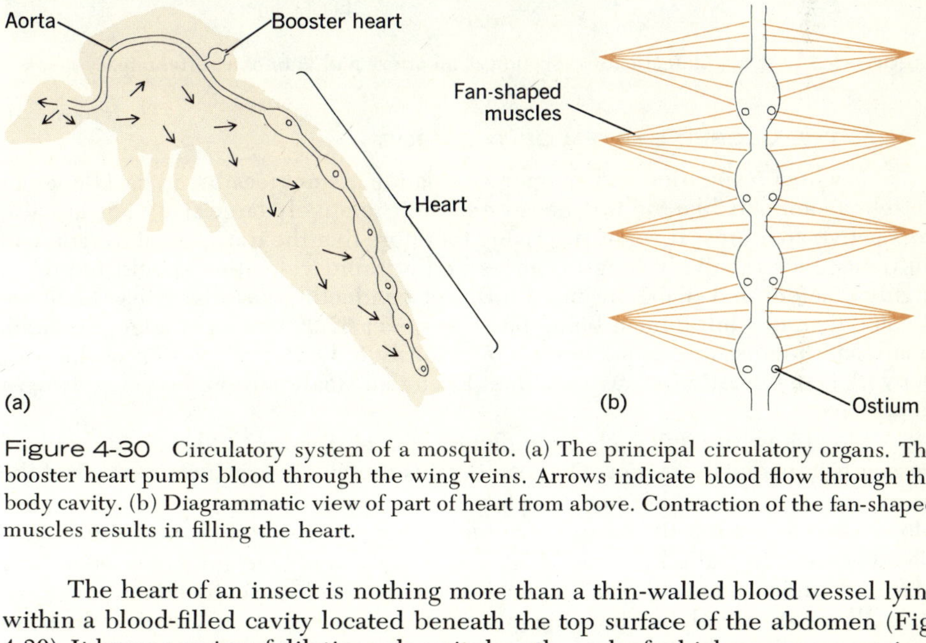

Figure 4-30 Circulatory system of a mosquito. (a) The principal circulatory organs. The booster heart pumps blood through the wing veins. Arrows indicate blood flow through the body cavity. (b) Diagrammatic view of part of heart from above. Contraction of the fan-shaped muscles results in filling the heart.

The heart of an insect is nothing more than a thin-walled blood vessel lying within a blood-filled cavity located beneath the top surface of the abdomen (Fig. 4-30). It bears a series of dilations along its length, each of which possesses a pair of lateral openings (*ostia*) equipped with valves. Each dilated portion is attached to the body wall by a pair of fan-shaped muscles. These muscles contract, enlarging the heart. This sucks the blood in through the ostia from the blood surrounding the heart. Muscles in the heart wall then contract, which close the valves of the ostia and push the blood forward in a blood vessel (*aorta*) that empties into the body cavity in the head. The body cavity is a continuous space surrounding all of the internal organs of the insect. It is filled with blood in which these organs are literally bathed. Cells and tissues are in direct contact with the circulatory fluid. Contracting muscles in the gut wall during digestion as well as those involved in respiration, body movement, and flight play a key role in propelling blood through the body and back to the cavity in which the heart is located. The more active the animal is the greater are its demands for nutrients and waste elimination and the more efficient the circulatory system becomes in meeting these needs.

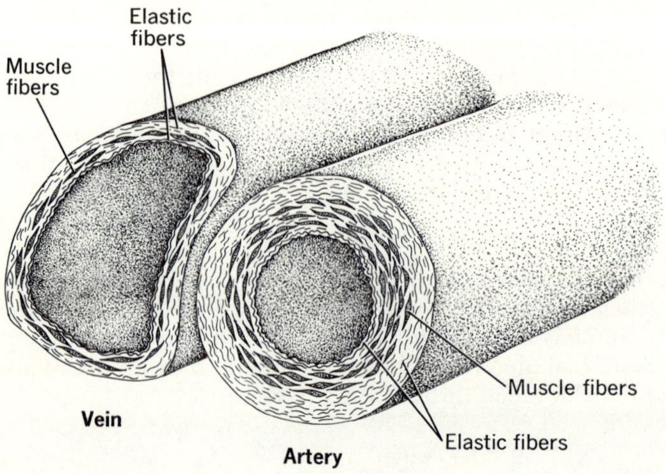

Figure 4-31 Diagrammatic cross section of an artery and vein of a vertebrate.

THE CLOSED SYSTEM OF THE FROG

Tracheal respiration and an open circulation are not feasible in backboned or vertebrate animals like the frog because of their relatively large size; they are also very active creatures. It is not surprising therefore that the transport of oxygen and nutrients over relatively long distances and at rapid velocities should require a highly efficient and elaborate mechanism of conducting vessels. It has been estimated that an adult human being possesses over 62,000 miles of large, medium, small, and microscopic blood vessels. Obviously only a small portion of this vast network is in operation at any one time. Even that small portion, however, is most impressive.

Blood flow through such a complex system of dividing and redividing blood vessels is impeded by friction. Overcoming this frictional resistance is the job of the heart. Since a great deal of pumping force is required to overcome this resistance, blood vessels leaving the heart carry blood under considerable pressure. These blood vessels are called *arteries* (Fig. 4-31) whose walls are thick, muscular, and elastic. As the rhythmically contracting heart forces a volume of blood into an artery, its walls are stretched, like blowing into a rubber balloon. Since valves in the heart prevent backflow, the stretched walls of such arteries impel the blood away from the heart and into the small arteries which branch from them. These small arteries are usually at some distance from the heart and supply specific regions of the vertebrate body (Fig. 4-32). Their walls are especially muscular. The contraction or relaxation of the muscular walls of these vessels changes their diameter, and hence regulates the flow of blood through them. It is by this mechanism that the vertebrate animal shunts its blood to active tissues and organs while restricting the flow to temporarily inactive regions (for example, the digestive tract during physical exertion).

The small arteries which service specific organs in turn divide into successively smaller vessels. The smallest of these are the *capillaries* whose walls are composed of one layer of flattened cells often less than one micron thick (Fig. 4-33). It is here that the bloodstream delivers nutrients and oxygen to the surrounding tissues and receives metabolic wastes from them. Diffusion occurs readily through the thin capillary walls into the tissue fluids in which all cells are bathed. Even

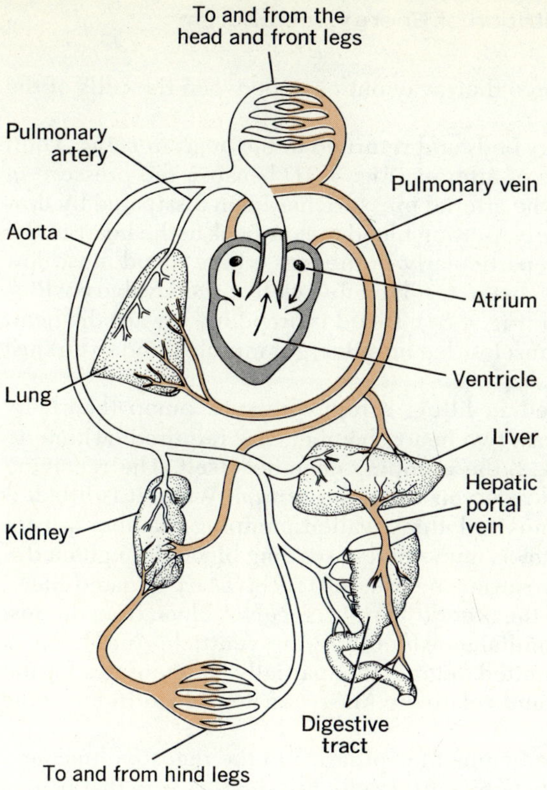

Figure 4-32 Schematic representation of the circulatory system of a frog. Arrows indicate direction of blood flow in the heart and the location of valves which prevent backflow. Veins and arteries are distinguished by color.

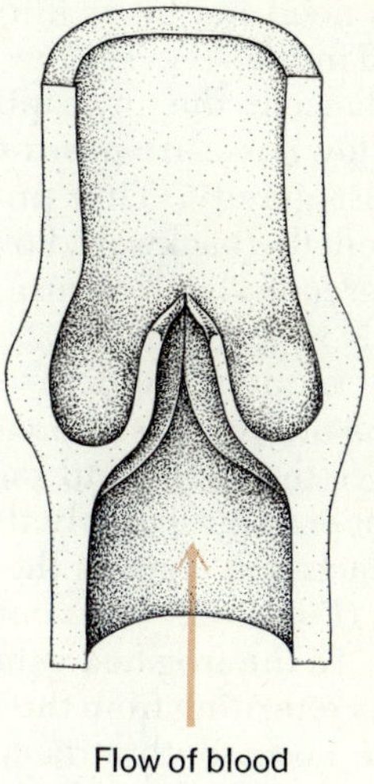

Figure 4-34 Valve in the vein of a vertebrate.

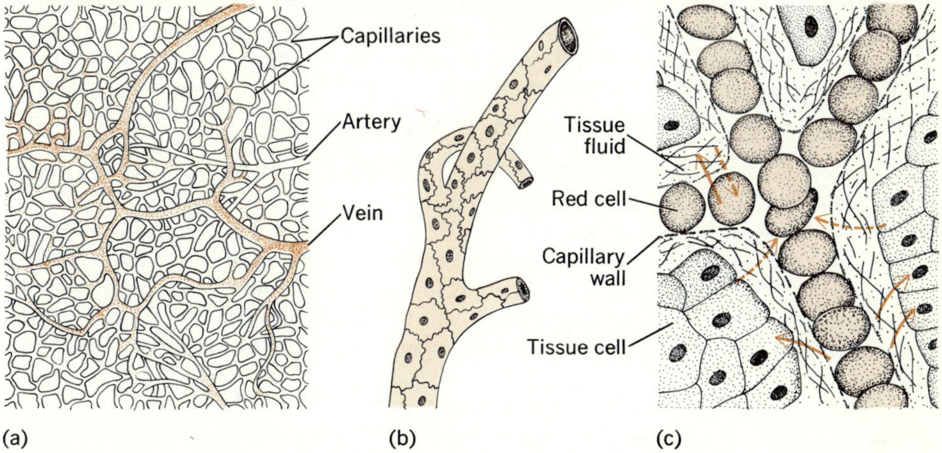

Figure 4-33 Capillaries. (a) Capillaries and the larger blood vessels which supply and drain them in the web of a frog's foot. (b) The structure of a capillary. (c) Capillaries and the transport of materials. Solid arrows indicate path of O_2 and nutrients; dashed arrows, the path of CO_2 and metabolic wastes.

amebalike white blood cells can force their way out or in between the cells of the capillary walls.

Blood is drained from capillary beds and returned to the heart in *veins*. Their walls need not be as thick as those of arteries (Fig. 4-31) because the pressure of blood in them is very low. Most of the arterial pressure has been dissipated by flow resistance in the tiny capillary vessels. Getting blood to flow back to the heart under very low pressure poses a problem particularly in the legs where blood must flow against gravity. This problem has been resolved by well-spaced valves which prevent the backward flow of blood (Fig. 4-34). Blood is propelled toward the heart by the contraction of surrounding muscles (leg muscles, gut muscles, and so on) just as it is in insects.

Another problem is presented in filling a high pressure pump (heart) by means of low pressure veins. At least two heart chambers are required: (1) one to receive the blood and push it into (2) the pumping chamber itself. The receiving chamber is a relatively thin-walled reservoir called an *atrium*. When it is filled, it contracts and charges the heavily muscled, thick-walled pumping chamber or *ventricle* (Fig. 4-32). The ventricle contracts immediately sending blood throughout the body. In the frog heart there are two atria, one (the left) receives oxygenated blood that is returning from the lungs and the other (the right) receives blood from the rest of the body. Both atria contract simultaneously, filling the ventricle. Inasmuch as frogs practice skin breathing to a limited degree, this partially compensates for the inefficiency of mixing oxygenated and relatively unoxygenated blood in the same ventricle.

It will be noted that one vein begins in capillaries in the digestive tract and ends in capillary beds in the liver (*hepatic portal vein*, Fig. 4-32). A vein that begins and ends in capillary beds is called a *portal* vein. This one, found in all vertebrates, is the means by which absorbed nutrients are transported to the liver. It is in this large and important organ where many of the metabolic pathways discussed earlier in this chapter take place, including the deamination of amino acids. The significance of the portal system to the frog's kidney is not known.

The tissue fluids to which we alluded previously are part of a separate *lymphatic circulation*. This system resembles the open circulation of the insect except that it is drained by innumerable lymph vessels which join and empty into veins by common lymphatic ducts. In the frog this is assisted by two pairs of lymph hearts.

CONTROL OF INTERNAL STABILITY

As we have pointed out before one of the unique characteristics of living organisms is their remarkable capacity to regulate their metabolic activities in their own self-interest. Regardless of changes which occur in the external environment of an animal, its internal environment is subject to very little variation: a steady state is maintained.[10] The circulatory system plays an important role in this phenomenon. The organs of excretion, such as kidneys, are devoted solely to this function. Why this should be so is one of the many fascinating examples in biology of one thing leading to another.

Most animals usually consume more protein than they require for biosynthesis.[11] The excess amino acids are deaminated to pyruvic acid. The free amino groups (NH_2^-) are converted into one of three different nitrogenous wastes: ammo-

[10] The technical term for this is *homeostasis*.

[11] This is necessary to assure the minimum intake of each of the several different amino acids essential to animal life.

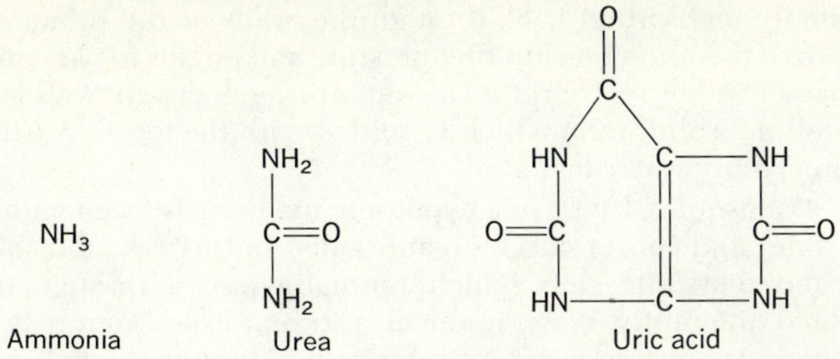

Figure 4-35 The structural formulas of three common nitrogenous waste products.

nia, urea, or uric acid (Fig. 4-35). Ammonia requires relatively little energy for its synthesis, but it is highly toxic and must be eliminated as a dilute solution in copious amounts of water. Urea takes more energy for its synthesis, but it is quite soluble and not nearly as toxic as ammonia. It can be eliminated with much less water loss. Uric acid is a relatively large molecule and requires the most energy for its production. It is, however, relatively nontoxic and only slightly soluble and can be excreted in crystalline form with scarcely any loss of water. The form in which nitrogenous wastes are excreted therefore has a great deal to do with the amount of water loss in the animal body. Body water content in turn affects the osmotic pressure of circulatory fluids; and the osmotic balance of body fluids is determined largely by those soluble inorganic salts that are essential to life. As a consequence of this chain of interrelationships excretory organs are fundamentally regulatory in nature. They are involved not only in the form in which nitrogenous wastes are excreted but also perform essential services in controlling the water and salt content of body fluids.

Let us examine how self-regulation of the internal environment is managed in two quite different animals.

THE MOSQUITO'S SOLUTION TO SELF-REGULATION

The excretory organs of mosquitos, like other insects, are long, slender tubules which empty into the intestine (Figs. 4-21 and 4-36). The nitrogenous end product of protein catabolism is uric acid, just as it is in birds and terrestrial reptiles. All of these animals share a common problem: water conservation in an environment in which a certain amount of water loss through respiratory surfaces, excretions, and even the skin cannot be avoided.

In the adult mosquito uric acid is secreted into the excretory tubules in a water solution. As this solution (urine) moves down each tubule toward the intestine,

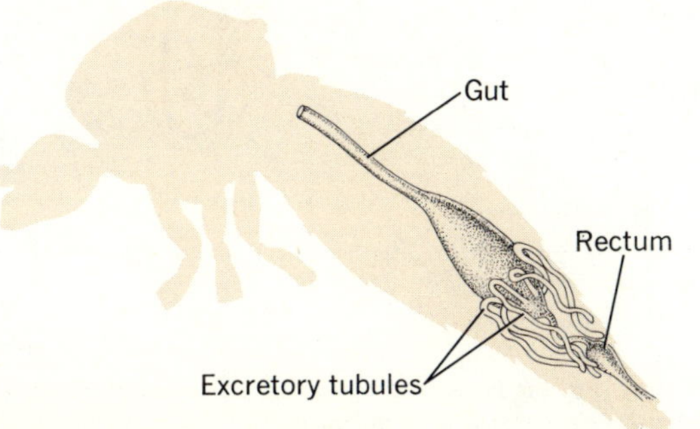

Figure 4-36 Excretory organs of the mosquito. See also Figure 4-21.

water is gradually reabsorbed back through the walls of the tubule and into the blood. By the time the urine reaches the intestine and begins to pass into the rectum it is a pasty mass. Further reabsorption of water through the gut wall leaves the uric acid in a crystalline (solid) form which is voided with the feces. Waste elimination results in scarcely any water loss at all.

The larval mosquito, living in a hypotonic medium, is faced with the problem of removing water and conserving inorganic salts. In this case the solution of uric acid, water, and inorganic salts which normally passes through the excretory tubules is treated differently. Water removal is accomplished simply by the reduced reabsorption of water in the excretory tubules and the elimination of fluid urine. Inorganic salts are conserved by their reabsorption through the walls of the rectum by active transport. Any condition, on the other hand, which tends to increase the salt concentration of body fluids would be compensated by decreased rectal reabsorption and the elimination of salts.

THE FROG'S SOLUTION TO SELF-REGULATION

The frog tadpole takes the easy way out and excretes its nitrogenous waste as ammonia. Living in fresh water the larva has no problem in water conservation and the synthesis of ammonia requires less energy than either of the other two nitrogenous wastes. The adult frog, on the other hand, spends a good part of its time out of water and is subject to considerable water loss through its moist skin and by evaporation from its lungs. As a result the frog takes the middle road and excretes urea. A moderate amount of water is used to flush out an excretory product whose synthesis requires only moderate amounts of energy.

The excretory system of the frog is quite different from that of the insect. The

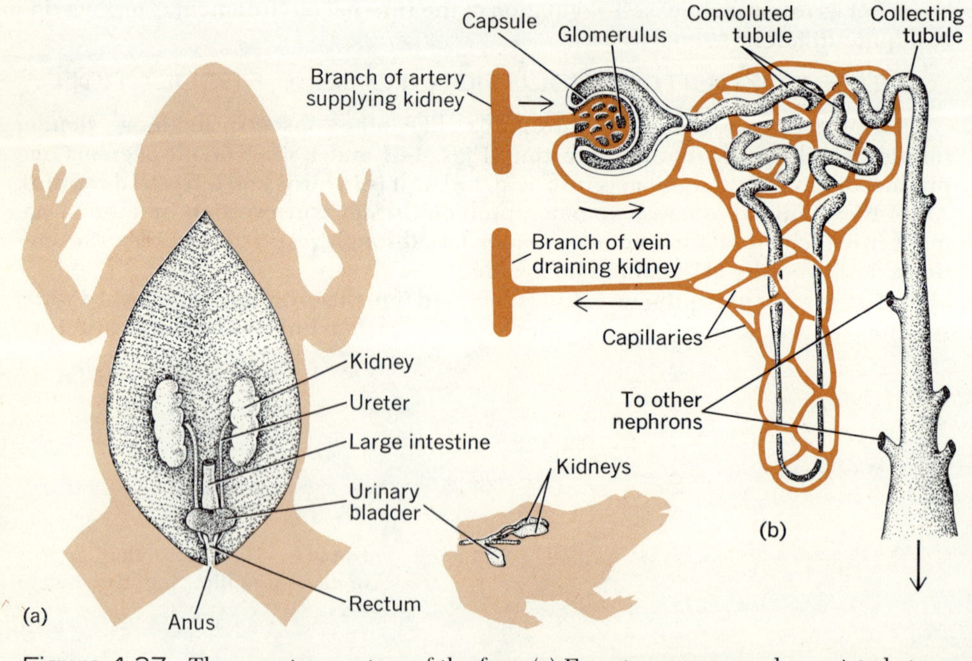

Figure 4-37 The excretory system of the frog. (a) Excretory organs and associated structures. (b) Diagrammatic representation of the nephron of a vertebrate.

difference in circulation between the two animals is partially responsible for this. The similarity in the basic process of excretion in the two organisms, however, is striking, as we shall see.

The excretory organs of the frog, as in all vertebrates, are a pair of kidneys (Fig. 4-37a). Each kidney is drained by a tube (*ureter*) which empties into the lower digestive tract (rectum). Urine passing down the ureters may be stored in the *urinary bladder,* a large pouch opening into the rectum near the entrance of the ureters. Each kidney is supplied with blood by a branch of the aorta (Fig. 4.32) and consists largely of many thousands of tubular excretory units called *nephrons* (Fig. 4-37b). The nephrons drain into common *collecting ducts* that ultimately discharge their contents into the ureter. One end of the folded nephron tubule is expanded into a cuplike capsule, which invests a tightly coiled ball of capillaries (the *glomerulus*). The tiny (but muscular) blood vessel leaving the glomerulus has a smaller bore than the small artery leading in. It immediately branches into a capillary network in which the remaining portion of the nephron (the *convoluted tubule*) is thoroughly entangled and in very close contact. The blood in this capillary bed is drained by a vein for its return to the heart.

Blood passing through the glomerulus is filtered through submicroscopic pores in the wall of the glomerular capillaries and into the cavity of the capsule. This filtrate is quite similar to whole blood except that it lacks blood cells and protein molecules which presumably are too large to pass through the minute openings in the capillary walls. Its composition therefore is comparable to the tissue fluids of the frog. As this filtrate passes into the convoluted tubule the same kind of process occurs as that observed in the excretory tubules of the insect. Large volumes of water are reabsorbed into the capillaries surrounding the tubules. The amount of water so transferred maintains constancy in the water balance. Ions of sodium, potassium, calcium, phosphate, and so on, are similarly treated, through selective reabsorption, to maintain stability in the salt balance. Glucose and amino acids are totally reabsorbed. Urea, synthesized by both liver and kidney, is only partially reabsorbed and is the major solute in the urine that is excreted.

These mechanisms, however, do not operate without guidance and control. An excretory tubule may reabsorb water in one case and not in another *only if it is somehow directed* to perform one or the other of these opposing functions. The complex muscular movements of a digestive tract do move food from one organ to the next *if they are coordinated.* An organism does indeed direct its internal affairs as well as its external affairs in its own best interest. We have so far, however, not given the matter of control and coordination much attention. In the next chapter we will attempt to rectify this.

QUESTIONS

1. Briefly discuss the interdependence of metabolism, integration, growth, and reproduction.
2. For a time a young bean plant sprouting from a seed is a heterotroph. Why is this?
3. Describe some of the metabolic events included in the category of biosynthesis.
4. The kinetic energy released by metabolism may be transformed in a variety of ways. Describe some of them.
5. What effect does energy transformation have on the efficiency of work performance?

6. The metabolic "packaging" of energy in ATP molecules has a number of advantages. What are some of them?
7. The release of free energy in a cell is regulated by the rate of energy utilization in that cell. How is this brought about?
8. Of the 38 molecules of ATP that are synthesized per molecule of glucose oxidized, how many are the result of oxygen-dependent reactions in mitochondria?
9. The intimate interrelationships of cellular respiration and biosynthesis are beautifully represented by reactions open to the end product of the "dark reactions." In broad terms outline some of these interrelationships.
10. In a similar way outline some of the interrelationships between cellular respiration and biosynthesis exhibited by end products of the noncyclic light reactions.
11. Describe as many ways as you can in which each of the following plant structures is adapted to serve the function indicated:

    (a) The leaf and photosynthesis
    (b) Guard cells and water conservation
    (c) Xylem vessels and water transport

12. Describe some of the different ways animals are adapted to securing food.
13. Describe as many ways as you can in which each of the following animal structures is adapted to serve the function indicated:

    (a) An earthworm's intestine for digestion and absorption
    (b) A tadpole's gills for gas exchange
    (c) A frog's lungs for gas exchange
    (d) The open circulatory system of an insect for the transport of materials.

14. What are some of the advantages and disadvantages of being a trachea breather?
15. Why is the regulation of the salt and water content of the animal body often intimately related with the excretion of nitrogenous wastes?
16. Although the excretory organs of a mosquito and a frog are quite different, the basic process of excretion is similar in both animals. In what respects are the regulation of the water balance and the elimination of nitrogenous wastes similar in the two species?

Suggested Readings

Burnett, A. L., and Thomas Eisner. *Animal Adaptation.* Modern Biology Series. New York: Holt, Rinehart and Winston, 1964.

Coult, D. A. *Molecules and Cells.* The Principles of Modern Biology. Boston: Houghton Mifflin, 1966.

Giese, A. C., *Cell Physiology.* Philadelphia: Saunders, 1962.

Griffin, D. R., and Alvin Novick. *Animal Structure And Function.* Modern Biology Series. New York: Holt, Rinehart and Winston, 1970.

Jellinck, P. H. *The Cellular Role of Macromolecules.* The Scott, Foresman series in undergraduate biology. Glenview, Ill.: Scott, Foresman, 1967.

Lee, A. E., and Charles Heimsch. *Development and Structure of Plants, A Photographic Study.* New York: Holt, Rinehart and Winston, 1962.

Lehninger, A. L. *Bioenergetics: The Molecular Basis of Biological Energy Transformations.* New York: Benjamin, 1965.

Levine, R. P. "The Mechanism of Photosynthesis," *Scientific American,* December 1969.

Loewy, A. G., and P. Siekevitz. *Cell Structure and Function.* Modern Biology Series. New York: Holt, Rinehart and Winston, 1969.

Racker, E. "The Membrane of the Mitochondrion," *Scientific American,* February 1968.

Ray, P. M. *The Living Plant.* Modern Biology Series. New York: Holt, Rinehart and Winston, 1965.

Sistrom, W. R. *Microbial Life.* Modern Biology Series. New York: Holt, Rinehart and Winston, 1969.

Steward, F. C. *Plants at Work.* Principles of Biology Series. Reading, Mass.: Addison-Wesley, 1964.

CHAPTER 5 Integration and Adaptive Behavior

THE IMPORTANCE OF INTEGRATION AND RESPONSIVENESS TO THE LIVING ORGANISM

A living organism is a dynamic system devoted to the maintenance of its highly organized state through the mobilization of energy. We examined in Chapter 4 some of the metabolic processes and mechanisms employed in this function—the machinery of life, so to speak. But an organism is much more than a mere machine. It is a highly coordinated system, quick to respond to its internal needs and bodily states. Digestion in animals, for instance, requires special detection devices sensitive to the progress of the digestive process. The information so monitored is conveyed to integrating centers, which in turn activate appropriate responding mechanisms. Only in such a fashion can the separate organs of the digestive system be expected to act in concert in the satisfactory digestion of a meal. The circulatory system must likewise be directed and coordinated in administering to the nutritive and oxygen needs of contracting muscles and secreting glands in the gut wall as well as in providing for efficient transport of the absorbed products of digestion. The metabolic processing of these same absorbed substances must in turn be regulated by a fine system of controls sensitive to the energy needs of the organism and the proper balance of water and inorganic salts in its body fluids.

Living things are equally responsive *as whole organisms* to the total environment in which they live. They carry on their activities in order to perpetuate themselves and the species to which they belong. They react adaptively to changes in the physical environment. Sand lizards burrow beneath the desert surface to seek shelter from the killing heat of the midday summer sun. Geese fly south to protect themselves from the rigors of cold. Plants turn their leaves toward the sun and send forth blossoms in response to seasonal changes in the lengths of days and nights. The behavior of living organisms is also both adjusted and adjustable to the biological problems of coexistence with other members of their species as well as the untold numbers of other creatures with which they come in frequent contact. Ants feast on the sugary excrement of aphids (Fig. 4-19). They also protect their benefactors from attack by predatory insects. The erratic flight of butterflies makes their capture by birds more difficult. The silent cunning of a cat stalking its prey favors the likelihood of a catch. Certain plants (sages and others) prevent crowding and competition from other plants by the production of secretions which interfere with seed germination and the growth of roots in seedling plants.

Life is indeed ingenious at any level. Our present purpose is to examine this ingenuity at the level of metabolic integration, response, and adaptive behavior.

MECHANISMS DEVOTED TO SENSITIVITY AND RESPONSE

Intracellular Mechanisms

All cells are *irritable:* they are capable of responding in varying degrees to certain changes in their environment. These responses are adaptive, that is, they favor survival. The environmental changes which evoke them are said to act as *stimuli.*

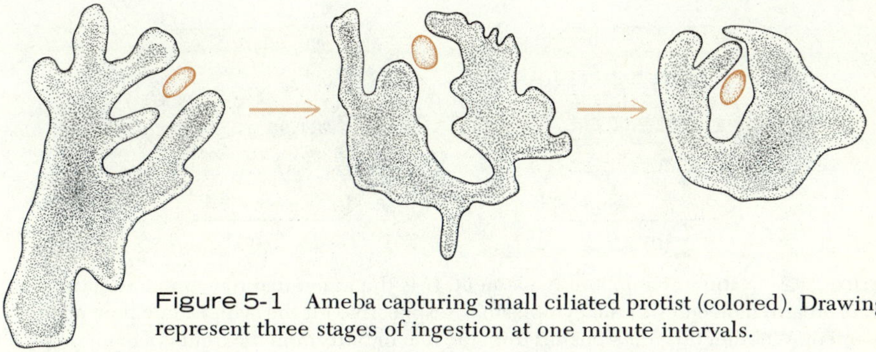

Figure 5-1 Ameba capturing small ciliated protist (colored). Drawings represent three stages of ingestion at one minute intervals.

The property of cellular irritability is best illustrated in unicellular organisms. The ameba (Fig. 5-1) possesses no specialized organelles either for receiving stimuli or for responding to them. Yet this protist will withdraw when touched by a probe but will engulf a small organism that touches it (Fig. 5-1). How it senses the differences between probe and food is not known, but chemical activators of some sort are presumably involved. How it carries on its flowing (ameboid) movement is also still a matter of some conjecture. The cytoplasm is clearly divided between an inner portion in a fluid, sol state and an outer layer in the form of a semisolid gel (Fig. 5-2). Cytoplasm in the inner sol flows forward, carrying the rest of the cell along with it. Upon reaching the forward end the fluid sol undergoes gelation, releasing a certain amount of water that passes to the rear. Here the outer gel solates, turns inward, and joins the forward-flowing sol.

It has been suggested that this differential movement of outer and inner layers is brought about by the contraction (folding) of fibrous protein molecules in the cytoplasm. Whether this is so or not we know virtually nothing about the metabolic activities responsible for these reactions or the manner in which they might be coordinated. We must frankly admit once more that the subcellular and macromolecular organization and coordination of the living cell is one of great complexity and subtlety. We must not lose sight of this at any point in the ensuing discussion.

Through the long course of evolutionary development, however, refinements have been superimposed on these more generalized patterns of cellular irritability. Specialization at the subcellular level has resulted in specific organelles designed to increase discrimination in detecting environmental changes (stimuli). The flagellated protist *Euglena,* for example (Fig. 5-3), has an eyespot which is sensitive to light. Because of this the organism is able to orient itself rather accurately with respect to light intensities. Specialization has also led to speedier conduction of cellular excitations over specific pathways which connect, in turn, to more efficient responding mechanisms. The unicellular ciliate *Euplotes* (Fig. 5-4) possesses

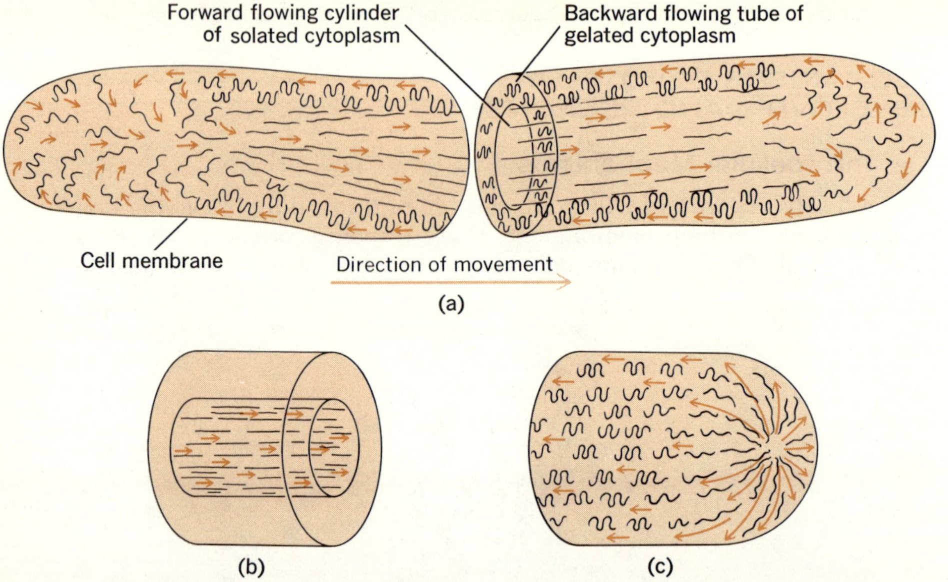

Figure 5-2 Nature of ameboid movement. It is theorized that the contraction (folding) of fibrous protein molecules in the cytoplasm is responsible for the cytoplasmic flow of ameboid movement. Whether this force pushes from the rear or pulls from the front is a moot point. This illustration suggests that it pulls from the front. (a) A diagram of a greatly simplified ameba. The advancing end shows unfolded protein molecules (short, straight lines) in the internal sol cylinder becoming folded (wavy lines), pulling solated cylinder forward, and forming a backward flowing tube of gelated cytoplasm on the outside. The opposite end shows folded molecules unfolding and joining the forward flowing sol. Small arrows indicate direction of cytoplasmic flow. (b) A three-dimensional view of a section of the sol cylinder. (c) A surface view of the advancing end and the formation of the backward flowing tube of gelated cytoplasm with its folded protein molecules.

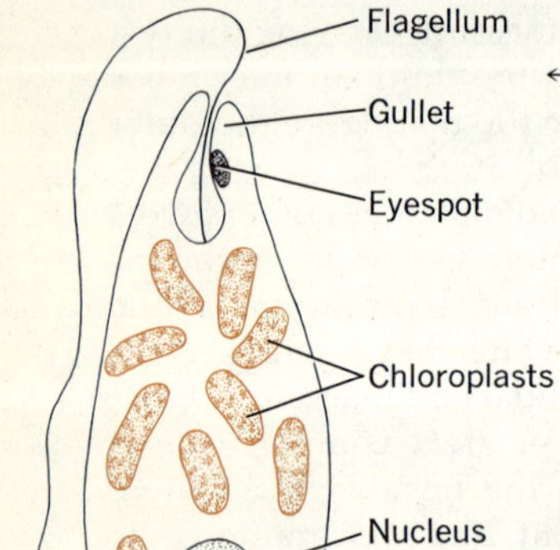

←Figure 5-3 *Euglena* possesses two organelles specialized for response: a light-sensitive eyespot and a locomotor flagellum.

Figure 5-4 *Euplotes*, a complex unicellular ciliate. Cutting the fibrils which connect five locomotor structures to a common coordinating center results in loss of coordinated movement.

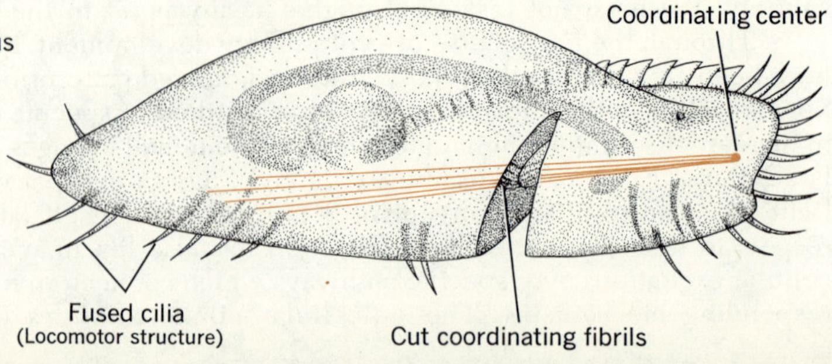

locomotor organelles consisting of fused bundles of cilia which enable the organism to crawl. These structures are connected to fibrils that emanate from a coordinating center. Figure 5-4 illustrates an experiment in which the fibrils to five locomotor organelles were cut, resulting in loss of coordinated movement. Incisions that did not bisect such fibrils produced no impairment in locomotion.

Nervous Mechanisms in Multicellular Animals

That some unicellular organisms find coordinating fibrils useful suggests that specialized conducting mechanisms of some sort would be a virtual necessity in the much larger many-celled animals. In fact, it should not come as any surprise to us to learn that animals have not only developed specialized cells, tissues, and organs for this function but have in many cases an elaborate and sophisticated organ system as well. This *nervous system*, as it is called, integrates the entire animal body, with its multitude of varied functions, into a harmonious working unit. Certain cells, tissues, or organs are especially sensitive to particular kinds of external or internal stimuli. These are called *sensory structures* or *receptors*. The *conductors*, on the other hand, are designed to receive, transmit, organize, and interpret all the biological signals discharged by receptors. The conductors activate appropriate responding mechanisms such as muscles or glands.

The latter, known as *effectors*, are found throughout the animal body in many different organs and organ systems (digestive, circulatory, locomotor, and so on).

CONDUCTORS AND INTEGRATION: CORRELATING ADAPTIVE RESPONSES WITH RECEPTOR SIGNALS

GENERAL ORGANIZATION OF THE CONDUCTOR SYSTEM

The cellular unit of structure in the nervous system is the nerve cell or *neuron*. It is much like any other cell except that one or more elongated processes radiate from it. In "simpler" animals such as the hydra, sea anemone, and jellyfish (Fig. 5-5) these neurons are arranged in a rather diffuse network throughout the

Figure 5-5 Nerve cells in this "simple" animal (*Hydra*) are arranged in a diffuse network throughout the body. (Eric Grave, Photo Researchers)

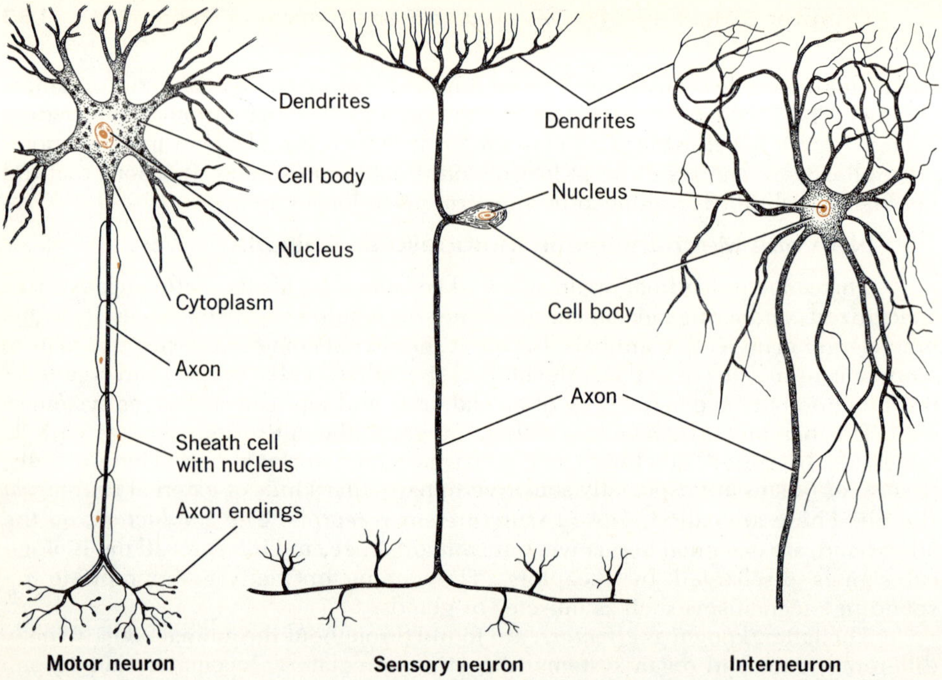

Figure 5-6 Three common types of neurons. Sensory neurons convey nerve impulses from receptors to the central nervous system (see text and Fig. 5-7). Motor neurons conduct impulses from the central nervous system to muscles and glands. Interneurons are restricted to the central nervous system and interconnect sensory and motor neurons. They also engage in complex interconnections among themselves.

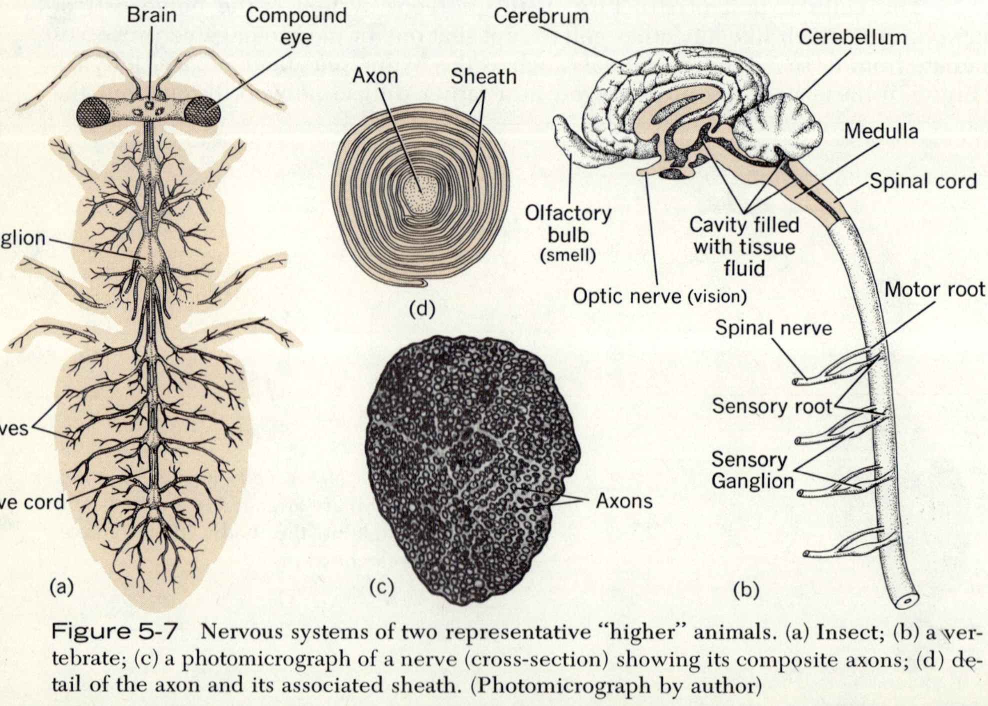

Figure 5-7 Nervous systems of two representative "higher" animals. (a) Insect; (b) a vertebrate; (c) a photomicrograph of a nerve (cross-section) showing its composite axons; (d) detail of the axon and its associated sheath. (Photomicrograph by author)

body. In animals like earthworms, insects, and vertebrates great variety in neuron structure is observed (Fig. 5-6). All of these cells, however, possess a prominent process which frequently may be of considerable length (long enough in man, for example, to reach from brain to fingertips). This *axon*, as it is called, is enclosed in a long series of *sheath cells*. The cell membrane of each of these cells is wrapped like a strip of adhesive tape around the axon, giving the axon the appearance of a string of sausages. The thinnest axons are less than 0.2 μ thick with sheaths of a single layer. Large axons, ranging from 30 to 300 or more microns in diameter, have sheaths of many layers. The axons of numerous neurons may be arranged side by side into a cablelike bundle called a *nerve* (Fig. 5-7).

The nerve cell body also gives off many short, profusely branched processes known as *dendrites*. Many cell bodies and their dendrites are commonly grouped together into a spherical mass called a *ganglion* (plural, *ganglia*), as shown in Figure 5-7. These are exceedingly important coordinating centers because it is here that contacts are made between the axon terminals of one neuron and the dendrites and cell bodies of other neurons. These intercellular contacts are called *synapses*. They are intimately involved in the subtleties of nervous function and are discussed in greater detail later. In view of these considerations it is logical that we should refer to the principal aggregations of ganglia within an animal as its *central nervous system*. The nerves which emanate from this system and ramify throughout the entire animal body constitute the *peripheral nervous system*.

In animals like earthworms and insects, the central nervous system is a double nerve cord lying below the digestive tract, which consists of axons interconnecting a series of paired ganglia. The front (anterior) end of the organism possesses particularly large ganglia called the *brain*. This is because of the concentration of receptors at the anterior end, which, in turn, is occasioned by the fact that these animals meet their world head-on (as is the case with practically all active animals). In vertebrates the central nervous system consists of an especially large and complex brain and a single spinal cord of considerable breadth which runs along the back of the animal. Relatively huge fiber tracts of axon bundles run up and down the spinal cord alongside masses of cell bodies more or less uniformly distributed throughout long, continuous columns. In the brain these fiber tracts terminate or originate in dense concentrations of interneurons (Fig. 5-6) in the outside layer (*cortex*) as well as in compact aggregations within. Nerves springing from the brain (*cranial nerves*) and spinal cord (*spinal nerves*) make up the peripheral nervous system.

THE OPERATION OF THE CONDUCTOR SYSTEM

Figure 5-8 is a schematic, greatly oversimplified illustration of a few of the nervous pathways involved when one responds to burning one's finger. The nerve impulses generated by receptors in the finger are transmitted over a *sensory neuron* into the spinal cord. This neuron almost immediately synapses with a *motor neuron* which carries impulses to muscles in the arm, and the hand is withdrawn. This simple in-and-out pathway is called a *reflex arc*.[1] Note, however, that the sensory axon also passes up the cord, synapsing first with an interneuron in the cord and then with an interneuron in the brain. The first neuron carries impulses to a motor neuron that stimulates muscles in the neck to turn the head and also reinforces impulses to the motor neuron going to the arm. The interneuron in the brain transmits

[1] Reflex action also involves the inhibition of antagonistic muscles (here those that would extend the hand) so that they do not contract. For simplicity this circuitry is not shown in the illustration.

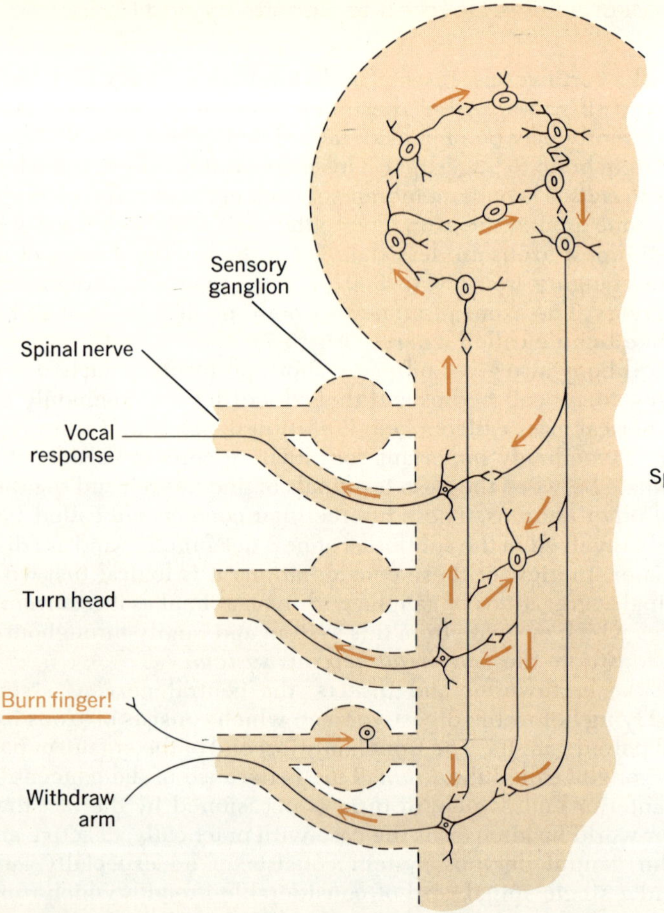

Figure 5-8 Hypothetical nervous pathways (schematic). Responses to a burned finger, such as withdrawing arm, turning head, and uttering a vocal response, involve a complicated circuitry.

impulses to a complicated circuitry in the brain developed through learning. Impulses then flow out of the brain over an interneuron that initiates a vocal response (which we will not quote) and provides additional reinforcement to the motor neurons already activated. All of these events occur, as we well know, with great speed and precision.

In actuality the pathways described embrace more than one neuron at each link in the circuit. The circuitry itself is much more complex. To appreciate the full impact of this complexity and its high order of efficiency we must examine the nature of the nerve impulse and the manner in which it is generated and transmitted throughout the system.

The Nature of the Nerve Impulse

As we have already learned, the nerve impulse is an action potential (Fig. 3-17). In the case of the neuron this action potential is carried along the membrane of the axon at a high velocity. The speed of impulse transmission is greater in warm-blooded animals than in cold-blooded animals and in large axons than in small axons. One of the slowest speeds has been recorded in the nerve net of the sea anemone (0.1 meter per second). In mammals nerve impulses travel at rates up to 160 meters per second (over 300 miles per hour). This rapidly advancing wave of depolarization involves only a few millimeters of axon length at any instant of time. Any point along the axon membrane within this short segment can obviously only be depolarized once until such time as the impulse passes on and polarity is restored at that point. Consequently, there is an upper limit to the number of nerve impulses that can be carried by an axon per unit of time. This maximum number lies between 250 and 2500 impulses per second in warm-blooded animals, depending on axon size.

The frequency with which nerve impulses are generated and transmitted by a neuron is the only variable in nerve cell conduction. Within a given neuron one action potential is like another: the kind and amount of energy released are identical from one impulse to the next. In any one neuron a nerve impulse is either generated and conducted or it is not: it is an all-or-none affair. Since it is produced by cellular activity, it travels the entire length of the axon without any loss in amplitude; when the axon divides, each branch carries its own undiminished action potential.

Variation in the frequency of nerve impulses is the means by which information is conveyed over the conductor system. Strongly stimulated receptors discharge a volley of impulses at high frequencies that are carried by neurons to muscles, for instance, and result in powerful muscular contractions. If the same receptors were excited by weak stimuli, very few impulses would be produced. These would elicit correspondingly mild muscular responses. In many such cases, however, there might not be any response at all. When one or very few impulses travel along an axon to a synapse with the dendrites or cell body of another neuron, they may or may not evoke impulses in the second nerve cell. The synapse acts as a kind of control "valve" that may or may not allow the transmission of low frequency nerve impulses. It also permits impulses to travel in one direction only[2]: from the axon of one cell to the dendrites or cell body of another. It plays a most important role in integration as we are about to see.

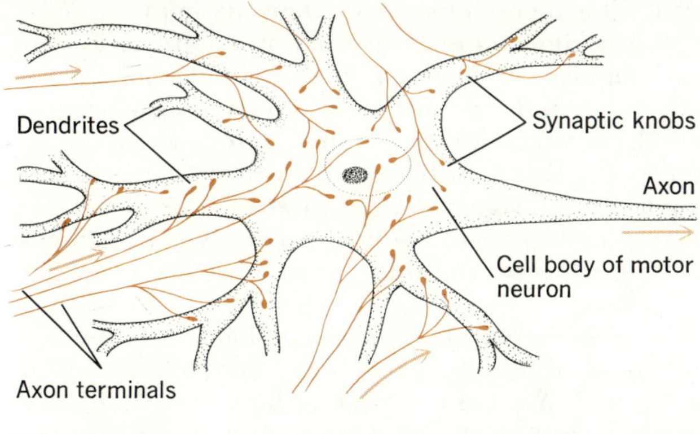

Figure 5-9 The synapse. Synaptic contacts are made on dendrites and cell body of motor neuron by axon terminals of other nerve cells. Arrows indicate direction of nerve impulse transmission.

The Synaptic Transmission of Nerve Impulses

An inspection of Figure 5-9 gives some suggestion of the intricate nature of synaptic connections. Every imaginable degree of complexity is possible. The terminals of many axons may converge on one nerve cell and the terminals of one axon may radiate out to many cells. The contact made by each *synaptic knob* of the multitude of axon terminals is actually not a physical connection between two cells; there

[2] An isolated axon will conduct an action potential in either direction.

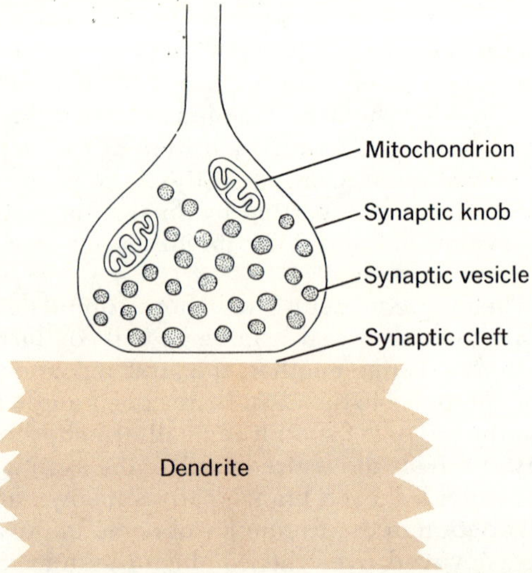

Figure 5-10 Detailed section of a synaptic "junction."

is a tiny space of about 20 mμ that separates the axon membrane from the membrane of the second cell. It is across this cleft that an impulse must be somehow either "projected" or generated anew so that it may continue its journey.

Regeneration appears to be the method used. It is believed that the synaptic knob stores a *transmitter substance* in large numbers of submicroscopic vesicles of uniform size (Fig. 5-10). When a nerve impulse reaches the synaptic knob, a certain number of these vesicles discharge their contents into the synaptic cleft. Thus freed, the transmitter substance somehow depolarizes the membrane of the second cell. When this depolarization reaches a certain minimum (threshold) value, an action potential develops in the second cell and the regenerated nerve impulse is passed along to the next synapse. This theory of synaptic transmission is not only supported by a growing body of experimental evidence but also helps to explain many interesting facets of nerve impulse conduction and animal behavior.

It has long been known that an impulse is delayed when crossing a synapse. About 0.6 millisecond (thousandths of a second) are required to regenerate the action potential. This is presumably the length of time it takes for transmitter substance to be released in sufficient amounts to reach the threshold for membrane depolarization. With continuous excitation over a conductor circuit synapses fatigue long before neurons do. Apparently the supply of transmitter substance in this instance has been temporarily depleted.

We are all familiar with the variable responses that are elicited by stimuli of varying intensities or by combinations of stimuli. A soft sound may not cause us to jump, but a loud noise frequently will. A noise or a flash of light perceived separately may be no cause for alarm; occurring together they commonly startle us. In essence these responses involve fundamental properties of the synapse itself. One or a few impulses reaching a synaptic knob often prove inadequate to excite the second neuron. The same neuron, however, will develop and transmit an action potential if the number of impulses is increased at one synapse or one or a few impulses reach it over each of two or more synapses (see Fig. 5-11a,b,c,d). This can be accounted for again on the basis of transmitter substance. One or a few impulses commonly fail to discharge enough transmitter substance to raise the second neuron

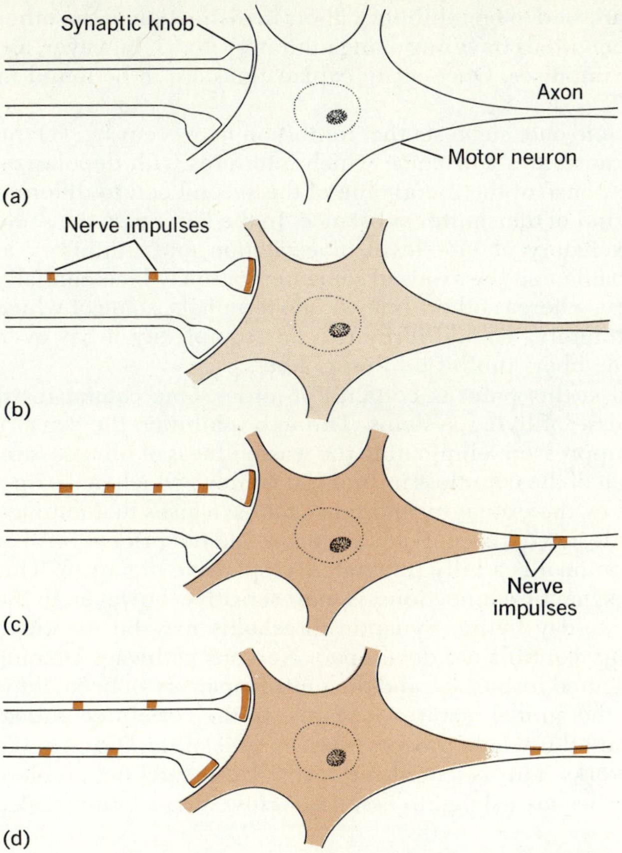

Figure 5-11 Excitation of a neuron by summation. An unstimulated neuron (a) may not be raised above a subthreshold level of excitation when stimulated by too few nerve impulses (b). Increasing the number of impulses by greater excitation at one synapse (c) or the accumulation of excitations in two or more synapses (d) will bring the neuron to the threshold level for depolarization and the discharge of nerve impulses.

above a subthreshold level of excitation. Increasing the number of impulses by (1) greater excitation at one synapse or (2) the accumulation of excitations in two or more synapses augments the release of transmitter substance to the threshold level for depolarization.

We are also familiar with a type of behavorial response that is quite the reverse of that described. We can, for example, stifle a sneeze (sometimes) by pressing against our upper lip. This involves *inhibition* of response. Many of the synapses on a given motor neuron restrict or entirely block the transmission of impulses arriving at other synaptic knobs (Fig. 5-12). Axons or axon terminals that ex-

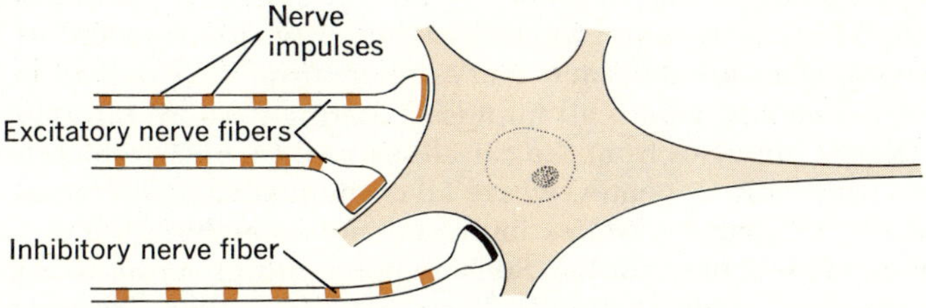

Figure 5-12 Restricting synaptic transmission by inhibition. Impulses along an inhibitory nerve fiber can block the transmission of impulses arriving at other synapses on the same motor neuron.

ercise this restrictive effect are said to be inhibitory fibers in distinction to the other excitatory ones. The action potentials traveling along inhibitory fibers, however, are no different from excitatory impulses. Once again explanations are to be found in the synapse.

Experimental evidence to date suggests that inhibition may occur by: (1) the release of another kind of transmitter substance which interferes with depolarization and/or (2) a variable response of the membrane of the second cell to different concentrations of the same kind of transmitter substance. In the latter case the same synaptic knob could be excitatory at one level of excitation and inhibitory at another. It also appears probable that the axons of some neurons may be essentially either excitatory or inhibitory, whereas others branch into terminals, some of which are excitatory and some inhibitory. To add further to this complexity it has even been demonstrated that some fibers inhibit inhibitory fibers.

A moment's reflection at this point is certainly in order. One cannot resist being impressed by the finesse of living systems. Through inhibition the sensory input from weak stimuli is suppressed, eliminating the wastefulness of unnecessary response. The activity of each of the countless millions of neurons which make up a nervous system is regulated by the excitatory and inhibitory synapses that impinge on it. Much of the elaborate design of this nervous system is laid out prior to birth or hatching, so that the young animal is a fully integrated, responsive organism. This circuitry, with its numerous synaptic connections, is most sensitive, however, to the sensory experiences of day-to-day living. Synaptic thresholds may be modified. New excitatory and inhibitory controls are developed. Nervous pathways become better channeled for more refined responses, and the entire repertory of behavior is enlarged and perfected. As the animal matures it learns. In the vertebrate animal with its particularly large brain this whole process takes on special significance. We would like to know how it works. This is a most interesting and searching problem indeed. It is also circular, for we are asking, in essence, "How does a brain think a brain works?"

A Brain Contemplates Itself

The human brain contains more than 10 billion interneurons (Fig. 5-13). There may be 10,000 synapses on one neuron, some of them excitatory and some inhibitory. This vast cellular assemblage has been modified and molded through the accumulation of memory stores and knowledge acquired by the system. The functioning of any one of these brain cells is intimately interwoven with all its countless fellows as well as the entire sensory input from receptors all over the body. Many of these cells will discharge impulses spontaneously even when isolated from incoming sensory stimulation.

Nerve impulses coming from the eye are generated by electromagnetic radiation, those from the ear by sound waves and those from touch receptors by pressure. Yet all these impulses are the same. Nerve fibers from the eye lead to a part of the brain (the visual center) where all impulses coming to it are *interpreted* as something seen. Nerve impulses from the ear are carried by fibers which terminate in a different area (the auditory center) where all of them are interpreted as sound.[3] Still another part of the brain *translates* impulses into something felt or touched. The brain is indeed the seat of sensation. Sensory perception is a *psychological interpretation* of external and internal stimuli. The relevancy of these interpretations to the

[3] If we could transpose the fiber tracts coming from ears and eyes, we would see thunder and hear lightning!

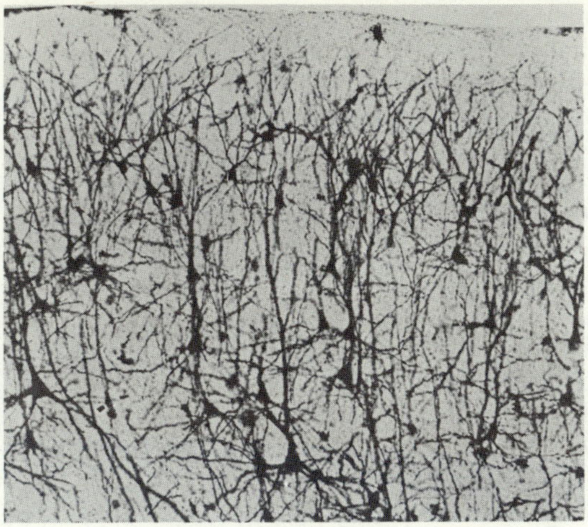

Figure 5-13 Some of the neurons in a typical section of the cerebrum (cortex) of a mammal. Many of the cells actually present are not visible because of the staining technique used in this preparation. (O. A. Sholl)

total environment enables the organism to direct its activities and behavior in order to maintain its biological integrity and enhance species survival.

How the brain performs these feats of data processing and then catalogs the gathered information into memory stores for future reference is one of the biggest puzzles facing modern biologists. Research in this area is very active and has led to many provocative observations relative to neuron physiology and learning. Laboratory rats (and goldfish) show significant brain changes when they undergo training in the performance of specific tasks. Both enzyme activity and protein synthesis in the brain are accelerated. The number of synapses is increased. In a study of marine, snail-like creatures (sea hares) the animals were trained over a period of several days in the laboratory to 12 hours of light followed by 12 hours of darkness. Certain ganglia were then removed from these animals and the nerve impulse discharge of isolated neurons was measured. Periodically, at a time coinciding with the time the lights had been turned on during the training period, the impulse rate increased quite significantly in the monitored neurons.

Some think that the firing patterns observed in this nerve cell "memory" are somehow related to RNA synthesis. RNA synthesis, of course, is known to be a prelude to the manufacture of protein. In other studies materials which had been synthesized in the cell body of a neuron were observed passing through tiny tubules in the axon and being discharged into a synaptic knob. A number of researches (on the octopus and several different vertebrates) have established that there are two distinct types of memory: short-term and long-term. Short-term memory presumably involves a rather quickly established nervous pathway which is transitory and impermanent. The ability to remember lasts only a few minutes to a few hours. If short-term memory is reinforced by repeated training, it becomes long-term memory—a more or less permanent memory store that can be remembered for weeks or years. Some studies suggest that drugs which interfere with protein synthesis prevent long-term memory when they are administered during the period of memory reinforcement (but not later); but they did not appear to affect short-term memory.

This recounting of scientific studies constitutes some of the primary evidence for a current hypothesis of the physiological mechanisms by which memory and learning take place. This hypothesis presumes that specific synapses are stimulated as a consequence of memory reinforcement. Synaptic stimulation results in enzyme induction (page 110) and the accelerated synthesis of protein. Increased amounts of protein pass from the cell bodies of interneurons in the circuit through the submicroscopic axon tubules toward those synaptic knobs being activated. New synapses are formed alongside those already present. The pathway is strengthened by additional synapses, and memory becomes long-term. The new synapses follow the old because some of the newly synthesized polypeptides are marker molecules unique to the neuron. Such marker substances could be relatively small molecules. A polypeptide chain of only 8 amino acids, if arranged in every possible combination of choices from 20 dissimilar amino acids, could yield 25,600,000,000 different kinds of molecules (20^8). This would be more than enough to provide each of the 10 billion interneurons in the human brain with two unique markers.

This hypothesis, even if found in the course of time to be true in all its details, leaves a multitude of questions unanswered (in fact, it creates some problems of its own). We live in an exciting period of discovery in biological research. This research is taking man to the edges of some ultimate frontiers. Such certainly is the case when a human brain stands before the mirror and asks, "What am I?" At the present time it is amazed and staggered by what it sees.

RECEPTORS: MONITORS OF ENVIRONMENTAL CHANGE

The astonishing performance of the integrative mechanisms just described is made possible in large measure by highly reliable and essential data-gathering devices previously identified as receptors. These are the sole means whereby an organism is apprised of the environmental conditions around it and within it. An animal responds only to those stimuli to which it is sensitive, and its behavior remains fully integrated and purposeful only when this capacity to detect stimuli is completely and continuously operative and unimpaired.

All receptors are either bare nerve endings (dendrites) or highly specialized neurons (Fig. 5-14). In the latter case they are often housed in a complex structure consisting of many accessory mechanisms (such as the ears or eyes). Each type of receptor is especially sensitive to but one kind of stimulus. This sensitivity reaches

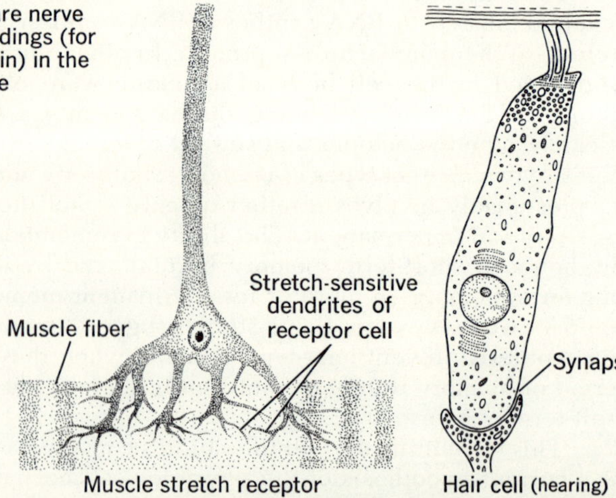

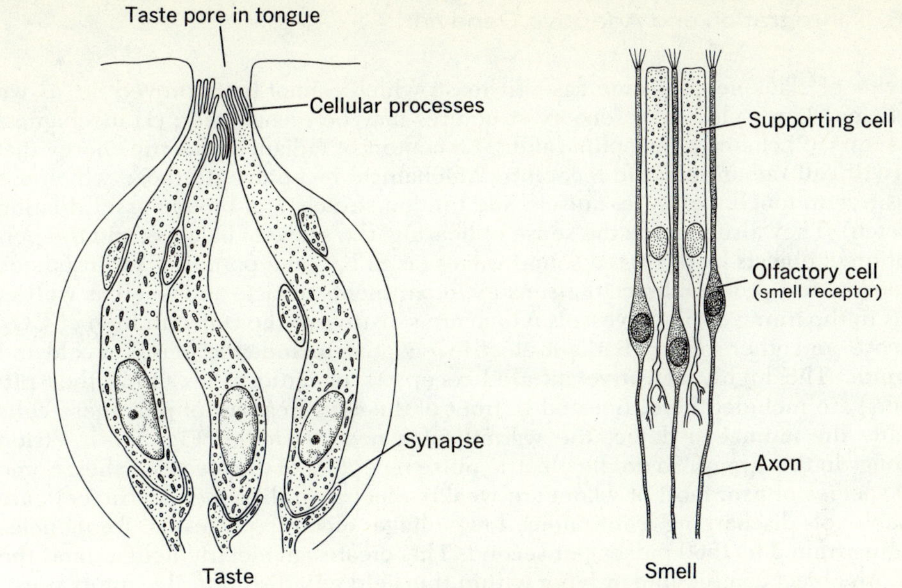

(c) **CHEMICAL RECEPTORS**

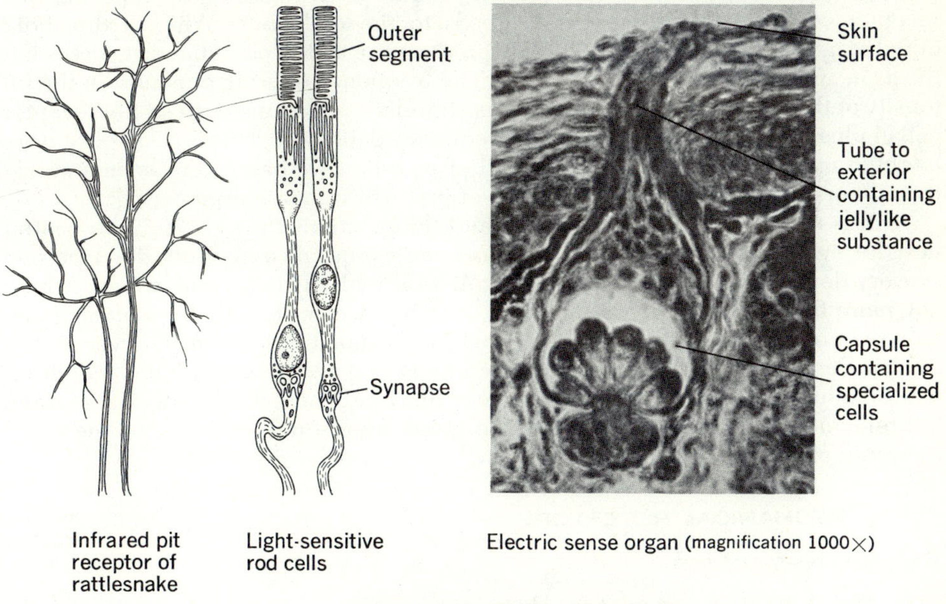

(d) **RADIANT-ELECTRIC RECEPTORS**

Figure 5-14 Animal receptor cells, (a) Pain receptors (bare nerve endings) in the eye. (b) Mechanical receptors; muscle stretch receptor in a crayfish on the left; human hair cell (hearing) on the right. (c) Chemical receptors; human taste receptors on the left; human receptors for smell on the right. (d) Radiant-electric receptors; infrared pit receptors of rattlesnake on the left; light-sensitive rod cells on the right. Photomicrograph is of the electric sense organ of an electric fish. This receptor is sensitive enough to detect current changes involving the movement of only 1000 electrons every 25 milliseconds (0.00000000000003 ampere). (H. W. Lissman, University of Cambridge)

heights of efficiency, in some cases at least, which cannot be improved on, as we shall see. Three classes of sensory structures may be recognized[4]: (1) mechanical receptors, (2) chemical receptors, and (3) receptors of radiant or electric energy that we will call radiant-electric receptors. Mechanical receptors are those which are sensitive to touch, pressure, muscle and tendon stretch, and blood vessel dilation (stretch). They also include the sense of hearing, the sense of balance, and the perception of objects by deflected sound waves (as in bats and porpoises for instance). Chemical receptors embrace the sensory departments of taste and smell as well as cells in the lining of blood vessels which are sensitive to the concentration of CO_2, glucose, and other solutes. Radiant-electric receptors include the senses of cold and warmth. The highly sensitive infrared receptors of rattlesnakes (and other pit-vipers) are included here. Located in front of the eyes in a pair of pits, these cells enable the animal to detect the warmth of a nearby victim (Fig. 5-14). Vision belongs in this group. So do the electric pulse receptors of the electric fishes. Some 500 species of fish, most of whom are weakly electric (unlike the electric eel), are capable of discharging continuous low-voltage electric pulses at frequencies ranging from 5 to 1500 pulses per second. This creates an electric field around the fish. An object coming into or lying within this field will distort it. Receptors sensitive to such distortions constitute both a guidance system and a means of capturing prey.

In all sensory structures the stimulus somehow depolarizes the dendrites of a receptor cell. This depolarization spreads to the cell body. When a threshold depolarization is reached, the cell body produces an action potential that passes out the axon and into the conductor system. The frequency of firing depends on the intensity of the stimulus. Under continuous stimulation the sensitivity of the receptor cell gradually diminishes and impulse frequency decreases. This process of *sensory adaptation* has been experienced by all of us when we have more or less accustomed ourselves to a strong odor which at first was rather overpowering.

It is difficult to discuss receptors and the reactions they elicit in the central nervous system in general terms, because one cannot meaningfully describe one sensory department in terms of another. A color really is not warm, sharp, or loud any more than it is salty, foul-smelling, or itchy. A color can be seen *only*. It can moreover be described solely by words whose connotations are *entirely* visual. We can use the word "warm" for red only if we make it synonymous with red and eliminate all allusions to temperature. Because of this fundamental attribute of sensation, the remainder of our discussion will be much more intelligible if confined to a sampling of specific sensory areas.

MECHANICAL RECEPTORS

Stretch Receptors

The abdominal region of the shrimplike crayfish is equipped with two sets of locomotor muscles. One set flexes (bends) and the other extends (straightens) the animal's abdomen and tail. Such movements, when repeated rapidly, enable the animal to swim backwards. As the tail extends, the powerful flexor muscles become stretched. This automatically results in their contraction by a receptor mechanism which is illustrated in Figure 5-15. Muscle fibers that make up a flexor muscle have stretch receptor cells whose sensitive dendrites rest on them. When the muscle

[4] Pain receptors are not included here since they are sensitive to varying stimuli at high intensities, for example, mechanical, thermal, chemical, and electrical.

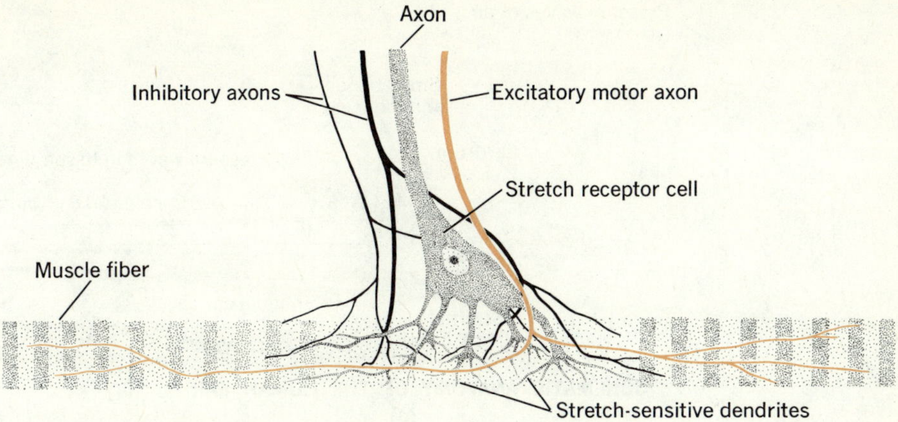

Figure 5-15 Stretch receptor of abdominal muscles of crayfish.

fiber is stretched, the dendrites are deformed. This results in depolarization of the dendrites. When the depolarization reaches a threshold value, the receptor cell begins discharging impulses. These are transmitted into the central nervous system and back out over motor neurons to the muscles and contraction ensues.

It will be noted in the illustration that there are inhibitory axons whose terminals synapse at many points on the dendrites of the stretch receptor cell. Impulses passing over these inhibitory fibers prevent firing of the receptor. In this way the central nervous system regulates sensory input by inhibition just as it does motor output. As a consequence the crayfish is able to keep its abdomen extended when, for example, it walks.

Similar mechanisms to these are found in our own bodies. (The knee jerk reflex results from stimulating stretch receptors in the knee-cap tendon.) Were it not for adjustments in the contraction of opposing muscles made possible by these sensory cells, we could not stand erect.

The Sense of Hearing

The receptor cells (Fig. 5-14b) whose impulses are interpreted by the human brain as sounds are stimulated in much the same fashion as those just described. The dendrites of these cells take the form of hairlike processes. The name, *hair cell*, refers to this feature. When these "hairs" are deformed by being bent, the receptor cell is depolarized.

The mechanism by which sound waves are converted into nerve impulses in man (and other mammals) is illustrated in Figure 5-16. Successive waves of compressed air (sound waves) produce corresponding vibrations in the *eardrum*. These vibrations are transmitted by a leverage system of three small bones to another membrane, the *oval window*. Here membrane movement is translated back into the original pressure waves, except now they are traveling in a tissue fluid rather than air. This tissue fluid is confined in three tubelike chambers (the *cochlea*). Two of these are continuous with one another, forming a hairpin-shaped duct. This duct begins at the oval window where compression waves are set up and ends at another membrane, the *round window*, which is pushed back and forth in synchrony with the vibrations of the oval window. These compression waves cause the *basilar membrane* of the third duct to move up and down at the same frequency. This membrane bears the hair cells on its upper surface. The "hairs" in turn are attached

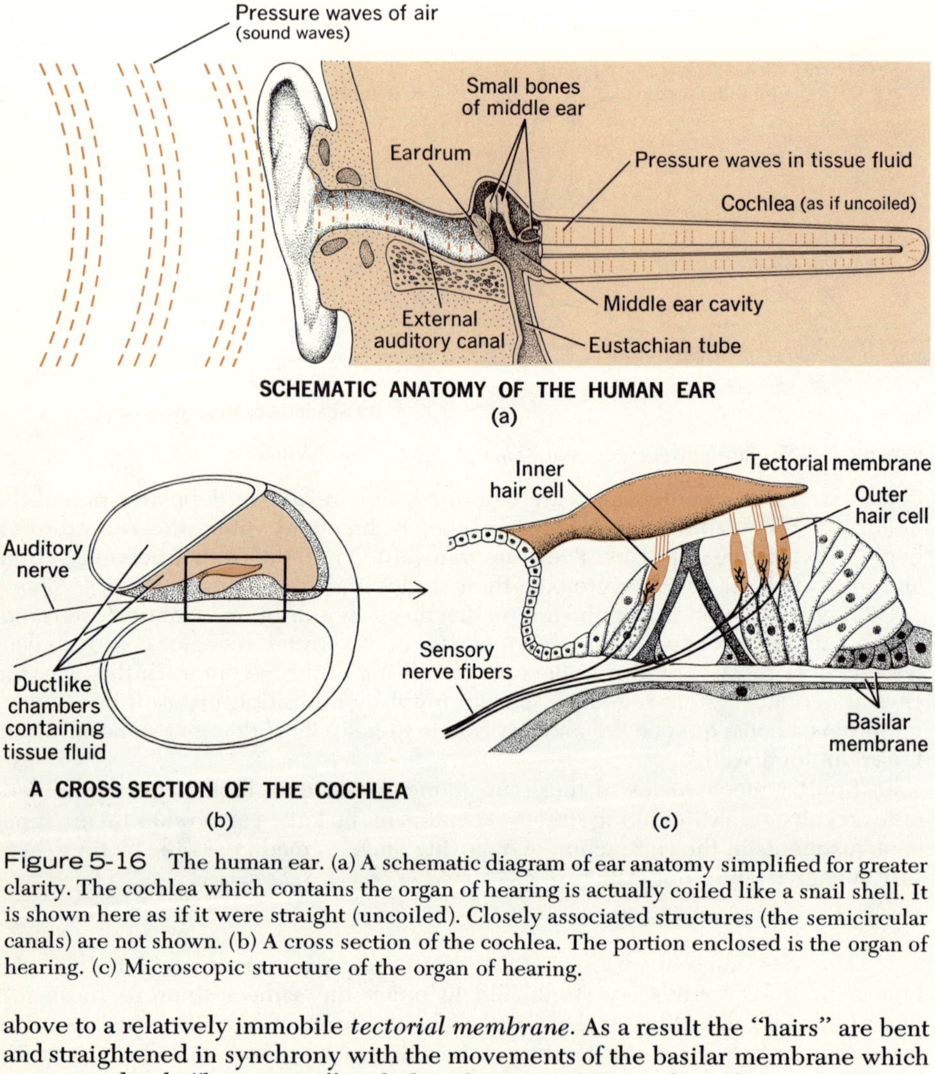

Figure 5-16 The human ear. (a) A schematic diagram of ear anatomy simplified for greater clarity. The cochlea which contains the organ of hearing is actually coiled like a snail shell. It is shown here as if it were straight (uncoiled). Closely associated structures (the semicircular canals) are not shown. (b) A cross section of the cochlea. The portion enclosed is the organ of hearing. (c) Microscopic structure of the organ of hearing.

above to a relatively immobile *tectorial membrane*. As a result the "hairs" are bent and straightened in synchrony with the movements of the basilar membrane which correspondingly "keeps time" with the vibrations of the oval window and eardrum.

The basilar membrane gradually widens from its narrowest end near the oval window (Fig. 5-17). As can be seen from the illustration any given frequency of sound waves elicits maximum movement of the basilar membrane at a specific resonance point.[5] Sensory neurons which synapse with hair cells at different locations along the length of the basilar membrane lead to different "pitch points" in the auditory center of the brain. One point interprets impulses as middle C, another as high C, and so on. Although large segments of the basilar membrane are set in motion by one sound wave frequency, most people can distinguish between 256 (middle C) and 259 cycles per second. This is possible because the sensory neurons from the cochlea synapse with a relay series of interneurons leading to the auditory

[5] A similar phenomenon can be demonstrated when middle C on a piano keyboard is kept depressed. If one hums middle C, the piano wires will vibrate in resonance.

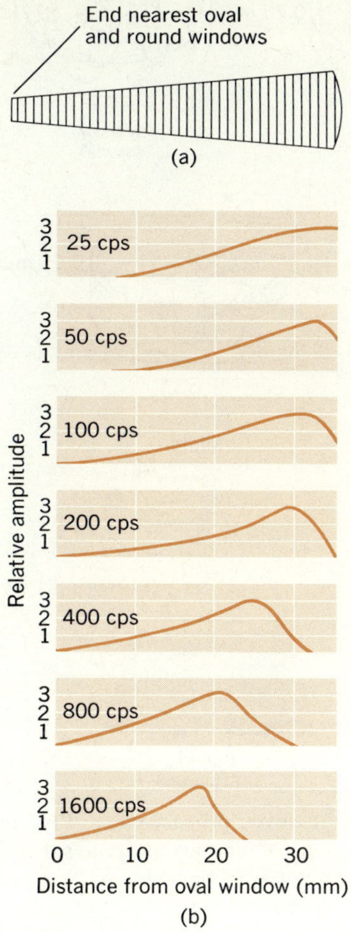

Figure 5-17 The basilar membrane. (a) From the end nearest the oval and round windows the fibers of the basilar membrane become longer like the strings of a harp or piano. (b) Diagrams showing basilar membrane movements in response to seven different sound frequencies. High frequency sounds move the basilar membrane near the oval and round windows, whereas low frequency sounds move the membrane at the opposite end. Sound wave frequency is given in cycles per second (cps). (Adapted from *Handbook of Experimental Psychology*, S. S. Stevens, Ed., John Wiley and Sons, New York)

center. At each synaptic step along this pathway the more rapidly firing neurons inhibit those which are firing less rapidly. The result is a progressive sharpening of the sensory messages. In fact, if our hearing were just a little more acute, we would be confused by the incessant roar created by molecular movement in the auditory apparatus itself. As it is the "hair" of a hair cell is sensitive to a distortion considerably less than the diameter of a hydrogen atom!

CHEMICAL RECEPTORS: THE SENSE OF SMELL IN VERTEBRATES

The olfactory (smelling) center of the brain receives nerve impulses from sensory axons whose cell bodies and dendrites (Fig. 5-14c)lie in the roof of the nasal passage of air-breathing vertebrates. In fishes they lie in the walls of dead-ended nasal sacs. The dendrites of these receptor cells terminate in specialized *olfactory hairs*. There is good reason to believe that these receptor cells are of several different types, each one representing a unique "primary odor" (equivalent in connotation to primary colors such as red or blue). These types differ from one another in that their sensory "hairs" possess receptor sites designed to accommodate volatile

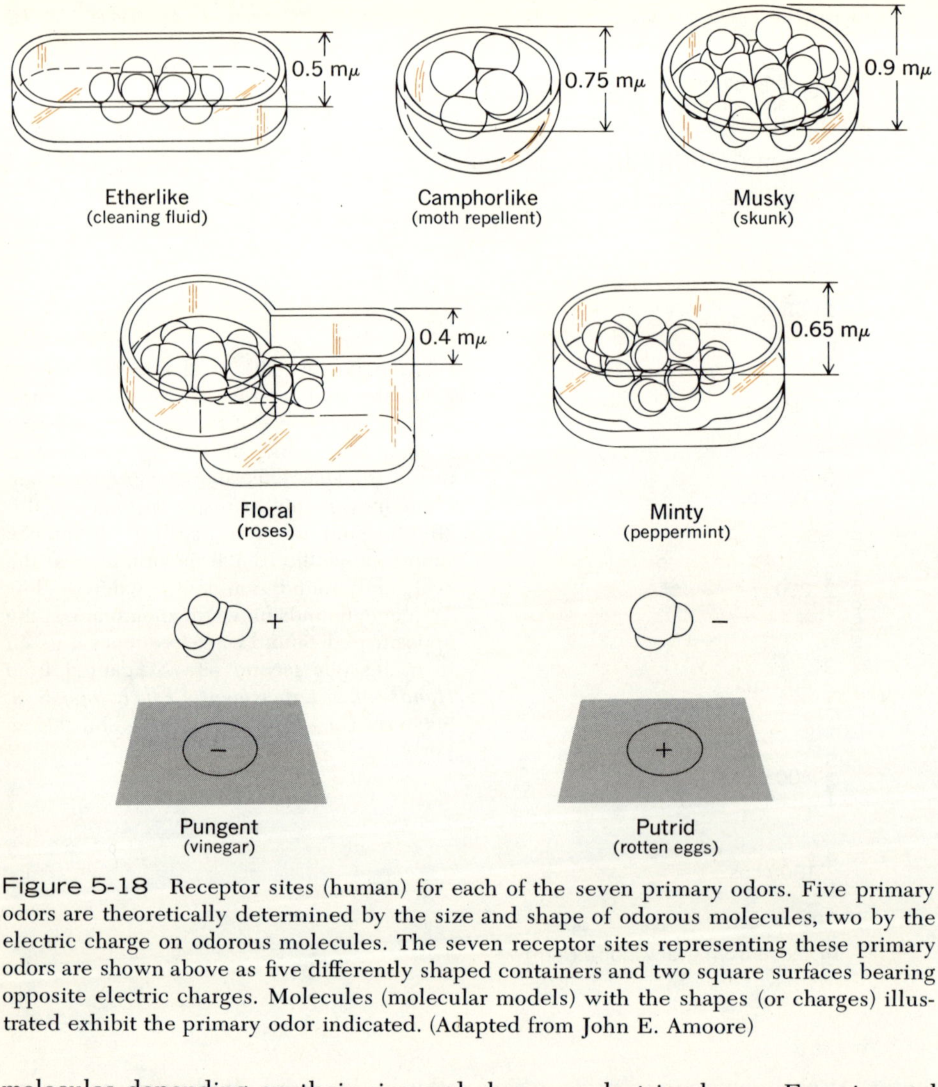

Figure 5-18 Receptor sites (human) for each of the seven primary odors. Five primary odors are theoretically determined by the size and shape of odorous molecules, two by the electric charge on odorous molecules. The seven receptor sites representing these primary odors are shown above as five differently shaped containers and two square surfaces bearing opposite electric charges. Molecules (molecular models) with the shapes (or charges) illustrated exhibit the primary odor indicated. (Adapted from John E. Amoore)

molecules depending on their size and shape or electric charge. Experimental studies on man indicate there are seven distinct kinds of such sites, five of which are predicated on the shape and size of odorous molecules and two on their electric charge (Fig. 5-18; camphorlike, musky, floral, pepperminty, etherlike, pungent, and putrid).

No combination of different odors will produce a primary odor, but combinations of primary odors may yield any complex scent desired. Molecules of similar size and shape that fit into only one specific receptor site and no other represent a pure primary odor. Even though these molecules may be quite dissimilar chemically they not only smell alike but smell quite differently from any other primary odor (Fig. 5-19). A molecule which may fit into more than one receptor site, either singly or in combination, has a complex odor, such as that of almonds (Fig. 5-20).

The activity of olfactory cells in the frog was studied by microscopic electrodes inserted into them. Different receptor cells were shown to discharge nerve impulses in response to different odors. This investigation revealed that the frog has about eight different receptors, five of which are camphorlike, musky, etherlike,

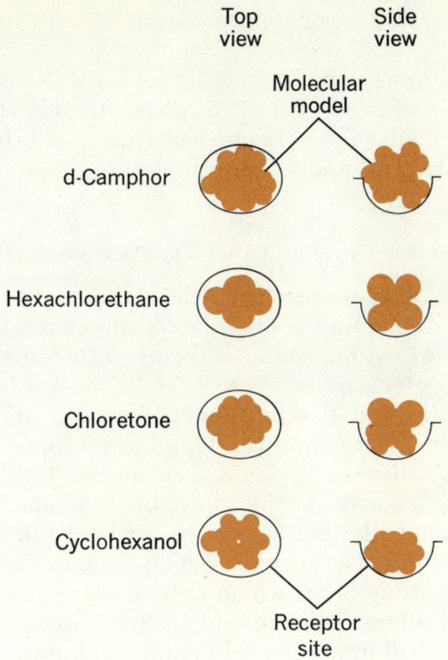

Figure 5-19 The colored shapes represent unrelated molecules. Because their size and shape are similar, they all fit the camphorlike receptor site and all smell like camphor. (Adapted from John E. Amoore)

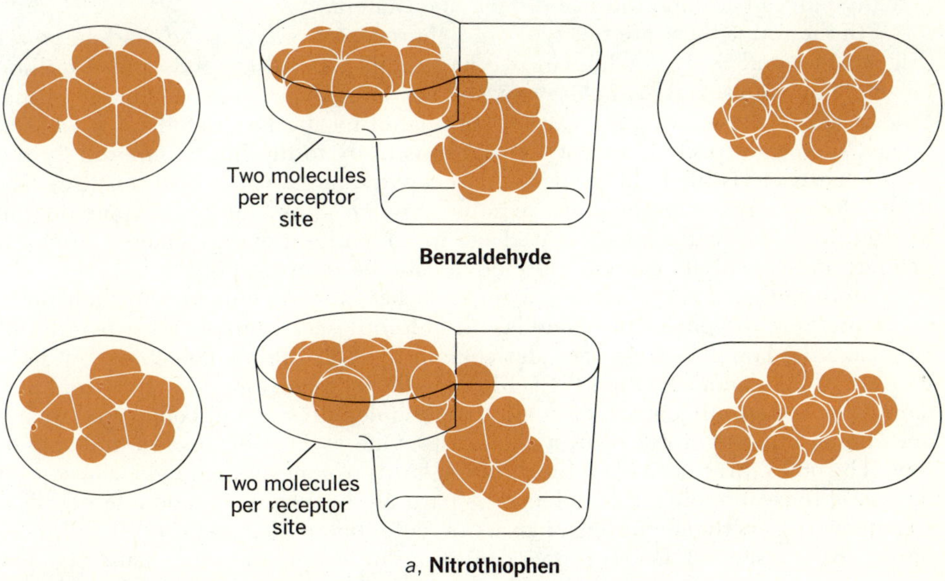

Figure 5-20 Two different molecules with similar shapes each fit more than one receptor site and smell like almonds (a complex odor). The almond odor is apparently a composite of camphorlike (left), floral (center), and pepperminty (right). (Adapted from John E. Amoore)

pungent, and putrid. It would seem that man and frogs are remarkably similar in this sensory department![6]

In the studies alluded to in Chapter 1 on fish migration (pages 15–16) it has been found that fish have an exceptional sense of smell. They can apparently detect the presence of just a few molecules of certain substances. In one test a group of fish were able to detect the presence of a substance diluted to the extent of one-half a thimbleful dissolved in a large lake.

RADIATION RECEPTORS: THE SENSE OF VISION IN VERTEBRATES

The eyes of the vertebrate animal (as well as those of the octopus and squid) are remarkably similar to a camera (Fig. 5-21a). A tough, outer supportive coat of fibrous connective tissue (the *sclera*) continues over the front of the eye as a transparent *cornea*. Several muscles attach to the sclera so that the eye can be moved in any direction. Within the sclera is a heavily and darkly pigmented layer (the *choroid*) that prevents the reflection of light inside the eye. This layer projects forward in front of the *lens* as a circular curtain, the *iris*, which is also colored (blue, hazel, or brown, for instance, in humans). The iris acts like the adjustable light aperture in a camera. Muscles in the iris arranged like the spokes in a wheel open (dilate) the aperture, or *pupil;* muscles in the form of a ring narrow (constrict) it. The double convex lens is held in place by suspending fibers which radiate out to connect with a circular muscle lying next to the sclera. Tension in the sclera, brought about by tissue fluids confined within the eyeball under pressure, pulls on the suspending fibers. Since the lens is passively elastic like a rubber balloon filled with air, this tension thins the lens for distance vision. When the circular muscle contracts, the scleral tension on the fibers is relieved and the lens thickens for near vision. In this way rays of light converge on the inner, light-sensitive layer, or *retina*, at the center of focus. (In some fishes the lens does not change in shape but actually moves back and forth as it does in a camera.)

In the retina there are two types of light receptor cells, the *rods* and the *cones* (Figs. 5-14d and 5-21b). A human eye has about 100 million rods and 5 million cones. They are packed to as close as 3 μ of one another in some parts of the retina. Since each of these receptors can be stimulated by the fine details of a focused image, such tight packing extends visual sensitivity to the limits imposed by the wavelengths of visible light itself. Some receptors, moreover, are so sensitive that they will discharge at least one nerve impulse when stimulated by only one photon of light (a photon is the smallest package in which light energy comes). Such efficiency and sensitivity cannot, quite obviously, be improved on.

Both rods and cones contain a series of flat membranous sacs in their outer segments, reminiscent of the lamellae in chloroplasts (compare Figs. 5-14d and 3-9). Like the lamellae these sacs also contain light-absorbing molecules, but here they are visual pigments rather than chlorophyll. There are four different visual pigments that apparently consist of one pigment molecule (*retinal*) conjugated to any one of four different protein (*opsin*) molecules. One of these opsins is found only in rods. The other three occur in three different kinds of cones. When light is absorbed by one of these proteins, the molecule is partially broken down and altered. This somehow triggers the beginning of an action potential in the receptor cell within 25 millionths of a second. Under proper conditions of illumination the visual pigment is rapidly resynthesized; the retinal and opsin are used over and over again.

[6] Honeybees, which smell with their antennae or "feelers," also appear to use the same kind of mechanism.

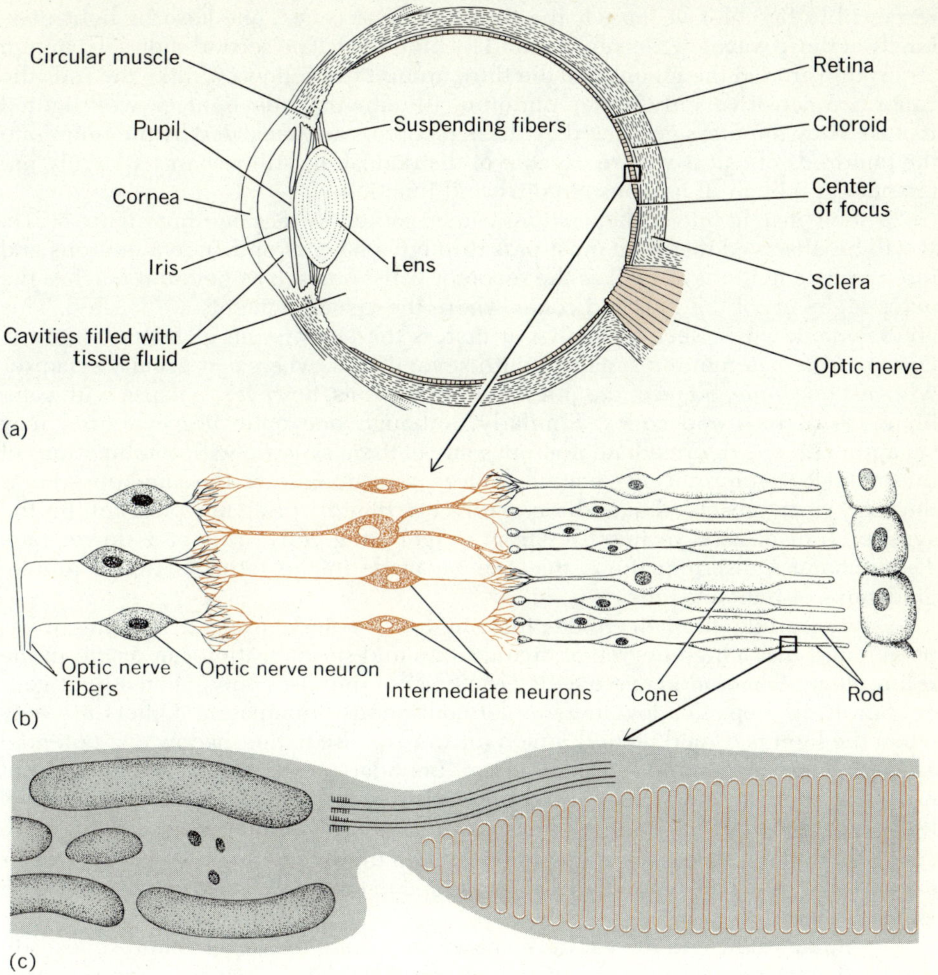

Figure 5-21 The human eye. (a) The anatomy of the eyeball. (b) The microscopic structure of the retina. (c) An ultramicroscopic view of a portion of a rod, showing stacked membranes of visual pigment.

Rods are scattered widely throughout the human retina. Their pigment molecules are highly sensitive to dim flashes of light. When from two to a dozen of them are stimulated simultaneously, a person will see a faint flash of light. Under strong illumination (daylight or a well-illuminated room) the visual pigment is broken down (bleached) rapidly. Rod vision is greatly impaired because resynthesis cannot surpass the rate at which bleaching is taking place. In a dark room, at night, or in red light there is little to no bleaching. Under such conditions the rate of synthesis exceeds the rate of bleaching. As the concentration of visual pigment increases the rods become more sensitive to the dim light. We say the eye has become *dark adapted.* In no case, however, are the rods sensitive to color: all impulses generated by them are interpreted as shades of gray.

Cones are the exclusive light receptor in the center of focus in the human eye. They diminish in number rapidly outside this retinal area. These receptors are responsible for color vision. Of the three kinds of cones, one absorbs light maximally at those wavelengths represented by blue light. The second shows maximum absorption around the greens and the third around the yellows. Unlike the rods the cones are sensitive only under optimum illumination (daylight or well-lighted rooms). How impulses generated by these receptors are translated by the brain into the hundreds of colors we are capable of distinguishing still remains a largely unfathomed problem in nervous circuitry and function.

Some insight into such questions can be gained by reexamining Figure 5-21b. It will be observed that light must pass through a layer of optic nerve neurons and intermediate neurons as well as the receptor cells themselves before it reaches the outer segments of the rods and cones where the visual pigments are located. This arrangement, which seems peculiar at first, is the basis for an elaborate circuitry. One intermediate neuron synapses with several rods, whereas it usually synapses with just one cone. Some of the intermediate neurons, however, synapse with combinations of rods and cones. Similarly, although one optic nerve neuron may synapse with one intermediate neuron, some of them synapse with combinations of intermediate neurons. These complex interconnections serve the same purpose as the relay series observed in auditory pathways. Rapidly firing receptors inhibit the synaptic transmission of impulses from neighboring cells firing at a slower rate. This enhances discriminations that can be made in the retina between minute gradations of light and shadow.

By inserting microelectrodes into single optic nerve axons in the frog, it has been found that a high degree of organization and specialization can reside in the retina alone. Some optic nerve cells fire bursts of impulses only when a light goes on, but firing drops to a low level under continuous illumination. Others fire only when the light is turned off or dimmed (or in response to the shadow of a potential enemy). Some respond to the sharpness of boundaries between light and dark but not to the contrast in darkness or lightness. Others respond to a boundary between light and dark only if it is moving. Some fire only under the stimulus of a curved boundary, stationary or moving, in which the area within the curve is darker than that outside. Such boundaries are provided in the frog's world by the insects on which it feeds.

Similar studies on the cat have shown that there are no special responses to boundaries in the retina. In the visual center of the brain, however, interneurons show amazing specialization: a great variety of kinds of boundaries, boundaries moving in specific directions, light patterns oriented in different ways, and so on. This versatility and richness of response are made possible by an elaborate and sophisticated central nervous system. The frog lives its modest, relatively uneventful but nonetheless effective life by considerably less investment in brain. The price it pays for this is greater expenditures for data-processing within the eye itself.

EFFECTORS IN MULTICELLULAR ANIMALS: AGENTS OF RESPONSE

The elegant machinery we have been describing gathers data and makes decisions. These decisions become useful to the organism only when they are put to work. Effectors, as the agents of response, complete the integrative system by which an organism regulates its activities and behavior. Effectors take a variety of forms. Already alluded to are the *electric organs*, a series of platelike structures representing modified muscle tissue in the electric fishes. Reference has also been made to the *luminescent organs* of fireflies. During the mating period males and females

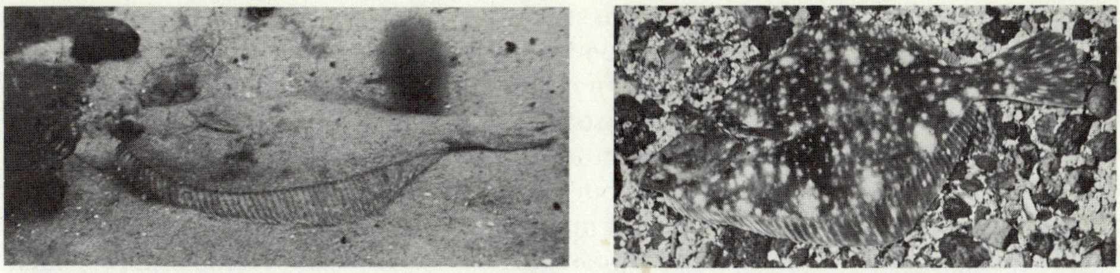

Figure 5-22 The flounder, like a number of other fishes, is able to alter its color to blend with the background. *Left:* Flounder against a background of sand. *Right:* Flounder against a background of varicolored gravel. (Russ Kinne, Photo Researchers)

are attracted to one another by their flashing lights. Luminescent organs also occur commonly in deep-sea fishes where they often serve as lures for capturing prey. Pigment-containing cells known as *chromatophores* are found widely in many different animals. The ability to control the distribution of pigment in each one of large numbers of chromatophores enables an animal to blend with its environment (Fig. 5-22). The vast majority of effectors, however, fall within two great groups: (1) muscles and musculoskeletal systems and (2) chemical secretions and glandular control systems. We will consequently focus our attention on these two areas.

MUSCLES AND MUSCULO-SKELETAL SYSTEMS

Movement of the body or some of its parts is a most important and, very frequently, the ultimate response of an animal. Although cilia are employed for this purpose, their use is of a rather specialized order, such as providing locomotion for certain invertebrate larvae or the creation of upward-flowing currents of mucus to cleanse the respiratory passages of vertebrates. Animal movement is largely accounted for by specialized cells that contain large amounts of fibrous, contractile protein. These are the muscle cells.

One of the simplest kinds of muscle cells are those that make up the *smooth muscle* of vertebrates (Fig. 5-23). These spindle-shaped, elongated cells contain fibrils of contractile protein molecules running in a lengthwise direction. Smooth muscle responds slowly to excitation, but it is capable of prolonged contraction and

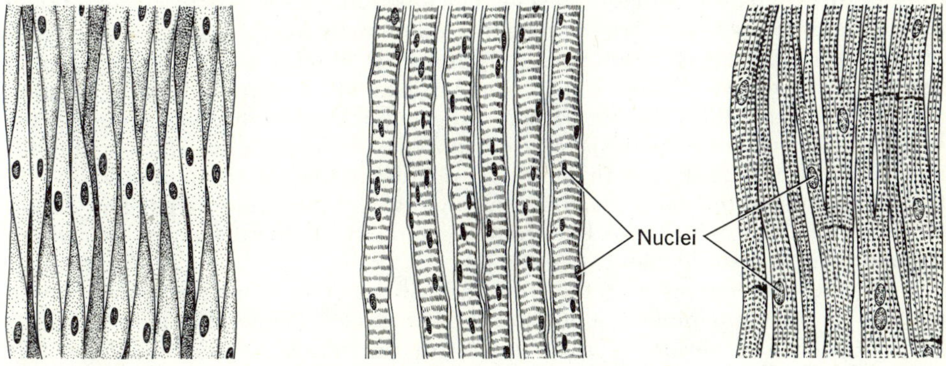

Figure 5-23 Common types of muscle cells.

does not fatigue readily. It is found most commonly in the walls of hollow organs such as the stomach, intestines, and blood vessels. In these organs the muscle tissue is frequently divided into two layers. In one layer the cells are aligned in a circular direction so that their contraction constricts the organ. They may even serve as a valve (between the stomach and intestine for instance). In the other layer the cells run lengthwise; their action brings about shortening and dilation. The coordinated, rhythmic, and wavelike contractions of muscle sets arranged in this way (whether of the smooth muscle type or not) make possible a varied repertory of movements such as one sees in a crawling earthworm or a giraffe in the act of swallowing or regurgitating. Their steady contraction, on the other hand, maintains a constant pressure on the contents of hollow organs and provides a means whereby tubular structures can alter their diameter. This feature is most important in the circulatory and digestive systems as well as in several others.

In a great variety of invertebrates including the earthworm, clam, octopus, crayfish and insect, and in all vertebrates, a highly specialized type of muscle tissue is found. Like smooth muscle the contractile fibrils are composed of a protein complex called *actomyosin.* Unlike smooth muscle, however, the actomyosin is structurally arranged so that it is visible under the light microscope as a series of cross-markings or striations. Figure 5-23 shows an illustration of the *striated muscle* found in vertebrate animals. This tissue consists of long, multinucleate fibers rather than discrete cells containing one nucleus. These fibers may reach lengths of an inch or more. Their diameter varies from about 10 μ in some fibers (a little more than smooth muscle cells) up to others which are as thick as a human hair (100 μ). Striated muscle is capable of exceedingly vigorous and rapid contractions. The wing muscles of some insects contract at rates in excess of 1000 cycles per second.

Backboned animals possess a third kind of muscle which is found only in the heart (*cardiac muscle,* Fig. 5-23). Although cardiac muscle is imperfectly striated, it is distinct from true striated muscle in many ways (its fibers are not multinucleate, the fibers branch and interconnect with one another, and so on). Cardiac muscle is intrinsically rhythmic and its contraction is smooth, powerful, and almost indefatigable. Our heart starts beating long before we are born and before any regulatory nerves have reached it. During a *restful day* it puts out 76,000 foot-pounds of work —enough to lift 150 pounds over 500 feet into the air. Yet this organ is only a little larger than your fist!

The macromolecular machinery of muscle contraction has been most fruitfully studied to date in striated muscle. The electron microscope has shown that a striated muscle fiber is in large part composed of many fibrils, each of which has a banded appearance (Fig. 5-24). Each fibril in turn is a highly ordered aggregation of tiny filaments of two distinct sizes. Six thin filaments are symmetrically grouped around each of the thicker ones (Figs. 5-24d,e and 5-25). The thicker filaments have been found by chemical analysis to be composed of the protein *myosin.* The thin filaments are a different protein, *actin.* A membranelike structure, the *Z-line,* is thought to anchor the actin filaments in some way at their midpoints. The darker bands of the fibrils (which look like striations under the light microscope) are where the actin and myosin filaments overlap.

In the presence of ATP the free ends of the actin filaments approach one another by somehow moving along the thicker myosin filaments (Fig. 5-25). This brings the Z-lines closer together and results in muscle contraction. How this occurs chemically is not at present clear. It is interesting that muscle fibers killed in glycerine and kept frozen for months will still contract in the presence of ATP when they are warmed. It is even more significant that contractile protein extracted from

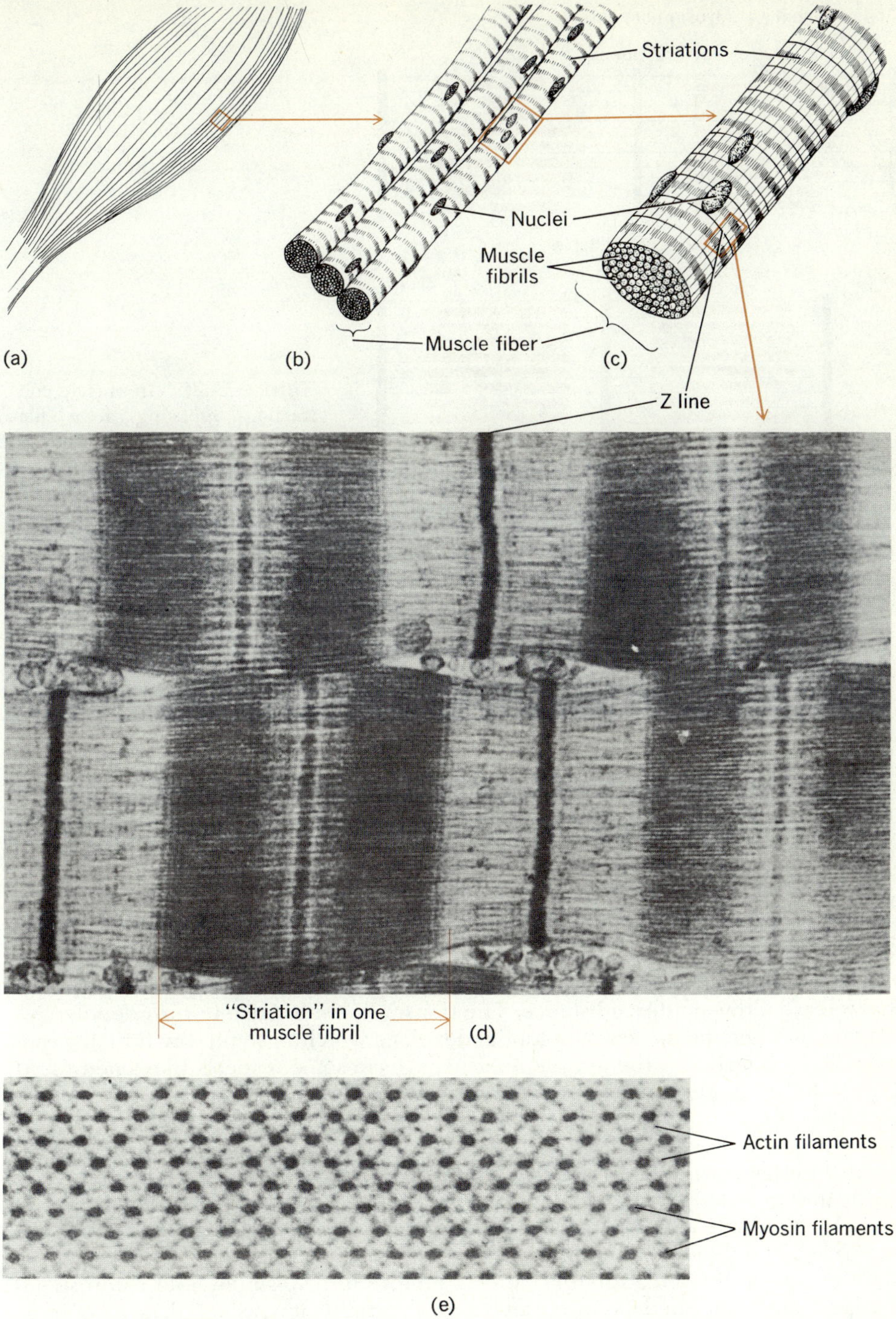

Figure 5-24 Striated muscle shown at progressively higher magnifications. (a) An entire muscle. (b) Three muscle fibers. (c) One muscle fiber and its many component fibrils. (d) An electron micrograph of rabbit muscle. (e) An electron micrograph (cross section) of insect muscle through region (striation) where actin and myosin filaments overlap (×250,000). [(d) and (e) Dr. H. E. Huxley]

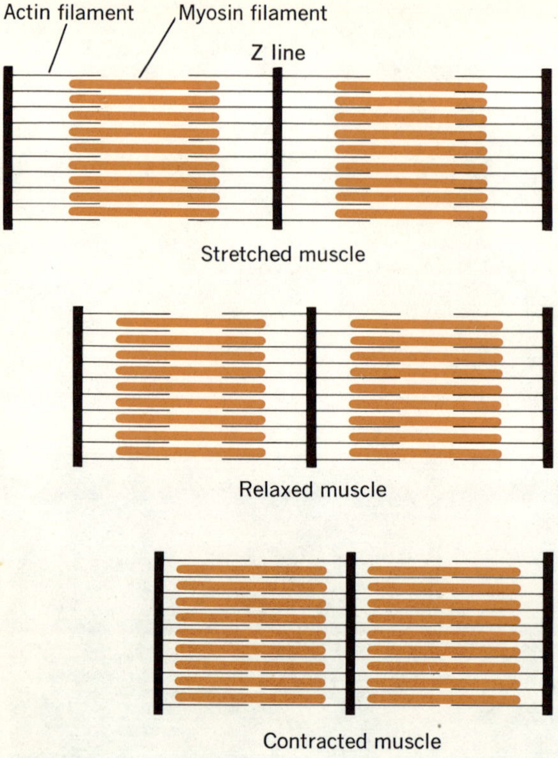

Figure 5-25 In muscle contraction opposing actin filaments approach one another by somehow moving along the thicker myosin filaments.

the amebalike slime molds and from the tails of sperm cells behaves remarkably similar to the actomyosin of insects and vertebrates. Could it be that movement throughout the animal and protist world depends on similar macromolecular mechanisms? If the answer to this turns out to be positive, it would indeed lend added support to the challenging thesis that life, amidst its great diversity, still possesses a basic unity of process and principle.

The contraction of striated muscle fibers is stimulated by motor neurons. The terminals of a motor axon end in a flattened, disclike expansion on the muscle fiber known as a *motor end plate* (Fig. 5-26). These end plates are synapselike junctions that release a transmitter substance. The transmitter substance in turn depolarizes the muscle fiber and an action potential develops which in some way initiates contraction. In vertebrates the muscle fiber behaves in some respects like a neuron. It even exhibits an all or none response by either contracting fully or not at all. The speed of contraction is greatly facilitated by an efficient transport system that includes not only an endoplasmic reticulum but a special system of ultramicroscopic tubules as well (white spaces and roundish structures, respectively, between fibrils in Fig. 5-24d).

In vertebrate animals the terminals of one motor neuron may serve anywhere from 5 to 200 different striated muscle fibers (Fig. 5-26). The group of muscle fibers and the neuron servicing them is called a *motor unit*. No single fiber can respond independently nor can the entire unit contract partially; it is, as noted, an all or none affair. Any given muscle is composed of many such motor units. Consequently, graded degrees of contraction are made possible depending on the relative number of these motor units in action at any one time. Nerve impulses normally reach motor units in a series of volleys at frequencies ranging up to several hundred per second. Since the firing rates of neurons serving different motor units are not necessarily the

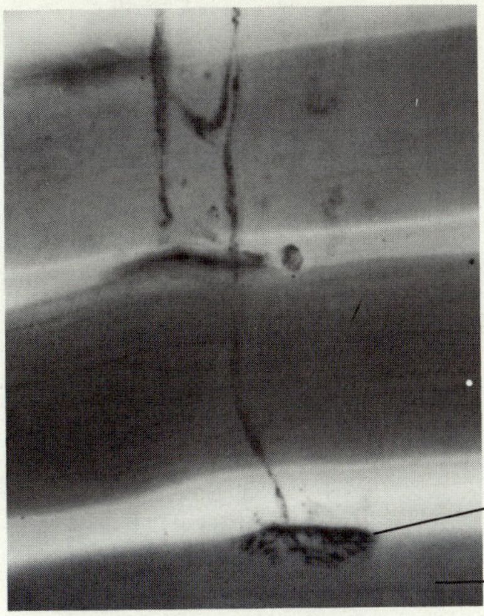

Figure 5-26 Synapselike junction (motor end plate) between motor axon terminals and striate muscle fibers. (Photomicrograph by author)

same, there is an overlap and fusion of contractions which are out of phase with one another. This results in the smooth muscular responses with which we are familiar.

Striated muscles are usually attached to bones or other skeletal elements like the thick, rigid cuticle on the outside of an insect or the calcareous shells that surround a clam. The interrelationships between skeleton and muscle are beautifully organized and constitute a *musculo-skeletal system* by which an animal is able to walk, crawl, dig, swim, or fly as well as perform a great variety of other movements of a highly specialized nature. To carry out these complex movements involves many muscles acting in concert. Teams of muscles are arranged in opposing sets (Fig. 5-27). A muscle pulling in one direction is matched by another pulling in the opposite direction. The contraction of one, the other, or both of these reciprocating

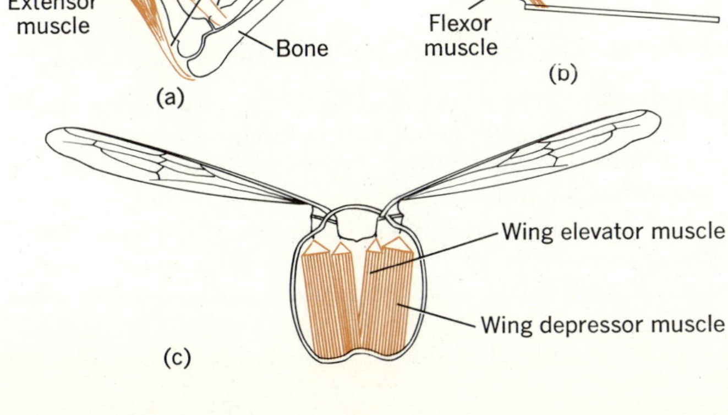

Figure 5-27 Typical arrangements of opposed muscles controlling the joints in the bony, internal skeleton of a vertebrate (a) and in the exoskeleton (outside skeleton) of an insect (b and c).

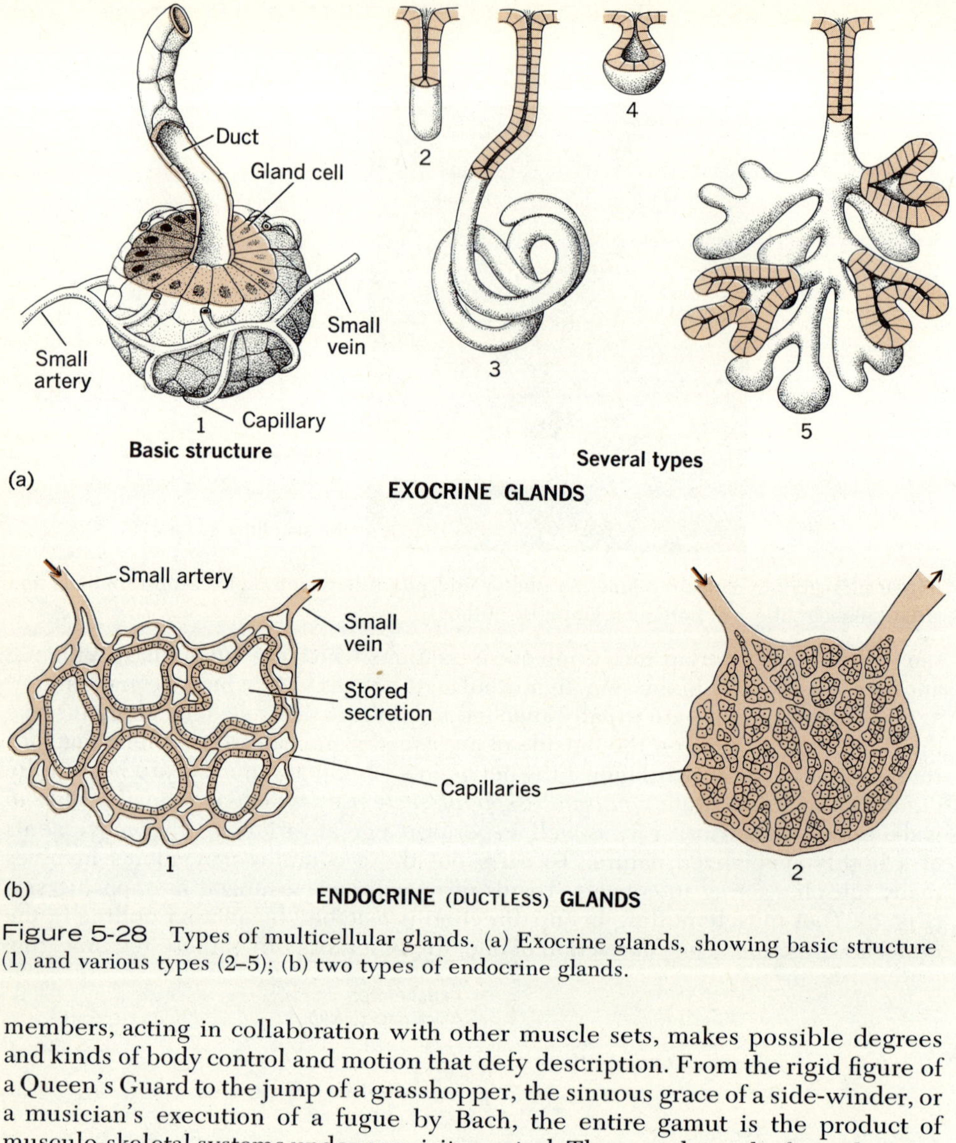

Figure 5-28 Types of multicellular glands. (a) Exocrine glands, showing basic structure (1) and various types (2–5); (b) two types of endocrine glands.

members, acting in collaboration with other muscle sets, makes possible degrees and kinds of body control and motion that defy description. From the rigid figure of a Queen's Guard to the jump of a grasshopper, the sinuous grace of a side-winder, or a musician's execution of a fugue by Bach, the entire gamut is the product of musculo-skeletal systems under exquisite control. The complex task of coordination falls to the central nervous system acting on information supplied by its monitoring receptors.

CHEMICAL SECRETIONS AND GLANDULAR CONTROL SYSTEMS

In addition to muscles, neurons activate many different glands whose secretions serve a great variety of metabolic, protective, and behavioral functions. Our bodies are cooled by the evaporation of sweat. Our eyes are cleansed and protected by tears. Poison glands act as protective and/or aggressive devices for frogs, insects, spiders, snakes, and so on. Chemical digestion is made possible by many glands which pour their secretions into the digestive cavity. Unicellular mucus glands lubricate the digestive lining and protect it from self-digestion, and so on. The list is most impressive.

The abundant multicellular glands of "higher" animals are of two distinct types: (1) the *exocrine glands* which discharge their secretions through ducts into cavities or onto the surface of the body (such as those described) and (2) the *endocrine* (ductless) *glands* which empty their secretions, called *hormones,* directly into the circulating blood (Fig. 5-28). In both types the active glandular cells selectively extract their raw materials from circulatory fluids.

The ability of an animal to regulate its metabolic activities and behavior is in large measure mediated by hormones. A fascinating group of exocrine secretions called *pheromones* similarly enables an animal to influence the physiological activities and behavior of other animals of its kind. The balance of our discussion will concentrate on these two groups of substances.

Hormones and Self-Regulation

Hormones are known to regulate the transformation (metamorphosis) of caterpillars into butterflies and to control growth in animals like crayfish and shrimp. The endocrine glands, however, have been much more intensively studied in vertebrates, so we will confine our attention to them.

It is likely that blood-transported endocrine secretions are involved in some way in the coordination of every aspect of metabolic activity. They are also influential in the development of behavior patterns (sexual responses, for example) and in predisposing individuals to such psychological states as nervousness or depression. Figure 5-29 illustrates a few of the endocrine glands found in man (and other vertebrates) and summarizes (in parentheses) *some* of the known regulatory activities engaged in by each of the glands shown. Only minute amounts of hormone are

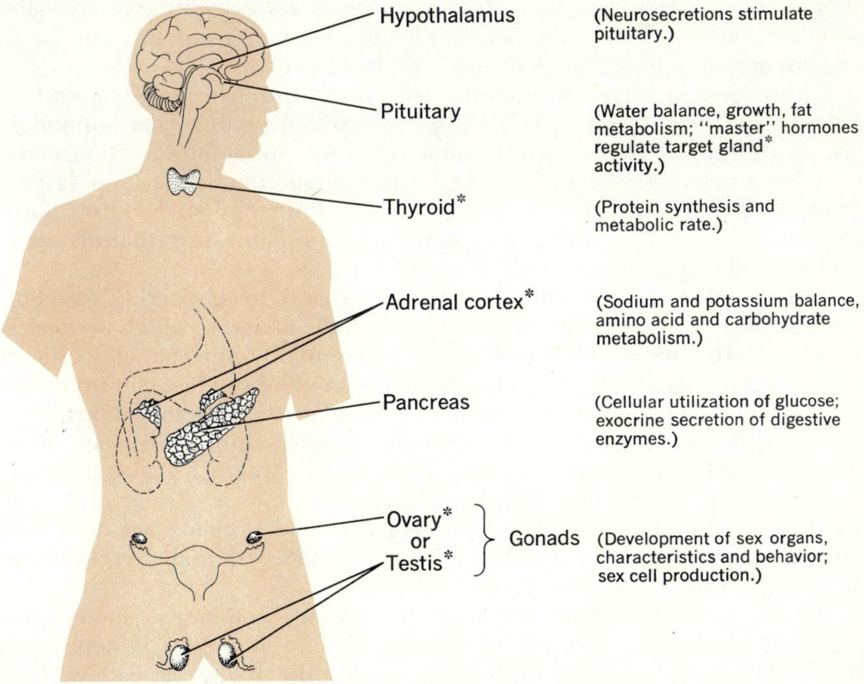

Figure 5-29 The endocrine system of man. The parenthetical notations list some of the metabolic processes and activities whose regulation is dependent on hormones secreted by the gland indicated. A number of endocrine structures are not shown.

required to be effective. In some cases these effects are widespread throughout the entire animal. *Thyroxin*, produced by the *thyroid gland*, increases the overall rate of metabolic activity and may contribute to general restlessness. *Insulin*, secreted by the *pancreas*, accelerates the cellular utilization of glucose and consequently lowers blood sugar levels. On the other hand, there are hormones that are much more localized in their effects, such as one of the *pituitary* secretions that regulates the tubular reabsorption of water in the kidneys.

The mechanisms by which endocrine secretion is coordinated vary from simple to complex. The simplest of these entails a direct feedback relationship between gland and bodily state in which not even the nervous system is involved. As a case in point insulin secretion is stimulated by high blood sugar levels (Fig. 5-30a). Secretion is suppressed by low blood sugar levels. Since insulin lowers the blood sugar content, this is a self-regulatory arrangement. A more sophisticated coordinating system utilizes the nervous system and the capability of neurons themselves to secrete hormones (*neurohormones*). *Neurosecretion* seems to be almost as characteristic of neurons as their role in nerve impulse conduction. This is particularly true of interneurons located in the *hypothalamus* (Fig. 5-29). Nerve impulses are routed to the hypothalamus from receptors in the eyes, ears, nose, skin, and so on, as well as receptors monitoring the blood supply in the hypothalamus itself (Fig. 5-30b). Stimulated interneurons produce neurohormones which are carried by a special portal vein (page 156) to the pituitary gland below. These hormones trigger the synthesis and release of specific pituitary hormones. Some of the pituitary hormones are known as "master hormones" because it is their function to activate certain other endocrine structures, the so-called target glands. These include the *thyroid, adrenal cortex*, and the *gonads* (*ovaries* or *testes*). Pituitary activation of target gland secretion establishes another feedback relationship, this time between the master gland (and/or hypothalamus) and its target.

Let us consider a typical example. The *gonadotropic* hormones stimulate the growth and activity of certain cells in the ovary that produce sex hormones (*estrogen*) as well as others that will become eggs. As estrogenic secretion increases and reaches a threshold value, further gonadotropin production is suppressed. The original stimulus for estrogen secretion is consequently removed and sex hormone production declines. As a result of these reciprocal relationships the release of sex hormones (and eggs) in the female is cyclic.

The spectrum of biologic activities integrated by endocrine effectors far exceeds the few examples of an essentially metabolic nature to which we have so far referred. Growth, reproduction, and behavior are equally subject to endocrine regulation. Varying secretory levels of the pituitary growth hormone can produce anything from a midget to a giant. Sex hormones stimulate the growth and development of the accessory sex organs (such as the penis and vagina). They are similarly responsible for the secondary sex characters that distinguish males and females (body hair, voice, muscle development in ourselves; antlers in deer, and so on). A castrated puppy never learns to lift his leg when he urinates nor does he ever develop a sexual interest in females. If he is injected with male hormone (*testosterone*), he becomes proficient in both capacities.

Behavior and hormones show an interesting reciprocity: some behavior depends on hormones, whereas the release of some hormones is dependent on behavior. A species of fish (stickleback) spends the winter in salt water. The increasing lengths of daylight during the spring result, through optic stimulation of the hypothalamus, in augmented secretion of thyroxin. The fish now prefers fresh water and migrates into streams where breeding takes place. In the female ring

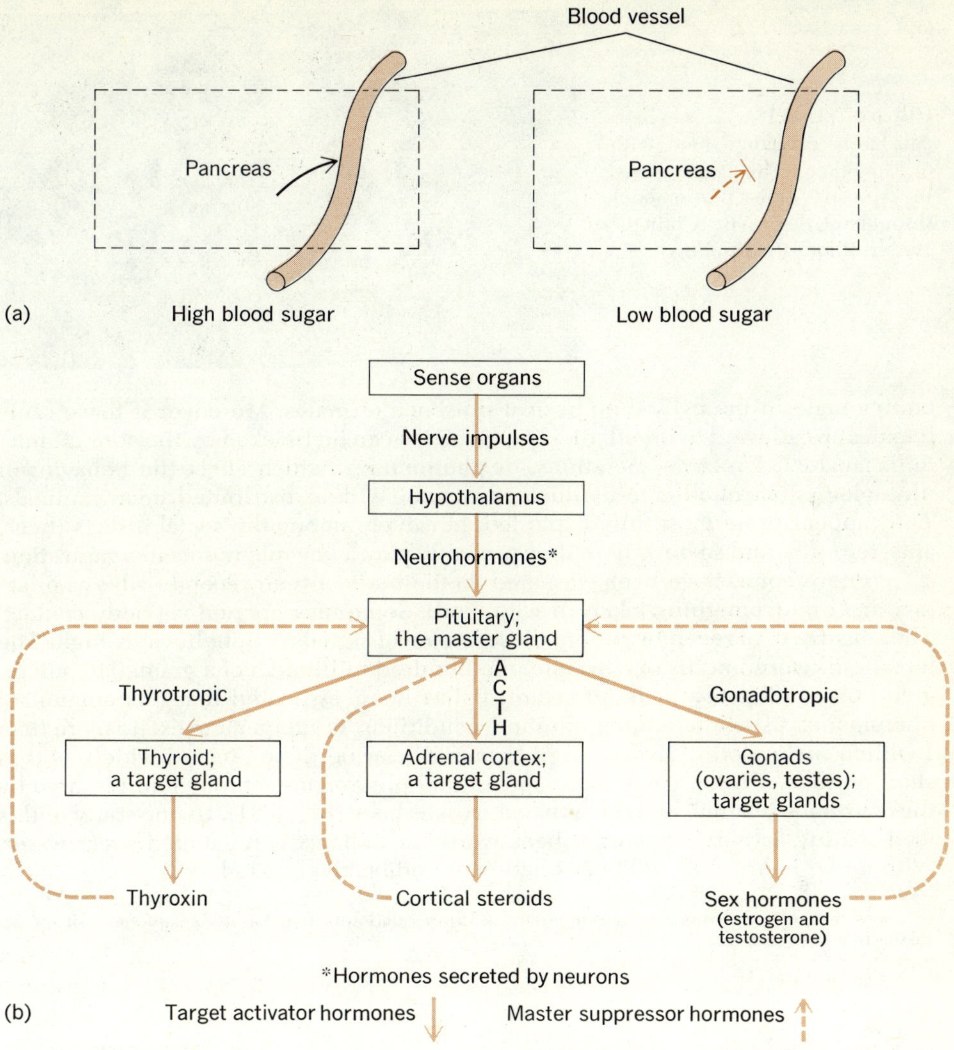

Figure 5-30 The two kinds of endocrine coordinating mechanisms. (a) Pancreatic hormone (insulin) secretion is stimulated (→) by high blood sugar and suppressed (⇢) by low blood sugar content. Since insulin depresses blood sugar level, this is a self-regulatory arrangement. (b) Hormone secretion of a target gland is activated by a series of dependent events beginning with sensory nerve impulses to the hypothalamus. Target gland secretion in turn controls its own output by suppressing (⇢) further pituitary secretion of target activator hormone.

dove sex hormone secretion depends on behavior. She will neither lay eggs nor incubate them unless a male bird is within sight.

Pheromones and Social Regulation

A phenomenon quite similar to that in doves is manifested in female mice. In this instance, however, reproductive behavior in the female is elicited by the odor rather than by the sight of males. Estrogen secretion and its attendent sexual responsiveness ("heat" or "estrus") are as much stimulated by wads of cotton con-

Figure 5-31 Honeybees constantly exchange food with one another. This also serves to spread oral pheromones throughout the hive. (Colin G. Butler, Photo Researchers)

taining male urine as by the physical presence of males. Moreover if these same females are allowed to breed, their pregnancies can be blocked by the odor of unfamiliar males![7] Exocrine secretions, or pheromones, which affect the behavior or physiologic state of other individuals seem to be widely distributed among animals. They appear to be most fully exploited, however, among the social insects (bees, ants, termites, and so on) where they probably play a key role in social organization.

Pheromones have been classified on the basis of their receptor sites as olfactory and oral (something taken in with food). Some may operate via body contact. The sensitivity of receptor sites to these chemical signals is unbelievably high. The female silkworm moth contains one one-hundred millionths of a gram (0.01 microgram) of an olfactory male attractant. It has been estimated that this amount of pheromone, distributed under optimum conditions, is adequate to excite more than 1 billion male moths. Even oral pheromones can be disseminated widely within short periods of time. Honeybees spread oral pheromones throughout the hive by the constant exchange of food between worker bees (Fig. 5-31). In one study of this food-sharing activity six worker bees were fed radioactively labeled sugar water. Within two hours over 7000 bees had been radioactively labeled.

[7] Mechanisms of this kind are important in the regulation of population density and will be discussed in a later chapter.

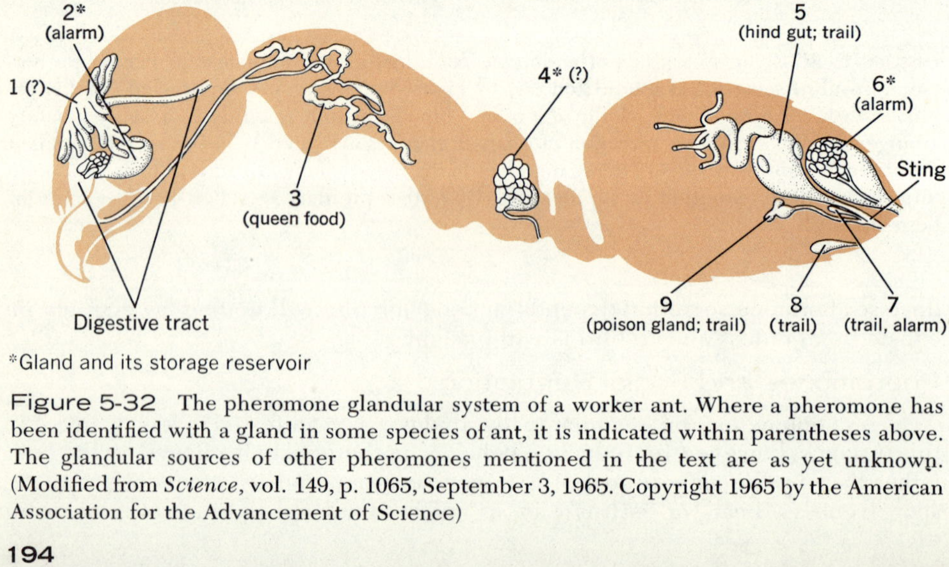

Figure 5-32 The pheromone glandular system of a worker ant. Where a pheromone has been identified with a gland in some species of ant, it is indicated within parentheses above. The glandular sources of other pheromones mentioned in the text are as yet unknown. (Modified from *Science*, vol. 149, p. 1065, September 3, 1965. Copyright 1965 by the American Association for the Advancement of Science)

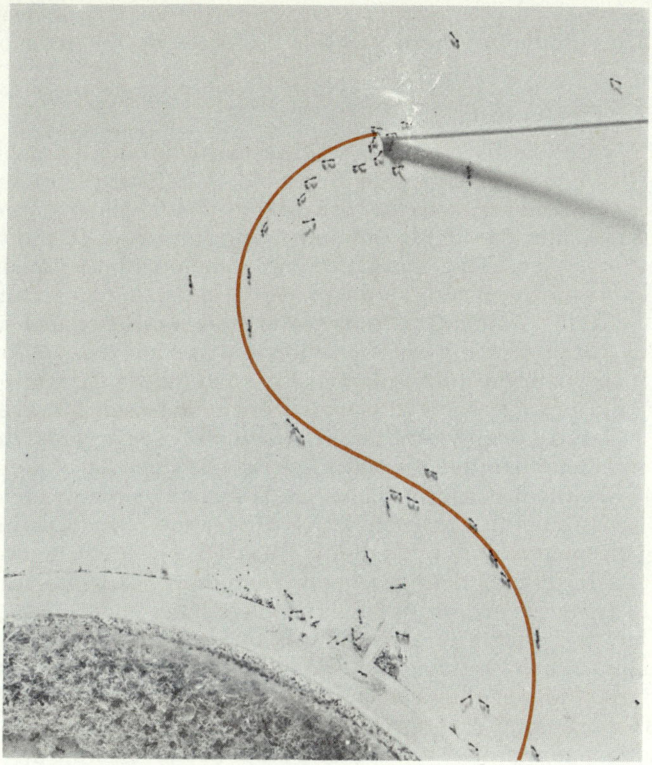

Figure 5-33 Worker ants follow artificial trail (colored line) laid down with a stick treated with the contents of gland #7 (Fig. 5-32) from a single ant. (Sol Mednick)

It is believed that as few as ten pheromones, acting singly or in combination, could mediate the entire organization of an insect society like that of the ant. These insects (like bees) are certainly well equipped with the requisite apparatus (Fig. 5-32). They also give evidence of a number of behavioral and physiological responses elicited by pheromones. Responses to a given pheromone vary with the concentration of the secretion and the context in which it is used. Some secretions (glands 2, 6, or 7) evoke alarm or digging reactions. In one species of ant, gland 3 secretes the oral pheromone by which the sterile female workers convert a (worker) larva into the reproductive queen of the colony. Olfactory pheromones, produced by glands 7, 8, and 9 and the hind gut, serve to gather workers and mark trails to food (Fig. 5-33). The trail pheromones are beautifully adapted for their purpose. Their volatility results in complete fading of the signal within a short time so that old trails cannot confuse new ones. They are also species-specific: each species of ant has its own language.

Several other chemical signals undoubtedly exist such as those for grooming and food exchange. In fact, even the decomposition products of a dead ant will trigger a kind of funereal response and workers will faithfully carry the departed to the refuse pile. If living workers are experimentally daubed with these decay products, they too are dutifully deposited at the colony "dump." On their return to the nest the "living dead" once again suffer the indignity of premature final services.

The process is repeated over and over again until the aura of death is finally removed from the bodies of these luckless creatures by the rigor of the ritual.

ADAPTIVE BEHAVIOR

Up to this point we have concentrated primarily on those organic structures and processes that enable organisms to direct and coordinate their internal and external activities. The internal activities are all those metabolic events discussed in Chapter 4 that maintain the highly ordered living state through the utilization of free energy in the system. The external activities include those equally important efforts expended in the quest and capture of energy and in the preservation and perpetuation of the living system. It is to these external activities that we apply the term "behavior." It is also through behavior that we are perhaps brought most clearly face-to-face with the full impact of life's conquest of energy: the dogged determination and perseverance with which organisms seek out every available source of life-sustaining energy—the adroit skill and ingenuity with which they use the world around them to further species survival. The sum total of these responses, and many more like them, constitute behavior. We would certainly be remiss by not including a discussion of this most characteristic feature of living things.

As noted elsewhere life is exceedingly frugal. Waste, inefficiency, and pretentiousness cannot be tolerated in the keen, competitive struggle for the energy market. Relatively simple organisms have relatively small investments in metabolic machinery. They consequently cannot afford, nor do they require, anything more than a modest inventory of behavioral responses. At the other extreme complicated organisms like the "higher" animals cannot afford to be without an elaborate and comprehensive behavioral repertory in order to bring in the amounts of energy needed to maintain and protect a metabolic plant of such complexity.

As we now proceed to look into various levels of behavior we should constantly bear the foregoing consideration in mind. Behavior is neither primitive and inefficient nor extravagant and superfluous. It is always adequate—no more, no less—for the organism and its needs.

Stereotyped Behavior

Many types of behavior are so inflexible and stereotyped that one can predict the response to a given stimulus. Many species of moths, for instance, are invariably attracted by a light source at night. Behavior of this kind is predominantly programmed or built into the organism. The light response of moths is largely predetermined by a nervous circuitry which is inherited in much the same fashion as the interconnected blood vessels in a circulatory system. Although the response is subject to some modification through learning, the inherited limits within which variation is possible are rather narrow. We will first consider levels of behavior such as these in which the stereotyped patterns established by inheritance are the predominant components.

BEHAVIOR IN PROTISTS

Actively motile protists demonstrate simple but effective responses to external stimuli. These may be roughly categorized as avoidance behavior and orientational behavior. An ameba withdraws when touched, an avoidance response noted previously. A paramecium backs away (withdraws) from an obstacle in its path, sets a new course, and takes off in a different direction. Orientational responses differ from avoidance in that they are not random movements. They are rather precisely aligned with the location or direction of the stimulus. A paramecium, for example, will usually head upstream into a gentle current. This type of response is called a *taxis* and is by no means restricted to protists. Some butterflies attempt to evade pursuing birds by flying into the sun. Honeybees locate food sources in relation to the sun. Even the behavior of plants is an orientational phenomenon except, of course, the plant reacts by growth responses rather than actual locomotion. Reactions of this sort are called *tropisms*.

BEHAVIOR IN PLANTS

Tropistic behavior in plants entails responses to external stimuli such as light, gravity, and physical contact with other objects (Fig. 5-34a,b,c). Plants grow toward the light that powers their photosynthetic machinery. They send their shoots upward into the air against the force of gravity and their roots downward into the nourishing soil. If we turn a young plant upside down, the shoot and root turn on themselves and continue to grow in their foreordained directions. A climbing vine will twist around an object it chances to contact. Some plants have tendrils (peas, squash, and so on) which will entwine themselves around a support in just a few moments.

(a)

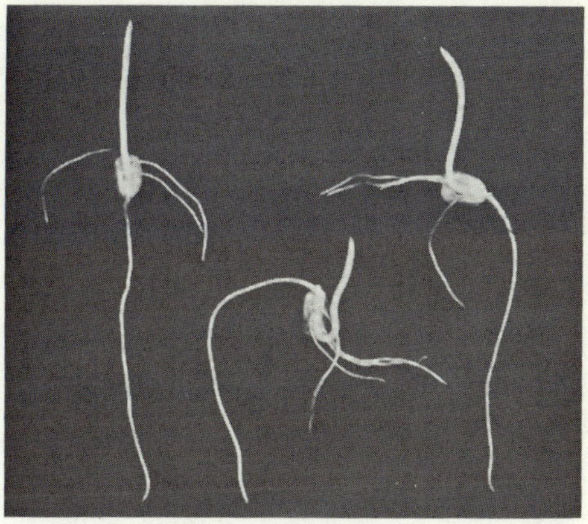

(b)

(c)

Figure 5-34 Three common tropisms in plants. (a) The stem of this geranium has bent in the direction of the light and the leaf blades are at approximate right angles to the light rays. (b) The seeds of these young corn plants were germinated in a right-side-up, vertical position (left), a horizontal position (right), and an upside-down vertical position (center); in each case the shoot grew upward and the root, downward. (c) This stout woody stem is wrapped in a tight spiral around the tree trunk which supports it. Such climbing woody plants are called lianas. [(c) Photograph by author]

Figure 5-35 Effect of growth hormones on the cabbage plant. *Left:* Untreated; *right:* Sprayed several times with a diluted growth hormone. (Dr. S. H. Wittwer, Michigan State University)

All of these behavioral responses are brought about by a controlled bending of some part of the root or shoot. This bending is actually the result of differential growth under the control of plant hormones. Plant hormones are produced in the tips of growing shoots and in leaves and are then transported down the stem. One of the three or more groups of hormones presumably produced by plants has the effect of accelerating growth by stimulating cell enlargement. These *growth hormones* have a pronounced effect on the rate of growth in plants (Fig. 5-35). The distribution of these hormones in the plant is affected by a number of stimuli. In a young growing stem, for instance, newly synthesized hormone will migrate from the lighted to the unlighted side of the growing tip and then descend down the unlighted side of the stem (Fig. 5-36a). Growth is consequently stimulated on the side opposite the light and the shoot bends toward the light. In older plants (Fig. 5-34a) such stem bending may result from the actual destruction of hormone on the illuminated side.

Gravity also has a decided effect on differential growth (Fig. 5-36b). When a shoot and a root are in a horizontal position, hormone accumulates on the lower side. This results in the shoot turning upward. The root, however, appears to be very sensitive to such concentrations of growth hormone.[8] Growth is consequently

[8] This sensitivity is not unusual. Concentrated solutions of growth hormones can not only stop growth but kill plants. Some commercial weed killers act in just this way.

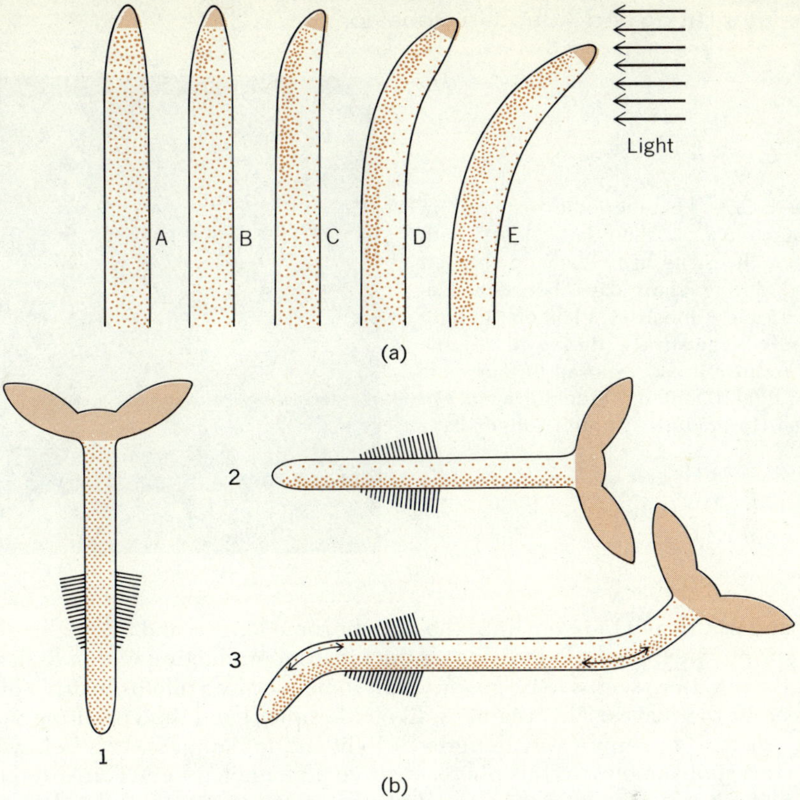

Figure 5-36 The effect of light (a) and gravity (b) on the growth hormone distribution in plants. The site of hormone synthesis is indicated by color, the relative concentration of hormone, by stippling. A seedling exposed to light concentrates its hormone on the shaded side of the stem (a). When a seedling is laid horizontally, hormone is transported to the lower side (b, 1 and 2). Arrows (b,3) show where cellular elongation takes place (see text). (Adapted from P. M. Ray, *The Living Plant.* Holt, Rinehart and Winston, 1965)

inhibited on the lower side and the root turns downward. The twining response in vines probably operates in a fashion similar to the stem reactions just described. Physical contact of the stem with a support results in accelerating growth on the opposite side of the stem so that the plant bends (coils) around the support.

Plants are by no means restricted to tropistic, orientational responses. We are all familiar with the dormant period of trees and shrubs during the autumn fall of leaves. This leaf fall may somehow be related to certain hormones affecting cell divisions at the bases of leaves. Still other hormones appear to be influential in the production of flowers and in stimulating fruit development in the flower. The blossoming behavior of plants is indeed a fascinating topic all by itself, and we have much yet to learn about it. The opening and closing of flowers in response to light or temperature may be due to differential growth or turgor changes on the upper and lower sides of flower parts. The periodic blossoming of plants every fall or every spring still remains a largely unfathomed problem. We are not, however, in total ignorance of some of the mechanisms at work.

Some plants, like corn, poinsettias, and chrysanthemums, are called *short-day plants* because they do not blossom until the days become shorter (and the nights longer) in late summer and fall (Fig. 5-37). *Long-day plants,* like radishes, holly-

Figure 5-37 Photoperiodic response in chrysanthemum, a short-day plant. Both plants are the same age. Plant on the left bloomed during short days between December 15 and February 1. Plant on the right grew only vegetatively during the same period because it was exposed to four extra hours of light. (Dept. of Floriculture and Ornamental Horticulture, Cornell University)

hocks, and iris, do not blossom until the days become longer (and the nights shorter) in the spring. Such photoperiodism (timing blossom production with a light period of specific duration) seems to be governed in some way by a photosensitive pigment produced in the leaves. This pigment, like chlorophyll and the visual pigments, is highly sensitive to certain wavelengths of light and to darkness. It is believed that when a threshold amount of this photosensitive pigment has been converted into its active form, it triggers a series of metabolic reactions that act as a kind of in-born "clock" in measuring night length. The sensitivity of the mechanism is remarkable. Some plants will flower in response to night length changes of only 10 to 15 minutes.

INSTINCTIVE BEHAVIOR IN ANIMALS

Instinct is the most complex (and least inflexible) level of stereotyped behavior. Although it is widely exploited by many kinds of animals, it is perhaps most dramatically developed in insects, spiders, fishes, and birds. A spider spins its web and a bird builds its nest on the basis of responses biologically programmed into the organism. These complex behavioral activities occur without benefit of prior training or observation. They are also species-specific. Two different species of birds will build their own unique kind of nests (Fig. 5-38). Some instinctive behavior, such as bird migration, is truly extraordinary. Many species migrate hundreds or thousands of miles while flying at night. Young birds commonly make these flights independently of their parents. Experiments with birds tested in planetariums indicate that they navigate by reference to the positions of stars in the heavens. The nature of the in-born celestial compass and clock requisite to this task is not at all understood.

The initiation of instinctive behavior depends on two requirements: (1) a physiological state of readiness within the organism and (2) a triggering or releasing stimulus provided by the environment. A female bird, for example, may start to build a nest only if her ovaries are in a state of reproductive readiness and an acceptable male bird of her species is present. Instinct has often been confused with learned behavior because the state of physiologic readiness commonly requires that individuals reach a certain stage of maturity. Birds fly instinctively but only after

Figure 5-38 Nest building by birds is unlearned behavior. Different species produce their own unique type of nest (oriole on the left; robin on the right). (Gerard Brown, Monkmeyer; Photo Researchers)

they have reached a certain stage of physical development. It has been shown that no learning, insofar as flight is concerned, takes place during this maturation period. We must not, on the other hand, assume instinct to be an automatic, robotlike behavioral display that is totally unlearned. It is capable of being modified by experience, sometimes to a very great degree. If the environment, for example, yields unusual, inappropriate, or inadequate releasing stimuli, instinctive behavior may be significantly altered, if indeed it is elicited at all. Learning in the honeybee is largely a matter of adapting instinctive responses to a variety of releaser stimuli, as we shall see later.

Learned Behavior

All behavior is dependent on and tailored to a strong hereditary component: the receptors, conductors, and effectors with which a given organism is equipped. Most, if not all animals, on the other hand, show adaptive changes at some time or another in their behavior that reflect the importance of past experience and learning. The behavioral machinery is like a musical instrument; it has an innate design which restricts its use in certain ways. Its repertory of performance, however, is limited only by the experience that plays upon it. Let us now focus our attention on these acquired and learned aspects of behavior.

HABITUATION

If stereotyped behavior is continuously evoked without purpose, there is a gradual elimination of the response. This is the simplest kind of learning, called *habituation.* If one vibrates a spot on a spider web, imitating the efforts of a trapped fly, the spider will run out to investigate. After several repetitions of this performance the spider no longer responds. It has become habituated to the stimulus. A change in the intensity or location of the vibrations, however, will immediately reinstate the original behavior.

IMPRINTING

In some animals at least (birds and mammals) it has been found that there is a certain critical period early in life when the experiences of the individual bring about significant behavioral effects which last for years if not throughout the life of the animal. Learning during this highly impressionable age has been called

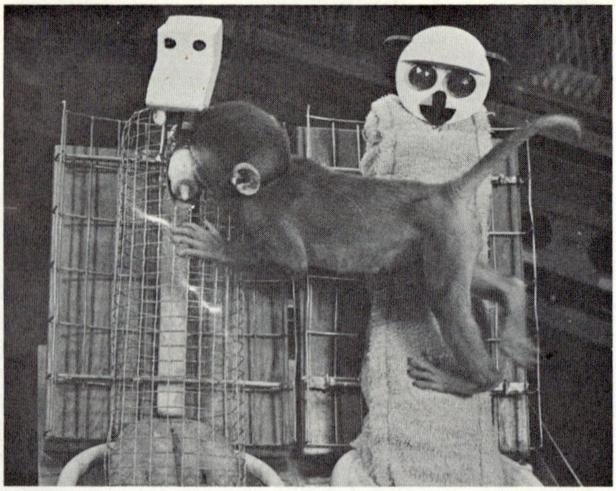

Figure 5-39 A baby rhesus monkey stands on its terry cloth "mother" while it nurses from the bottle of the wire "mother." (Wisconsin Regional Primate Research Center)

Figure 5-40 A baby monkey frightened by a strange object (teddy bear in foreground) flees to its substitute terry cloth "mother" rather than to the wire "mother." (Winsconsin Regional Primate Research Center)

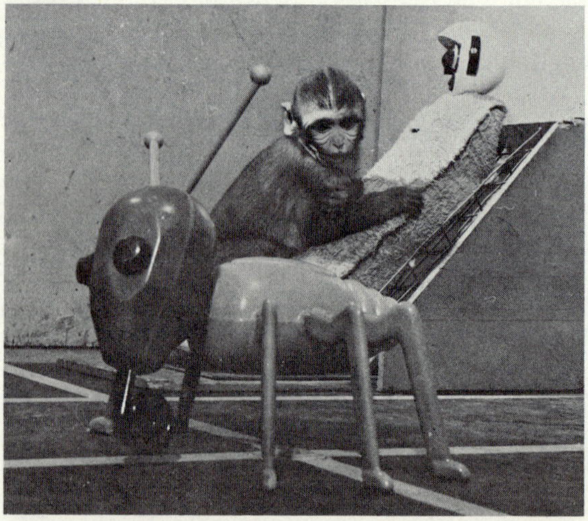

Figure 5-41 Securely clinging to its substitute "mother," the baby looks on a strange stimulus with curiosity rather than fright. (Wisconsin Regional Primate Research Center)

imprinting. In baby monkeys the first six months of infancy seem to be particularly critical in the development of behavioral patterns which persist into adult life. A satisfactory relationship between mother and child appears to be of the utmost importance during the first half of this period. Monkeys raised without real or substitute mothers develop a variety of behavioral abnormalities both as youngsters and as adults. They sit for hours staring into space. Rather than jabbering and screaming at a strange man who approaches their cage they attack themselves by biting and pinching their own bodies. As they become sexually mature the males are unable to copulate with even normal sexually excited females. Females which conceive and bear young are indifferent if not abusive mothers.

The fourth to the sixth months of infancy seem to be an even more critical imprinting period. During this interval it appears that play activity with other infants is more important to normal behavioral development than the prior mother-child relationship. Monkeys raised with their mothers but isolated from other monkeys develop severe abnormalities in both social and sexual behavior. On the other hand, infants allowed only 20 minutes a day of association and play with other infants develop normally even when they have been raised with imitation mothers composed of cloth.

In a study of baby monkeys being reared in the absence of their mothers it was discovered that the infants formed a strong attachment to the folded gauze sheet on the floor of their cages. The attachment resembled that of a human infant for its teddy bear (or blanket). This prompted the researchers to investigate the significance of the feel of soft materials to the establishment of a mother-infant bond. They therefore built two artificial, substitute "mothers." One was made of wire with a crude wooden head and a support to hold a nursing bottle in its "chest" (Fig. 5-39). The other was made like the "wire mother" except that it was covered with soft terry cloth. The terry cloth mothers proved vastly superior. Regardless of which mother held the food, the infants clung to their terry cloth mothers for about 18 hours every day (Fig. 5-39). If an infant was frightened by a strange object in its cage, such as a mechanical teddy bear that walked and beat a drum, the baby immediately ran to the terry cloth mother rather than the wire mother (Fig. 5-40). When a baby monkey was removed to strange surroundings, with toys scattered about, it would become typically terrified, throwing itself to the floor and rocking its body to and fro with its head clasped between its hands. When the terry cloth (but not the wire) mother was placed with the baby, the terror soon vanished and the infant looked upon the strange monsters more with curiosity than with fright (Fig. 5-41).

MOTIVATION

Receptors in the hypothalamus are known to monitor the concentration of water in the blood as well as the relative amounts of glucose and possibly many other solutes. A deficiency detected here is translated into some need or drive that we label by names such as hunger and thirst. This drive generates or *motivates* whatever behavior is required to satisfy the physiologic need. It must not be assumed, however, that drives are restricted to physiologic need. They are not. They are, rather, aimed at species fitness and survival. Young cats with full bellies stalk imaginary prey. Monkeys, apes, and people work at puzzles for no other reason than to solve them. Whether one lives by stealth or by cunning, survival demands that the art be practiced well. And this it truly is. Animal play is a biological device by which behavior patterns conducive to species survival are practiced with zest and vigor.

CONDITIONING

Some animal psychologists feel that all learning arises through the conditioning of motivated organisms. If such is the case habituation, trial-and-error learning, maze learning, habit—even the practice of reason itself—are all levels of behavior based on the conditioned response as their fundamental unit. The process of conditioning can perhaps best be understood by describing a typical laboratory design used in conditioning studies. A box easily large enough to accommodate a pigeon without crowding is equipped with a food hopper. If the bird pecks a readily accessible operating key, the hopper opens long enough for the animal to pick out a kernel of grain. The key is wired to an automatic monitor that records how many times the key has been pecked per unit of time. The bird is motivated by being deprived of food before it is put in the box.

At first the pigeon performs seemingly random acts such as preening itself, cooing, spreading its wings, and pecking at the box. Sooner or later the key is pecked and the hopper opens. The bird picks out a kernel of grain and the hopper closes. Key pecking now begins to increase at a faster and faster rate. Food is *reinforcing* (rewarding) pecking behavior. Through augmented pecking responses directed at the key we say the pigeon is being *conditioned* to key operation. Learning so induced in the pigeon is quite persistent. Without any "refresher courses" these birds retain their conditioned response for at least six years. If, once a bird has been conditioned to peck the key at a high frequency, the hopper is adjusted so that it no longer opens, the bird continues to peck at a rapid rate. It requires several hours before the frequency of pecking finally drops back to the level before reinforcement. During this period of behavioral *extinction* the bird occasionally flaps its wings, coos, and turns away from the key with some emotional distress (not greatly unlike the rest of us under comparable circumstances).

The nature of conditioned behavior can be manipulated to a considerable degree by scheduling reinforcement in various ways rather than automatically reinforcing every peck at the key. We can, for example, adjust the hopper so that it opens only after a predetermined number of pecks (like piecework pay for human labor). In this case the pecking rate becomes exceedingly rapid. Fatigue is the only apparent restriction to response rate. We can, on the other hand, adjust the hopper so that it opens at a set time interval after the first peck, regardless of the number of pecks (like the payment of a monthly salary). Although this results in stable behavior over a period of time, the pecking rate becomes quite slow just after the last reinforcement and then speeds up in anticipation of the next. One of the most effective schedules is when reinforcement occurs at random time intervals (like "one-armed bandits" and other games of chance). Animals, including man, are natural gamblers because nature rewards her creatures at random. In the interest of survival animals have evolved a great sensitivity to random reinforcement schedules. Pigeons placed on a random time schedule that averaged one reward every 5 minutes pecked two to three times per minute steadily for 15 hours.

Food (or water) is by no means the only reinforcer of conditioned responses. Monkeys will press a key to get a view of another monkey outside the box. Chimpanzees will press a lever in order to work a puzzle. The behavior of a variety of animals can be reinforced by removing an unpleasant stimulus, such as stopping a loud noise or warming a cold animal. Punishment appears to have little value. Mild punishment seems to be relatively ineffective in changing behavior permanently; strong punishment is effective but is accompanied by undesirable emotional side effects. Monkeys that were forced to learn by the threat of painful electric shock often died of stomach ulcers, whereas control animals receiving the same electric shocks continued to be healthy.

COMPLEX CONDITIONING

Conditioned responses can be linked together in sequences of varying degrees of complexity. This occurs when animals are required to make trial-and-error responses to different kinds of multiple choice situations. Animals vary in their skill at learning under such conditions; the performance of earthworms is fair, that of insects and octopuses good to excellent. Monkeys and the apes rank exceptionally high. The chimp in particular is capable of solving novel problems by reasoning.

Maze learning is commonly used as a test of problem-solving capabilities. A maze is an enclosed corridor or tunnel that may be a simple T-shape, such as is used to condition earthworms, or it may be quite complicated. In a study designed to compare the maze-learning ability of ants and rats, a sophisticated maze was used (Fig. 5-42). The animals begin at "start" and must wend their way through the maze to the "exit" where the reinforcers are located. Every wrong turn is counted as an error. Although rats learn faster than ants, both species reach a high level of competence after a number of trial runs.

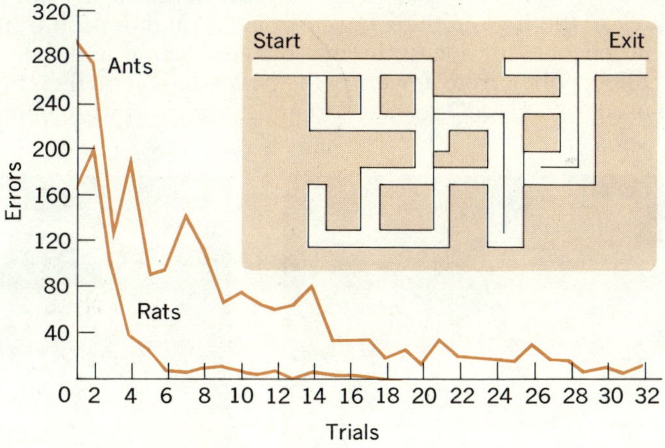

Figure 5-42 A maze used to compare the learning abilities of ants and of rats. The rats ran to food; the ants to their nests. The learning curves show that in this test rats learn faster than ants, but both species eventually reach a high level of performance. (Adapted from T. C. Schneirla in K. Roeder, Ed., *Insect Physiology*, John Wiley, 1953)

Another test is more subtle and involves a symbolic kind of counting. Food is placed in one of two identical hoppers placed side by side. The hoppers are charged on successive trials according to a set sequence, such as right-right-left-left (RRLL). Rats utterly fail this test. Cats after extended training can master a simple RRLL without error. Raccoons are able to extend the series to RRLLRR. Monkeys are successful up through RRLLRRLL.

Caged chimpanzees will obtain a banana that is out of their reach by piling boxes on top of one another. They may also make a long pole by fitting short lengths of bamboo into one another. In one such test the chimp led an investigator by the hand to a spot below the banana, at which point one primate thoroughly surprised the other by climbing on his shoulders to retrieve the delectable morsel.

Social Behavior

The majority of animals are social at least to the extent that they mate and reproduce. Many species live in social aggregations or societies. Whatever the extent to which individuals interact in a social context the cohesiveness and continuity of the social unit are predicated on the behavior which collaborating individuals direct at one another. This *social behavior* may have its base in relatively inflexible, instinctive patterns or in responses which are essentially learned. We will investigate both. We trust the reader will observe, however, that even stereotyped behavior is by no means stereotyped. It is fraught with spontaneous responses designed to accommodate unforeseen circumstances. Without learning honeybees as well as chimpanzees could not survive.

THE SOCIAL BEHAVIOR OF HONEYBEES

Man provides honeybees with ready-made hives in which to live (Fig. 5-43). These structures contain frames of man-made wax sheets on both sides of which the bees produce their combs by secretions from their wax glands (Fig. 5-43b). Some of the cells in these combs are used by the reproductive queen bee as brood chambers in which she lays her eggs (Fig. 5-43c). Others are used as storage chambers for a kind of bee food we call honey. Foraging bees collect the sugary nectar from flowers, carry it back to the hive in their *honey stomach,* and deposit it in comb cells. Here the nectar changes to honey through enzyme action and dehydration. The foragers also collect pollen from flowers in the *pollen basket* on their hindmost legs. Pollen is also used as food and is cleaned from the foragers by grooming bees in the hive.

(a)

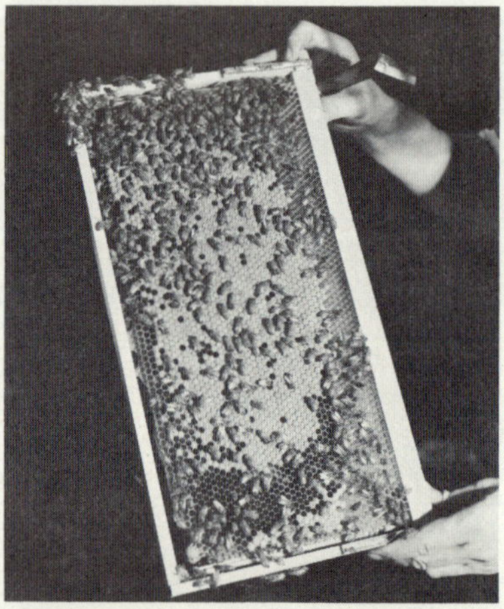

(b)

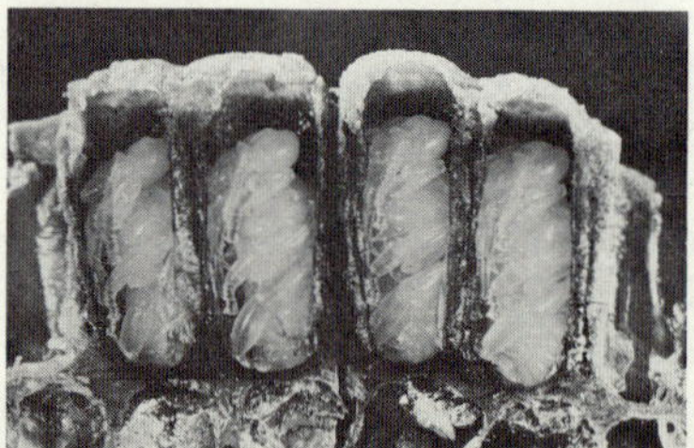

(c)

Figure 5-43 Beehives (a) contain vertically stacked frames of wax sheets on both sides of which bees build their combs (b). The cells of the comb are used for the storage of honey and as brood chambers for the developing young (c). [(a) and (b) U.S.D.A.; (c) Lynwood Chace, from National Audubon Society]

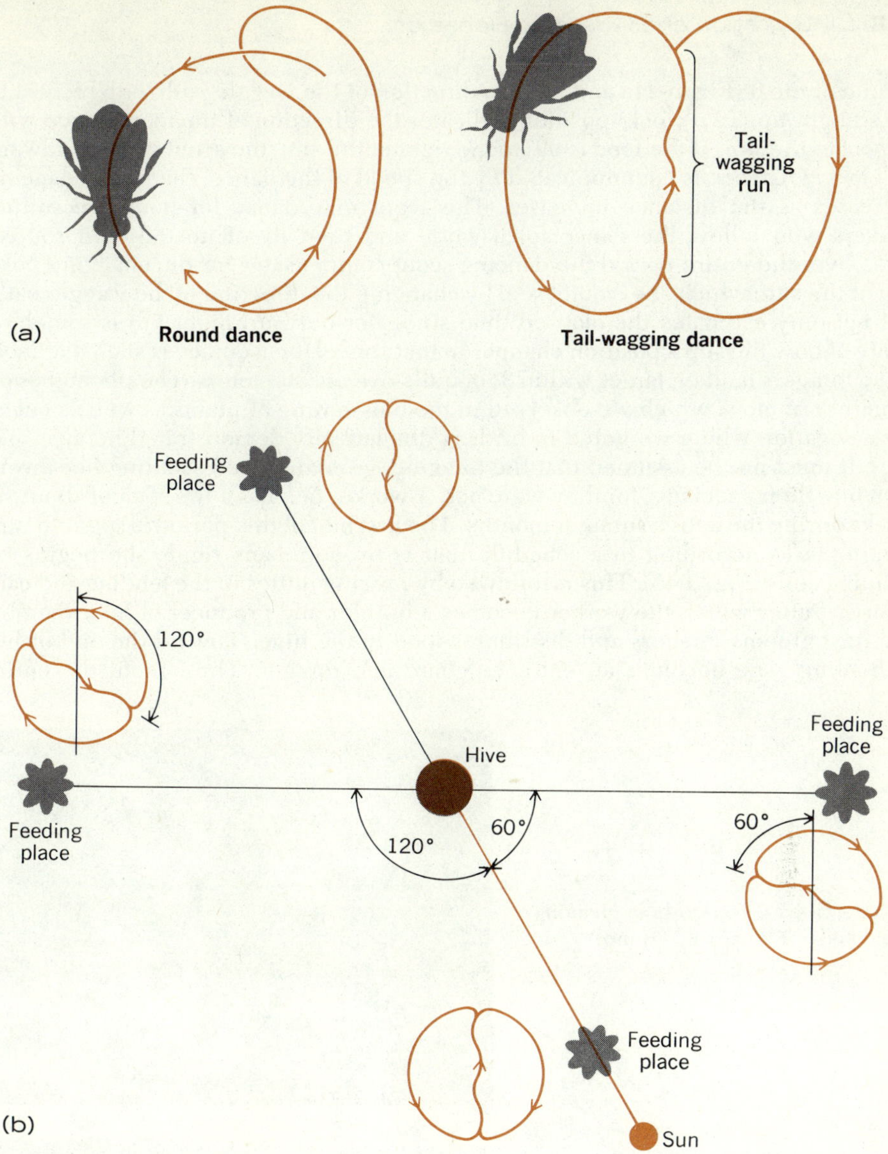

Figure 5-44 Bees use a dance by which they communicate the location of a food source. The round dance (a, left) is for nearby food sources, the waggle dance (a, right) for more distant food sources. In the waggle dance the waggle path indicates the direction of the food with respect to the position of the sun. It is danced on a vertical surface in which straight up ("noon") represents the position of the sun (b). (Adapted from Karl von Frisch, *Bees, Their Vision, Their Chemical Senses, and Language*. Copyright 1950 by Cornell University)

Food is located in the field by scout bees which return to the hive to relay their information to prospective foragers. This is accomplished by a fascinating bee dance usually executed on the vertical face of a honeycomb. If the food is closer than 50 to 100 meters, the scout performs a round dance of alternating clockwise and counterclockwise circles (Fig. 5-44a). The scent of nectar on her body gives the clue as to the nature of the food source. If the food is farther than 100 meters, the scout describes a figure 8, waggling her abdomen along a straight path through the middle of the 8 (Fig. 5-44a and b). The direction of the waggle path with respect to the straight up (12 o'clock) position indicates the direction of the food source with respect to the sun. If the food is 90° to the right of the sun, the scout waggles toward 3 o'clock. Distance is communicated by the speed of the dance; the rate of dancing decreases as the distance increases. The scout may dance for hours, recruiting workers who follow the dance for a while and then fly off to the food source.

Over the entire period the dancing scout compensates for the changing position of the sun, which she cannot see, by changing the direction of the waggle path. She not only estimates the elapsed time since her outward flight but is somehow aware of how the sun's position changes in that time. Her accuracy is such that most of the foragers land on target within 3° of bulls-eye at 250 meters. The phenomenon of an inborn clock which we observed in the blossoming of plants as well as celestial navigation which we noted in birds is dramatically demonstrated here.

It must not be assumed that the foraging, grooming, and scouting bees were born into their specialty, for they were not. A worker bee has a life span of about six weeks during the active summer months. The first half of this period is spent in performing tasks according to a schedule that is by no means rigid. She begins by cleaning cells (Fig. 5-45). This is followed by nursing duties in the feeding and care of larvae, after which the worker becomes a builder and producer of honeycombs. She next grooms foragers and distributes food in the hive. Toward the end of her "schooling" she becomes an entrance guard and forager. Throughout the entire

Figure 5-45 Worker bees cleaning brood cells. (Richard F. Trump)

Figure 5-46 Workers ventilating the hive by beating their wings at the hive entrance. (Treat Davidson from National Audubon Society)

period the worker patrols the hive, carrying out debris, dead workers, and so on. Interestingly enough she apparently does absolutely nothing for about 40 percent of the time. One day (the 17th) in the life of a worker included cell cleaning, honeycomb building, guarding the hive entrance, foraging, several play flights and, we presume, resting.

Life in the hive is particularly dynamic during the summer months, not only because of foraging activities but also because of critical temperature levels within the hive. When the temperature rises above 35.5° C, workers go to the hive entrance and propel air into the hive by using their wings (Fig. 5-46). Nursing bees spread watery honey over the brood cells to evaporate. They beg neighboring bees for more honey water, which is freely given, and a wave of begging spreads throughout the hive. Grooming bees that relieve the returning foragers of their bounty seek out those with the most dilute sugar. A forager with good rich food of lesser water content can scarcely find a worker to take her load. The foragers now begin to bring in diluted nectar. Some even carry water only. After the hive has cooled sufficiently, the situation is reversed. A forager returning with rich nectar is now welcomed by several workers. Those bearing the watered-down product are effectively ignored.

The picture painted here is one of bustling activity and great adaptive spontaneity. By no stretch of the imagination is it one of a totally unlearned, inflexible society.

THE SOCIAL BEHAVIOR OF WILD CHIMPANZEES

The highly structured order of a honeybee society has its direct antithesis in a wild chimpanzee "community." These denizens of equatorial Africa are nomadic in nature. Wherever their feeding activities take them at the end of a day they build their tree nests and bed down for the night. A given area harbors perhaps 100 chimpanzees which are dispersed in highly variable groups. Each of these groups has between 6 and 30 individuals, but their numbers are not constant from day to day.

The only stable unit in this society is the female and her infant, although older children may also be included. Figure 5-47 is a photograph of a family group studied by Jane van Lawick-Goodall. The mother ("Flo") was over 30 years old and had an infant ("Flint") and a three-year-old daughter ("Fifi"). A six-year-old female and a twelve-year-old adult male may also have been Flo's children because all five

Figure 5-47 A family group of chimpanzees including the mother, her 3-year-old daughter and infant son. (Baron Hugo van Lawick © National Geographic Society)

Figure 5-48 Male chimps commonly greet one another with an embrace. (Baron Hugo van Lawick © National Geographic Society)

Figure 5-49 *Right:* A juvenile chimp using a grass stalk trimmed as a tool for collecting insects. (Baron Hugo van Lawick © National Geographic Society)

Figure 5-50 Chimpanzee with a club attacking a stuffed leopard. (J. van Zon)

animals commonly engaged in grooming one another (picking dried skin, weed seeds, and so on, from the others' bodies).

Chimps show a remarkable attachment to their youngsters and great aptitude in their care. An adult male patted Flint on the head in a fatherly fashion. Fifi would often become too persistent in trying to take over the care of her baby brother. On these occasions Flo would engage her in distractive behavior by tickling or grooming her, or chasing her around a tree. In over 2000 hours of on-the-spot observation of chimps, Jane Goodall has observed a punishing response on only ten occasions, and in all of these reassurance immediately followed in the form of an embrace.

The interpersonal relationships of adult animals are equally intriguing. When two male chimps meet they may bow or nod. Sometime they embrace (Fig. 5-48). Two males startled by a sudden noise will reach out and touch one another on the shoulder. One ape often seeks reassurance from another by extending a hand to be touched. This commonly occurs when females and juveniles approach a male who has just vented his frustrations by a temper tantrum of branch pulling. A male chimp has been observed removing a foreign body from the eye of his female companion.[9] In answer to her whimperings he crouched before her and, peering into her eye at close range, manipulated the eyelids with his two forefingers. Male chimps commonly hunt and kill other animals like monkeys and the young of baboons, bushpigs, and antelopes. After the hunter has had his fill, he will hand morsels of meat he has torn from the carcass to the outstretched hands of females and juveniles gathered around him.

Chimpanzees are primarily vegetarians, their diet consisting mostly of fruits, leaves, flowers, seeds, stems, and bark. Insects, however, seem to be a staple article in their daily menu (unlike the intermittent consumption of meat). These include two species of termites (eaten at different times of the year), ants and caterpillars. In collecting ants and termites chimps prepare special sticks by breaking and trimming a stalk of grass or a twig (Fig. 5-49). The tool made for ant collecting is considerably longer than that for termite collecting. The pruned stick is inserted into the opened insect nest and after a moment the stick is withdrawn. The insects clinging to it are promptly eaten. Young chimps learn to "termite" by watching adults. Baboons watch the procedure also but do not make insect sticks even though ants and termites are a delicacy to them.

Chimps make a variety of tools and engage in remarkably humanlike behavior. They crush leaves and use them as sponges in soaking water from crevices. They also use leaves as toilet paper and as towels in cleaning and drying themselves. Adriaan Kortlandt conducted a fascinating study of chimps native to an area of forest growth interspersed with open plains. He rigged a stuffed leopard with a windshield-wiper mechanism so that the head and tail would move. The mock leopard was placed in an area occupied by about 30 chimps. In short order "Hell broke loose" as one member of the expedition put it. During the next 20 minutes there was a pandemonium of hooting and yelling in which the leopard was attacked methodically. Innumerable objects were thrown both overhand and underhand with remarkable accuracy. Males broke off stout branches and stripped them of their twigs and leaves (Fig. 5-50). These clubs were used with great vigor. The animals charged in erect and literally beat the leopard to pieces. The humanlike quality of the entire affair was striking. As the attacking "warriors" rejoined their "troops" there was much embracing—and handshaking!—all the way around.

[9] These animals were long-time companions in an outdoor enclosure at the Yerkes Laboratories of Primate Biology in Florida.

QUESTIONS

1. What are receptors, conductors, and effectors? How do they relate to one another and to integration in the organism?
2. In general terms describe the organization and circuitry of the conductor system in animals.
3. In what manner is the synaptic transmission of nerve impulses believed to take place?
4. What are inhibitory fibers and how is inhibition thought to occur?
5. Discuss a current hypothesis of the physiological mechanisms by which learning takes place.
6. How many different kinds of receptors can you name?
7. In what ways are the receptor cells for stretch, hearing, smell, and vision similar? How do they differ?
8. Discuss the kinds of muscular response found in hollow organs and the arrangement of smooth muscle cells which makes such movement possible.
9. Describe how muscles are organized in the musculoskeletal system of man or an insect.
10. In ring doves, seeing a male initiates egg production in a female. Describe the physiological chain of events which lead to egg-laying in this case.
11. What are pheromones? What are some of the ways they are known to regulate social organization and behavior?
12. Distinguish between and illustrate:
    (a) Avoidance behavior, taxis, and tropism
    (b) Dormancy of plants and photoperiodism
    (c) Instinct and motivation
13. Discuss this statement: "Instinct is a totally automatic and unlearned form of behavior."
14. Recall at the end of Chapter 1 that the question was asked, "How does a gull know what it looks like?". Could you now answer this question hypothetically by utilizing the phenomenon of imprinting?
15. In animal conditioning random reinforcement schedules are often very effective. Give a reasonable explanation for this.
16. Although much of insect behavior is instinctive, these animals are quite competent at problem solving. Substantiate this statement.
17. Do you wonder why many consider chimpanzees to be our closest living subhuman relatives?

Suggested Readings

Adler, J. "Chemoreceptors in Bacteria," *Science,* **166,** November 12, 1969.
Agranoff, B. W. "Memory and Protein Synthesis," *Scientific American,* June 1967.
Amoore, J. E. *et al.* "The Stereochemical Theory of Odor," *Scientific American,* February 1964.
Barrington, E. J. W. *The Chemical Basis of Physiological Regulation.* Undergraduate Biology Series. Glenview, Ill.: Scott, Foresman, 1968.
Dethier, V. C., and E. Stellar. *Animal Behavior.* Foundations of Modern Biology Series. Englewood Cliffs, N. J.: Prentice-Hall, 1964.
Eccles, Sir John. "The Synapse," *Scientific American,* January 1965.
Frisch, K. von. *Dancing Bees.* New York: Harcourt, Brace and World, 1965.

Gardner, R. A., and B. T. Gardner. "Teaching Sign Language to a Chimpanzee," *Science,* **165,** August 15, 1969.

Griffin, D. R., and A. Novick. *Animal Structure and Function.* Modern Biology Series. New York: Holt, Rinehart and Winston, 1970.

Hailman, J. P. "How an Instinct Is Learned," *Scientific American,* December 1969.

Hayashi, T. "How Cells Move," *Scientific American,* September 1961.

Hendricks, S. B. "How Light Interacts with Living Matter," *Scientific American,* September 1968.

Hess, E. H. "Imprinting in Birds," *Science,* **146,** November 27, 1964.

Huxley, H. E. "The Mechanism of Muscular Contraction," *Science,* **164,** June 20, 1969.

Katz, B. "How Cells Communicate," *Scientific American,* September 1961.

Kloot, W. G. Van der. *Behavior.* Modern Biology Series. New York: Holt, Rinehart and Winston, 1968.

Lorenz, K. Z. *King Solomon's Ring.* New York: Apollo Editions, 1961.

Marler, P. R., and W. J. Hamilton, III. *Mechanisms of Animal Behavior.* New York: Wiley, 1966.

McGill, T. E. *Readings in Animal Behavior.* New York: Holt, Rinehart and Winston, 1965.

Michael, C. R. "Retinal Processing of Visual Images," *Scientific American,* May 1969.

Miller, W. H. *et al.* "How Cells Receive Stimuli," *Scientific American,* September 1961.

Ray, P. M. *The Living Plant.* Modern Biology Series. New York: Holt, Rinehart and Winston, 1965.

Schaller, G. B. *The Mountain Gorilla: Ecology and Behavior.* Chicago: University of Chicago Press, 1963.

Scott, J. P. *Animal Behavior.* The Natural History Library. New York: Doubleday, 1963.

Smith, D. S. "The Flight Muscles of Insects," *Scientific American,* June 1965.

Tinbergen, N. "The Evolution of Behavior in Gulls," *Scientific American,* December 1960.

Wilson, C. L., and W. E. Loomis, *Botany.* New York: Holt, Rinehart, and Winston, 1967.

Wilson, D. M. "The Flight-Control System of the Locust," *Scientific American,* May 1968.

Wilson, E. O. "Chemical Communication in the Social Insects," *Science,* **149,** September 3, 1965.

CHAPTER 6 Growth and Development

The discussion on behavior has demonstrated the significance of development and change to living systems. We cannot take the organism for granted as if it is what it is because it has never been anything else. A living thing is what it is, just as it behaves like it behaves, by reason of its developmental history. A species is the historical product of millions of years of evolutionary development. A population of organisms is the historical product of many generations of reproducing parents. Each individual, in turn, has a history tracing back to the reproductive products from which it sprang. A progressive series of integrated metabolic events gradually unfolds that we call growth and development. When this unfolding has reached a more or less stable state, we say the individual is fully grown. Our concern so far has been essentially with this fully grown, more or less independent individual. But what of the free-living mosquito larva (Fig. 4-21a)? It is no less independent than the winged adult.

It is worthwhile at this point to broaden our outlook and investigate the developmental aspects of living systems. Life is a process in which change and modification are constant ingredients. Growth and development are not restricted to larval or juvenile organisms. A pruned shrub will grow new shoots and blossoms. A lizard that has lost its tail may grow a new one. We humans continue to grow new skin until we die. Following the surgical removal of most of our liver we can even regenerate a new one in a short time. Life and development go hand in hand. Being alive entails a continuous rebuilding of structures already built. The ever-changing character of the organism's personal history will be the center of our attention in this chapter.

THE NATURE OF GROWTH

Growth refers simply to an increase in mass. In the unicellular protists like *Paramecium* and *Euglena*, this is brought about by the synthesis of additional organelles (endoplasmic reticula, mitochondria, and so on), resulting in a gradual increase in cell size. In plants and animals growth occurs through cell division and the subsequent increase in size of the resulting daughter cells. These, in turn, on reaching a certain optimum mass, divide and a second generation of daughter cells is produced that also increase in size. This may be repeated indefinitely as long as the organism is alive, although the rate of cell division usually declines with age. Many plants grow in this manner so that their size is roughly proportional to their age. Some plants and most animals, on the other hand, reach a certain optimum size which they do not usually exceed. Growth does not cease at this point, however. Throughout the remainder of the lives of such creatures growth is restricted to the replacement of worn-out, damaged, or degenerated parts.

The mechanics of cell division shows a remarkable uniformity among plants and animals as well as many protists. We will direct our attention to some of the details of this important process.

THE MECHANICS OF CELL PROLIFERATION: MITOSIS

We learned in Chapter 3 (Fig. 3-5) that the nucleus of a cell, when viewed with a conventional light microscope, appears to consist in large part of a loose network called the chromatin net. This chromatin material, when photographed under the electron microscope, has actually been shown to be highly attenuated, threadlike chromosomes which are dispersed throughout the nucleus. We also learned (Fig. 3-28) that these chromosomes are in part composed of DNA molecules which embody the hereditary units or genes, which in turn control the metabolic activities of the cell.

Any given species of organism has a constant number of chromosomes in its body cells. This constant number is distributed in paired sets of chromosomes. The members of a pair are usually similar but differ from all other pairs. Man, for example, has a total of 46 chromosomes in each of his body cells.[1] These are distributed as 23 pairs of distinct chromosome sets. The broad bean plant has a total of 12, or 6 different chromosome pairs.

One chromosome in each of the 23 homologous chromosome sets of a human being came from the person's father. The other chromosome member in each pair is like its mate but came from the individual's mother. Each pair therefore consists of a paternal and maternal chromosome. In the same way, the bean plant possesses 6 paternal and 6 maternal chromosomes which it derived from its parents and which are distributed in 6 homologous pairs. We will not concern ourselves at this time with how these parental contributions are apportioned to the offspring. This lies in the province of sexual reproduction and is considered in Chapter 7.

Although the chromosomes cannot be seen in most nondividing cells under the light microscope, they become discernible during the process of cell division. Usually just before division the cell begins to synthesize chromosomal DNA as well as RNA. During this period the DNA content of the nucleus is precisely doubled. The progressive changes which occur after this result in the production of two daughter cells. The process is called *mitosis.* Although the entire sequence is an uninterrupted series of events, for descriptive convenience it is divided into four arbitrary stages or phases called *prophase, metaphase, anaphase,* and *telophase* (Fig. 6-1). The metabolically active, nondividing cell, up to the point at which the nuclear DNA content is doubled, is said to be in *interphase.* For clarity the semidiagrammatic drawings show only one pair of homologous chromosomes, one member of which is black and the other colored. At the side of the drawings are a number of photographs of representative stages of the mitotic process. Those on the left were taken of cells in the growing root tip of a broad bean plant; those on the right were taken of cells in the embryo of a whitefish.

Interphase

Late in interphase, but prior to division, the chromosomes replicate (Fig. 6-1b). In animal cells the two pairs of centrioles begin to separate. As they do so protein fibers appear between them forming what is known as the *spindle.* Fibers also begin to radiate from each centriole pair like the spokes of a wheel, producing a starlike *aster.* Centrioles and asters are not visible in plant cells.

Prophase

The threadlike chromosomes now start to coil into a tight helix (Fig. 6-1c,d). As they thicken and shorten, they are more readily seen. Since each chromosome

[1] There are exceptions to this which need not concern us in this brief account.

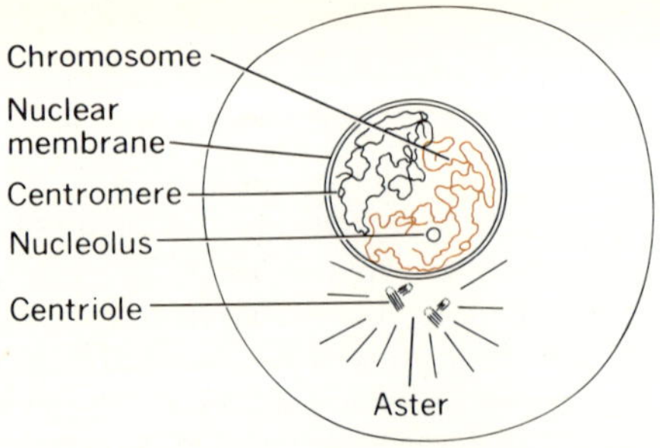

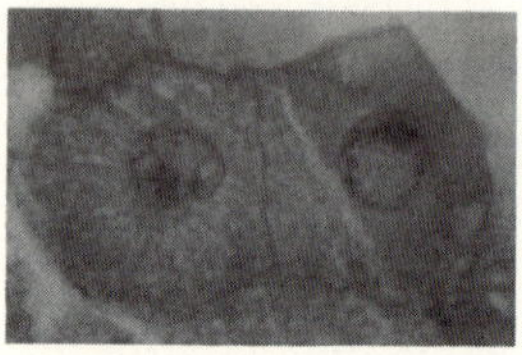

Interphase
(a)

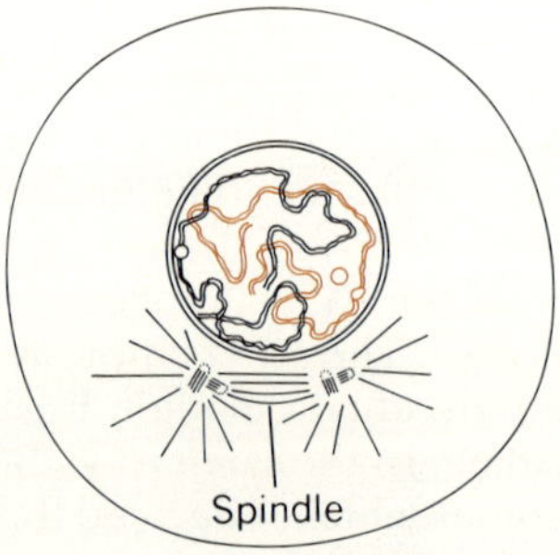

Late interphase
(b)

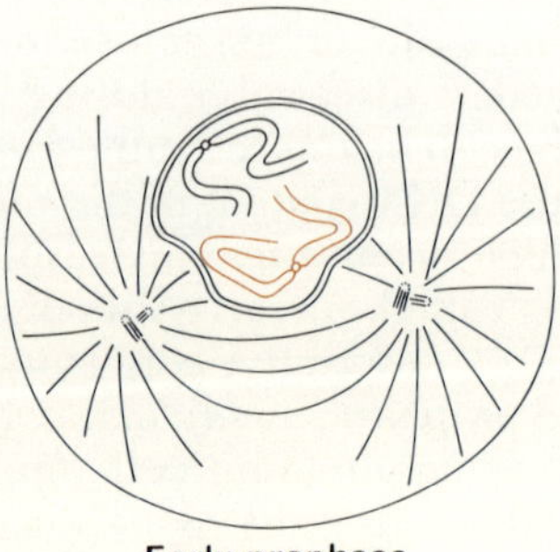

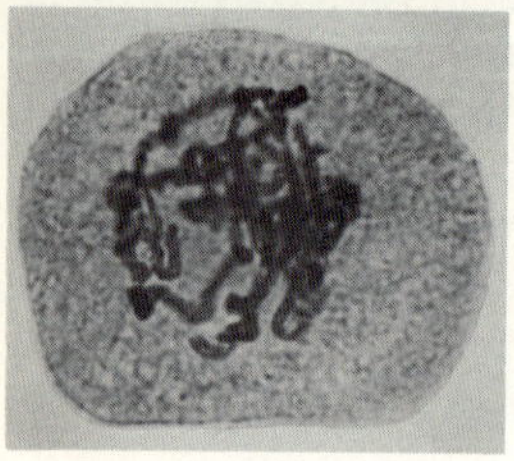

Early prophase
(c)

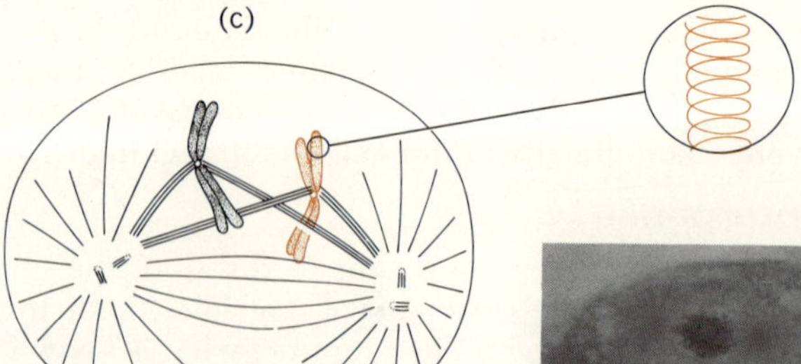

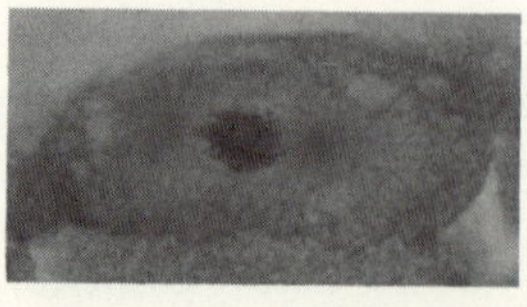

Late prophase
(d)

Figure 6-1 Mitosis. Schematic drawings of an animal cell illustrate the process. Only one pair of chromosomes is shown, one member of which is black, the other colored. (a–b) Middle and late interphase; (c–d) early and late prophase; (e) metaphase; (f) anaphase; and (g–h) early and late telophase. During prophase the chromosomes become condensed through coiling [shown by inset in (d)]. To the left of the drawings, representative photomicrographs of mi-

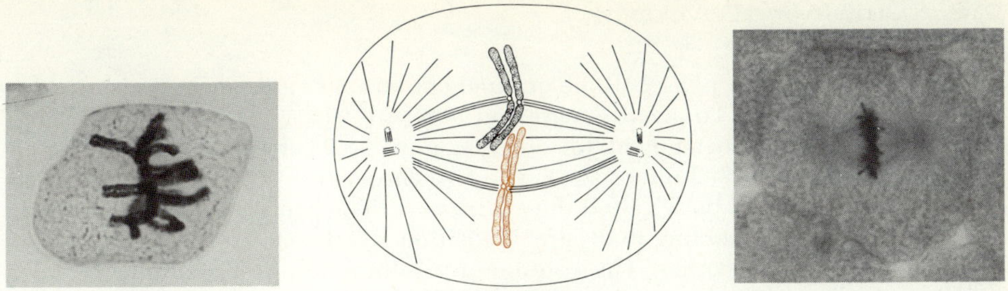

Metaphase
(e)

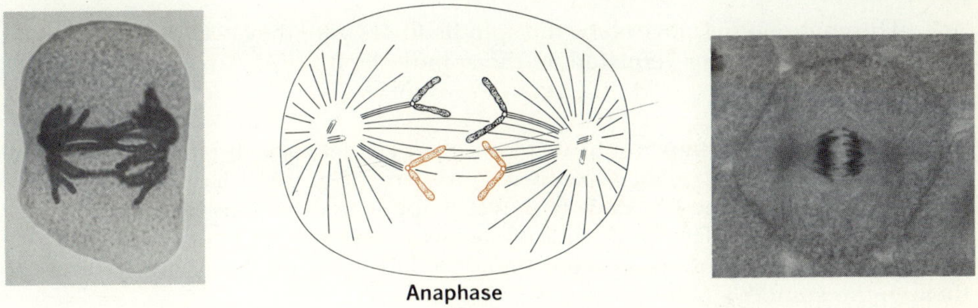

Anaphase
(f)

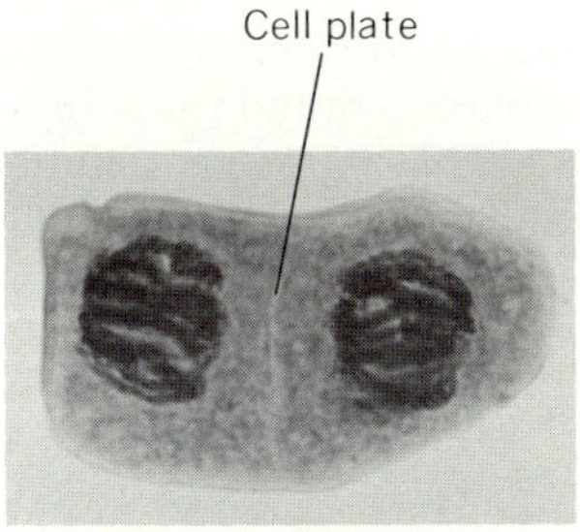

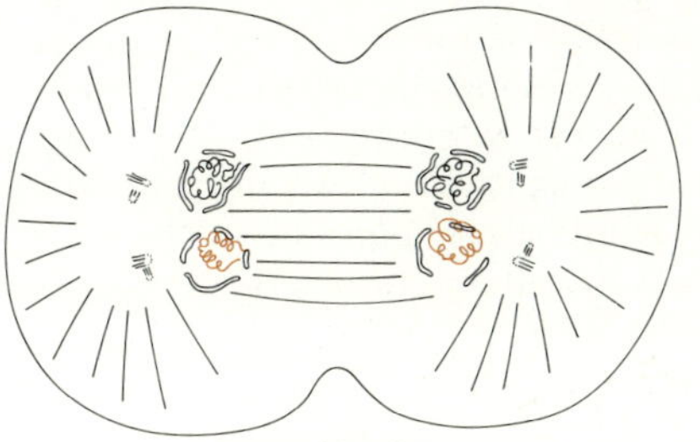

Early telophase
(g)

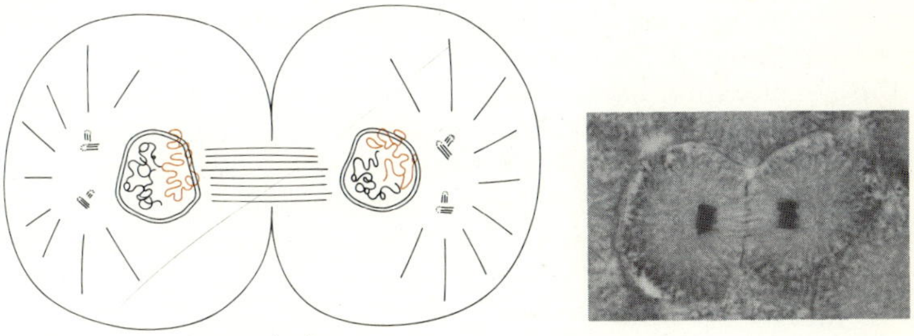

Late telophase
(h)

tosis in the broad bean plant show middle prophase, metaphase, anaphase, and telophase. To the right, representative photomicrographs of mitosis in the whitefish embryo illustrate interphase, late prophase, metaphase, anaphase, and late telophase. (Bean plant by Dr. C. D. K. Cook, University of Zurich; whitefish by author)

replicated itself in interphase, all the original chromsomes are now represented by identical duplicates. To avoid confusion we call these identical duplicates *chromatids;* the broad bean plant, for example, now has 24 chromatids representing two complete sets of 12 chromosomes. The two chromatids in a set are attached to one another by a small body called the *centromere.* The centrioles move apart and the spindle enlarges, coming to occupy a position which extends from one side or pole of the cell to the other. The centromeres joining duplicate chromatids now become attached to the fibers of this spindle. While these events are occurring the nuclear envelope and nucleolus are gradually disappearing.

Metaphase

The centromeres move along the spindle fibers until they occupy a position in the middle periphery or "equator" of the spindle figure (Fig. 6-1e).

Anaphase

The centromeres now replicate themselves so that each chromatid is attached to spindle fibers by its own separate centromere (Fig. 6-1f). Each chromatid or duplicated chromosome proceeds to move in opposite directions toward the poles of the spindle figure. The loose ends of each chromosome are bent backwards in a V-shape as if the centromeres were being pulled upon by the fibers connecting them to the spindle's poles.

Telophase

Each complete set of chromosomes reaches its respective pole and begins to uncoil (Fig. 6-1g,h). The sequence of events at this point is almost the reverse of prophase. Nuclear envelopes and nucleoli gradually reappear. Each centriole produces a new centriole in animal cells and a cytoplasmic constriction develops which will gradually separate the two nuclei and the cytoplasm around them into separate cells. In plants a *cell plate* takes the place of a cytoplasmic constriction. The synthesis of a cell wall on either side of this plate signals the completion of cell division.

THE SIGNIFICANCE OF MITOSIS

Chromosome behavior in the mitotic processes of plants and animals is not only similar but also remarkably uniform. Since chromosomes contain the genes which are the blueprints and directions for cellular activity and development, this is not surprising. Mitosis is a device that provides not only for the exact duplication of this genetic information but also for its precisely equal apportionment between the cells that result from division. Every daughter cell is assured of receiving the same chromosome complement possessed by the cell from which it came. Biological integrity is guaranteed. The multicellular organism controls the nature of additions to its mass when it grows and the character of replacements for its maintenance and repair.

DEVELOPMENT: CHANGE IN FORM

Considerably more is involved in the growth of a living thing than the mere division of identical cells and their subsequent increases in size. A multicellular organism is no mere aggregation of identical cells. It has not simply gotten bigger and bigger. It has *developed.* As long as it is alive, it maintains itself and undergoes certain changes in form. It has passed through a highly ordered series of transformations before it reaches an adult stage. *Development* by no means stops here. The organism must continue to repair itself. Structures whose use is intermittent through the year must be restimulated or be produced anew. The sex organs of seasonally breeding vertebrates must be reactivated. Many plants yield an entirely new "crop" of leaves and blossoms every year. These cyclic phenomena, together with the continuing necessity for the repair and maintenance of organic structure, sooner or later begin to assume formidable proportions. The organism finds itself aging and death finally ensues. Life has no *status quo.* It is a developmental phenomenon whose hallmark is ceaseless change.

In order to gain some insight into these dynamic processes, we will consider two quite diverse organisms. We trust the reader will note that even in two very dissimilar creatures the fundamental components of development are remarkably uniform.

Early Development of a Frog Tadpole

The early embryonic development of a frog tadpole from a fertile egg cell can be divided into three stages: (1) the initial period of cell proliferation (Fig. 6-2a); (2) a period of continued cell division characterized by cell movements which result in organized layers and masses of cells (Fig. 6-2b); and (3) the progressive appearance of organs whose component cells gradually differentiate into distinct cellular types or tissues (Fig. 6-2c). Growth is largely restricted to the last stage.

The egg cell, or *ovum,* contains a large amount of yolk, most of which is located in one-half of the cell called the *vegetal hemisphere.* The yolk nourishes the developing embryo. Although the remaining *animal hemisphere* contains some yolk, the nucleus and cytoplasm comprise most of its contents. During the initial period of development the fertile ovum undergoes a series of mitotic divisions which yield a round ball of many small cells (Fig. 6-2a). The yolk-laden cells in the vegetal hemisphere are relatively large because the yolk presumably impedes the rate at which mitosis occurs. A cavity gradually develops in the animal hemisphere toward the end of this first phase of cellular divisions, and the embryo is then called a *blastula* (Fig. 6-2a; 4,5).

The more rapidly dividing cells of the animal hemisphere now begin to migrate downward over the yolk-laden cells. An inward turning front of cells produces a crescent-shaped depression called the *blastopore,* which marks the approximate location of the future anus (Fig. 6-2b; 1). This depression increases in size with continued cell division and movement, extending forward as a cavity (the *archenteron*) which will become the digestive cavity and its associated structures in the tadpole. The growing archenteron gradually eliminates the earlier cavity (Fig. 6-2b; 2–5). During this period of cell movements beginning with blastopore formation the embryo is referred to as a *gastrula.*

A thickened tongue-shaped plate now gradually forms along the top of the embryo. The lateral edges of this plate grow into folds (Fig. 6-2c; 1,2). The crests of these folds grow toward one another, meeting and fusing in the midline to form a closed tube (Fig. 6-2c; 3). This *neural tube* will form the spinal cord. Its anterior end is expanded to form the brain. An outpocketing in the lateral wall of the brain signals the beginning of a pair of eyes (*optic vesicles*). The blastopore closes and the anus starts to form below it. A deepening depression in the floor of the archenteron will ultimately give rise to the liver. Below this, masses of cells start to hollow out to form the heart. The component cells gradually *differentiate* into cardiac muscle tissue (Fig. 6-2c; 4,5). Other cell masses anterior to the heart and below the brain

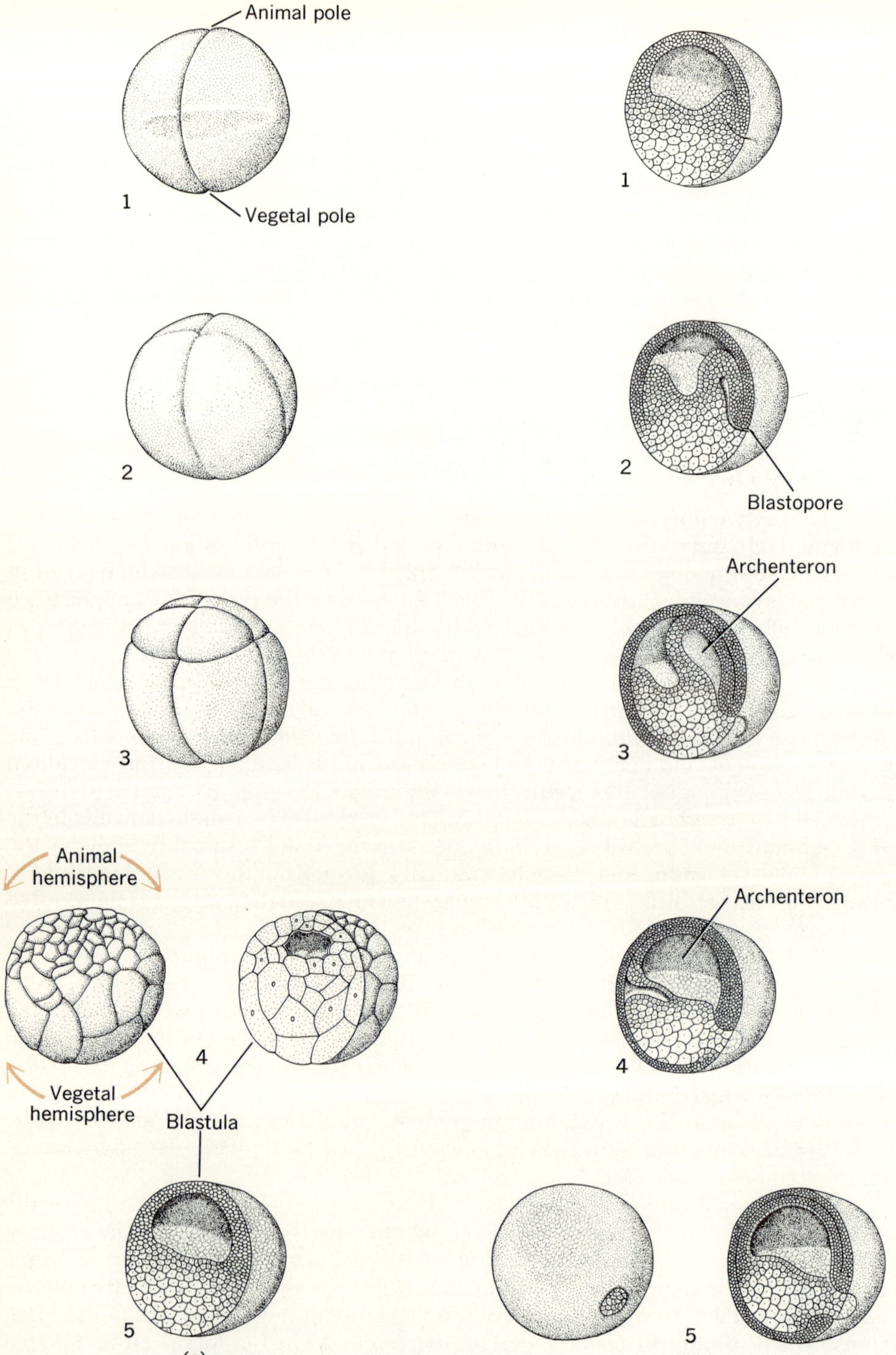

Figure 6-2 Embryonic development of the frog tadpole. (a) From early mitotic divisions of the fertile egg to the blastula stage; (b) from early to late gastrulas; (c) from early organ formation to tissue differentiation (see also Fig. 4-23).

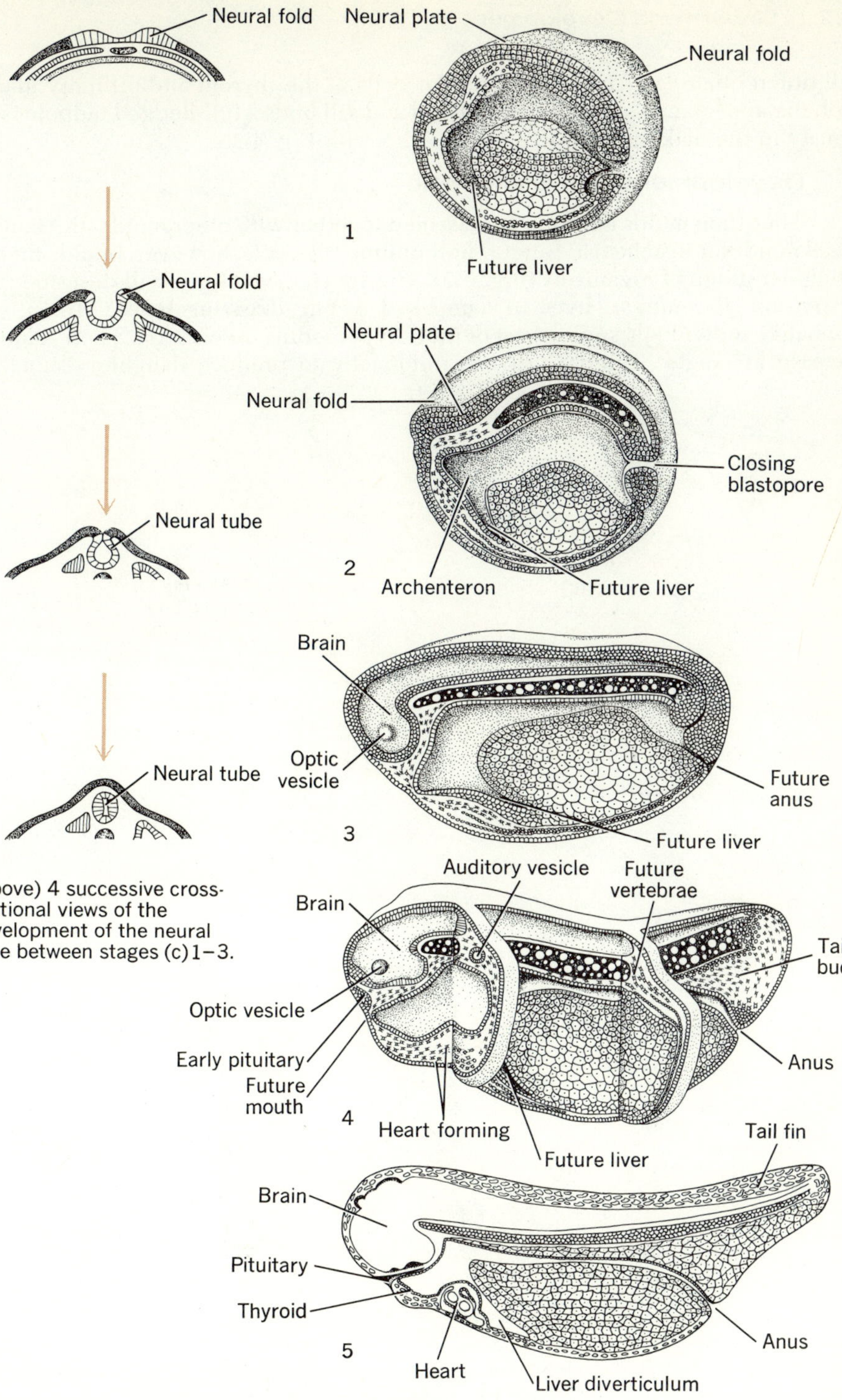

(Above) 4 successive cross-sectional views of the development of the neural tube between stages (c)1–3.

(c)

will differentiate later into the glandular cells of the thyroid and pituitary glands. With the appearance of a growing tail bud and gill buds a full-fledged tadpole is obviously in the making (compare Fig. 6-2c; 5 with Fig. 4-23).

Development of a Slime Mold

The slime molds are usually classified together with other molds (the familiar bread molds for instance) as fungi. Their unique life cycle, however, brands them as a singular group of organisms (Fig. 6-3). The species which we will describe (*Dictyostelium discoideum*) lives in damp soil among decaying leaves. Unicellular amebalike individuals spend most of their lives feeding on bacteria. Their numbers increase as single "amebas" divide mitotically to produce daughter "amebas."

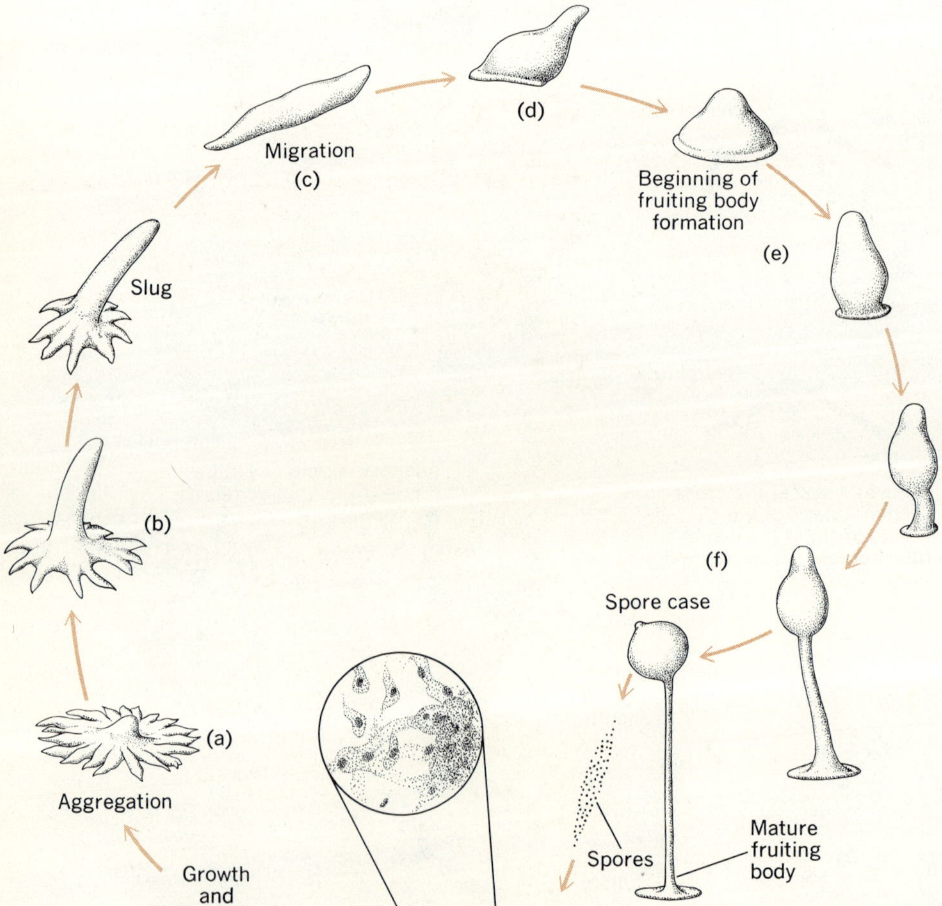

Figure 6-3 Life cycle of a slime mold. Spores become ameboid cells which aggregate into a star-shaped mass (a). A moving sluglike form is produced (b-c) which, after several hours, rounds up (d) and is transformed into a fruiting body (e–f) consisting of a spherical spore case, stalk, and base. Spores produced in the spore case are released and each may germinate to form an ameboid cell.

These daughter cells feed and grow and in turn divide. After several generations of this the total number of amebas may become quite large. At this point an intriguing phenomenon may take place. Amebas start to aggregate in a star-shaped mass as streams of cells from all directions converge on a common gathering site (Fig. 6-3a).[2] The growing multicellular mass of protoplasm (which is a bit slimy and gives the group its name) now begins to move as an integrated unit. The center of the mass rises into an elongated cone as its radiating "roots" shorten (b). The cone-shaped body now turns on its side and continues to migrate about like a minute garden slug (c). These movements last for several hours, after which the slug rounds up (d) and undergoes a dramatic transformation into a type of fruiting body found in many fungi (e–f). Some of the cells make up the base of the structure; others form the stalk. Still others become differentiated into reproductive spores within the spherical spore capsule. The unicellular spores, on release from the spore capsule, may germinate, each one producing an "ameba," and our story ends where it began.

In retrospect we can see in this curious organism, far removed from a frog, a series of three developmental stages which run parallel to those we observed in the tadpole: (1) an initial period of mitotic cell divisions followed by (2) cell movements which result in an organized mass of cells (a, b, and c) and (3) the gradual appearance of structures whose component cells differentiate progressively into distinct cellular types (d, e, and f). With minor variations, progressive changes of this nature are typical of the developmental history of all multicellular organisms as well as a host of unicellular ones.

From our investigation thus far the following question gradually emerges (and has undoubtedly already occurred to the reader). How are these developmental changes brought about? When we recall that mitosis guarantees the faithful transmission of hereditary material from one generation of cells to the next, an even more penetrating question can be asked. How do cells containing the same genetic information in their nuclei become different? Let us investigate these questions, for they strike at the heart of conditions and processes which govern and control development.

THE REGULATION OF CHANGE IN FORM

The Basic Point of Reference

We learned in Chapters 3 and 4 that the cell is both the producer and the product of a great multitude of metabolic reactions which distinguish the living from the nonliving. The genetic information coded chemically in the nucleus embodies the blueprints which specify the synthesis of enzymes. These enzymes mediate metabolic reaction sequences that span the spectrum of anabolic and catabolic events. These same metabolic events are highly sensitive, in turn, to the presence or absence of certain substrates and end products in the cellular environment, as we saw in Chapter 3 (Fig. 3-28). According to the Jacob-Monod hypothesis the chemical milieu has a great deal to do with determining whether the repressor substance of a given regulator gene is in its active or inactive state. Differing patterns of enzyme induction and suppression, dependent on varying metabolic environments, may yield significant deviations in the ultimate destinies of once-similar cells.

There are, of course, many situations in which the physicochemical setting for gene-controlled events may be expected to vary. We will confine ourselves to two

[2] This behavior can be elicited in laboratory populations by allowing the food supply to become depleted.

which have been found to exert far-reaching effects on developmental patterns: (1) interdependencies within the cell between nucleus and cytoplasm and (2) interactions between cells involving time and space.

Interdependencies between Nucleus and Cytoplasm

A CASE STUDY IN SEA URCHIN EGGS

The sea urchins are a group of marine invertebrates related to the starfishes. They commonly inhabit rocky tidepools where they may be found frequently in large numbers. They have proven to be very useful organisms for embryological studies because they are readily adaptable to laboratory life in an aquarium and their sexual products (ova and sperm) lend themselves admirably to experimental manipulation.

One species in particular was selected with great foresight for what has become a classical series of researches. The eggs of this creature possess a band of red pigment granules below the equator (Fig. 6-4). The nucleus may occupy any position in the cell with respect to the band. Since this band identifies a specific portion of the egg, it serves as a cytoplasmic marker which can be followed during the course of development. As can be seen from Figure 6-4 the red marker band can be traced through the early mitotic divisions of the fertile egg through the blastula to the lining of the archenteron of the gastrula and larva. It would appear from this series of observations that the ultimate destiny of some of the cells in the larva at least is determined by the particular region of egg cytoplasm from which they were originally derived.

This tentative conclusion has been tested experimentally in a number of

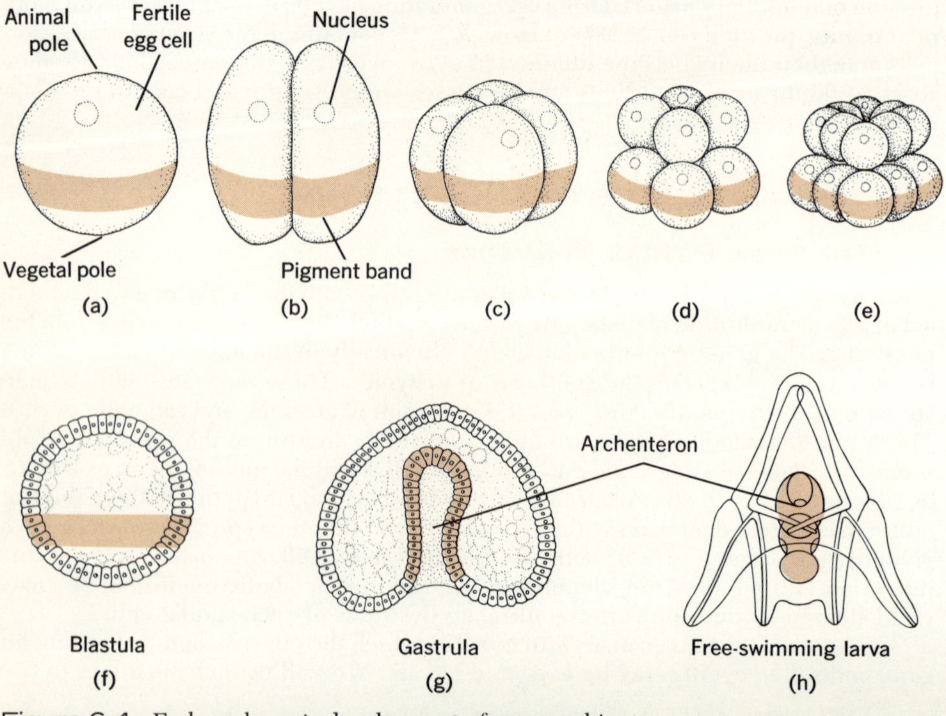

Figure 6-4 Early embryonic development of a sea urchin.

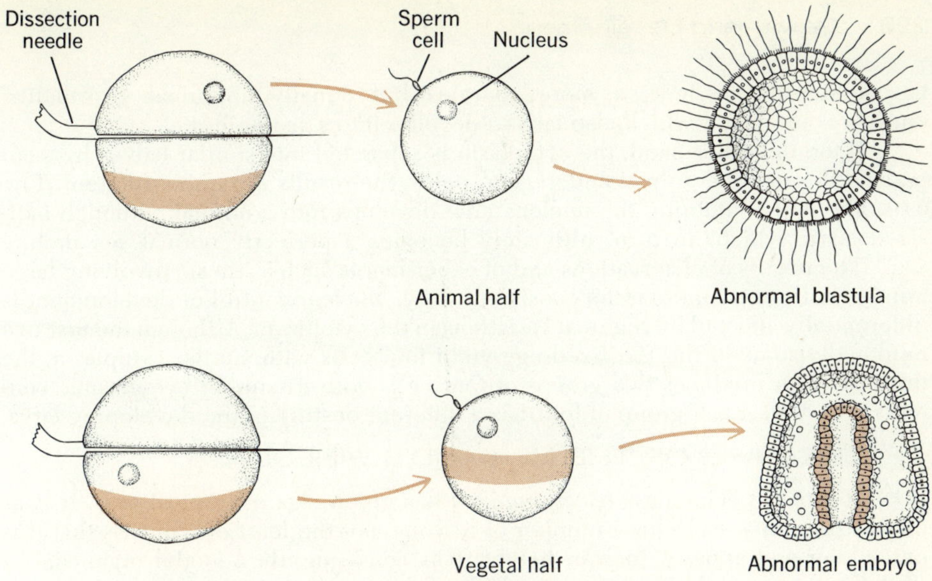

Figure 6-5 When the cytoplasm of a sea urchin ovum is divided into dissimilar halves by a cut separating the animal and vegetal poles, the half containing the nucleus develops abnormally following fertilization (union of sperm and egg cells).

ways. One method was to cut the ovum into halves by a fine glass needle. Since the nucleus may be found in different cells either in the animal or vegetal hemispheres, various kinds of halves may be cut. The half containing the nucleus is then *fertilized* (*fertilization* is the union of egg cell and sperm cell nuclei) and its development followed.

The cytoplasm may be divided into dissimilar halves, for instance, by a cut separating animal and vegetal poles (Fig. 6-5). In this case if the "animal" half happens to contain the nucleus, the resulting embryo is an abnormal, ciliated blastulalike form which swims about and finally dies. If the "vegetal" half chances

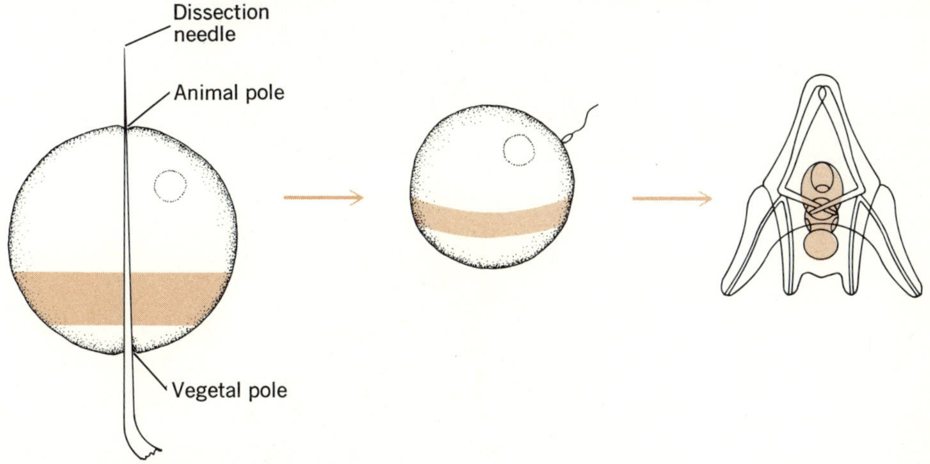

Figure 6-6 When the cytoplasm of a sea urchin ovum is divided into similar halves by a cut through the animal and vegetal poles, the half containing the nucleus develops into a half-sized but normal larva.

to contain the nucleus, a more complex but equally abnormal gastrulalike embryo is produced which also fails to develop into a sea urchin.

If, on the other hand, the cytoplasm is separated into similar halves by a cut passing through the animal and vegetal poles, the results are quite different (Fig. 6-6). The half containing the nucleus now develops into a normal, although half-sized larva, which, in turn, ultimately becomes a perfectly normal sea urchin.

The results of observations and of experiments such as these, involving large numbers of ova, seem to tell a consistent story. Nuclear control of development is differentially affected by regional variations in the cytoplasm. Although the first two mitotic divisions of the fertilized egg yield four cells with similar cytoplasm, the third division produces two groups of four cells with dissimilar cytoplasmic contents (Fig. 6-4). Each group of four has a different destiny in the developing larva.

THE CASE OF THE MERMAID'S WINEGLASS

Mermaid's Wineglass (*Acetabularia*) is a green alga of tropical seas. It is as interesting as its name for a number of reasons, not the least of which is that it is unicellular and about 1 inch in height. It is consequently a model organism for studying the interrelationships of nucleus and cytoplasm.

This seaweed consists of a rootlike base in which the nucleus is located, a long vertical stalk, and a cap in the form of a shallow cup (Fig. 6-7). The cap develops at about the time cell growth ceases. At this point the nucleus is quite large. Up to the time the cap matures to become a reproductive structure, the organism is a single giant cell. When the cap is fully formed, the enlarged nucleus undergoes repeated divisions and many tiny daughter nuclei migrate up the stalk into the cap, converting it finally into a mature fruiting body.

From the observations just made it would appear that there is some sort of interrelationship between the state of development of cap cytoplasm and nuclear behavior. We noted that (1) the nucleus is quite large when the cap is developing,

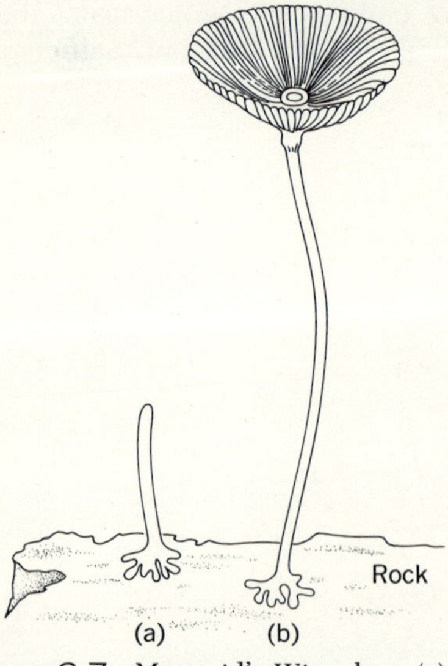

Figure 6-7 Mermaid's Wineglass. (a) An immature, growing cell; (b) a fully developed, one-celled individual, with cap, stalk, and a base containing the nucleus.

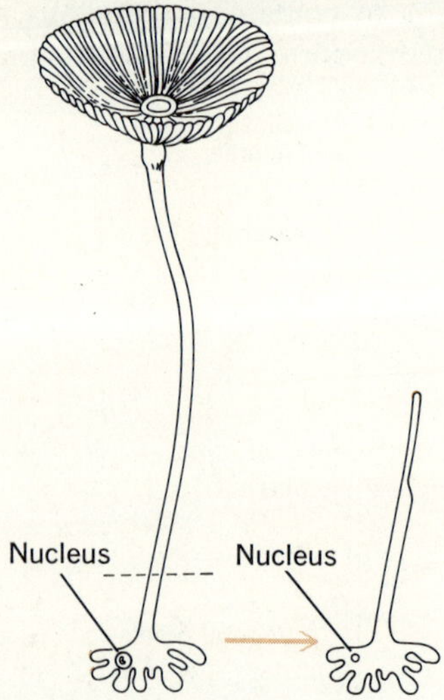

Figure 6-8 When the Wineglass cell is decapitated, the nucleus shrinks quite perceptibly.

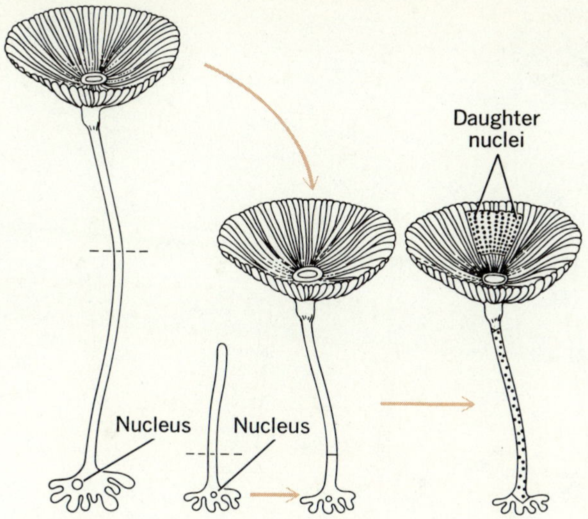

Figure 6-9 If a fully formed cap is transferred to the base of a young cell, the "juvenile" nucleus passes through nuclear divisions characteristic of its mature counterpart.

but (2) the nucleus divides repeatedly after the cap is fully formed. Since this alga is readily capable of regenerating lost parts, it is quite amenable to the surgical removal of portions of the cell as well as the transplanting of these parts to other individuals. It is possible therefore to investigate the influence of cap cytoplasm on the nucleus by experimental surgery. If, for instance, a mature cap is removed from a cell, the nucleus shrinks perceptibly (Fig. 6-8). Evidence suggests that vegetative regrowth at the cut end of such a decapitated cell is brought about by messenger RNA synthesized in the nucleus and transported to the regenerating stalk. If, on the other hand, a fully developed cap is transferred to the base of a young, growing cell, the small "juvenile" nucleus passes through the same nuclear divisions characteristic of its mature, enlarged counterpart (Fig. 6-9). It would indeed seem that cap cytoplasm does influence nuclear behavior in a significant fashion.

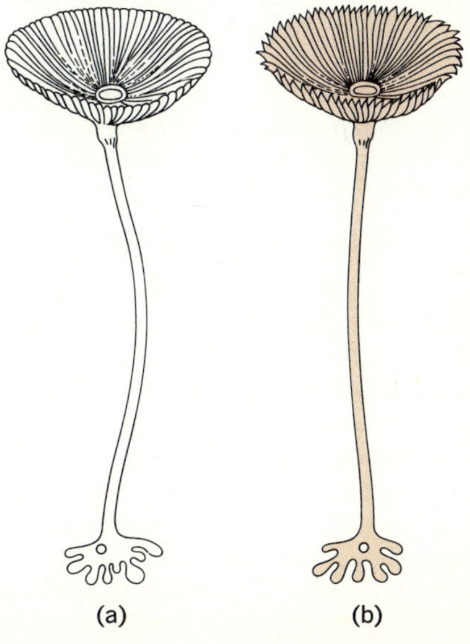

Figure 6-10 Species of *Acetabularia* are distinguished by the structure of their caps. (a) A species with a smooth-rimmed cap; (b) a species with a notched-rim cap.

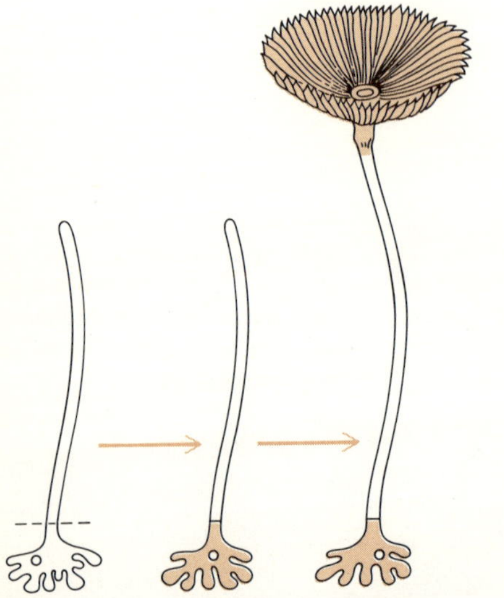

Figure 6-11 Grafting a stalk from a "smooth" species onto the base of a "notched" species ultimately results in the production of a notched cap.

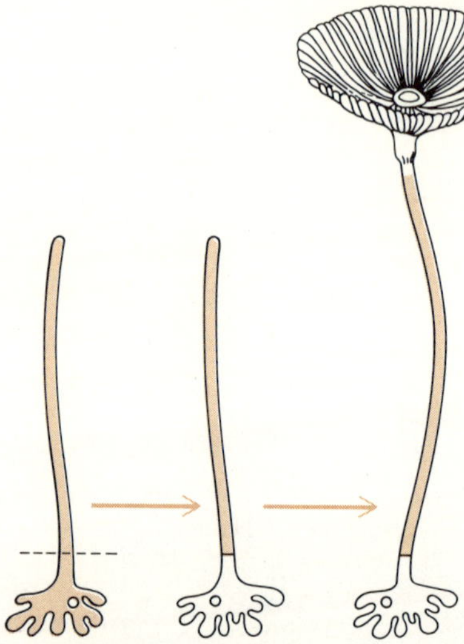

Figure 6-12 Grafting a stalk from a "notched" species onto the base of a "smooth" species ultimately results in the production of a smooth cap.

The reciprocal effect of the nucleus on cap development is also beautifully demonstrated in this organism. Several different species of *Acetabularia* are easily distinguished by the structure of their caps. One of them differs from the species we have been describing by having a notched rather than a smooth rim (Fig. 6-10). Interesting results are obtained when the nuclei of these two species are interchanged. When the growing stalk of a young smooth individual is grafted to the base of a notched organism, the stalk ultimately produces a cap characteristic of the notched species (Fig. 6-11). We use the word "ultimately" here because the regenerating cap is often intermediate between the two species. Sometimes the transplanted stalk apparently contains enough residual mRNA from the original "smooth" nucleus that the resulting cap is the joint effort of two different nuclei. If this hybrid cap is cut off, however, the next cap is typical of the notched species. Comparable results are obtained when the transplantations are reversed by using a base from the smooth species (Fig. 6-12). The nucleus eventually determines the type of cap that will develop.

Interactions Between Cells

We have seen that gene-controlled events within a cell are subject to varying modifications as they are exposed to differing cytoplasmic environments within that cell. We will now observe that gene-governed metabolic processes are also affected by conditions that lie outside the cell and have their origin in the intercellular relationships that come to exist in developing communities of cells. In the late blastula of the frog, for example (Fig. 6-2a;5), some cells are deep within the yolk-laden mass of the vegetal hemisphere. Others line the blastular cavity. Still others lie at the embryo's outer surface. The relative concentrations of O_2 and CO_2, pH, and other critical factors vary in these differing cellular environments. Metabolic activity begins to become diversified according to embryonic regions and locations. Masses of cells start to produce specific metabolic products and by-products which diffuse to neighboring cells. Many of these same products presumably activate or inactivate the repressors of metabolic reactions. Organs begin to develop and tissues differentiate as adjacent cell communities become integrated in a common function. Different but interdependent reaction sequences of this general nature gradually spread throughout the entire embryo as the development of a new individual progressively unfolds.

CELLULAR INTERACTION IN FROG EMBRYOS

It is possible to determine the potential destiny of some part of an early embryo by staining it. The stained tissue can then be identified and followed through development. A certain portion of an early gastrula near the animal pole has been thus treated and is known as the *presumptive eye region.* During the course of normal development presumptive eye tissue in the early gastrula becomes a differentiated eye in a later embryo (Fig. 6-13a). If, however, the presumptive eye tissue is surgically removed from a gastrula and placed in an isotonic salt solution the consequences are quite different (Fig. 6-13b). An irregular mass of undifferentiated cells is produced that finally dies. Apparently presumptive eye tissue has no capacity of its own to differentiate properly.

If we now transplant presumptive eye tissue from several gastrulas to various positions on older embryos, even more interesting results are obtained (Fig. 6-13c). Now the grafted cells are exposed to differentiating structures in the older embryos. If the graft is adjacent to that part of the embryo where a brain, eye, and olfactory organ are developing, the graft faithfully follows suit and gradually differentiates into brain, a pair of eyes, and the beginning of an olfactory organ. Differentiation in the transplant is similarly under the control of the older, host embryo when it adjoins areas that will become auditory vesicles and gills. By placing such grafts in various locations it has been found that presumptive eye tissue can be molded or induced to develop into almost any structure in the body.

A graft may also induce differentiation in a host (Fig. 6-14). Removing a piece of tissue from the upper (dorsal) lip of the blastopore in an early gastrula and transplanting it to a different location in another early gastrula induces the development of a second, partial embryo at the site of the graft. This secondary embryo is composed mostly of host tissues induced to differentiate along directions not normal for their location. This kind of interaction whereby one group of cells seems to bring about differentiation in another group of cells is referred to as *embryonic induction.*

The broad spectrum of potential destinies open to presumptive tissue is soon lost in the course of development. Sometime after the early gastrula stage but prior to the first appearance of neural folds (between (b)2 and (c)1 in Fig. 6-2) the presumptive eye region ceases to be presumptive and becomes *determined.* That is, the developmental potential of this tissue is now fixed; it will differentiate into an eye only. If we transplant this determined tissue from an embryo similar to (c)2 in Figure 6-2 into an older embryo, it produces an eye no matter where placed (Fig. 6-15).

It appears from these experiments that *time* and *place* are exceedingly important factors in patterns of embryonic development. Certain topographic regions in the early gastrula will progressively differentiate according to specific potencies *as long as their natural cellular environment is not disturbed.* After a period of time has elapsed, during which masses of cells undergo extensive movement, these same topographic regions continue to differentiate according to their respective potencies *even though their natural cellular environment is disturbed.* Determined tissues gradually become differentiated and a tadpole begins to take form.

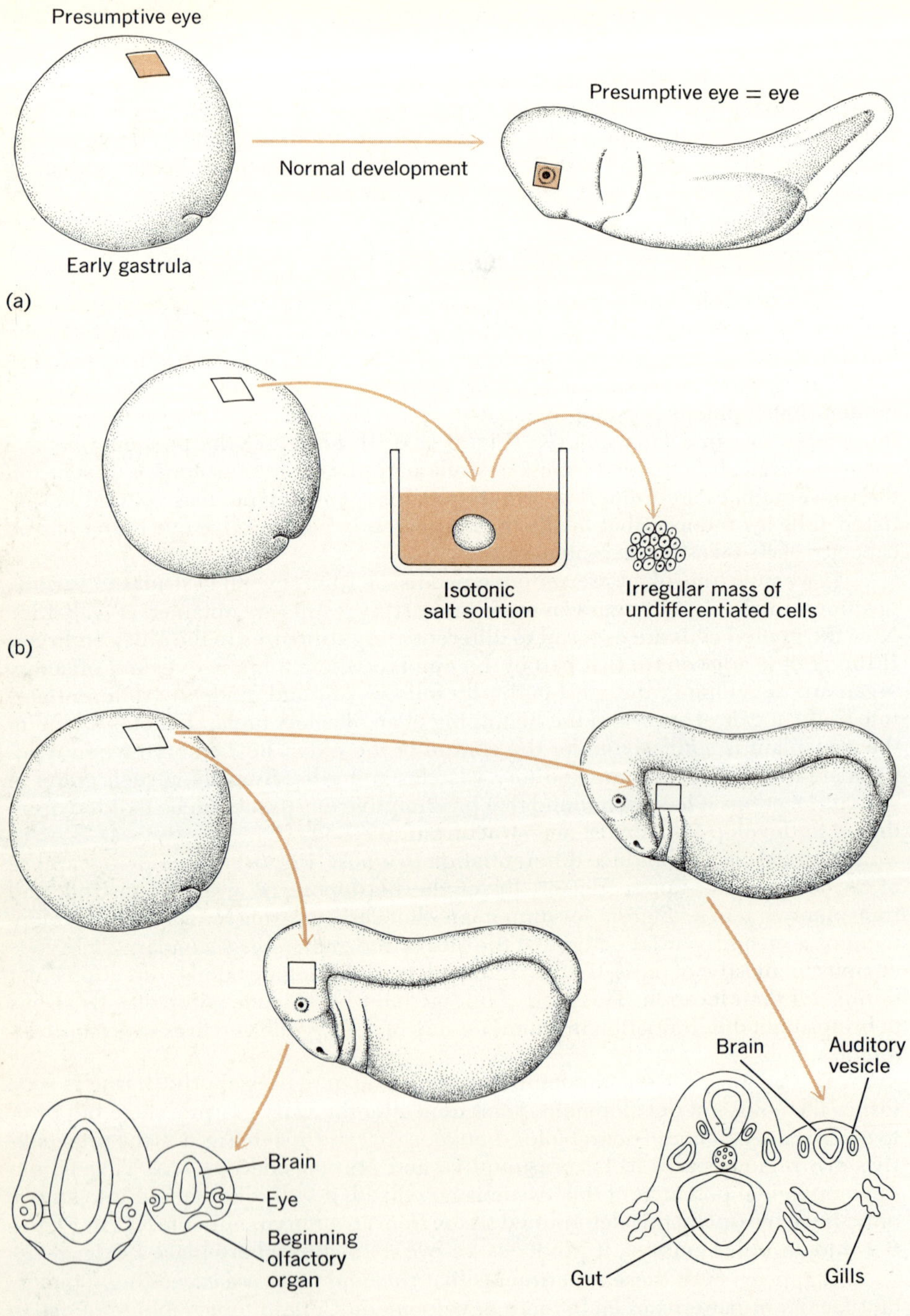

Figure 6-13 A region in the early gastrula of a frog embryo normally develops into an eye (a). If this region is removed from the embryo and kept alive, it fails to differentiate (b). If it is transplanted to an older embryo, it develops into different structures depending on where it is placed (c). (See also Fig. 6-2.)

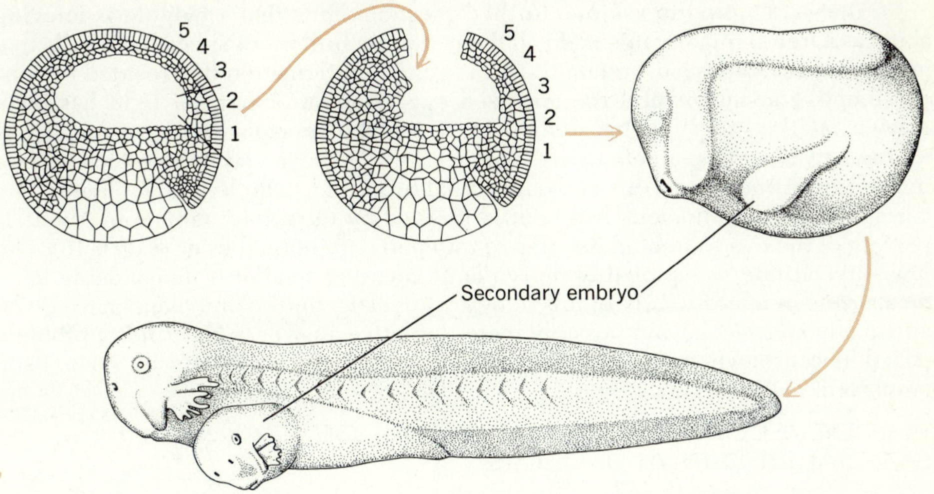

Figure 6-14 A piece of tissue from regions 1 and 2 in an early gastrula, transplanted to another early gastrula, induces the host to form a secondary embryo at the graft site. The opening made in the host embryo heals.

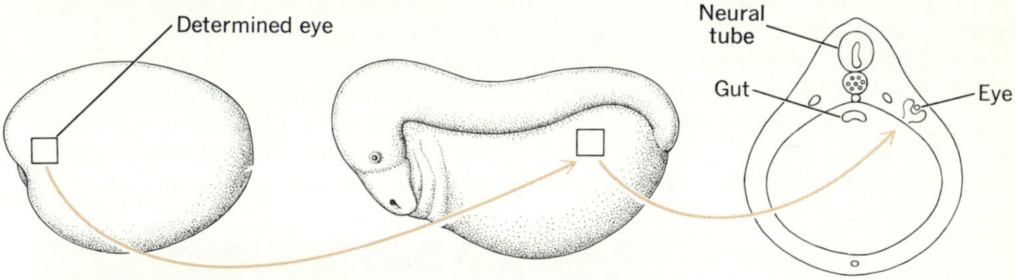

Figure 6-15 The presumptive eye tissue of the early gastrula becomes a determined eye region by the time the neural folds are developing. Transplanting determined eye tissue to a later embryo results in the development of an eye regardless of where it is placed.

Figure 6-16 A South African clawed frog raised from an ovum whose nucleus had been replaced by one taken from an intestinal lining cell of a tadpole. (J. B. Burdon)

One is tempted to ask one further question. Now that a tadpole is foraging about as a free-living organism, do all the countless billions of specialized cells that make up this creature still retain the same genetic information that resided in their presumptive ancestors and the fertilized egg? A great amount of light has been thrown on this question by studies involving the transplantation of a nucleus into an egg cell whose nucleus has been destroyed or removed. In one such study a nucleus from one of the cells of the intestinal lining of a tadpole was implanted into a frog's egg whose nucleus had been destroyed by ultraviolet radiation. About $1\frac{1}{2}$ percent of the ova so treated developed into perfectly normal frogs (Fig. 6-16). The low survival rate was probably due in large measure to almost unavoidable technical errors in microsurgery of this delicacy. Investigations using plants have given an equally spectacular but possibly more definitive answer to the kind of problem posed because whole intact cells from completely mature tissues have been employed.

THE DEVELOPMENTAL POTENTIAL OF SPECIALIZED PLANT CELLS

Pieces of specialized tissues may be surgically removed from leaves, stems, or roots and kept alive in sterile nutrient media containing inorganic salts, sugars, and possibly other additives. Isolated cells loosened from this tissue and thus released from the metabolic restrictions of their normal environment may develop into fully formed mature plants. This has been successfully accomplished using tissues from ferns and lilac, both wild and cultivated carrot and tobacco plants. In one such study of the tobacco plant pieces of tissue were removed from the pith of a fresh stem and kept alive in a nutrient (culture) medium. Single cells were then dissociated from these tissue pieces and isolated from one another in separate cultures. Some of these cells gave rise, through repeated divisions, to a growing mass of cells (Fig.

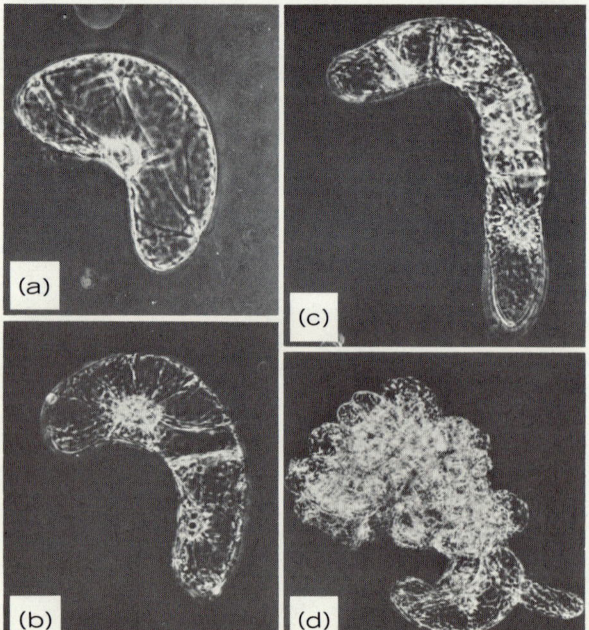

Figure 6-17 Development of a mass of cells from a single cell isolated from the fresh-stem pith of a tobacco plant. (a) A single cell (×750); (b) two-cell stage (×600); (c) a filament of five cells (×470); (d) mass of cells (×375). (Vilma Vasil and A. C. Hildebrandt, University of Wisconsin.

Figure 6-18 Differentiation and plantlet formation from cultured cell colonies of tobacco. (a) Undifferentiated mass of cells ($\times\frac{3}{2}$); (b) Differentiation of shoots with leaves ($\times\frac{2}{3}$); (c) Plantlets with roots and shoots ($\times\frac{2}{3}$). (Vilma Vasil and A. C. Hildebrandt, University of Wisconsin)

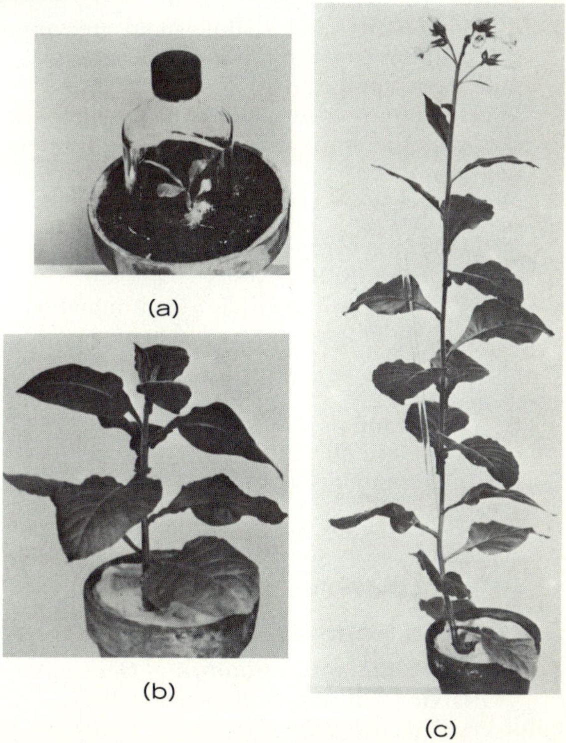

Figure 6-19 Subsequent development of tobacco plantlets in Figure 6-18 after transfer to soil in pots. (a) $\times\frac{1}{3}$; (b) $\times\frac{1}{3}$; (c) $\times\frac{1}{6}$. (Vilma Vasil and A. C. Hildebrandt, University of Wisconsin)

6-17). When such a colony of cells was transferred to another (solid) medium, it differentiated roots and shoots with green leaves (Fig. 6-18). These rooted shoots, in turn, when planted in soil became mature tobacco plants with normal roots, stems, leaves, and flowers (Fig. 6-19).

The answer to our previous question is in the affirmative. The differentiated, specialized cells of many adult organisms do retain the full complement of their hereditary endowment. They are the exclusive beneficiaries of a well-guarded heritage handed down through a cellular pedigree of countless mitotic generations dating back to the fertilized egg.

The fertilized ovum through its master plan of chromosome-bound volumes in DNA code develops and grows into a self-directed organism according to specifications. This new creature is but a further extension of ceaseless activity and modification. Its never-ending conquest of energy entails continuous change. The highly ordered state of life demands constant biosynthesis in order to guarantee the necessary flow of free energy in the system. No organism maintains this living state forever. Maturity turns to senility which is brought to an end by death. The dynamic flow of life is by no means impeded by this event, but rather served. Long before the advent of death individual organisms prepare for the replacement of the metabolic system they cannot sustain indefinitely. They provide for their continuity through reproduction. New individuals gradually come into being as the dynamic and endless cycling of life completes one turn.

QUESTIONS

1. Compare growth in unicellular and multicellular organisms.
2. What is the significance of mitosis?
3. Although a frog and a slime mold are very dissimilar organisms, their basic stages of development follow remarkably similar patterns. Describe and illustrate these similarities.
4. In general terms describe how it is hypothetically possible for cells with indentical genetic information to become different during the course of embryonic development.
5. Describe a study or experiment in which:

   (a) Differing cytoplasmic environments within a cell modified the course of gene-controlled events

   (b) Differing intercellular environments within an embryo modified the course of gene-controlled events

6. Distinguish between the following:

   (a) Presumptive tissue

   (b) Determined tissue

   (b) Differentiated tissue

7. Cite evidence for the belief that the diverse and specialized cells of a mature multicellular organism possess the same hereditary material as that in the fertilized egg from which the organism developed.

Suggested Readings

Alston, R. E. *Cellular Continuity and Development.* Undergraduate Biology Series. Glenview, Ill.: Scott, Foresman, 1967.

Balinsky, B. I. *An Introduction to Embryology.* Philadelphia: Saunders, 1965.

Barth, L. J. *Development: Selected Topics.* Principles of Biology Series. Reading, Mass.: Addison-Wesley, 1964.

Bodemer, C. W. *Modern Embryology.* New York: Holt, Rinehart and Winston, 1968.

Bonner, J. T. "Hormones in Social Amoeba and Mammals," *Scientific American,* June 1969.

Bonner, J. T. *Morphogenesis: An Essay on Development.* New York: Atheneum, 1963.

Ebert, J. D. and I. M. Sussex. *Interacting Systems in Development.* Modern Biology Series. New York: Holt, Rinehart and Winston, 1970.

Gibor, A. "Acetabularia: A Useful Giant Cell," *Scientific American,* November 1966.

Gilchrist, F. G. *A Survey of Embryology.* New York: McGraw-Hill, 1968.

Gurdon, J. B. "Transplanted Nuclei and Cell Differentiation," *Scientific American,* December 1968.

Jacobson, A. G. "Inductive Processes in Embryonic Development," *Science,* **152,** April 1, 1966.

Konigsberg, I. R. "The Embryological Origin of Muscle," *Scientific American,* August 1964.

Mazia, D. "How Cells Divide," *Scientific American,* September 1961.

Moore, J. A. *Heredity and Development.* New York: Oxford University Press, 1963.

Overbeek, J. van. "The Control of Plant Growth," *Scientific American,* July 1968.

Saunders, J. W., Jr. *Animal Morphogenesis.* Current Concepts in Biology. New York: Macmillan, 1968.

Sussman, M. *Animal Growth and Development.* Englewood Cliffs, N.J.: Prentice-Hall, 1964.

Vasil, V., and A. C. Hildebrandt. "Differentiation of Tobacco Plants from Single, Isolated Cells in Microcultures," *Science,* **150,** November 12, 1965.

Wessells, N. K., and W. J. Rutter, "Phases in Cell Differentiation," *Scientific American,* March 1969.

CHAPTER 7 Continuity and Diversity: reproduction

We observed that the growth and the development of organisms are under the direction of hereditary material passed on to them from their parents. Our concern has been with the nature of that development and the manner in which it takes place. The "personal" history of individual organisms is quite germane to biology because after all it is only through the energy mobilized by fully operative living beings that life can be sustained at all. Equally essential to the phenomenon of life, however, is the necessity for its perpetuation. No living creature can endure forever as a metabolic "booster pump" maintaining the flow of life. Genetic information must not only contain the requisite directions for producing an organism, it must also provide for its successful transfer so that there will be succeeding generations of "brand new booster pumps." The continuity of life can be assured only through reproduction as DNA directions are recycled through one organism and on to the next. We learned in Chapter 3 (page 57) that this can be done in two ways—asexually and sexually.

ASEXUAL REPRODUCTION

Asexual reproduction is accomplished without the aid of specialized sex cells. All parental chromosomes are replicated and apportioned equally among the progeny so that offspring are identical to the parent from which they were derived.

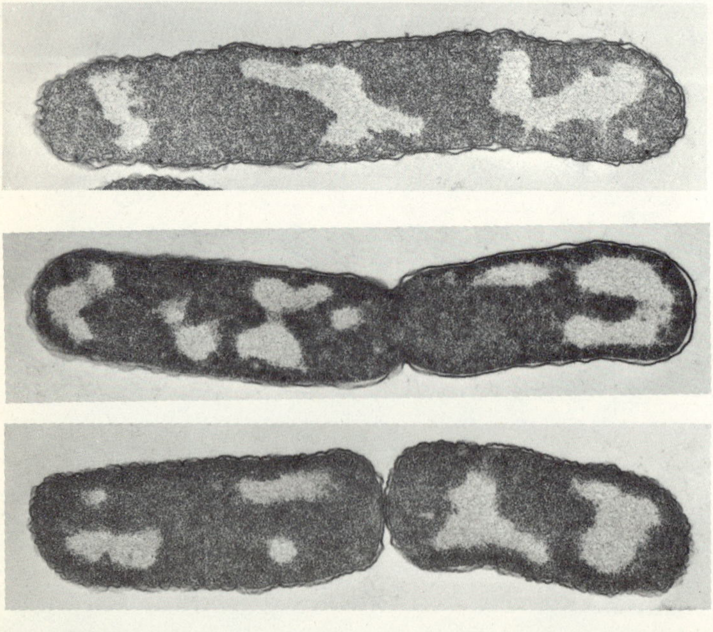

Figure 7-1 Fission in a bacterium (the colon bacillus, *Escherichia coli*). The lighter portions of the bacteria are the nuclear areas. (×20,800) (Oak Ridge National Laboratory)

This method of reproduction is used extensively among monerans, protists, plants, and the "lower" animals and assumes a variety of forms.

Among unicellular species (bacteria, *Ameba*, *Paramecium*, and so on) the organism may divide into two approximately equal halves through fission (Fig. 7-1). In bacteria, the single chromosome is simply replicated. In protists, the chromosomes are assorted by mitosis. Simple fission is commonly compounded in the protists, resulting in a whole series of divisions and the elaboration of many cells (or nuclei) from one original cell or nucleus (Fig. 6-9).

Among animals (for example, certain worms) a type of asexual reproduction occurs which bears a superficial resemblance to fission. The worm begins to constrict in its midsection as if it were tightening an imaginary belt. Two worms are subsequently formed approximately one-half the length of the original. Each new worm regenerates the parts it is missing and grows to the approximate size of the original, undivided worm, at which time the process of fragmentation may be repeated.

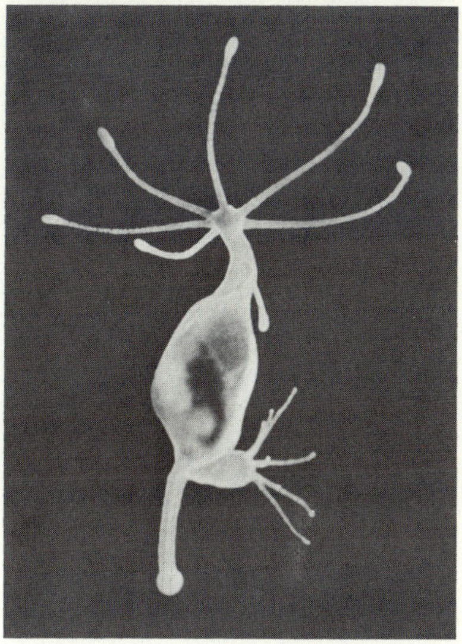

Figure 7-2 Reproduction by budding in *Hydra*. The buds will later separate from the parent organism. (See Fig. 5-5.) (Laurence Perkins, Photo Researchers)

Many of the "lower" animals resemble plants in their manner of asexual propagation. *Hydra* (Fig. 5-5) may reproduce by the formation of buds along its column, as shown in Figure 7-2. These buds later separate from the parent and lead separate lives. In other organisms related to *Hydra*, however, the buds do not separate. The end result of continual budding is a branched, shrublike colony of organisms.

Comparable vegetative methods of propagation are seen in the sprouting potato (Fig. 4-1) or the "suckers" that spring from the roots of the banana plant (Fig. 7-3). The wild strawberry plant may multiply asexually by producing new plants from its horizontal stems (Fig. 7-4). A great many plants can even propagate themselves from leaves (Fig. 7-5). The plant kingdom has exploited asexual reproduction most extensively. New plants produced this way grow quickly because they derive much of their sustenance during the early stages of growth from the established parent organism.

Figure 7-3 The banana plant propagates itself by "suckers" which spring from the roots.

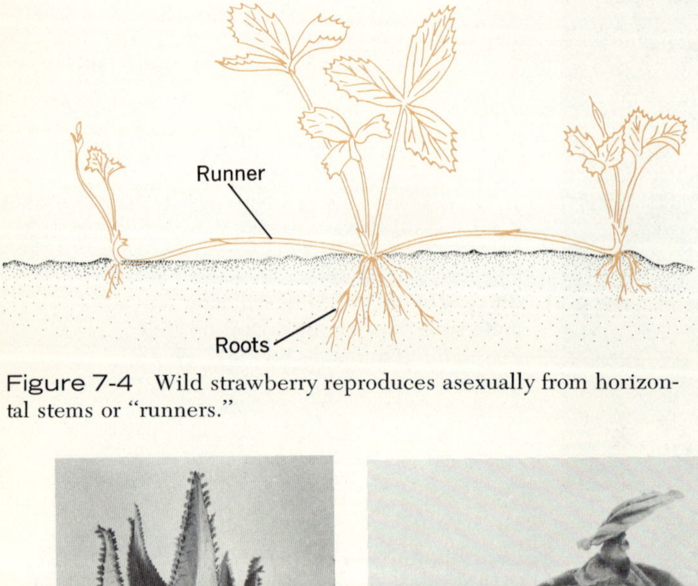

Figure 7-4 Wild strawberry reproduces asexually from horizontal stems or "runners."

Figure 7-5 (a) The ornamental air plant buds off tiny plantlets at the margins of the leaf. (b) If the stem of an African violet leaf is planted in soil, small plants may be produced at the leaf base. (T. H. Everett)

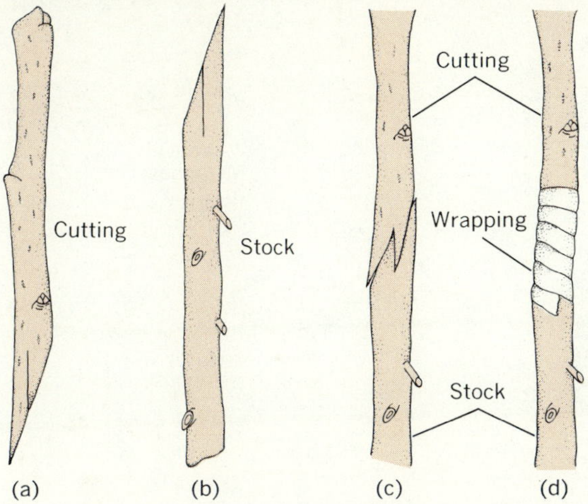

Figure 7-6 Graft used to propagate apple trees. (a) Cutting from the stem of the desired variety. (b) Section of stock with roots pruned. (c) Clefts in cutting and stock forced together. (d) Completed graft with wrapping.

The facility with which plants propagate vegetatively has made it possible for man to graft a cutting from one woody plant (tree, shrub, or vine) onto another plant of the same or related species. One type of graft is illustrated in Figure 7-6. Most of the shoot of a young tree is removed and a cutting from another tree is then inserted into the cleft of the cut stump (the stock) and securely wrapped. The shoot that develops will be typical of the tree from which the cutting was taken. This practice, which has been in use for centuries, has many applications. The roots of many desirable fruit-bearing varieties of plants are susceptible to certain diseases. Crop production is greatly increased by grafting cuttings from these strains onto stocks whose roots are disease resistant. A number of different kinds of tree fruits (for example, the navel orange and McIntosh apple) have been grown on grafted trees whose cuttings date back in some cases a hundred years ago or more to a single tree. These particular fruit trees could not be grown from the seed of the fruit because the seed does not "breed true." Seeds are produced by sexual reproduction in which two *different* parents participated.

SEXUAL REPRODUCTION

Sexual reproduction is very commonly employed by practically all organisms from bacteria and protists to plants and animals. Although blue-green algae have never been observed to reproduce sexually, very few other species fail to exhibit sexuality in some form at some time in their lives.

Sexual reproduction is distinguished in particular by the fact that it entails the *reorganization of nuclear material.* Two nuclei fuse in the act of fertilization (page 225). Each of these nuclei is usually carried in a separate sex cell or *gamete* (in animals either a sperm or ovum). The result of this union is a fertilized egg (or *zygote*). As noted in Chapter 6 this fertilized egg or zygote, *with its full complement of chromosomes,* develops into a new individual. A moment's reflection at this point is in order. If the chromosome number of a sexually reproducing organism is to be kept constant, some mechanism is necessary to reduce the number of chromosomes in gametes by one-half. Otherwise every time fertilization took place the chromosome number would be doubled. This would be a paradoxical situation in light of the fact that mitosis maintains the constancy of the chromosome complement so faithfully throughout the entire course of development.

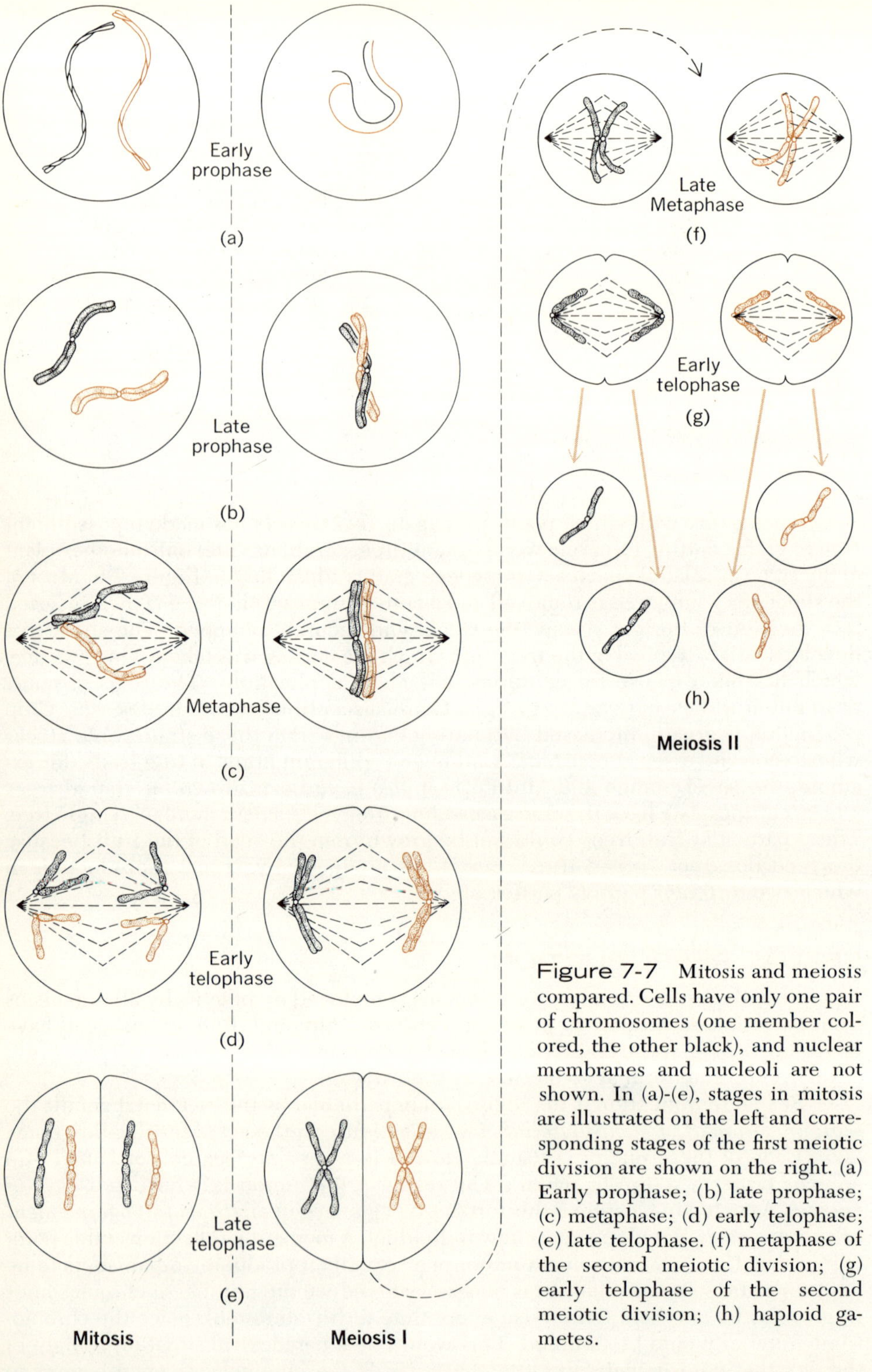

Figure 7-7 Mitosis and meiosis compared. Cells have only one pair of chromosomes (one member colored, the other black), and nuclear membranes and nucleoli are not shown. In (a)-(e), stages in mitosis are illustrated on the left and corresponding stages of the first meiotic division are shown on the right. (a) Early prophase; (b) late prophase; (c) metaphase; (d) early telophase; (e) late telophase. (f) metaphase of the second meiotic division; (g) early telophase of the second meiotic division; (h) haploid gametes.

Mature sex cells do, in fact, contain exactly half the number of chromosomes found in the zygote. A gamete possesses one each of all the different kinds of chromosomes that the fertilized egg harbors in pairs. The human zygote has 46 chromosomes distributed in 23 different homologous pairs (page 215). A human sperm (or ovum) has 23 different chromosomes. For convenience the paired number characteristic of a fertilized egg is said to be the *diploid* number (usually abbreviated $2n$). The unpaired number found in gametes is called *haploid* (abbreviated n). In man $2n$ (diploid) is 46; n (haploid) is 23.

The process by which the chromosome number in gametes is reduced is called *meiosis*. Although this process bears many similarities to mitosis, it is also distinctly different in some respects, as is evident by studying Figure 7-7. In the formation (maturation) of sex cells two consecutive cell divisions occur. Each of these two meiotic divisions can be separated into phases comparable to those in mitosis. During prophase of the first meiotic division homologous chromosomes become intimately attached to one another along their length in a process known as *synapsis* (a). (For simplicity only one pair of chromosomes is illustrated: $2n = 2$.) Each chromosome then duplicates itself, forming a tightly knit group of four chromatids, called a *tetrad* (b). Each chromosome now consists of two identical chromatids joined by a common centromere. The chromosome pair is correspondingly represented by a tetrad (four attached chromatids). It is in tetrad formation (and its consequences) that meiosis and mitosis differ most sharply.

In metaphase the single centromere connecting two chromatids attaches to a spindle fiber (c). Each centromere then moves, during anaphase, toward opposite poles carrying its doublet of chromatids with it (d). Telophase consequently produces two cells, each of which contains a single chromosome (consisting of two identical chromatids as shown in e). One of these chromatid doublets represents the original paternal member of the chromosome pair (color), the other the original maternal member (black).

The second meiotic division is similar to mitosis. The centromere moves onto the equatorial "plate," becomes attached to a spindle fiber, and replicates itself (f). Each chromatid now goes to opposite poles of the cell as if pulled by its centromere (g). The second meiotic division is completed (h), resulting in four sex cells, each of which contains the haploid number of chromosomes (two with one maternal chromosome and two with one paternal chromosome).

Not only has the number of chromosomes been reduced by one-half but the nuclear material has been reorganized as well. This becomes particularly evident if the four gametes (h) are allowed to pair off at random and fuse with one another (a process known as *self-fertilization* which is possible in some organisms). Three dissimilar zygotes could be formed, two of which would differ from the original diploid "mother cell" which gave rise to the gametes. Identifying them by their chromosome content they are (1) maternal-maternal (black-black), (2) maternal-paternal (black-colored), and (3) paternal-paternal (colored-colored).

Although this description of meiosis has been purposely oversimplified to avoid confusion, it is adequate for our immediate needs. We can now examine sexual reproduction in several different representative organisms. As the discussion develops we will be in a stronger position to probe deeper into meiosis and consider more fully the far-reaching consequences of sexuality for biological phenomena.

SEXUAL REPRODUCTION IN UNICELLULAR ORGANISMS

Conjugation: Sexual Reproduction without Gametes

Species of *Paramecium* and many other ciliates can reproduce either asexually by simple fission or sexually by a process known as *conjugation*. Isolated populations of one species of *Paramecium* may be reared in the laboratory through innumerable generations in which fission is the sole method of reproduction. If individuals from a different population of this species are introduced into the culture, however, the paramecia begin to reproduce by conjugation. An individual of one strain becomes attached to an individual of the other strain (Fig. 7-8). One species of *Paramecium* has at least eight such mating types or "sexes." In no case will paramecia of the same mating type conjugate, but any combination of two different strains will. Cases of this sort involving multiple sexuality necessitate a revision in the traditional notion that sexual reproduction is restricted to two kinds of mating individuals.

Conjugation is further distinguished by the fact that gametes are not employed in the process. Fertilization results from an exchange of nuclei between two conjugating individuals. Figure 7-8 depicts this process in a paramecium (*P. caudatum*) in which each individual has two nuclei. The small diploid *micronucleus* is directly engaged in conjugation, whereas the large *macronucleus* is not. As two individuals pair off, the macronucleus in each cell begins to dissipate while the micronucleus undergoes two meiotic divisions (a, b, and c). Three of the four haploid micronuclei in each organism then degenerate (d). The remaining micronucleus in each conjugant then divides mitotically (e). One of the two haploid micronuclei in each cell is now exchanged between the conjugants (f) and fertiliza-

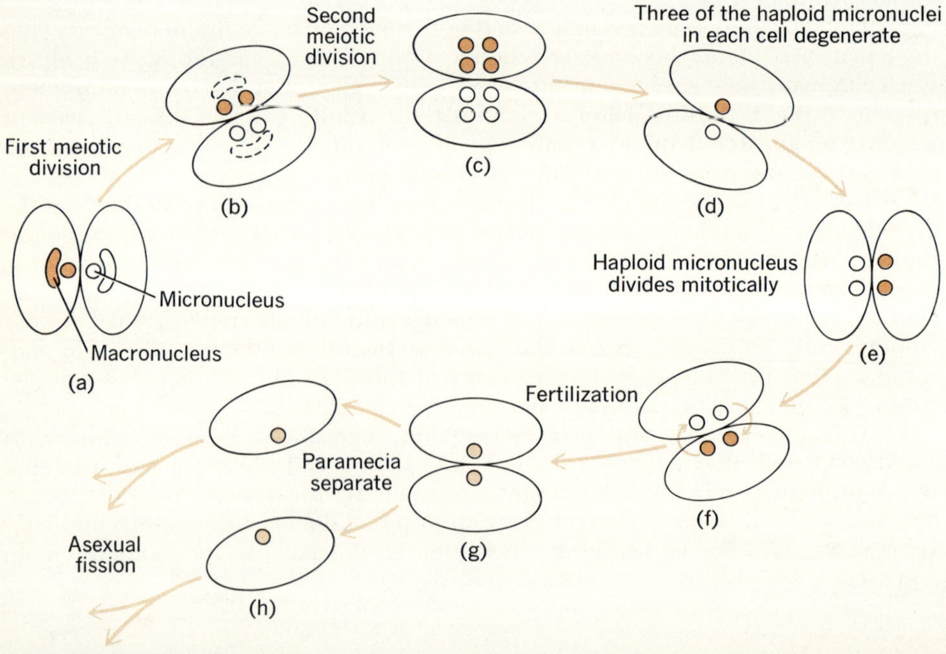

Figure 7-8 Conjugation between two paramecia of different mating strains (*P. caudatum*).

tion occurs as the migrating micronucleus from one cell fuses with the resident micronucleus of the other (g).

The zygote micronucleus resulting from this union contains the diploid number of chromosomes in which a haploid chromosome complement was contributed by each of the two conjugating organisms. The paramecia now separate (h) and each individual undergoes a series of mitotic nuclear and cellular divisions in which the macronucleus is reconstituted and four new paramecia are produced. The details of this period of development and proliferation are not relevant to us here.

An interesting type of self-fertilization may occur within a mating strain after a number of successive generations has been produced asexually. *Single* paramecia will pass through stages which are essentially comparable to conjugation (a–e). At the point where two haploid micronuclei have been formed (e) they fuse. Consequently, a new zygote micronucleus is formed that is the result of nuclear reorganization (meiosis) and fusion *within one individual.*

Sexual Reproduction Involving Gametes

Many of the unicellular green algae also reproduce either asexually or sexually. Sexual reproduction, however, involves the fusion of sex cells. The flagellated alga *Chlamydomonas* is a typical example of one of these organisms. This alga is commonly found in fresh water ditches and ponds and is distinguished by two flagella of equal length, a cup-shaped chloroplast that occupies most of the cell, and a cellulose cell wall (Fig. 7-9a). Its nucleus differs from that in a one-celled protist like *Paramecium* in that it contains the haploid number rather than the diploid number of chromosomes.

Sexual reproduction in *Chlamydomonas* is thought to be induced by the depletion of nitrogenous substances in the surrounding water. In any event it does not occur unless individuals of opposite mating types come in contact with one another. These types are usually designated by the symbols + and −, as indicated in Figure 7-9. A haploid vegetative cell divides mitotically four to five times, pro-

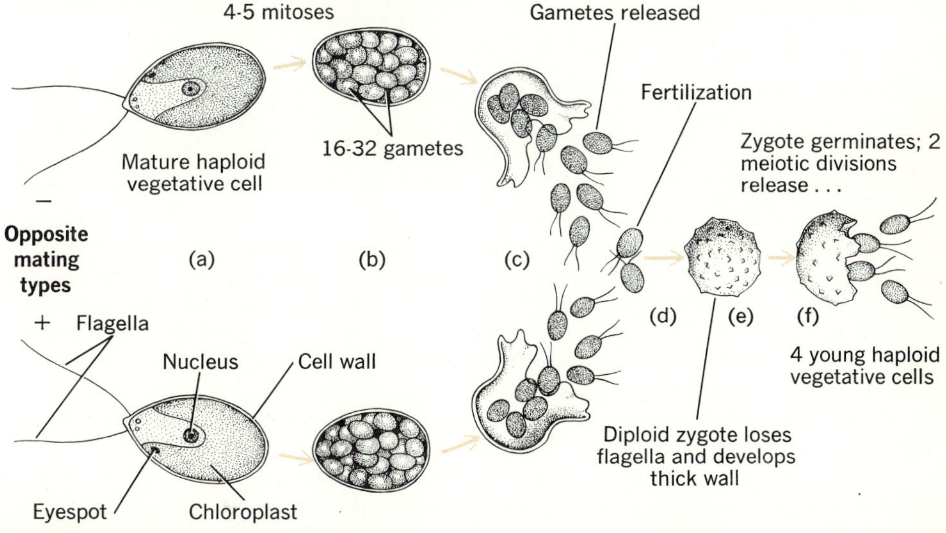

Figure 7-9 Sexual reproduction between opposite mating types in the alga, *Chlamydomonas.*

ducing 16 to 32 small sex cells (a, b). These gametes, contained within the cell wall of the original vegetative cell (a), soon develop their own flagella and cell walls. The cellulose case which encloses them then bursts, releasing the gametes (c). The gametes shed their cell walls and members of opposite mating types eventually pair off, attaching to one another at their flagellated ends (d). Their cytoplasms gradually merge and their nuclei unite, forming a diploid zygote. The zygote loses its flagella and develops a thick protective coat (e). In this state the organism is able to withstand the cold of winter or the dryness of summer. When conditions become favorable, the zygote germinates. Two meiotic divisions result in the production of four small haploid cells which are released to the surrounding water (f). These in turn grow into mature vegetative cells, completing the cycle.

Chlamydomonas thus differs considerably from *Paramecium* in the details of sexual reproduction. Over the entire life cycle only the zygote is diploid. True gametes are produced. They are, however, indistinguishable from vegetative cells and may be regarded essentially as small vegetative cells that have a tendency to unite under certain circumstances and behave like gametes. In a few species of *Chlamydomonas* the gametes, although similar in form, are produced in two different sizes. Other flagellates (for example, the colonial flagellate *Volvox* shown in Plate 9 of Chapter 3) carry sexual differentiation much further. Gametes are produced that differ in form as well as in size. Some are small and actively motile, others large and nonmotile—types we recognize as sperm and ova. Some individual colonies of *Volvox* produce only sperm cells; others produce only eggs. Individual organisms can be distinguished, in other words, as males or females.

Sexual differentiation and specialization reach their highest state of development in the multicellular plants and animals. As we consider these "higher" forms the reader should remember that the fundamental requirements for sexual reproduction are no less perfectly met in conjugating paramecia than they are in copulating rabbits. As organisms increase in size and complexity greater degrees of specialization are required to achieve the same reproductive efficiency found in simpler forms of life. Regardless of the sophistication of reproductive mechanisms, we will note that sexual reproduction inevitably "boils down" to a nuclear reorganization made possible by meiosis.

SEXUAL REPRODUCTION IN PLANTS

It is evident at this point that sexual reproduction involves a cycle in which fertilization and meiosis must follow one another in an alternating pattern. The meiotic division of diploid cells produces haploid gametes whose fusion restores the diploid condition. In plants the interdependence of these two processes is quite clearly marked by two distinct generations of plants (or portions of plants) that alternate with one another. One generation is diploid and begins with the zygote. As this diploid form matures it produces spores (minute reproductive structures that are usually unicellular). This spore-producing plant form is called the *sporophyte.* When spores released by the sporophyte germinate, they develop into a haploid, gamete-producing plant form known as the *gametophyte.* If sperm and eggs produced by the gametophyte join in fertilization, the resulting zygote marks the beginning of the next sporophyte generation.

This *alternation of generations* is nicely demonstrated in ferns (Fig. 7-10). What most of us recognize as a fern is actually the sporophyte generation (a). Spore-forming structures develop usually on the underside of the leaf or frond. Within these bodies diploid spore mother cells divide by meiosis producing four haploid

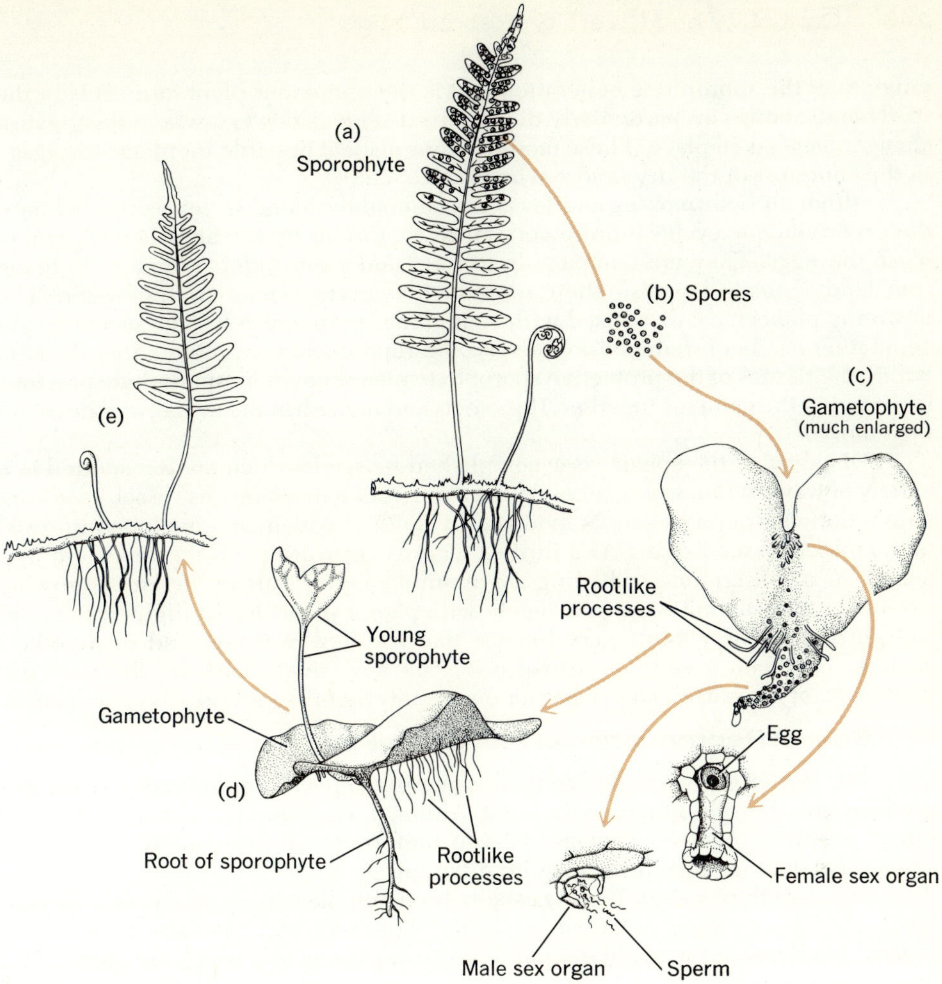

Figure 7-10 Life cycle of a fern. (a) Sporophyte; (b) spores; (c) gametophyte (much enlarged) showing male and female sex organs; (d) gametophyte bearing a young sporophyte (much enlarged); (e) new, independent sporophyte.

spores each. These meiotic divisions mark the beginning of the gametophyte generation. Spores are released from the sporophyte (b). If the wind-borne spores are carried to moist, shaded soil, they germinate, developing into delicate, commonly heart-shaped gametophytes about $\frac{1}{4}$ inch long (c). The lower surface of the haploid gametophyte bears both male and female sex organs. The male organs give rise to flagellated, motile sperm which swim to the eggs of other gametophytes if adequate moisture is present. Self-fertilization generally does not occur because the male and female organs of one gametophyte tend to mature at different times. In a given gametophyte only one zygote matures. Reproduction of gametes signals the end of the haploid generation. The embryo sporophyte which develops from the zygote secures its nourishment from the autotrophic gametophyte until it becomes a self-sufficient young plant with roots, stem, and leaf, (d), after which the gametophyte dies and decomposes (e).

Modifications of the pattern of alternating generations described in ferns occur throughout the plant kingdom. In the mosses it is the gametophyte generation rather than the sporophyte generation that is the prominent plant form. It is in the seed plants, however, particularly the groups that bears flowers, where the greatest changes have taken place. These modifications made it possible for plants to engage in the conquest of the dry land environment.

Although both mosses and ferns are commonly found in terrestrial habitats, their reproductive cycles require very moist conditions for the flagellated sperm to reach the eggs. They are comparable to frogs and salamanders which miss being true land animals because their reproductive cycle occurs in the water. The flowering plants have dispensed with swimming sperm entirely. The gametophyte generation has been reduced to microscopic reproductive structures that develop within the tissues of the protective sporophyte plant body. The sporophyte provides for bringing the gametes together. It protects and nourishes the zygote and developing embryo.

Dispersal of the species is accomplished by seeds which are transported in a variety of ways (wind, water, animals, special ejection mechanisms). Each seed contains a dormant embryo and its food supply, both of which are enveloped in resistant, protective seed coats. As a further security measure even the seeds are protected. As a consequence offspring are assured a better start in life and a growing site where they do not have to compete with parent plants for sunlight, water, and soil nutrients. The evolutionary lineage that resulted in these (and many other) changes has been a very successful one. The flowering plants are the most numerous plants on earth today, whether one counts them as individuals or as species.

Sexual Reproduction in Flowering Plants

The flowering plants derive their name from special reproductive structures produced by the sporophyte plant body. Although considerable variation exists in flower anatomy between species, a basic uniformity of structure is found even among flowers that appear to be quite different.

A typical flower (Fig. 7-11) is usually borne on the expanded end of a specialized stalk. This terminal enlargement, called the *receptacle*, gives rise to four floral organs. The first are the leaflike *sepals*. Next come the *petals* which are often white or colored and commonly the most conspicuous part of the flower. The *stamens* are located just within the ring or whorl of petals and are also arranged in a circle. Each stamen consists of a terminal *anther* (containing the *pollen*) supported on a slender *filament*. The center of the flower has a vase-shaped *pistil* whose enlarged base, the *ovary*, contains an *ovule*[1] (forerunner of the seed). The slender stem of the pistil, the *style*, bears a cap known as the *stigma*, which may be variously shaped depending on the species and is often coated with a sticky secretion. The stamens and the pistil are directly involved in sexual reproduction and are therefore considered essential parts of the flower.

The events leading to gamete formation and fertilization are best described by beginning with the anther. A section through an anther from a young flower bud shows that it contains four chambers (Fig. 7-12a). Within these chambers are many diploid spore mother cells. As the anther grows each diploid cell undergoes meiosis, producing four haploid spores (b). Each spore then divides unequally by mitosis forming a larger *tube cell* and a small *generative cell*. The cell wall thickens around these two cells producing a *pollen grain* with a very resistant, usually roughened coat (c). This pollen grain is actually *a male (unisexual) gametophyte*. At

[1] The ovaries of most flowers contain several ovules. The watermelon flower has a very large number.

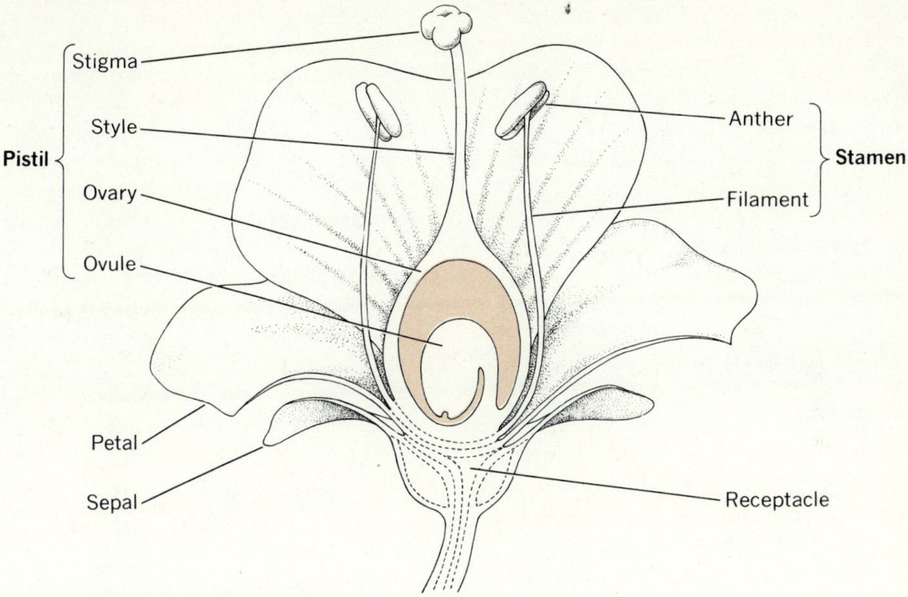

Figure 7-11 Diagram of the organs of a flower.

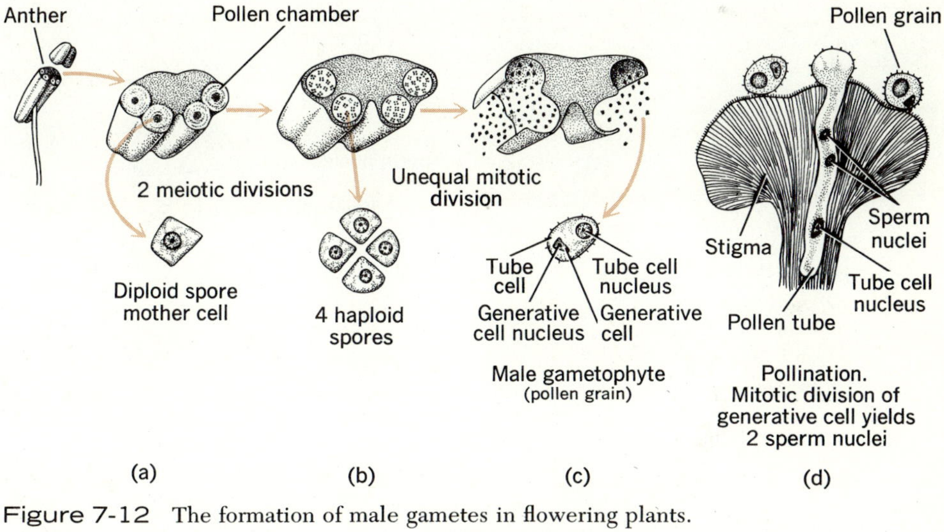

Figure 7-12 The formation of male gametes in flowering plants.

this point the anther splits releasing its powdery pollen. If pollen contacts the stigma of the flower or is carried to the stigma of another flower of the same species (by wind, insects, birds, and so on), it germinates, forming a *pollen tube* which starts to grow down into the style (d). Sometime during *pollination* (the transfer of pollen from anther to stigma) the generative nucleus divides mitotically, yielding two haploid *sperm nuclei* that will be carried by the growing pollen tube to the ovule (d).

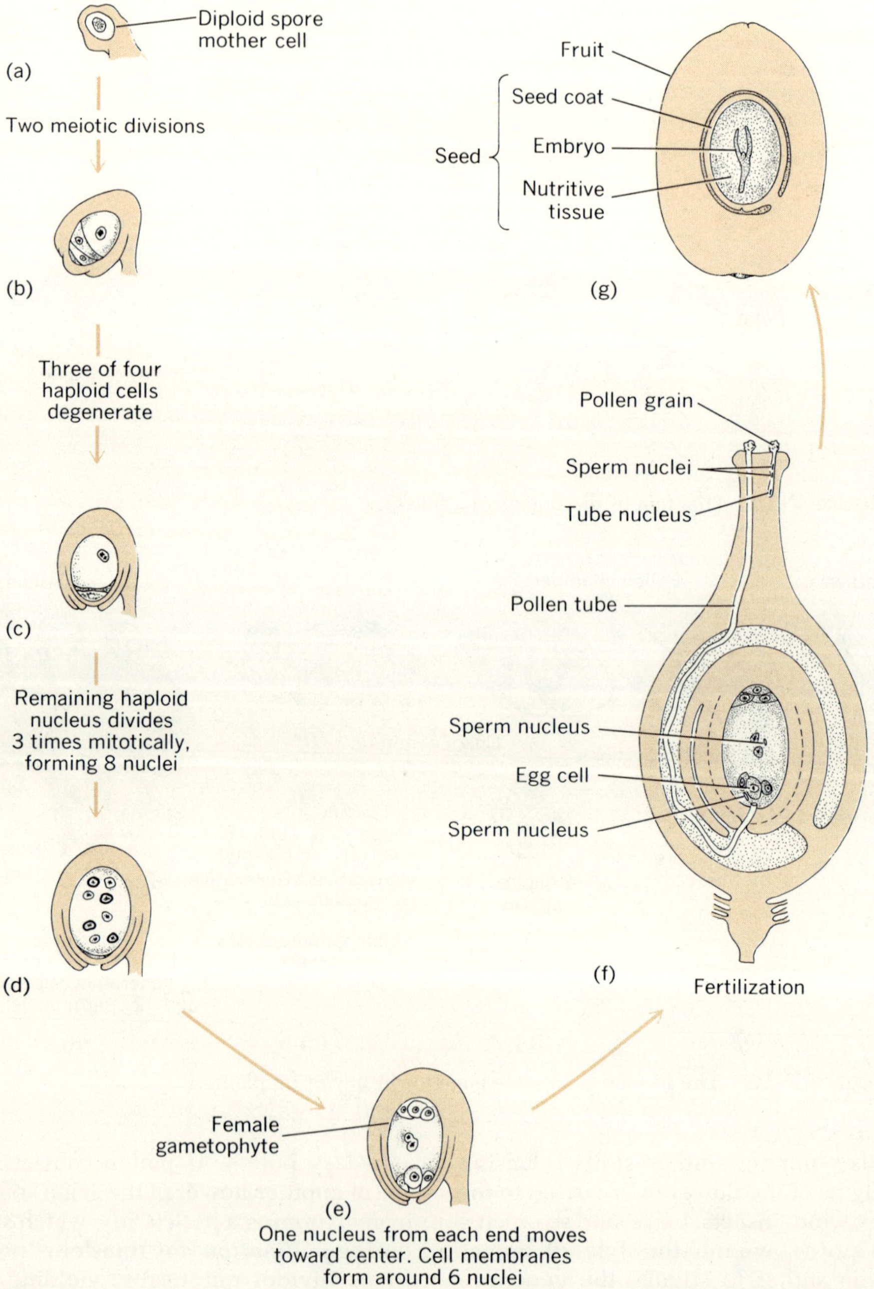

Figure 7-13 The life cycle of a flowering plant from female gametophyte formation to fruit development.

Within the ovule corresponding changes lead to the formation of the female gamete (Fig. 7-13). The young ovule usually contains a single, large diploid spore mother cell (a). This cell passes through two meiotic divisions, yielding four haploid spores. Three of these spores, however, begin to degenerate (b). The nucleus of the large surviving spore now divides by mitosis three times, producing eight nuclei which become distributed in two groups of four at opposite ends of the "cell" (c-d). One nucleus from each end moves toward the center of the eight-nucleate structure as cell membranes form around each of the remaining six nuclei (e). This seven celled body is a *mature female gametophyte*. One of the three cells in one end of the gametophyte is the egg cell. A pollen tube (male gametophyte) which has grown through the style and penetrated the ovule now releases its two sperm nuclei. One of these nuclei fuses with the egg nucleus producing a zygote (f). The other unites with the two nuclei not enclosed by a cell membrane, forming a unique triploid nucleus with chromosomes in sets of three ($3n$). The other five cells in the female gametophyte degenerate, as does the tube nucleus. The triploid nucleus passes through repeated mitotic divisions, eventually giving rise to a nutritive tissue (to nourish the embryo plant). As the zygote develops into an embryo, the outer portions of the ovule become converted into a seed coat. The seed (embryo, nutritive tissue, and seed coat) is further invested within tissue contributed by the ovary itself. The entire assemblage is called a *fruit* (g).

It should not be thought that the sepals and petals are of little use in reproduction. One, the other, or both of these structures serve important functions in pollination. The sugary nectar that is so attractive to hummingbirds and many different insects is commonly secreted at the bases of petals and sepals. The color imparted by these floral organs is a very effective lure (along with odor) to animal pollinators. The form and arrangement of petals sometimes determine whether the flower will be self-pollinated or cross-pollinated. (A stigma receiving pollen from anthers of another flower is said to be *cross-pollinated;* pollen from the same flower would result in *self-pollination*).

Even though cross-pollination is less direct than self-pollination, most plants are modified to encourage cross-pollination. This is done in a variety of ways. Sometimes the stamens and pistils of a given flower do not mature at the same time. In other instances, floral structures are so arranged that it is unlikely that pollen can pass directly from anther to stigma. Many species are self-sterile; pollen fail to germinate on a stigma of the same plant. There are even some species in which a given plant either produces flowers without stamens (pistillate flowers) or flowers without pistils (staminate flowers). We again note a most interesting feature which we have previously observed in *Paramecium*, *Chlamydomonas*, and the ferns. *Sexual reproduction distinctly favors the exchange of gametes between different individuals.*

SEXUAL REPRODUCTION IN ANIMALS

Although the fundamental process of sexual reproduction is the same in both plants and animals, we would expect considerable variation in detail between two such dissimilar groups. In animals, for instance, meiosis leads directly to the formation of gametes rather than to the development of a haploid generation which produces gametes. Although some animals, such as certain jellyfishes (Fig. 7-14), are said to reproduce by an alternation of generations, both generations are diploid. A sexual diploid generation alternates with an asexual diploid generation. Only the gametes produced by the gonads of the diploid male and female jellyfish are haploid.

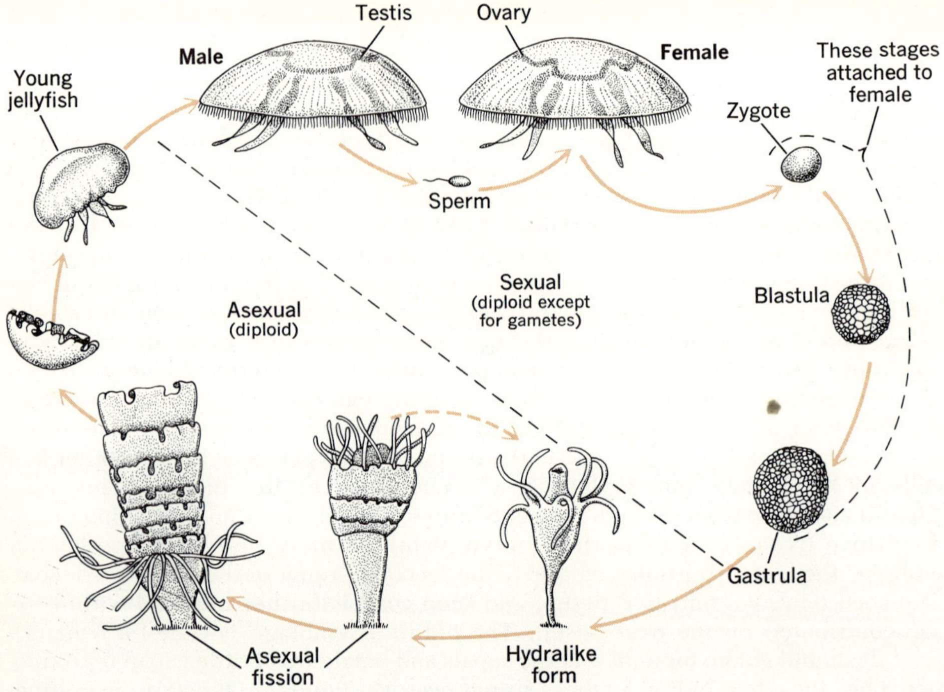

Figure 7-14 Alternation of generations in the life cycle of a jellyfish.

The sex cells of all animals are distinctly different. Despite their usually microscopic size, the female gametes or ova are commonly one to several thousand times larger than the male gametes or sperm. They are well supplied with rich cytoplasmic food reserves to nourish the developing embryo. They are produced in an *ovary* (female gonad) from undifferentiated diploid sex cells, as shown in Figure 7-15. After an initial period of mitotic proliferation each diploid cell begins to grow. Two meiotic divisions occur, in each of which practically all of the cytoplasm is apportioned to one daughter cell. The other daughter cell, scarcely more than a nucleus, is called a *polar body.* The first polar body may or may not divide during the second meiotic division, but in either event all polar bodies degenerate, leaving one relatively large egg cell. One diploid cell has produced one haploid cell in two meiotic divisions.

Sperm are produced in a testis (male gonad) in a similar fashion (Fig. 7-15). The cytoplasm is apportioned equally, however, during meiosis. The final phase of maturation is characterized by cellular differentiation in which each gamete develops a tail (flagellum) and loses most of its cytoplasm. From each diploid cell four motile sperm have been produced that are essentially haploid nuclei equipped with tails.

In most animals the ovaries and testes are located in different individuals. Those organisms, such as the earthworm, that carry both male and female gonads are called *hermaphrodites.* Even hermaphroditic species, however, seldom practice self-fertilization. Their reproductive systems are often so structured that fertilization can only occur by an exchange of sperm between two individuals. In other instances the two gonads commonly mature at different times in the same individual or (as in many plants) the organism is self-sterile.

Fertilization in animals encounters some of the same problems faced by plants. Motile sperm require a fluid medium in which to reach the eggs. Delicate unicellular gametes are also easily killed by drying. Aquatic animals, however, do not usually experience these difficulties. In many fish (for example, trout and salmon) the male merely discharges sperm over a mass of eggs laid on the stream bed by its mate. Such *external fertilization* is not possible with terrestrial animals like insects, reptiles, birds, and mammals. These organisms practice *internal fertilization.* A male and female engage in a temporary physical contact called *copulation* by which sperm are transferred directly from the body of the male into the body of the female. Fertilization takes place within the protective, fluid environment of the female reproductive system.

Among species which practice internal fertilization embryonic development may follow one of several patterns. In birds, many species of reptiles, and insects, the yolk-laden fertilized egg cells are invested in a relatively impervious coating or shell and then laid by the female in some protected place, where they undergo development and hatch into young. In some species (for example, certain lizards and snakes) the eggs are kept within the body of the female and hatch there. These creatures consequently give birth to free-living young. In most mammals (and a few

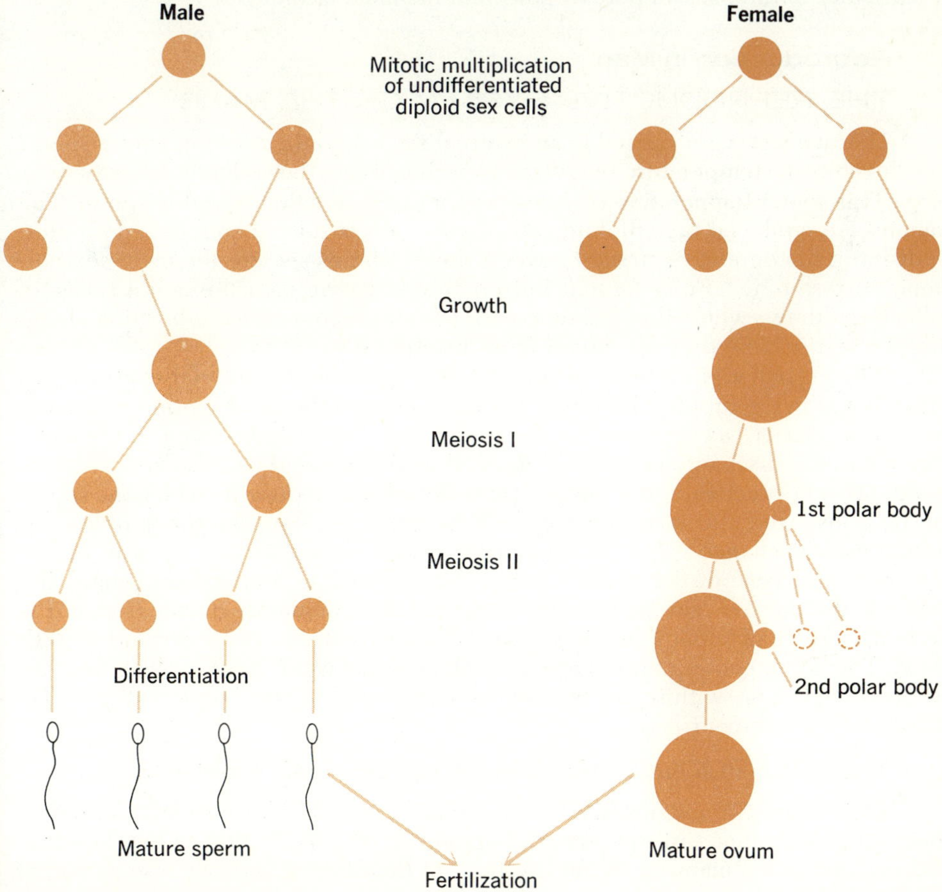

Figure 7-15 The maturation of gametes in animals.

other vertebrates) the developing young are not only carried inside the mother but derive their nourishment almost entirely through an intimate connection with the maternal bloodstream rather than solely through nutrient stored in the egg cell.

The care and protection which animals give to developing young have a great deal to do with egg production and reproductive efficiency. In fishes like the eels which do not protect their young, one female will lay millions of eggs in one breeding season. The female stickleback fish, on the other hand, lays only about 100 to 150 eggs in a season. The young are protected in a nest built and guarded by the male.

Similar variations in parental care exist among insects. Care of the young is particularly well developed in social insects such as honeybees (as previously noted), termites, and ants. Birds and mammals, among vertebrates, have carried parental protection to its highest degree of development. The young are continuously cared for, in most cases, until they are capable of fending for themselves. It is perhaps in the placental mammal, however, where there is the greatest assurance that one zygote will successfully develop into a new individual. Like the flowering plants these mammals protect and nourish their embryos within the ample body of the parent organism. Let us therefore turn our attention to placental mammals and in particular to the most important placental mammal of them all.

Reproduction in Man

THE REPRODUCTIVE SYSTEM OF THE MALE

The two testes are carried in an external sac called the *scrotum* where sperm may develop at a temperature below that which exists in the abdominal cavity (Fig. 7-16). This cooler temperature is necessary for the production of viable sperm. The gametes are produced from diploid cells in the walls of several convoluted tubules. Cells lying in connective tissue between these tubules secrete the male sex hormone, testosterone.[2] Fully formed but nonmotile sperm pass through a series of collecting tubules where they are delivered to a single convoluted tube lying alongside the testis. Here they are stored in an inactive state.

When ejaculation (release of sperm) occurs, wavelike muscular contractions, which begin in the tubules draining each testis, carry millions of sperm up through the sperm duct or *vas deferens*. As the sperm pass below and behind the bladder they receive a discharge of *seminal fluid* from the *seminal vesicles* and *prostate glands*. This fluid activates the sperm. It also serves as a medium by which the sperm can be transferred and in which they will be protected from the shock of normal environmental changes. The mixture, now called *semen*, passes by way of the *urethra* to the base of the *penis* where strong muscular contractions propel the semen to the outside. The penis is well adapted as a copulatory device in that it is made up mostly of spongy tissue whose cavernous spaces can be distended with blood. The organ consequently becomes enlarged and rigid. As a result semen can be deposited deep within the female copulatory organ (the *vagina*, Fig. 7-17) during sexual intercourse.

THE REPRODUCTIVE ORGANS OF THE FEMALE

Each of the two ovaries in a human female possesses an outer layer or cortex containing many thousands of potential eggs cells (Fig. 7-17). Only a few hundred of them ever reach maturity in her lifetime. At the onset of *puberty* (the period of

[2] Both the gamete-producing and hormone-producing tissues of the testes are regulated by gonadotropic hormones, as in the female (refer to Fig. 5-30b).

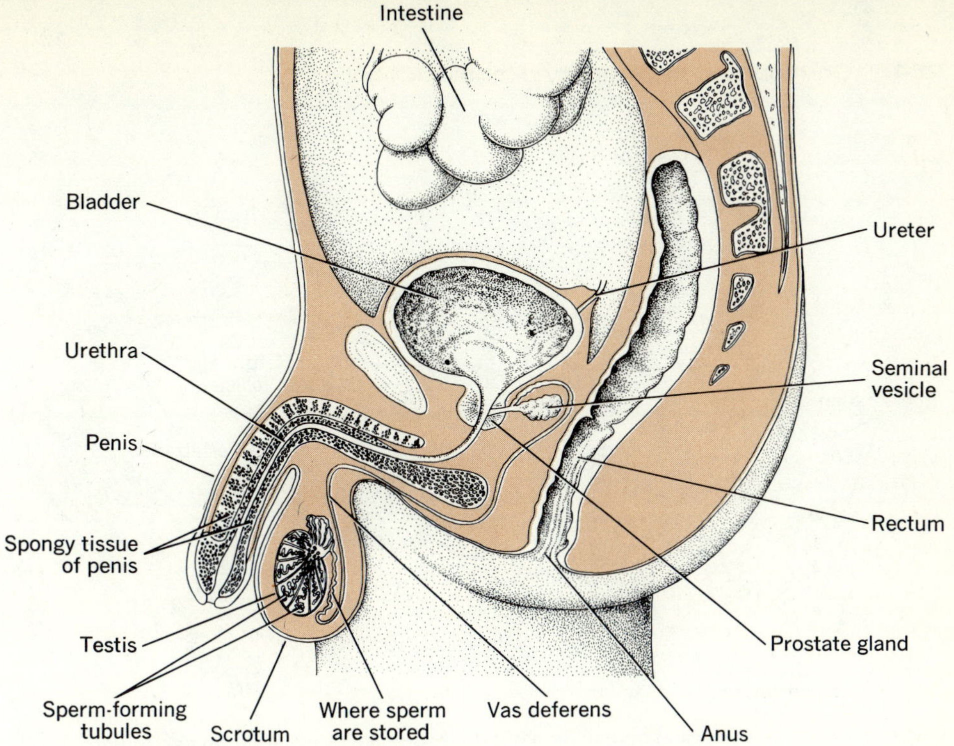

Figure 7-16 Reproductive organs of the human male (semidiagrammatic). The vas deferens follows a less direct route to its junction with the duct of the seminal vesicle than that shown.

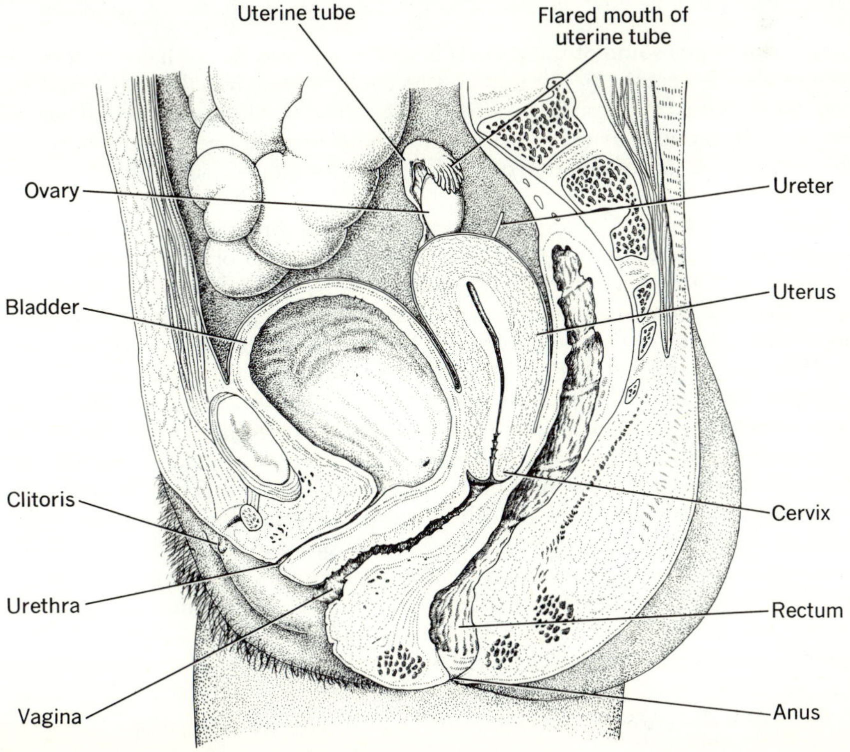

Figure 7-17 Reproductive organs of the human female.

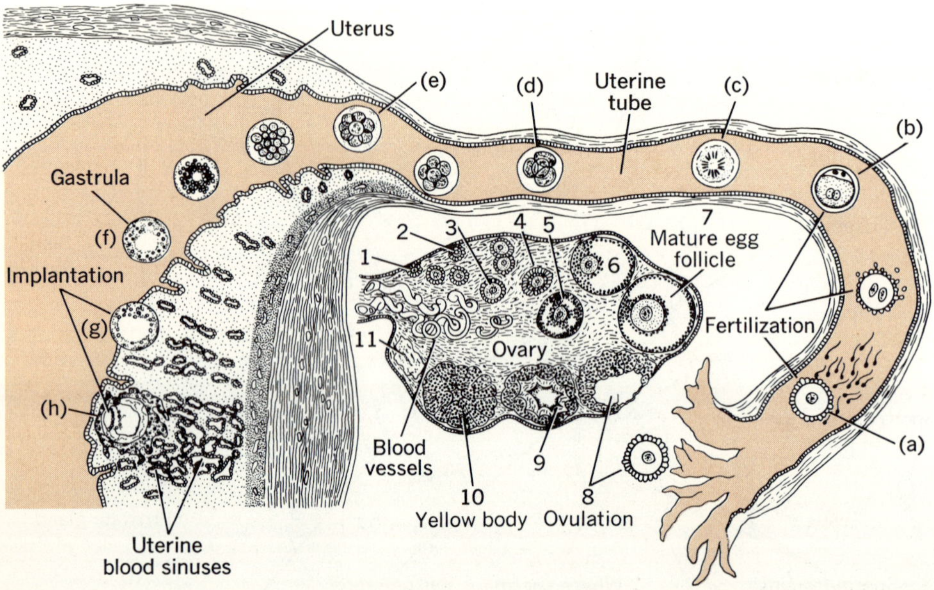

Figure 7-18 Schematic diagram showing the sequence of events from egg follicle development to ovulation, fertilization and early embryonic development: (1–7) stages in egg follicle development; (8) ovulation; (9–11) young, mature and degenerated yellow body; (a–b) fertilization; (c–f) formation of the gastrula; (g–h) implantation in the uterine wall. The expanded end of the uterine tube actually surrounds the ovary at the time of ovulation.

sexual development) some of these cells begin to mature. As we have observed the maturation of ova is cyclic. In the human this cycle is known as the menstrual cycle[3] and requires approximately 28 days to run its course. Although several egg cells may begin a given cycle, usually only one succeeds in completing it and becoming a mature ovum. Let us follow the course of development of one such successful ovum.

Under the influence of a gonadotropic hormone one of the cells at the outer edge of the ovary gradually becomes surrounded by a layer of small cells (Fig. 7-18, 1–3), forming a young *egg follicle*. The follicle cells around the ovum increase in number. Fluid gradually accumulates between these cells and a follicular cavity develops. After about 14 days a mature egg follicle is produced and the first meiotic division has been completed (Fig. 7-18, 4–7). During this period certain cells in the ovary (presumably cells in the connective tissue) secrete estrogen. This hormone suppresses further secretion of the follicle-stimulating gonadotropin (called FSH for follicle-stimulating hormone). Estrogen also brings about an increase in the blood supply to the uterus (womb) and a thickening of the uterine lining.

Around the 14th day the egg follicle ruptures, releasing the egg into the expanded end of the uterine tube (Fig. 7-18, 8). This process, called *ovulation*, ushers in a new pattern of events. The secretion of estrogen falls off and the cells of the ruptured follicle become larger. These large cells increase in number as a new structure called the *yellow body* begins to form in the old follicle (Fig. 7-18, 9). The yellow body functions as an endocrine gland. For several days it secretes a hormone

[3] This is paralleled in most subhuman mammals by a cycle of recurring periods of sexual excitement ("heat") called the *estrus cycle*.

called *progesterone.* This hormone continues the development of the uterine lining and the suppression of FSH secretion. If fertilization does not occur, however, the secretion of progesterone diminishes and the yellow body starts to shrink. On or about the 28th day of the cycle *menstruation* begins. The lining of the uterus deteriorates, releasing a certain amount of blood and tissue through the vagina. The pituitary, relieved of the suppressive effects of estrogen and progesterone, begins secreting FSH once more and another cycle begins.

INTERCOURSE, FERTILIZATION, AND EMBRYONIC DEVELOPMENT

The *clitoris* (Fig. 7-17) of the female is the counterpart of the penis in the male and also contains spongy, erectile tissue. Like the penis it commonly evokes sexual excitement when stimulated. During the period that the ovarian follicle is growing the estrogen level of the blood rises. This is not uncommonly associated with increased sexual interest, particularly about the time of ovulation when sexual intercourse is most likely to result in conception (fertilization). Since sex interest in the human male is more or less continuous, the reproductive and integrative mechanisms of man are nicely designed, as they are in all organisms, to assure procreation of the species.

The culmination of sexual excitement during intercourse is marked in both sexes by an intense emotional response called an *orgasm.* In the male this is accompanied by a discharge of semen deep in the vagina at the mouth of the *cervix* (Fig. 7-17). Between 200 and 500 million sperm begin a three-hour journey through the cervix and uterus and on into the upper reaches of the uterine tubes. Although many expire along the way millions complete the trip. If an egg is not awaiting them on their arrival, they will retain their viability (fertilizing capacity) for a period of about 24 to 48 hours. If an egg was ovulated prior to intercourse, it remains viable for an estimated 18 to 24 hour interval. If sperm and ovulated egg meet within these approximate time limits, fertilization may take place.

Although usually only one sperm actually penetrates the ovum, thousands of them are apparently required to make this possible. As shown in Figure 7-18, 8, the ovulated ovum is protected by a covering layer of cells. Each sperm carries a tiny quantity of enzyme that breaks down the cementing substance holding these cells together. It is believed that several thousand sperm are required to provide enough enzyme to loosen these cells sufficiently for fertilization to take place. Once this occurs a single sperm penetrates the egg (a), whereupon a change occurs in the egg cell membrane so that no additional sperm can enter. The egg cell now passes through its second meiotic division. The haploid sperm and egg nuclei approach one another and fertilization is complete (b). Cilia lining the uterine tube, aided by muscular contractions, gradually move the developing embryo toward the uterus (c–e). After several days the embryo reaches the uterus in the gastrula stage (f). The embryo now becomes *implanted* in the uterine wall. It actively invades the uterine lining until it is completely buried (g–h). Membranes produced by the embryo send rootlike processes into the uterine wall which is now engorged with blood. These rootlike processes become surrounded by microscopic lakes of blood (uterine blood sinuses). This intimate union of embryonic and maternal tissues is called the *placenta* (Fig. 7-19).

The placenta provides the means whereby the embryo receives nutrients and oxygen from, as well as delivers metabolic wastes to, the maternal bloodstream. When the embryo develops its own circulatory system, the blood is circulated to

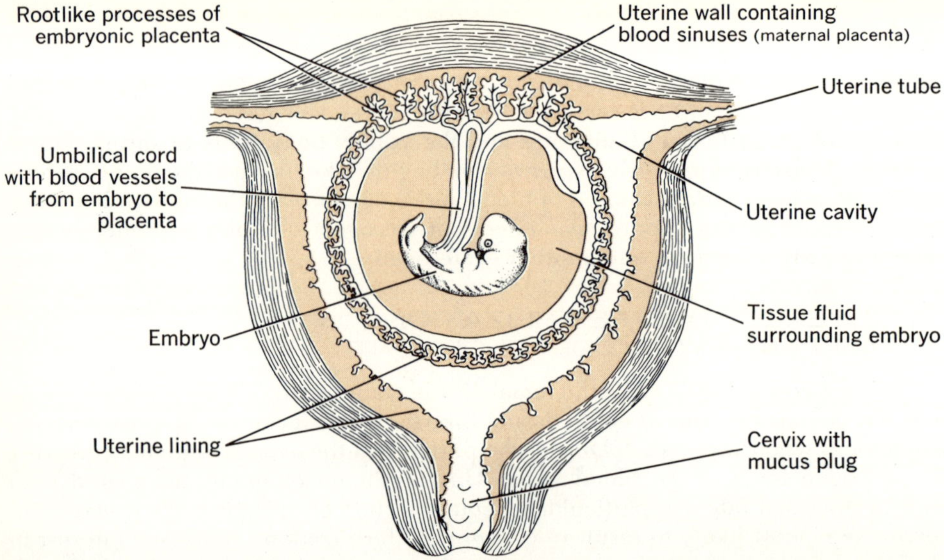

Figure 7-19 Human uterus with placenta and embryo at about the sixth week of pregnancy.

and from the placenta by way of blood vessels in the *umbilical cord.* The two bloodstreams are entirely separate. The exchange of materials between them is accomplished by diffusion and active transport.

Early in the course of embryonic development the yellow body in the ovary becomes a fully formed structure (Fig. 7-18, 10). It maintains the uterine conditions characteristic of pregnancy for several months through progesterone secretion. After a time the placenta itself begins to take over progesterone secretion and the yellow body gradually degenerates, leaving a scar in the ovary (Fig. 7-18, 11). After about 9 months of pregnancy the placental secretion of progesterone declines. The muscular walls of the uterus begin to contract in wavelike movements ("labor pains") and the infant is born. Shortly thereafter the placenta ("afterbirth") is expelled and pituitary hormones initiate the secretion of milk by the mammary glands.

On occasion multiple births occur. These are usually the result of more than one egg having been ovulated at the time of conception. If two eggs are ovulated and fertilized, the resulting children are called *fraternal twins.* Since they came from separate sperm and eggs, they are no more alike than ordinary brothers and sisters. A less frequent type of multiple birth is the one-egg twins. These individuals are genetically identical. They are derived from the division of a very early embryo (possibly as early as the two-celled stage) and consequently come from the same sperm and egg. They are really produced asexually from a common embryo.

UNISEXUAL REPRODUCTION IN PLANTS AND ANIMALS

It should not be assumed that it is always necessary for gametes to fuse before the development of a new individual begins. Unfertilized frog eggs can be stimulated to develop by pricking them with a needle dipped in frog blood. Varying stimuli (such as changes in temperature, pH, and osmotic pressure) have proved similarly effective in experimentally activating the development of unfertilized eggs in other species. In one case a fatherless female rabbit was reputed to have been produced. This phenomenon is known as *parthenogenesis,* which means virgin (*partheno*) birth (*genesis*). It occurs naturally in a number of different invertebrate animals (including many insects such as honeybees) and also in a small percentage of unfertilized turkey eggs. The turkey, however, is the only vertebrate for which there is scientific evidence of the natural occurrence of parthenogenesis.

Parthenogenetic offspring may be either haploid or diploid. In the honeybee they are haploid. The fertilized eggs laid by the queen develop into diploid female workers (or queens), whereas the unfertilized eggs become haploid males. Diploid offspring can be produced in two ways: (1) the unfertilized egg behaves as a zygote at the end of the first meiotic division (the second meiotic division is eliminated) or (2) the chromosomes of the mature, haploid ovum replicate twice while the cytoplasm of the cell divides only once. The rabbit referred to was a diploid individual experimentally produced (presumably) by the first of these methods.

Figure 7-20 Plantlets emerging from the anther of a tobacco plant. (J. P. Nitsch)

Unisexual reproduction is by no means restricted to animals. Several different kinds of plants produce fruits (usually seedless) in the absence of pollination and fertilization. The best known of these in which the process occurs naturally are seedless grapes and the cultivated varieties of pineapple and banana. The most striking example of this phenomenon, however, is provided by the tobacco plant. Uninucleate pollen grains (see Fig. 7-12b), when grown on a culture medium, have been shown to be capable of developing into haploid tobacco plants. The plantlets, shown in Figure 7-20, mature and flower normally, but do not produce seed.

The process of unisexual reproduction poses some intriguing questions. Since only one gamete is involved, is it truly sexual reproduction, which presumably requires a union of gametes? If our definition merely stipulates that some sort of nuclear reorganization has taken place, then unisexual reproduction is sexual reproduction. Let us abandon arbitrary labels and delineations at this point and reexamine the essence of sexual phenomena from the standpoint of its biological significance.

SEXUAL REPRODUCTION: EXPEDITOR OF BIOLOGICAL DIVERSITY

Although many organisms reproduce both sexually and asexually and a very few appear to reproduce by asexual methods only, the overwhelming majority of living creatures reproduces sexually. At first sight this seems unusual. Asexual reproduction is simple, direct, and efficient. Sexual reproduction, on the other hand, entails an alternating pattern of meiosis and fertilization which is never simple and often necessitates elaborate mechanisms, particularly in multicellular organisms as we have seen. The reason for this preference in sexual reproduction must lie in the nature and quality of the offspring produced. There is certainly a fundamental difference in the kind of young reproduced by the two methods. Asexually formed young are usually identical to one another and to the parent from which they came. No two sexually reproduced individuals, on the other hand, are ever exactly alike. Let us first see why this is so before we consider the biological implications of its being so.

We have made repeated reference to the fact that sexual reproduction entails a nuclear reorganization. The hereditary material of a sexually reproducing parent is somehow reshuffled in its gametes by reason of the process of meiosis itself. This reordering of hereditary material takes place as a result of two characteristic features of meiosis: (1) the random reassortment of chromosomes and their genes and (2) the recombination of genes on the same chromosome through what is called *cross-over*. Let us investigate each of these sources of novelty and change in turn.

The Independent Assortment of Chromosomes and Their Genes

Figure 7-21 represents schematically a simple case of meiosis involving just two homologous pairs of chromosomes. For simplicity they are represented diagrammatically, one pair being round and the other square in cross section. The two

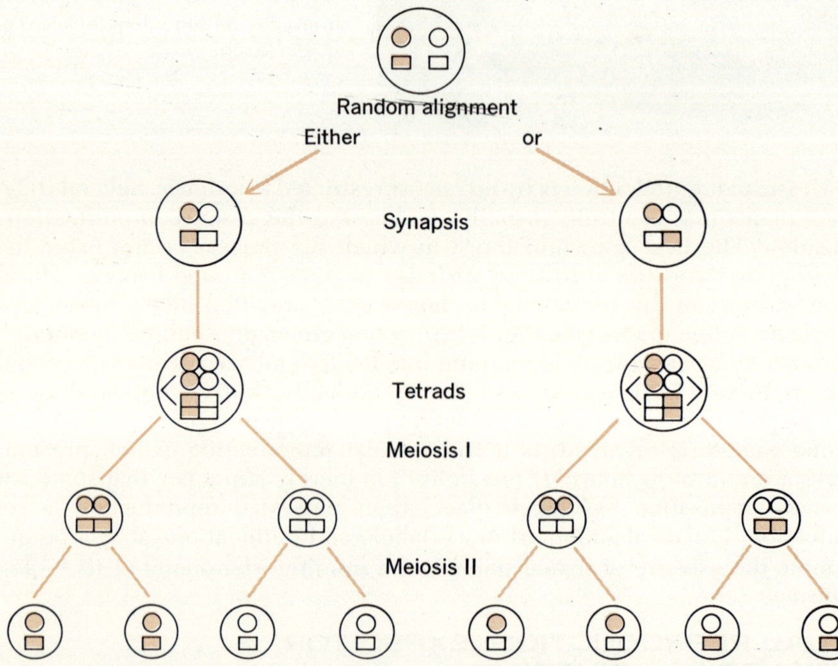

Figure 7-21 A schematic representation of the random reassortment of maternal and paternal chromosomes during meiosis. Different sets of homologous chromosomes align themselves on the equatorial plate independently of one another. With two pairs of chromosomes there are four possible combinations of maternal and paternal chromosomes in mature gametes. (Paternal chromosomes colored, maternal white).

chromosome sets align themselves on the equatorial plate of the spindle independently of one another. The paternal chromosomes (colored) of the two homologous sets may be on the same or opposite sides of the equatorial plate. In any one cell the alignment is strictly according to chance. With just two pairs of chromosomes there are only two different ways the two tetrads can be arranged with respect to one another; any other arrangement would result in separating the chromatids that represent each chromosome. This occurs in the second, not the first meiotic division.

As a consequence of this independent alignment and assortment of chromosomes (and their genes) meiosis may result in every possible haploid combination of maternal and paternal chromosomes, in this case four. The probability of any specific combination participating in fertilization would be one chance in four. Had we used three pairs of chromosomes rather than two there would be eight possible combinations ($2 \times 2 \times 2$ or 2^3). With 23 pairs, as in man, the possible number of arrangements is enormous (2^{23}). It is interesting to note that even if diploid young were reproduced by parthenogenesis as a result of the meiosis illustrated in Figure 7-21, none of them would have the same chromosome complement as the parent. They would all have matched sets of either maternal or paternal chromosomes.

The genes on a pair of homologous chromosomes are located along the length of each chromosome in corresponding linear arrangements. Each chromosome pair therefore represents a group of gene pairs. The members of any given gene pair are not necessarily identical replicas of one another. They are similar in that they affect the development of the same hereditary characteristic. This characteristic may have two alternative expressions, such as dark eye color and light eye color in humans. Basic eye color is controlled by a pair of alternative genes, one for the development of dark eye color, the other for light eye color. We will go into these matters in more detail in Chapter 8. These brief remarks should suffice for our present purpose.

Let us suppose that Figure 7-21 depicts two of the 22 pairs of chromosomes in the domestic rabbit. One pair of genes determines whether the rabbit's coat color will be gray or black. Let us assume that this gene set is located on the round chromosome and that the paternal (colored) chromosome carries the gene for gray and the maternal (white) chromosome carries the gene for black. Another pair of genes, located on a different pair of chromosomes, determines whether the rabbit's hair will be long or short. Let us suppose that the paternal (colored) square chromosome carries the gene for long hair and the maternal (white) square chromosome carries the gene for short hair. Since these genes must go wherever their chromosomes take them, there may be four different gene assortments in the gametes produced: (1) gray, long; (2) black, short; (3) gray, short; and (4) black, long. Since only one of these combinations will be involved in any specific act of fertilization, the opportunity for variation in offspring is excellent.

The Recombination of Genes through Cross-Over

Since chromosomes are passed along during meiosis as whole units, it might be presumed that all the genes on a given chromosome would always be transmitted together as a "package deal." This would be true were it not for the fact that two chromatids within a tetrad often break during synapsis and exchange their broken segments. This process, called *crossing-over,* occurs randomly between any two chromatids in a tetrad and in anywhere from one to several places. Figure 7-22 shows crossing-over taking place at one point only (indicated by the arrow) between a paternal (colored) and maternal (white) chromatid. Two gene locations are shown on each chromatid as reference points: *a* and *B* on the paternal chromatid and their alternative counterparts *A* and *b* on the maternal chromatid. Crossing-over

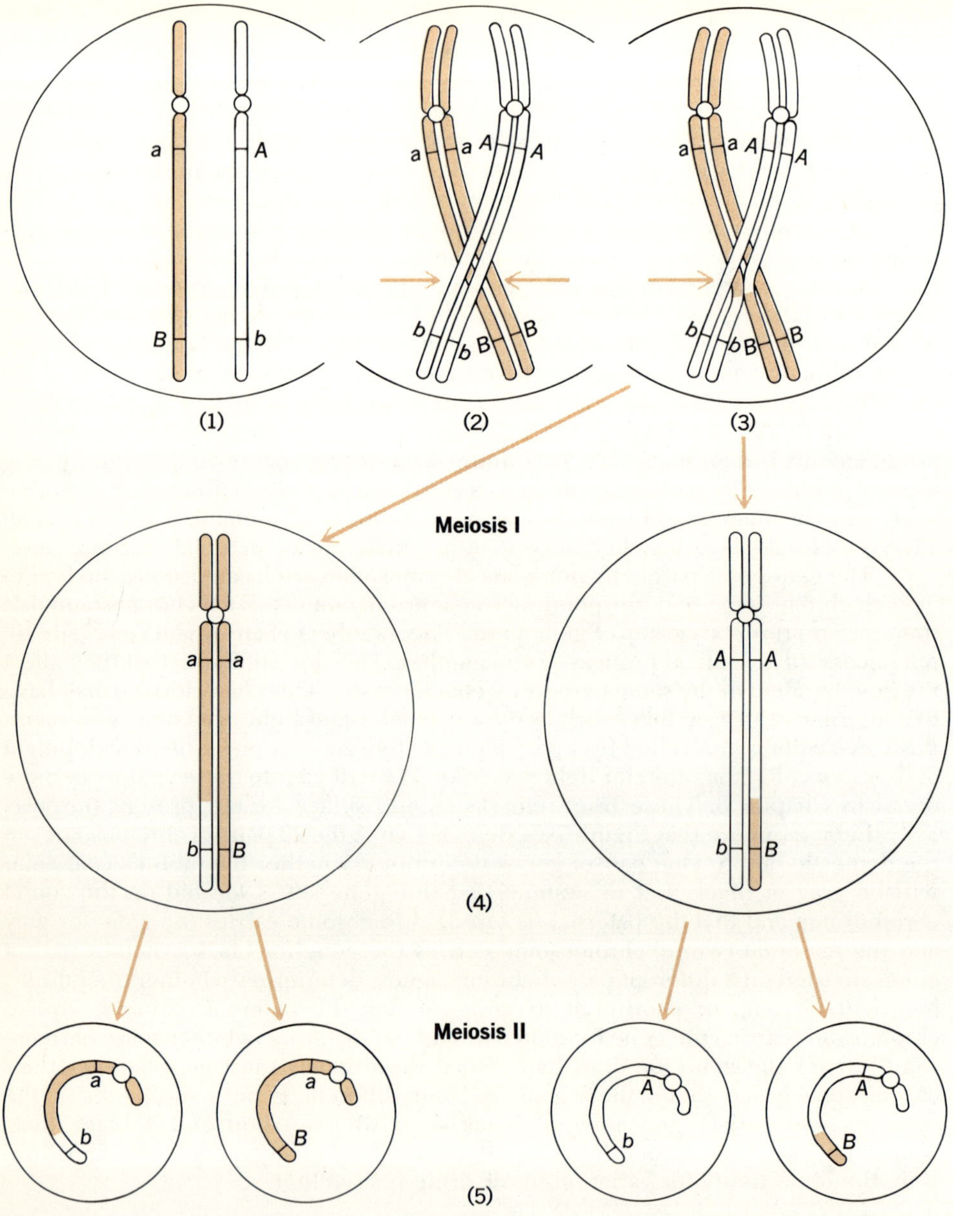

Figure 7-22 Crossing-over (arrow) and the recombination of genes on one chromosome. Two gene locations are shown on each chromosome as reference points: *a* and *B* on the paternal chromosome (colored), and their counterparts *A* and *b* on the maternal chromosome (white). (1–4) Meiosis I; (4–5) meiosis II.

is depicted between these gene locations on two chromatids but not on the other two chromatids. Consequently, either the *A* or *a* gene can become linked with either the *B* or *b* gene. Four different kinds of gametes can be produced, two through cross-over (*A-B* and *a-b*) and two without cross-over (*a-B* and *A-b*).

A case such as this can again be demonstrated in the domestic rabbit. A gene set for spotted coat versus unspotted coat occurs on the same chromosome with the gene set for short hair versus long hair. Because of gametes produced with and without cross-over every combination of these coat characteristics is possible: spotted, long; spotted, short; unspotted, long; and unspotted, short.

The Significance of Biologic Diversity

In short, meiosis fosters reproductive variety not only through the *reassortment* of genes on different chromosomes but also through the *recombination* of genes on the same chromosome. The potential for biological diversity becomes staggering when one tries to imagine the different possible genetic combinations between the male and female gametes of two organisms, let alone the combinations possible in a *population* of reproducing individuals.

The capacity for variation in sexually reproducing populations is of the greatest adaptive significance. Without variety organic evolution could not take place. Without variability organisms could not have maintained their adaptiveness over eons of time to environments that refused to remain stable. Lacking the capacity to change, living things would have become extinct long ago. Indeed life itself would never have evolved in the first place. Natural selection can operate only within diverse populations where survival is predicated on biologic differences and the relative degrees of fitness and unfitness of the individuals within those populations to all kinds of conditions. Biologic diversity is the raw material out of which the beautifully adapted creatures of this planet have been made.

QUESTIONS

1. Describe several different forms of asexual reproduction as it occurs in monerans, protists, plants, and "lower" animals.
2. Compare and contrast the processes of mitosis and meiosis.
3. Compare sexual reproduction in *Paramecium*, *Chlamydomonas*, and *Volvox*.
4. Describe the alternation of generations in ferns.
5. Describe some of the adaptations which enabled flowering plants to invade dry land environments so successfully.
6. What are the female parts of a flower? What are the male parts?
7. How does fertilization occur in a flower?
8. Of what importance are sepals and petals to reproduction?
9. Sexual reproduction favors the exchange of gametes between different individuals. How is this accomplished in paramecia? How is it done in ferns, flowering plants, and animals?
10. Describe the adaptive mechanisms which make internal fertilization possible in man.
11. How does fertilization alter endocrine secretion in the human female and what are some of the consequences?
12. Clearly explain why sexually reproduced offspring are always dissimilar.
13. Even though asexual reproduction is simple and efficient whereas sexual reproduction often necessitates elaborate mechanisms, the overwhelming majority of species reproduce sexually. Why is this?

Suggested Readings

Arey, L. B. *Development Anatomy*. Philadelphia: Saunders, 1965.

Johnson, W. H. *et al. Principles of Zoology*. New York: Holt, Rinehart and Winston, 1969.

Grollman, S. *The Human Body*. New York: Macmillan, 1969.
Ham, A. W. *Histology*. Philadelphia: Lippincott, 1969.
Lee, A. E., and C. Heimsch. *Development and Structure of Plants: A Photographic Study*. New York: Holt, Rinehart and Winston, 1962.
Muller, W. H. *Botany*. New York: Macmillan, 1963.
Nitsch, J. P., and C. Nitsch. "Haploid Plants From Pollen Grains," *Science*, **163,** January 3, 1969.
Sistrom, W. R. *Microbial Life*. Modern Biology Series. New York: Holt, Rinehart and Winston, 1969.
Vickerman, K., and F. E. G. Cox. *The Protozoa*. Introductory Studies in Biology. Boston: Houghton Mifflin, 1967.
Wilson, C. L., and W. E. Loomis. *Botany*. New York: Holt, Rinehart and Winston, 1967.

SECTION III The Responsive Population: source of stability and change

PROLOGUE

Intact living organisms are the smallest units in which the full ensemble of living processes is manifested. They are what they are by reason of the parental generations that produced them and are the products of a population that extends backwards in time. To an equal extent they are what they are because *those not like them* have been eliminated. Living things are the selected products of contemporary populations that have survived the tests for biologic fitness. Life can be understood only as a population continuum extending into time and space.

A species is a population of individuals, most of whom differ genetically in varying degrees from one another. Such a population presents many models for acceptance or rejection. Like a well-managed investment corporation, life, in its conquest of energy, diversifies its holdings as a hedge against an indefinite future. Models rejected by environmental pressures as unfit are eliminated from the inventory. Accepted models become the parents of the next generation. The population responds to the environmental direction imposed upon it.

It is here, in the dynamic ebb and flow of living populations, that the drama of life unfolds. It is also here, and only here, that the biologist can expect to get definitive answers to his ultimate questions of why and what for.

Mature white-bark pine "trees" are dark-colored shrubs at 12,000 ft in the Sierra Nevada Mts. (Photograph by the author)

CHAPTER 8 Genetics: hereditary foundations for stability and change

THE FUNDAMENTALS OF HEREDITARY TRANSMISSION

Mendelian Single-Factor Inheritance

An Austrian monk by the name of Gregor Mendel was the first person to give the study of heredity and variation (*genetics*) a firm scientific foundation. In 1866 (seven years before chromosomes were even known to exist) he published a paper describing breeding tests with garden peas he had begun ten years earlier. Although this report came to the attention of some of the best biologists of that time, it received scant notice. The scientific world did not yet have the background to appreciate its real significance. It was not until 1900 that three biologists,[1] working independently of one another, reached conclusions similar to those of Mendel. They were unaware of Mendel's work at that time. As happens so often in science, when a field is "ready," discovery is virtually certain. Even if a modest Austrian monk had never become fascinated by the problem of hereditary transmission, it is doubtful whether the development of genetics as a science would have been seriously affected.

Mendel chose garden peas in most of his work for several reasons. Flower structure is such that cross-pollination does not usually take place. It is therefore relatively easy to develop and maintain true-breeding varieties. Cross-pollination can be performed experimentally by the surgical removal of immature anthers followed later by artificial pollination of the stigma by hand. He was also impressed by the fact that the plant possessed a number of characteristics (seven that he was aware of) that were expressed in two contrasting forms. Four are listed in Figure 8-1. Traits of this nature lend themselves to objective analysis. A plant is either tall or short. Its flowers are either red or white. Barring environmental variables, one should not be confronted in breeding tests with the prospect of making subjective judgments concerning relative height or degrees of redness or whiteness, at least in the parents one selected to begin with.

Mendel first established true-breeding strains for the various traits. He then made crosses between strain pairs as indicated in Figure 8-1. The offspring in all cases resembled just one of the parents. The characteristic present in the offspring he referred to as *dominant*. That which did not appear he called *recessive*. Of the four pairs of traits shown, round and yellow seeds, red flowers, and long stems are obviously dominant.

He then allowed all the hybrid offspring of these crosses to self-pollinate, producing a second generation of plants. Interestingly the recessive characters reappeared in this population in the ratio of three dominants to one recessive. No individuals were intermediate between or a blend of dominant or recessive traits: there were no pink-flowered or medium-sized plants.

[1] Correns (Germany), DeVries (Holland), and Tschermak (Austria).

PARENTS	OFFSPRING	2ND GENERATION CENSUS	RATIO
1. Round × wrinkled seeds	All plants round	5474 round	
		1850 wrinkled	2.96:1
2. Yellow × green seeds	All plants yellow	6022 yellow	
		2011 green	3.01:1
3. Red × white flowers	All plants red	705 red	
		224 white	3.15:1
4. Long × short stems	All plants long	787 long	
		277 short	2.84:1
	Totals	12,988:4352	2.98:1

Figure 8-1 A summary of four of Mendel's crosses. Parent plants differing in one characteristic, for which they bred true, were cross-pollinated. The offspring of each of these crosses were then allowed to self-pollinate, producing the second generation.

Mendel concluded from studies of this nature that hereditary factors are separate entities that do not mix when they are transmitted to offspring. They remain unaltered as they pass from one generation to the next. Each pea plant has two factors for each characteristic. When gametes are formed, these factors become segregated so that a sex cell receives only one factor for each trait. Offspring consequently receive one factor for a given trait from each of their parents.

The remarkable conformity between Mendel's conclusions and the behavior of chromosomes during meiosis is quite obvious. As we have observed, any one kind of gene (Mendel's factor) is located on a particular chromosome and at a certain place on that chromosome. It is believed that each of the chromosomes of a homologous pair carries genes for the same characteristics. Consequently, a pair of chromosomes constitutes a battery of many different gene pairs representing a variety of potential traits. The two members of a gene pair determining the development of a particular character may be present in any possible combination in a diploid cell. In other words, they may exist in a dominant-dominant, dominant-recessive, or recessive-recessive combination. During meiosis, with the separation of maternal and paternal chromosomes, there will be a corresponding segregation of hereditary factors carried by these chromosomes.

At this juncture it will be helpful to become familiar with some useful techniques and terms. First, it is much simpler to refer to a gene by a letter of the alphabet than by a name. The letter that represents a gene set is capitalized for the dominant and in lower case for the recessive gene. Thus *R* represents the gene for red flowers and *r* the gene for white flowers. The alternative forms of a gene set, such as *R* and *r*, are said to be *alleles*. In the case of *R* and *r* these alleles may exist in any one of three possible combinations in the diploid cell: *RR*, *rr*, or *Rr*. The first two of these gene combinations, or *genotypes*, represent a double dose of the same allele. Such genotypes are said to be *homozygous*. *RR* is a homozygous dominant genotype and *rr* is a homozygous recessive genotype. *Rr*, carrying one of each allele, is a *heterozygous* genotype. These genotypes affect the physical appearance of pea plants by producing two kinds or *phenotypes* of individuals: (1) red-flowered plants (genotypes *RR* and *Rr*) and (2) white-flowered plants (genotype *rr*).

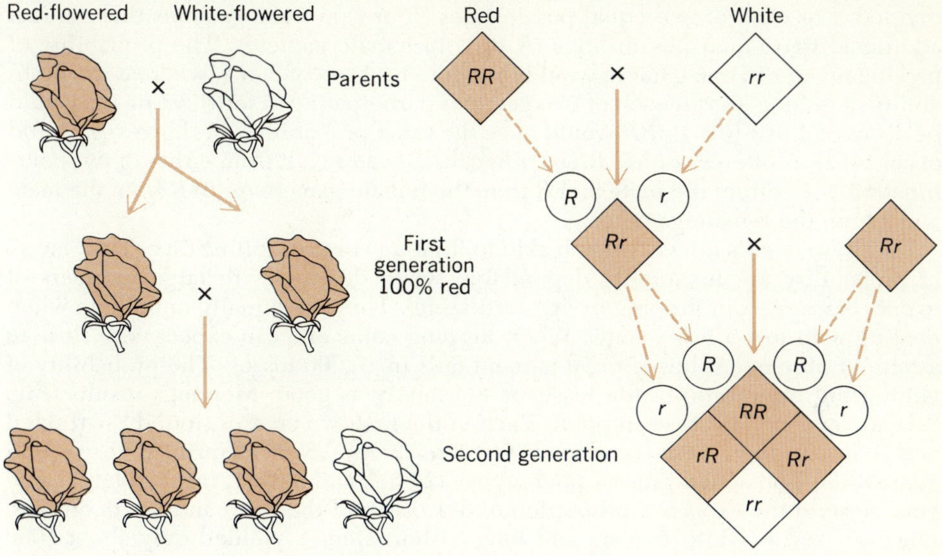

Figure 8-2 The second generation progeny of grandparents which were homozygous for red flowers (*RR*) and white flowers (*rr*) exhibit at 3 : 1 ratio of red-flowered to white-flowered plants. Although the first generation carries the recessive allele *r*, it is not expressed phenotypically; all of this generation are red-flowered (*Rr*). Compare gene segregation in the first generation cross to chromosome behavior in meiosis (Fig. 7-7).

With this background in vocabulary to assist us let us now trace the course of genetic events in Mendel's red-flowered × white-flowered crosses (Fig. 8-2). True-breeding (homozygous) red-flowered plants, *RR*, are crossed with true-breeding (homozygous) white-flowered plants, *rr*. Since the gametes of the red-flowered parents can carry only the *R* allele and those of the white-flowered parents only the *r* allele, all of the offspring will have to be heterozygous (*Rr*). Because of dominance they will develop red blossoms.

The gametes produced by these hybrids, however, will be of two types. Half of them will carry the *R* allele and half will have the *r* allele because of the segregation of maternal and paternal chromosomes in meiosis (refer to Fig. 7-7). As these plants are left to self-pollinate one or the other of the two kinds of sperm nuclei will unite in fertilization with one or the other of the two kinds of egg cell nuclei strictly according to chance. The resulting genotypes can be easily determined by the checkerboard method as shown in Figure 8-2. Four gene combinations are possible: *RR*, *rR*, *Rr*, and *rr*. Since *rR* and *Rr* are genetically indistinguishable, the ratio that these genotypes bear to one another is 1 *RR* : 2 *Rr* : 1 *rr*. Since three out of the four genotypes contain at least one *R* allele, $\frac{3}{4}$ or 75 percent of these plants should be red-flowered. One-quarter are *rr* and should bear white blossoms. Insofar as the phenotypes of these offspring are concerned the ratio should be 3 red-flowered : 1 white-flowered. As we can see from Figure 8-1, Mendel's actual results came quite close to these predicted calculations (705 red to 224 white or 3.15 : 1).

It should be emphasized that the gametic combinations which occur during fertilization are the result of chance. Half the male gametes of the first-generation hybrids carried allele *R*, the other half allele *r*. The female gametes were similarly divided. For each of two equal possibilities (*R* or *r*) in male gametes there are an additional two equal possibilities (*R* or *r*) in female gametes. The probability of picking up an *r* in one gamete would therefore be 1 out of 2 or $\frac{1}{2}$, whereas the probability of getting an *r* in each of two gametes participating in fertilization (*rr*) would be 1 out of 4 or $\frac{1}{4}$ ($\frac{1}{2} \times \frac{1}{2}$). *RR* would have the same probability. *Rr*, however, would occur twice as often as either *RR* or *rr* because it can result from either of two combinations: (1) *r* from the male and *R* from the female gamete or (2) *R* from the male and *r* from the female gamete.

These events are very much akin to flipping coins or rolling dice. The "laws" of probability are inexorably dependable when dealing with large numbers of events (such as coin flipping or egg fertilizing). They are equally unreliable when dealing with just a few events. When flipping coins one can expect very close to a ratio of 50 percent heads to 50 percent tails in 10,000 tosses. The probability of getting the same ratio in ten tosses is not nearly as good. Mendel's results (Fig. 8-1) are an excellent case in point. Each of the first two crosses (round × wrinkled and yellow × green seeds) produced between 7000 and 8000 offspring in the second generation. The actual ratio of phenotypes (2.96:1 and 3.01:1, respectively) came very close to the expected proportion of 3:1. Each of the third and fourth crosses, however (red × white flowers and long × short stems), yielded only about 1000 second-generation offspring. The observed ratios here showed somewhat greater deviations from that predicted (3.15:1 and 2.84:1, respectively).

Genetic Reassortment

Mendel was not only concerned with the inheritance of single characters but also with the manner in which a combination of characters might be transmitted. Are different characteristics passed along together or are they inherited as separate, independent entities? It is interesting in this regard that the seven traits he used happened to be controlled by genes which are apparently located on seven different chromosomes (and the garden pea has just seven pairs of chromosomes!). One of Mendel's approaches to this problem, illustrated in Figure 8-3, was to cross-pollinate plants that were homozygous for round and yellow-colored seeds (*RRYY*) with plants homozygous for wrinkled, green seeds (*rryy*). As might be expected, all the hybrid offspring displayed the dominant traits. Since one parent produced gametes, all of which carried *R* and *Y* (on different chromosomes) and the other produced gametes all of which carried *r* and *y* (on different chromosomes), the first generation was *RrYy* (round and yellow-seeded). It was the second generation of this cross that convinced Mendel (who, after all, knew nothing of chromosomes) that the hereditary factors for different traits are assorted independently. *RY* and *ry* do not necessarily go together. There must also be *Ry* and *rY* gametic combinations because there were four kinds, not just two kinds of plants in the second generation (Fig. 8-3). Out of 556 plants he observed round yellow, round green, wrinkled yellow, and wrinkled green phenotypes in a ratio that closely approached the theoretical 9:3:3:1 ratio predicted on the basis of independent assortment (Fig. 8-4).

In the test crosses described, as well as many others, Mendel firmly established several important concepts which are basic to an understanding of inheritance.

1. The alleles constituting a gene pair are segregated into separate gametes. As we know now, this occurs as a consequence of the reduction in chromosome number during meiosis.

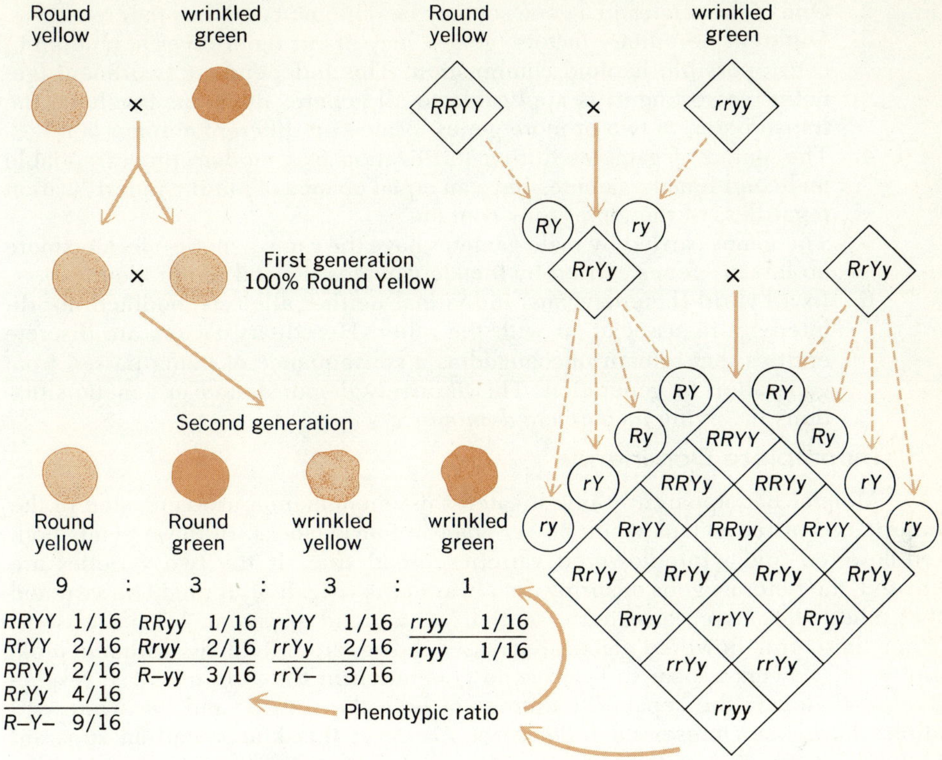

Figure 8-3 The genetic reassortment of genes carried on different chromosomes. The second generation descendants of a grandparent which was homozygous round- and yellow-seeded (*RRYY*) and one which was homozygous wrinkled- and green-seeded (*rryy*) exhibit a phenotypic ratio of $\frac{9}{16}$ round, yellow: $\frac{3}{16}$ round, green: $\frac{3}{16}$ wrinkled, yellow: $\frac{1}{16}$ wrinkled, green. The alleles for seed shape and color reassort at random. Compare gene reassortment in the first generation cross with chromosome reassortment in Figure 7-21.

		(*R-Y-*) ROUND YELLOW	(*R-yy*) ROUND GREEN	(*rrY-*) WRINKLED YELLOW	(*rryy*) WRINKLED GREEN	
Theoretical	Ratio	9	3	3	1	
	Plant numbers	313	104	104	35	= 556
Observed	Ratio	9.1	3.1	2.9	0.9	
	Plant numbers	315	108	101	32	= 556

Figure 8-4 In his test of the cross illustrated in Figure 8-3, Mendel obtained 556 plants in the second generation. His observed results closely approached the theoretical 9:3:3:1 phenotypic ratio predicted on the basis of independent gene assortment.

2. One of the alleles in a gene set may be dominant and the other recessive.
3. Different hereditary factors (genes) may assort themselves at random in every possible haploid combination. This independent assortment (genetic reassortment) is applicable to all genetic situations involving the transmission of two or more genes located on different chromosomes.
4. The union of gametes during fertilization is a random process. Viable male and female gametes have an equal chance of joining in fertilization regardless of the genes they contain.
5. The genes carried by male gametes have the same genetic effect, no more no less, as genes carried by female gametes (*Rr* and *rR* are identical).
6. In a hybrid (heterozygous) individual neither allele is modified nor diluted by its association with the other. Hereditary factors are discrete entities that remain unchanged as a consequence of being passed from generation to generation. This is nicely demonstrated in genetic situations involving *incomplete dominance.*

Incomplete Dominance

Mendel had apparently investigated this phenomenon as manifested in the inheritance of red versus white flower color in four-o'clocks. In these plants both red-flowered and white-flowered varieties breed true. If the two varieties are crossed, the heterozygous offspring are always pink (Fig. 8-5). It might be assumed that inheritance here consisted essentially of a blend of parental characteristics. When these pink-flowered plants are crossed, however, individuals with red, pink, and white flowers appear in the second generation in the ratio of 1:2:1, respectively. Obviously the separate identities of both the red (R) and the white (R') alleles have been preserved fully intact. Cases of this kind entail an apparent blending of parental traits because neither allele is dominant to the other. Hetero-

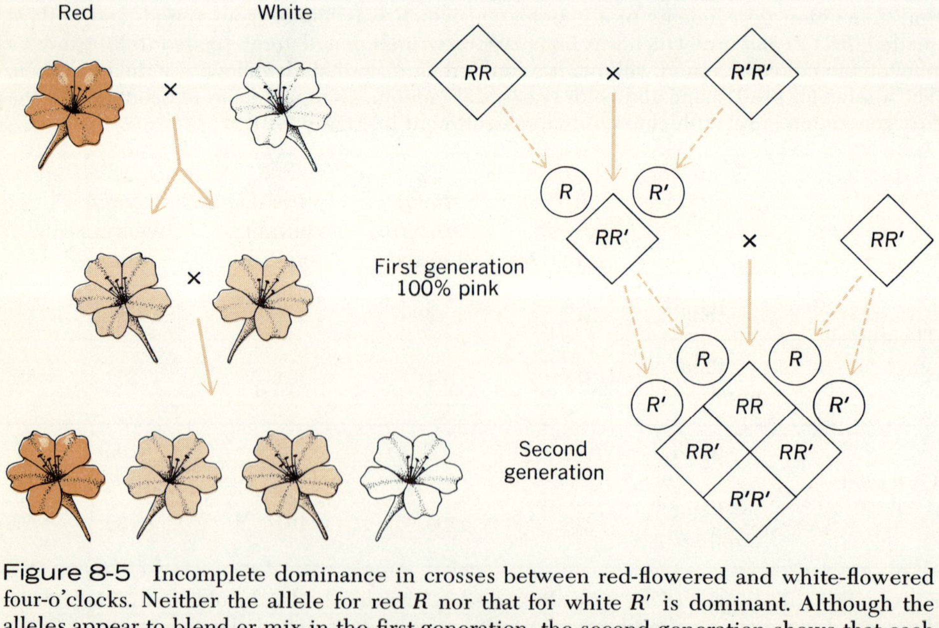

Figure 8-5 Incomplete dominance in crosses between red-flowered and white-flowered four-o'clocks. Neither the allele for red R nor that for white R' is dominant. Although the alleles appear to blend or mix in the first generation, the second generation shows that each allele has maintained its separate and distinct identity.

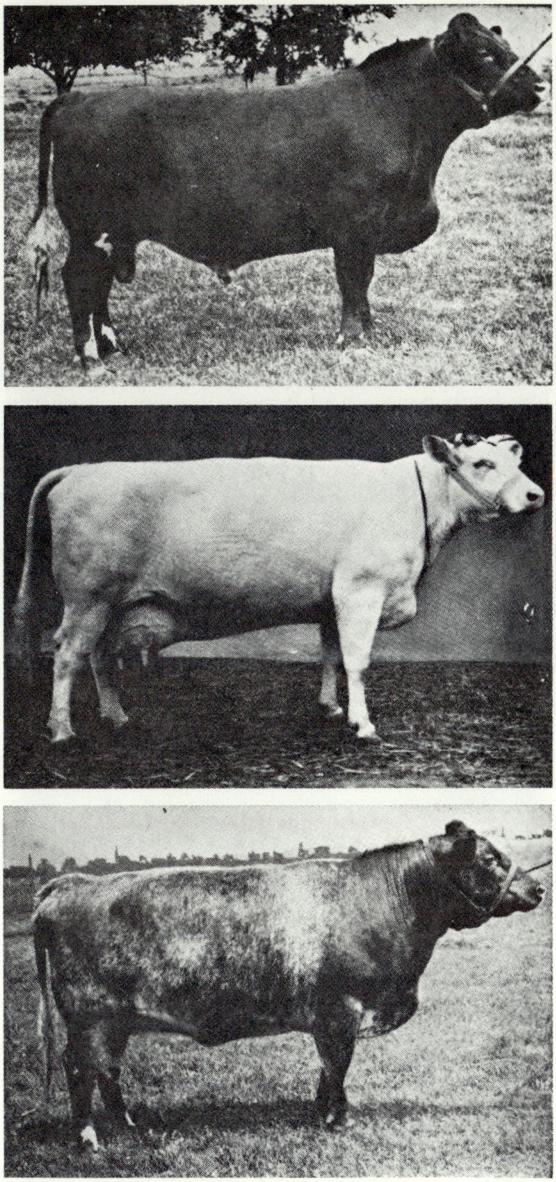

Figure 8-6 Incomplete dominance in a cross between a red Shorthorn bull and a white Shorthorn cow. The offspring is a mixture of red and white called roan. (L. H. Snyder, *Principles of Heredity*, D. C. Heath, 1946)

zygous individuals (RR') are consequently intermediate in phenotype because of of the phenotypic influence of two alleles, each of which is *incompletely* dominant to the other.

Examples of incomplete dominance are common throughout the living world. Figure 8-6 illustrates a classical example in the inheritance of coat color in Shorthorn cattle. A red × white cross produces calves that are roan (a mixture of red and white). Roan by roan crosses produce offspring in the ratio of 1 red : 2 roan : 1 white (in a large population of such offspring).

PHENOTYPES IN TERMS OF GENE ACTION

It will be useful at this point to relate the phenotypic results of genetic crosses to the nature and action of genes themselves. How do the alleles of a gene set differ biochemically? What is dominance and recessiveness at the level of enzyme action and protein structure? The reader may profit by referring back to pages 102–110 where enzyme synthesis and its regulation through gene action were discussed. We learned then that a gene, through a linear series of four different bases (A, T, G and C), encodes the primary structure of a specific protein molecule. If one of these bases were changed to one of the others, the corresponding codon in mRNA would be altered so that a different amino acid might be specified. The resulting protein molecule might therefore deviate from the normal by a substitution of one amino acid. This is often adequate to alter the biological activity of a protein significantly even though a change has been made in only one of several hundred amino acid monomers making up the molecule. It is this sort of thing that distinguishes the alleles in a gene set.

An explanation for many cases of recessiveness lies in the fact that some alleles result in the synthesis of biologically inadequate or inactive enzymes. Dominant alleles, on the other hand, commonly yield biologically active enzymes. A heterozygous individual would presumably produce one-half the amount of enzyme that a homozygous dominant one would. If one-half the amount of a particular enzyme performs a given metabolic function adequately, we would have a case of complete dominance. If it does not, incomplete dominance is involved. One must not assume, however, that recessive alleles are always the inactive

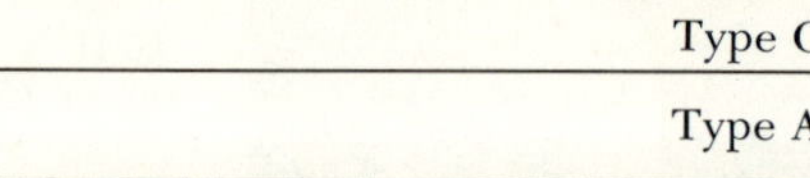

(a)

A PERSON WITH THESE GENOTYPES	WILL HAVE THESE BLOOD GROUPS (PHENOTYPES)
ii	Type O blood
I^AI^A I^Ai	Type A blood
I^BI^B I^Bi	Type B blood
I^AI^B	Type AB blood

(b)

Figure 8-7 Multiple alleles in human blood groups. The ABO blood groups in man are classified by the presence or absence of two blood group factors, A and B, carried in red blood cells. Allele I^A controls the development of factor A, allele I^B controls factor B and allele i produces no blood group factor. (a) I^A and I^B are codominant with respect to one another, and are dominant to the recessive i allele. (b) The six possible genotypes produce four phenotypes: blood types O, A, B, and AB.

forms of a gene. It is possible for a recessive allele to yield an enzyme that converts a given substrate into some useful product. The dominant enzyme might be so modified that it has a higher affinity for the substrate than normal enzyme but produces end products of little metabolic value (certain hereditary diseases of man may be of this nature).

Multiple Alleles

We have so far discussed patterns of inheritance in which there are but two alleles in a gene set. There are many instances, however, in which more than two alternative alleles occupy the same position on a given chromosome and affect the same genetic character. Such multiple alleles, as they are called, may exist in considerable variety (up to 12 or more). Any one individual, of course, is some combination of two of the various alleles. A simple example of this sort dealing with human blood groups is illustrated in Figure 8-7. The ABO blood groups in man are classified by the presence or absence of two blood group substances (designated the A and B substances)[2] carried by red blood cells. There are three alleles involved. Allele I^A controls the development of substance A and allele I^B controls substance B. Neither of these alleles is dominant to the other. A third allele, *i*, produces no blood group substance and is recessive to the others. Of the six different combinations of two of these three alleles, four blood groups (phenotypes) are possible: O (with neither substance A nor B), A (with substance A), B (with substance B), and AB (with both blood group substances).

Single Genes with Multiple Effects

Since genes actually operate at the macromolecular and metabolic level (in the biosynthesis of enzymes, blood group substances, and structural proteins such as hemoglobin), the phenotype of an organism is the end product of a very elaborate developmental process. Certainly metabolic pathways and interacting feedback systems such as the Krebs cycle are complex enough (refer to page 131). Environmental influences on gene-mediated events during embryonic development are of matching complexity. A single chemical reaction can have diverse effects because of the intimate interdependencies of metabolic events.

Confronted by these considerations one is led to the inescapable conclusion that a gene may produce several phenotypic results (for example, the gene responsible for the enzyme converting $D \rightarrow E$ in Fig. 3-21). In fact, it is not at all unlikely that *every* gene controls the development of more than one trait. Many examples of this nature are known. In man a hereditary disease (phenylketonuria) is characterized by the absence of an enzyme that converts one amino acid (phenylalanine) into another (tyrosine). The recessive allele responsible for this (when homozygous) not only brings about high concentrations of phenylalanine in the blood but also occasions the development of a number of collateral effects: (1) failure to produce normal amounts of pigment in the skin; (2) tendency toward epileptic convulsions; and (3) severe mental retardation. By restricting the dietary intake of phenylalanine early in life the symptoms of this genetic malfunction can be largely, if not entirely, abated. The phenotypic influence of this allele therefore is not

[2] These substances are differing glycopeptides: molecules composed of a polysaccharide combined with a peptide.

only versatile but varied. Depending upon the environment (diet), the phenotype may be affected in several ways or not at all.

Multiple Genes with a Single Effect

As stated, a single enzyme (controlled by a gene), within a long sequence of interrelated and dependent metabolic reactions, can influence several different end products. These end products in turn are ultimately responsible for the development of certain phenotypic characters. It is equally true that many different enzymes (controlled by many different genes) within a long metabolic sequence are responsible for the synthesis of a common end product (for example, substance Z in Fig. 3-21). The phenotypic trait which develops as a consequence of this product is the result of many different genes acting in concert. Instances of this nature, in which one pair of genes may modify the expression of another pair of genes, involve *genetic interaction.*

Let us examine gene interaction and other patterns of inheritance which deviate from the relatively simple mechanisms investigated by Mendel.

THE INTERRELATIONSHIPS OF GENES IN TIME (INTERACTION) AND SPACE (LINKAGE)

The Interaction of Nonallelic Genes

From what has been said we can conclude that no gene acts by itself insofar as its phenotypic effects are concerned. An organism is the product of countless integrated reactions mediated by enzymes, which in turn are controlled by genes. Furthermore, the course of these reactions as well as their ultimate phenotypic consequences are sensitive to varying environmental conditions. The development of an individual is therefore dependent on the total genotype and all the environmental variables that may modify the phenotypic expressions of that genotype.

Let us probe into the nature of these dynamic events by first considering a simple type of genetic interaction.

Simple Interactions Involving Two Nonallelic Genes

In Mendel's test cross in which he allowed heterozygous round and yellow-seeded plants (*RrYy*) to self-pollinate it will be recalled that the phenotypic ratio of the offspring was 9:3:3:1 (Fig. 8-3). This ratio is typical of crosses involving complete dominance in which two gene sets assort independently and each effects a different character. When two different sets of alleles located on different chromosomes effect the *same* character, however, the 9:3:3:1 ratio will become modified in some way. How it is modified depends on the nature of the interrelationship between the two nonallelic genes. Let us examine a specific case.

A number of plants are known to produce cyanide (hydrocyanic acid, HCN). White clover is one of these. Some genetic strains have a high cyanide content, whereas others show very low yields. In the synthesis of cyanide in white clover the metabolic reaction illustrated in Figure 8-8(a) is believed to take place. Enzyme 2 (specified by allele *B*) converts a substrate into cyanide. The substrate, however, is produced from a precursor substance by the action of another enzyme (enzyme 1 specified by allele *A* of another independently assorting gene set). Two dependent reactions must occur (both enzymes must be present) for cyanide syn-

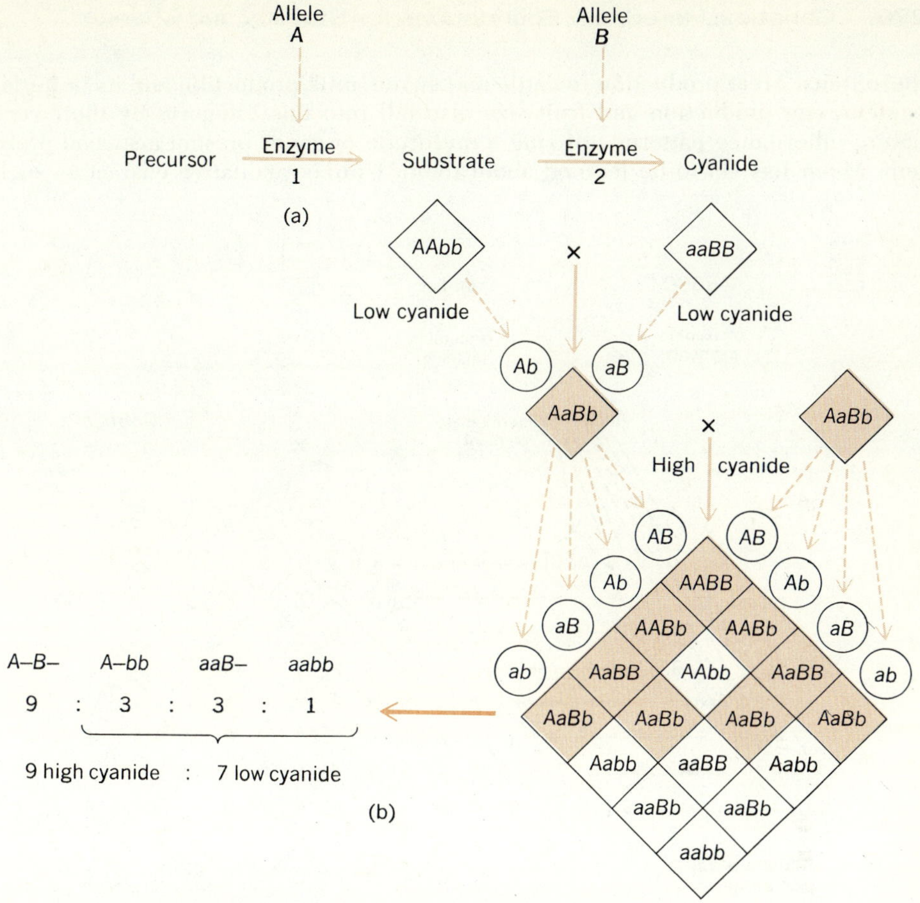

Figure 8-8 Gene interaction in white clover. (a) Two enzymes (1 and 2) specified by two genes (*A* and *B*, respectively) convert a precursor to cyanide. (b) Since both enzymes are necessary for cyanide production and the recessive alleles (*a* and *b*) do not specify active enzyme, interesting genetic results can take place. Two different strains homozygous for low cyanide produce heterozygous high cyanide progeny. The ratio of high to low cyanide progeny in the second generation is 9:7.

thesis to take place. Since the recessive alleles in both sets (*a* and *b*) do not specify an active enzyme, only those genotypes processing at least one *A* and one *B* (*A*—*B*—) will produce a high cyanide phenotype (Fig. 8-8b). Unexpected results sometimes occur in these genetic situations, as shown in Figure 8-8. A cross between two different strains homozygous for low cyanide content (*AAbb* × *aaBB*) will produce offspring all of which have high cyanide yields (*AaBb*). The 9:7 phenotypic ratio in the second generation is merely a modification of a 9:3:3:1 in which the 3:3:1 portions of the population are phenotypically indistinguishable.

For the reader interested in pursuing gene interaction further, additional examples are in the questions at the end of this chapter.

Complex Multiple-Gene Inheritance

A great many hereditary characteristics are influenced by a number of nonallelic genes, particularly those traits in which we as human beings are often most interested. Intelligence, height, and degree of skin pigmentation in man are among these traits. Meat production in cattle and swine, milk production and its butterfat content, egg production and fruit size also fall into this category. By their very nature inheritance patterns utilizing a multitude of genes present a special problem. Much has yet to be learned about them. Unlike qualitative characters such

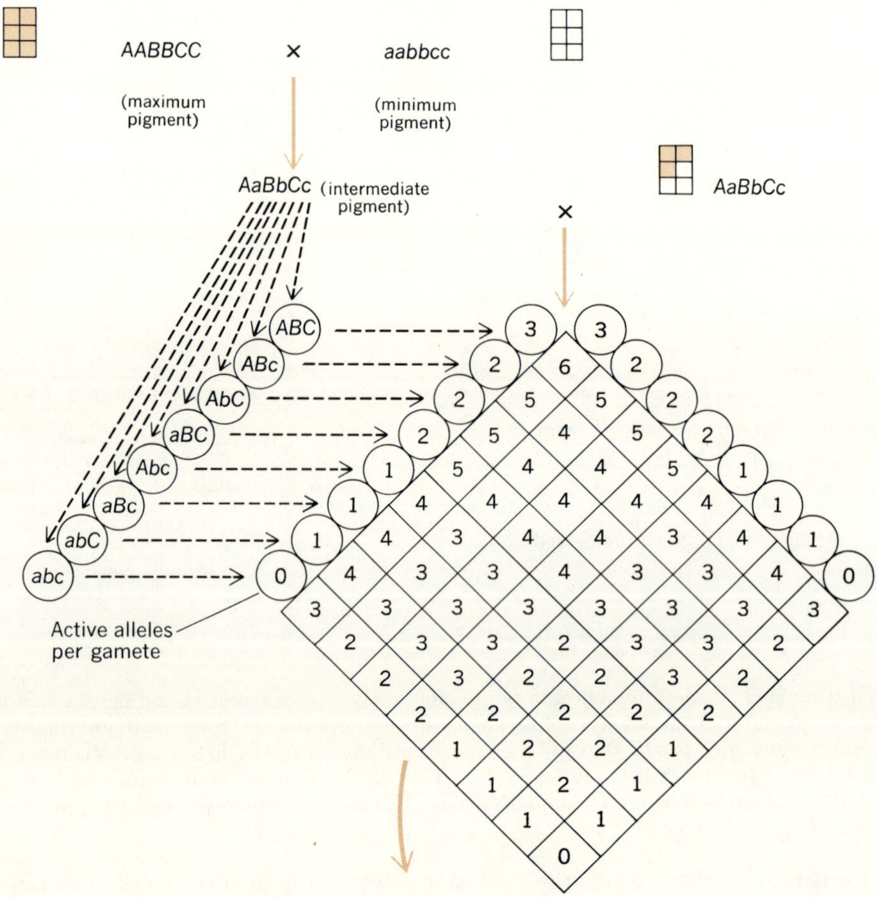

6 Active alleles (*AABBCC*)	5 Active alleles	4 Active alleles	3 Active alleles	2 Active alleles	1 Active allele	0 Active alleles (*aabbcc*)
1/64	6/64	15/64	20/64	15/64	6/64	1/64
Very dark	Dark	Dark intermediate	Intermediate	Light intermediate	Light	Very light

Figure 8-9 The inheritance of human skin pigmentation. This simplified model assumes that three pairs of incompletely dominant nonallelic genes are involved. Each *A*, *B*, or *C* allele is active and contributes one unit of pigment (■). Each *a*, *b*, or *c* allele is inert and contributes no pigment (□).

as red versus white flowers, multiple-gene traits do not fall into distinct phenotypic groups. Rather than being black versus white, people show almost infinite gradations between very dark and very light. Variation is gradual and continuous rather than abrupt and discontinuous. Such characteristics are said to be quantitative rather than qualitative.

A hypothetical model will perhaps help us gain some appreciation of multiple-gene inheritance and quantitative traits. It is believed that many pairs of genes are involved in human skin pigmentation. For simplicity's sake let us assume there are only three pairs of independently assorting genes, as illustrated in Figure 8-9 (*A-a*, *B-b*, and *C-c*). All gene sets show incomplete dominance. Each *A*, *B*, or *C* allele is active and contributes one unit of pigment (■). Each *a*, *b* or *c* allele is inert and contributes no pigment (□). The minimum pigment possessed by very light people (such as in freckles) is assumed to be the product of another set of genes. According to such a model the progeny of very dark and very light parents would be intermediate in color. The offspring of such intermediate parents would fall into seven groups differing from one another by one unit of pigment and span-

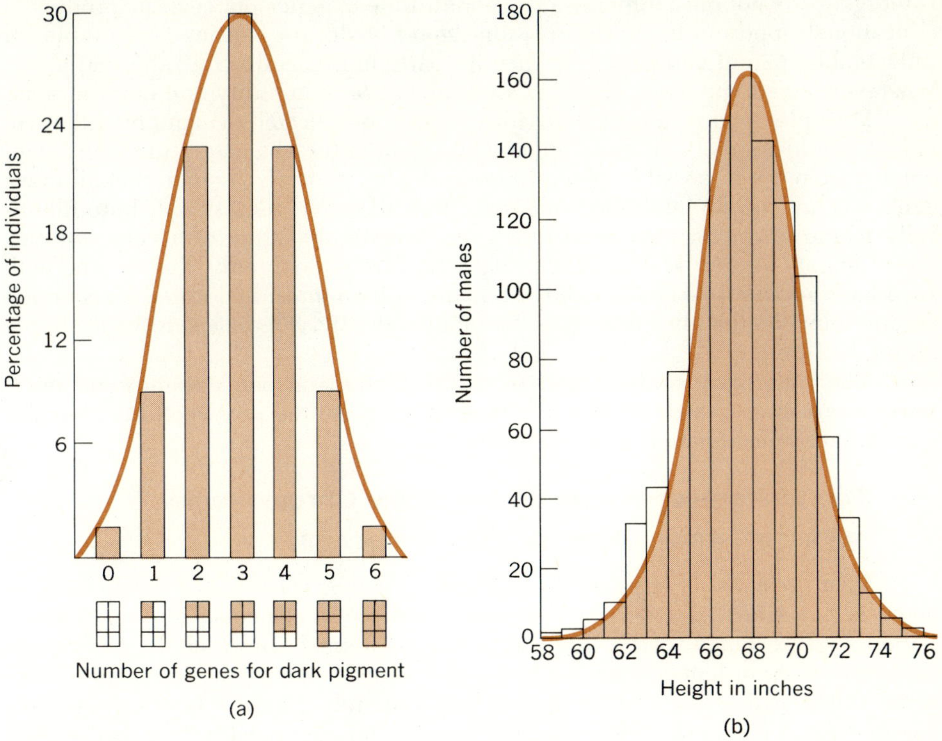

Figure 8-10 The variation found in traits controlled by many genes follows a normal curve of distribution, especially when allowance is made for environmental influences. (a) A bar graph and curve of human skin color variation based on the phenotypic ratio in the second generation of Figure 8-9. The phenotypic bars are separated to accommodate environmental intergradations resulting from varying exposure to the sun. (b) A similar bar graph and curve showing height in a random sample of 1083 men. Height is influenced by many different gene sets in man. [(a) Modified from Cockrum and McCauley, *Zoology*, Saunders, 1965, and (b) from Villee, *Biology*, Saunders, 1962]

ning the color spectrum from very dark to very light. Over 75 percent (50/64) of this second generation, however, would fall within an intermediate range (light intermediate to dark intermediate). The extremes (dark and light) trail off at either end of the preponderant middle group.

When one superimposes the effects of environment on such a phenotypic distribution, the distinction between color groups becomes further obscured. Individuals with genotypes for light to intermediate pigment will be darker when exposed to the sun than when not. When we consider the effects of environment, and that many more than three pairs of genes are involved (and consequently many more than seven genetic groupings), we can readily understand the continuous variation that is observed in quantitative traits like skin color. Figure 8-10 illustrates this. When we take the hypothetical phenotypic ratio of Figure 8-9 and convert it to a bar graph and curve, we find it agrees remarkably well with a similar graph and curve representing height distribution in an actual sampled population. Both height and skin pigmentation are controlled by many genes.

Environment plays an important role in the development of quantitative (as well as qualitative) traits. Height is affected by diet and general state of health. Intelligence is not only influenced by a multitude of genes but is also a product of educational opportunity (or deprivation), good health (or sickness), favorable (or unfavorable) parent and peer relationships early in life, and many other variables. Whatever phenotype is expressed is a product of *both* heredity *and* environment.

Multiple-gene systems are obviously more complicated than our hypothetical model would imply. The development of quantitative characters may very well employ from ten to possibly several hundred gene pairs. To assume that all these gene sets are incompletely dominant and that all active alleles contribute identically to the character expressed is a gross oversimplification of a very complex phenomenon. Various kinds of gene interaction are undoubtedly at work. And with such large numbers of genes involved, many of them must be located on common chromosomes so that they cannot assort independently, at least according to simple Mendelian principles.

The hereditary transmission of genes located on one chromosome poses some interesting questions. We will therefore devote the next section of our discussion to a consideration of this subject.

The Linkage of Genes on the Same Chromosome

SEX DETERMINATION AND LINKAGE

That presumably large numbers of genes may be carried on a single chromosome is beautifully demonstrated in the hereditary mechanism for sex determination. In a great many of the "higher" plants and animals, in which the individuals are either male or female (but not both), the chromosome complements of the two sexes differ. Either the male or the female (depending on the kind of organism) has one pair of chromosomes which differ not only genetically but also in their size and shape.[3] Because of the fundamental role this pair of chromosomes plays in sex determination they are called the *sex chromosomes.* In mammals and many other animals it is the male that possesses the dissimilar pair. One chromosome carries many genes and is called the *X* chromosome. Its structurally distinct "mate" carries only a few genes and is called the *Y* chromosome. The sex chromosomes in

[3] The group of insects to which the ants and bees belong is an exception. The males are haploid and the females diploid, as the reader may recall.

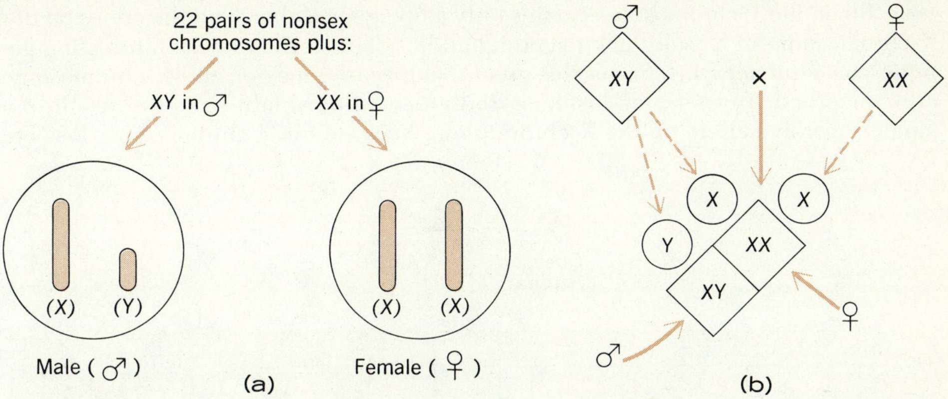

Figure 8-11 Sex in man is determined by a pair of chromosomes, the sex chromosomes. (a) In the male (♂) the members of the pair are dissimilar and are called the *X* and *Y* chromosomes. The female (♀) possesses two similar *X* chromosomes. (b) As a result of meiosis the *X* and *Y* chromosomes are segregated into separate sperm. Fertilization by *Y*— carrying sperm will result in male offspring, by *X*— carrying sperm, female offspring.

females constitute an homologous set (*XX*). These relationships are shown schematically in Figure 8-11a with respect to human sex chromosomes.

During meiosis half of the sperm will receive an *X* chromosome and half will receive a *Y* chromosome (Fig. 8-11b). All the eggs, however, will possess only an *X* chromosome. (Both sperm and eggs, of course, will carry one each of the 22 other nonsex chromosomes.) When fertilization occurs the chances therefore will be about equal that the egg will be fertilized by an *X*-bearing sperm or a *Y*-bearing sperm. Fertilization by a sperm with a *Y* chromosome results in a zygote which develops into a male (*XY*). Fertilization by a sperm with an *X* chromosome results in a zygote which becomes a female (*XX*). The sex of an individual then is normally determined at fertilization, predicated on which type of sex chromosome the fertilizing sperm carries.

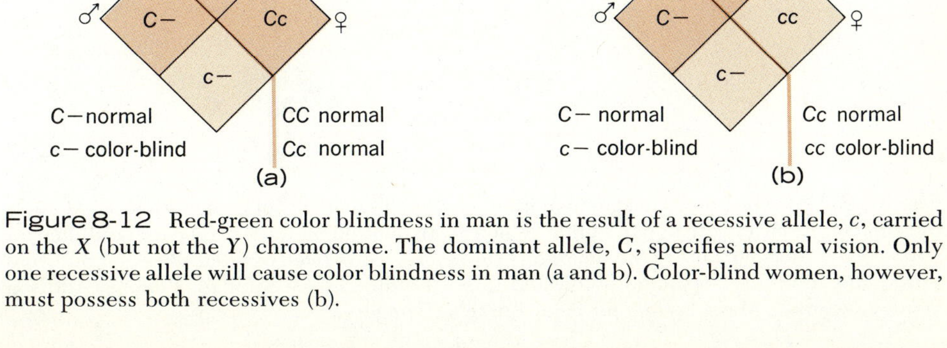

Figure 8-12 Red-green color blindness in man is the result of a recessive allele, *c*, carried on the *X* (but not the *Y*) chromosome. The dominant allele, *C*, specifies normal vision. Only one recessive allele will cause color blindness in man (a and b). Color-blind women, however, must possess both recessives (b).

Since the *Y* chromosome carries only a few genes, almost all the genes on the *X* chromosome of a male lack matching alleles. This leads to very interesting genetic consequences insofar as the great majority of genes on the *X* chromosome are concerned. In men, for instance, red-green color blindness is the result of a single recessive allele on the *X* chromosome because the *Y* chromosome does not

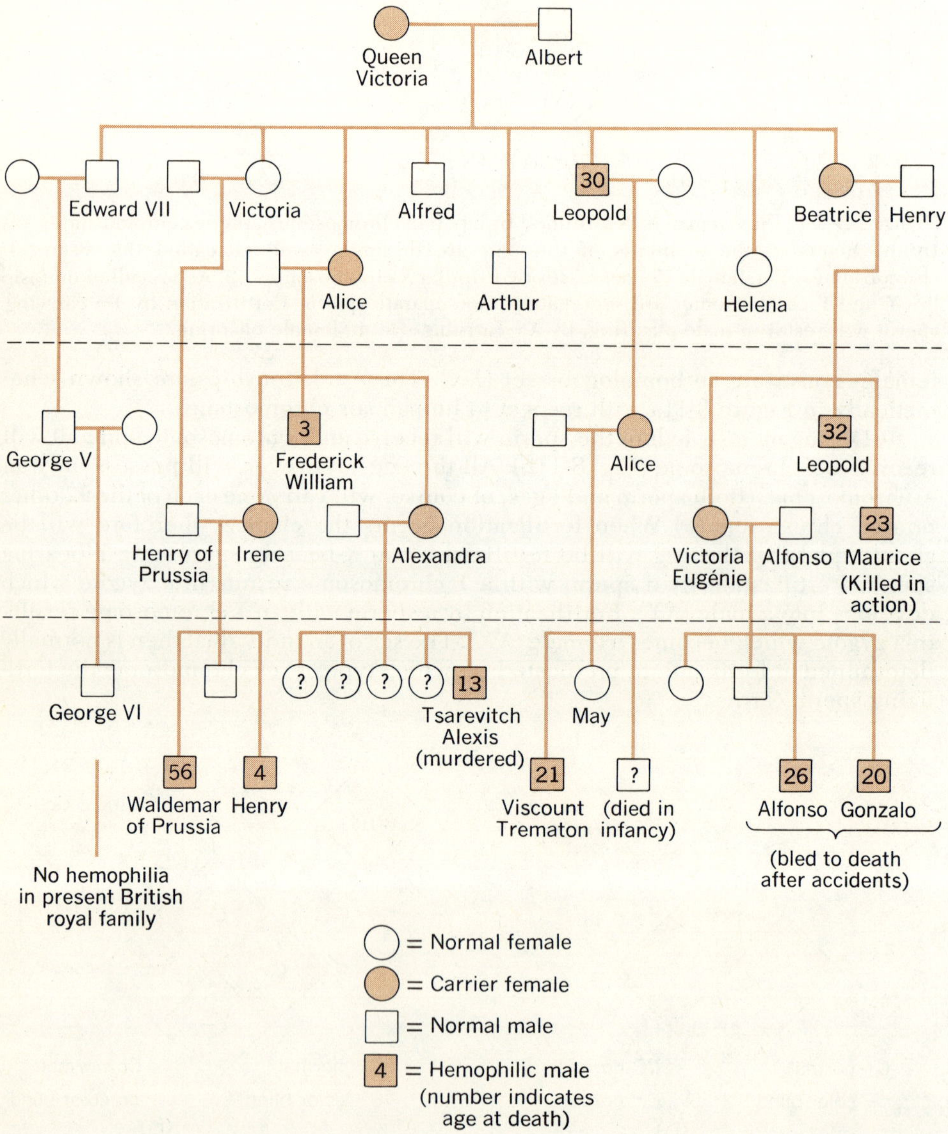

Figure 8-13 The inheritance of hemophilia in the descendants of Queen Victoria. Many of the grandchildren and greatgrandchildren have been omitted because the recessive allele was not received by them. One of the daughters (Louise) has also been omitted. First, second, and third generations are separated by dashed lines. (Modified from Kimball, *Biology*. Addison Wesley, Reading, Mass. Source for data: Burke's *Peerage*)

carry this gene. Although some women are known to have this trait, it is much more common in men. The reasons for this become evident on inspecting Figure 8-12. A male need inherit just one recessive allele (*c*) from his mother to be color blind (Fig. 8-12a). A color-blind woman, on the other hand, must have received a recessive allele from her mother as well as from her father because she must be homozygous recessive to show the trait (Fig. 8-12b). Color-blind women, in other words, must have color-blind fathers and mothers who are either color blind or are normal-sighted heterozygous carriers of the gene for color blindness (*Cc*). Genes such as this that are located on a sex chromosome and consequently result in different phenotypic distributions in the two sexes are said to be *sex-linked.*

A striking example of the disproportionate number of males who exhibit sex-linked traits is seen in the incidence of a form of human hemophilia. This is a disease in which the blood fails to clot normally and is the result of a recessive sex-linked allele. Figure 8-13 shows the occurrence of this disease in the descendants of Queen Victoria who, although not hemophilic herself, was a (heterozygous) carrier of the allele. Hemophiliacs or "bleeders" have a rather short life expectancy (particularly without medical assistance) because of their inability to stop bleeding. The hemophilic Waldemar of Prussia, however, lived to be 56. Of the ten known hemophiliacs in the pedigree all were males.

Linkage Groups and Crossing Over

With the multitude of genes required to produce an organism (there may be a million or more in man) it is evident that each of the different chromosomes must bear large numbers of genes. Since all the genes on a chromosome must normally be transmitted together, they constitute what is called a *linkage group.* There are as many linkage groups as there are chromosome sets. In man there are 23 linkage groups, and in the garden peas studied by Mendel, 7. As we saw in Figure 7-22, however, the linkage of genes on the same chromosome may be broken through *cross-over,* thereby making recombinations of genes possible between homologous chromosomes.

We have already stated the belief held by biologists that each gene has its own specific location on a particular chromosome and that all the genes on a given chromosome are arranged in a precise linear order from one end of the chromosome to the other. The evidence supporting this belief comes from breeding experiments and observational studies of the chromosome itself.

The common fruit fly (*Drosophila*) lends itself beautifully to linkage experiments because it reproduces rapidly, has only four pairs of chromosomes, and shows a wide variety of easily distinguished contrasting traits. The normal (wild type) fly has long wings (dominant allele *V*) and a gray body (dominant allele *B*). Alternative traits are vestigial wings (recessive allele *v*) and black body (recessive allele *b*). If a homozygous black, vestigial male is mated with a homozygous gray, long female, we would expect all the offspring to be long-winged and gray. If females from this cross are now mated to a homozygous recessive (black, vestigial) male, we would, on the basis of independent assortment, expect the results shown in Figure 8-14. There should be four kinds of individuals in the second generation in approximately equal numbers: gray, long; gray, vestigial; black, long; and black, vestigial. The actual results, however, are quite different: 82 percent of the offspring are gray, long, and black, vestigial, whereas only 18 percent are gray, vestigial, and black, long.

Such results can be explained on the basis of linkage and cross-over, as shown in Figure 8-15. These particular genes for body color and wing length are located

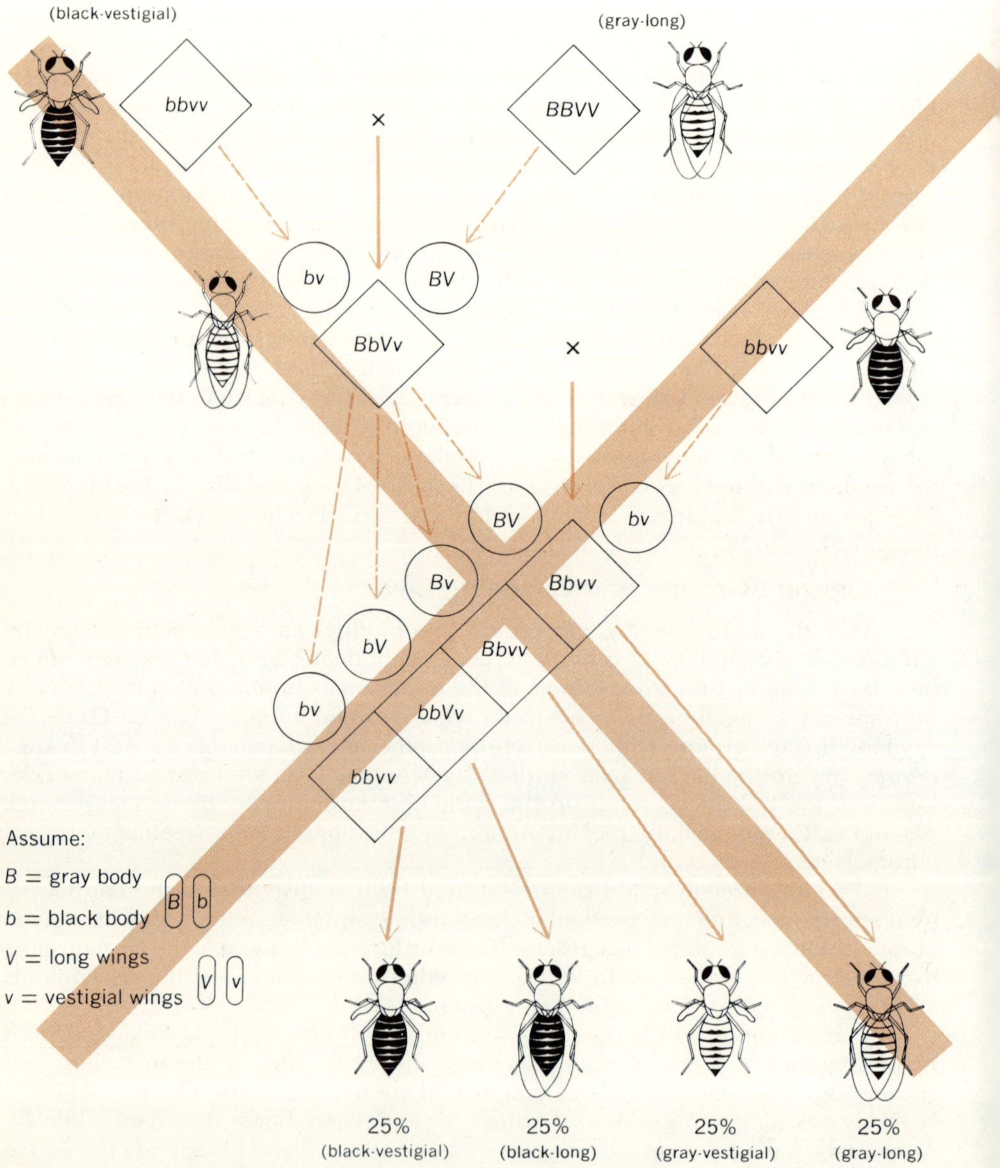

Figure 8-14 If crosses of fruit flies having gray bodies and long wings with flies having black bodies and vestigial wings followed the principle of independent assortment, the above test crosses would yield a 1:1:1:1 phenotypic ratio in the second generation. *This is not the case.*

on the same chromosomes. In 82 percent of the gametes cross-over did not occur between these two genes, but in 18 percent of the cases it did. It is of particular note that in all crosses of this kind the percentage of cross-overs between these genes is remarkably constant whenever sufficient numbers of progeny are involved.

There is good reason to believe that the percentage of cross-overs between any two genes on the same chromosome reflects with a fair degree of accuracy the

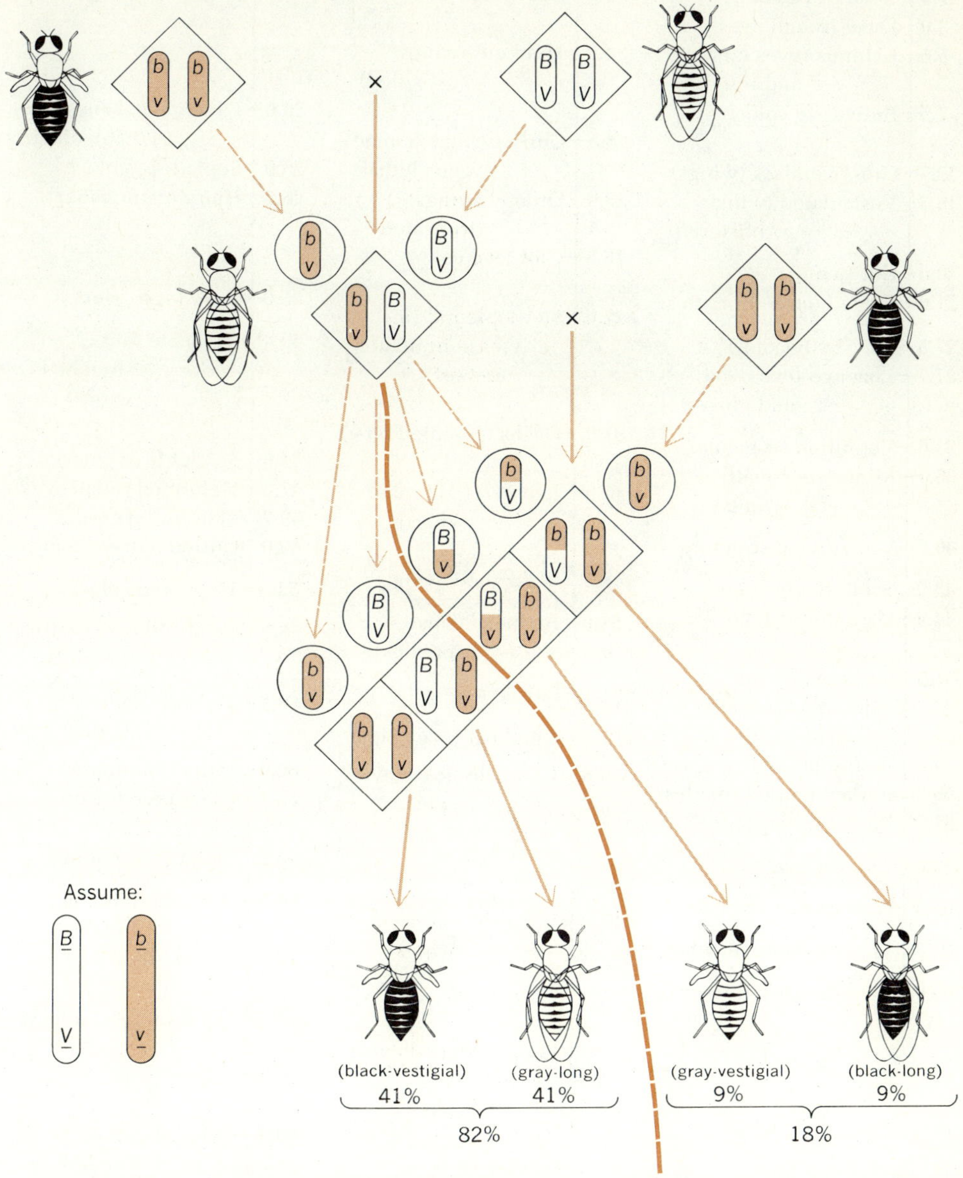

Figure 8-15 If gray-long by black-vestigial crosses involve gene linkage and cross-over, the above test crosses would yield a preponderance of gray-long and black-vestigial individuals in the second generation. *This is the case.*

relative distance separating the genes on the chromosome. Certainly the likelihood of a chromosome break occurring between two widely separated genes is greater than that between two which are close together. If one assumes that a chromosome is equally susceptible to breakage at any point along its length (which by no means is always true), then the percentage of cross-over between two linked genes will be proportional to their distance apart. Utilizing this assumption, geneticists have

Figure 8-16 Chromosome map of *Drosophila melanogaster.* Chromosome I is the *X* chromosome. Chromosome IV is very small. Gene locations are not to scale. (Modified from Johnson *et al., Essentials of Biology,* Holt, Rinehart and Winston, 1969)

prepared chromosome maps in which one unit of map distance is defined as the distance within which cross-over occurs 1 percent of the time.

Figure 8-16 represents such a map of some of the genes that have been located on the four chromosomes of the fruit fly. The numbers to the left of the characters represent map units. Note that chromosome II carries the black (48.5) and vestigial (67.0) alleles. Black and vestigial (or gray and long) are 18.5 map units apart ($67.0 - 48.5 = 18.5$) which is also their cross-over frequency. The relationships of black and vestigial to other genes on chromosome II are determined in a similar fashion. The cross-over between black and sternopleural is about 26 percent ($48.5 - 22.0 = 26.5$), whereas that between sternopleural and vestigial is higher. One can conclude, therefore, that (1) sternopleural cannot lie between black and vestigial because its cross-over percentage with both black and vestigial is greater than the cross-over percentage between black and vestigial and (2) sternopleural must lie closer to black than to vestigial because it has a lower cross-over frequency with black.

Corroboration of this precise linear ordering of genes has come through observing the giant chromosomes from the salivary gland of *Drosophila* (Fig. 8-17a). These chromosomes display a series of dark and light bands. The dark bands have been shown to bear a close relationship with the location of genes. One band, however, usually represents the position occupied by several genes. Chromosome mapping based on these bands gives a truer picture of the actual spatial arrangement of genes than cross-over frequency does. (Notice in Figure 8-16 that the total map units in chromosomes II and III exceed 100, whereas chromosome I falls short of 100, indicating that chromosomes are not equally liable to breakage at all points.)

The occurrence of chromosome puffing (Fig. 8-17b) also attests to the linear order of genes. The puffs are separated strands of DNA which have been shown to represent sites of active DNA or RNA synthesis. These puffs develop at varying places along the length of a chromosome, depending on cellular activity. Along the same chromosome where RNA synthesis has occurred in different puffs at dif-

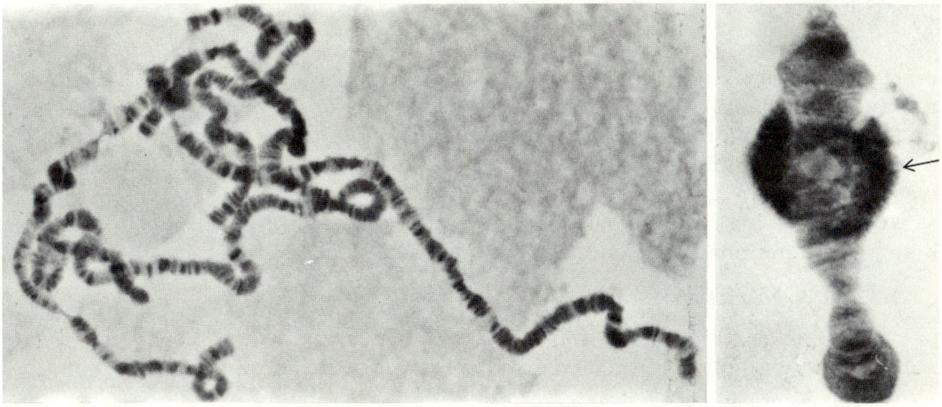

(a) (b)

Figure 8-17 (a) Giant chromosomes from salivary gland of *Drosophila*. Apparently chromatids replicate repeatedly without accompanying cell division, forming huge chromosomes consisting of many identical strands lying side by side. (b) Chromosome puff (arrow) in giant chromosome. Puffing indicates active genes in process of RNA synthesis. [(a) Photomicrograph by author; (b) Dr. W. Beermann]

ferent times, it has been found that the RNA synthesized in one puff differs from that manufactured in another.

BIOLOGIC NOVELTY: FOUNDATION OF ORGANIC EVOLUTION

We have reviewed some of the hereditary processes underlying the continuity of life. In so doing we noted that the same genetic mechanisms that enable "like to beget like" also guarantee variety and diversity in that which is begotten. The *reassortment* and *recombination* of thousands upon thousands of genes assure that every gametic union will contain elements of genetic novelty not found in other zygotes. As Charles Darwin observed over a century ago, no two sexually reproduced organisms are ever exactly alike.[4] It should not be assumed, however, that the unlimited potential for variation in living things is solely the result of genetic reassortment and recombination. Fundamental changes in both chromosome structure and number are possible as well as changes in the genes themselves. These changes, called *mutations*, are normally of rare occurrence, but when rare events of this nature are considered in the context of the large breeding populations in which they occur they become powerful forces in organic diversity and change.

A given mutation may take place only once in a million gametes, so that a species numbering 10 million individuals would produce this particular mutation 10 times in every generation. However, since countless thousands of different mutations are possible (different genes subject to change), such a population would yield untold thousands of mutant individuals in each generation.

Now that we have become familiar with basic genetic principles through studying specific test crosses, we can discuss the more realistic hereditary phenomena at work among natural breeding populations. We will first examine mutation before considering population genetics.

Mutation

Mutation may involve entire chromosomes (*chromosomal mutations*) or specific genes (*gene* or *point mutations*). Chromosomes and their genes are stable entities. This stability, however, is relative. The hereditary mechanism is subject under natural circumstances to infrequent but nevertheless perceptible change. The incidence of this change can be greatly accelerated by many products of human technology (a growing number of mutation-inducing or *mutagenic* chemicals as well as ionizing radiation such as X rays and ultraviolet light). We reserve these latter considerations for discussion in a later chapter.

CHROMOSOMAL MUTATIONS

Chromosomal mutations fall into two categories: (1) gene rearrangements resulting from chromosome breakage and (2) changes in chromosome number.

CHROMOSOMAL REARRANGEMENTS

Figure 8-18 illustrates some examples of chromosomal rearrangements. These include genic reorientation within the same chromosome set (duplication, deletion, and inversion) and genic exchanges between nonhomologous chromo-

[4] So-called identical twins actually develop by the asexual division of a sexually produced embryo (see page 256).

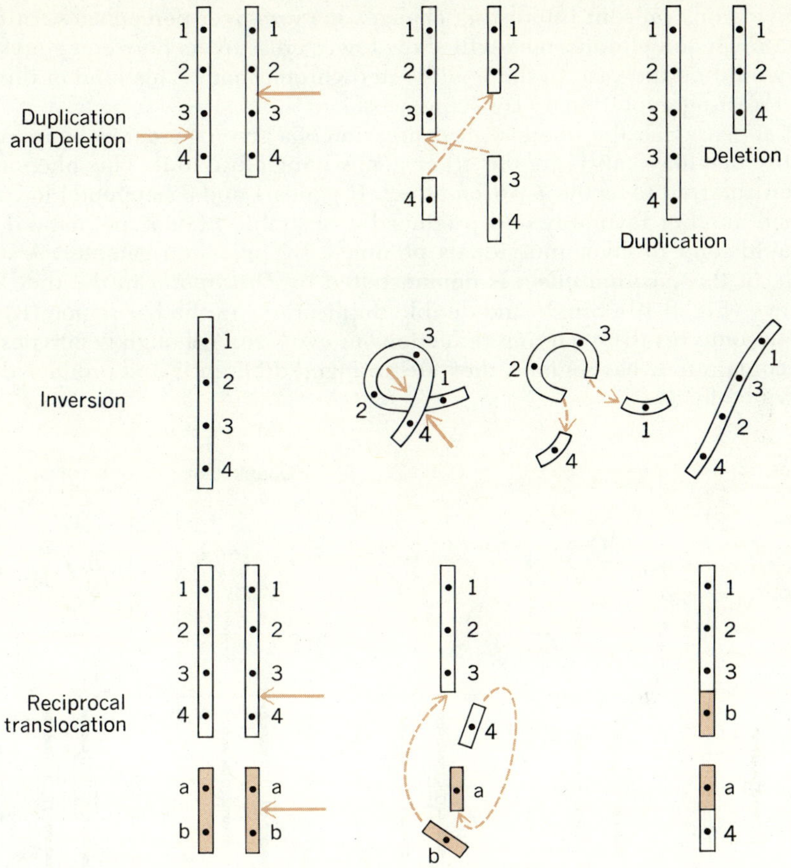

Figure 8-18 Types of mutations due to chromosome breaks. On infrequent occasions chromosomes may break (solid arrows), producing segments which can reattach themselves in abnormal ways (dashed arrows), resulting in various chromosomal rearrangements. Translocation entails exchanges between two nonhomologous chromosomes. Numbers and letters represent the locations of selected genes as points of reference.

somes (translocation). When chromosomes break, the separated fragments may become reattached to one another in unusual ways. Any chromosome fragment, on the other hand, which fails to reattach is almost certain to be lost in succeeding cell divisions if it happens to lack the centromere.

Interesting phenotypic effects accompany mutations of this kind. A zygote resulting from two gametes, one of which carried a chromosome deletion, would be haploid for the genes not supplied by that gamete (in Fig. 8-18 gene 3 and all its immediate neighbors on either side). Any single recessive alleles among this group carried by the normal gamete would express themselves phenotypically (as with sex-linkage in man). If the zygote was homozygous for the same deletion, it most likely would not develop normally (if at all) because of the absence of necessary genes. Another obvious phenotypic modification would result from the change in gene linkages. In a translocation certain genes from two nonhomologous chromosomes become associated in a new linkage group.

Inversions present interesting changes in cross-over percentages. In Figure 8-18 the inversion chromosome will show fewer cross-overs between genes 1 and 3 than would be the case in the noninverted chromosome. This kind of thing can have interesting evolutionary consequences.

It appears that the phenotypic expression of a given gene may be influenced (favorably or unfavorably) by the other genes lying close to it. This phenomenon has been referred to as the *position effect.* If genes 1 and 3 happened to complement one another favorably and produced a desirable phenotype, natural selection would tend to favor individuals produced by inversion gametes. A striking example of the position effect is demonstrated by *Drosophila* in the trait known as bar eye (Fig. 8-19). Single and double duplications in the bar region (B) of the *X* chromosome result in varying reductions in eye size. Although genotypes 3 and 4 both contain four bar regions, they are arranged differently and produce distinct phenotypic effects.

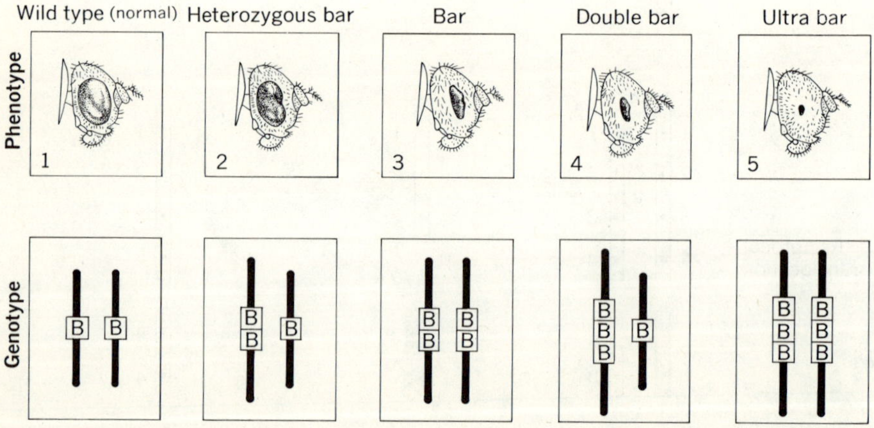

Figure 8-19 Effects of a chromosome duplication in *Drosophila.* Single and double duplications in the bar region (B) of the *X* chromosome result in varying reductions in eye size. Additional modification also depends on whether the mutations are homozygous or heterozygous.

CHANGES IN CHROMOSOME NUMBER

Occasionally a meiotic cell division may follow one of several anomalous courses. The members of one pair of chromosomes, for example, may fail to move toward opposite poles of the cell. They stay together and pass into one of the daughter cells, as shown in Figure 8-20b. This results in two daughter cells, one of which has an extra chromosome, whereas the other has one less than it should. If these two daughter cells were human gametes, one would have 24 chromosomes and the other 22. Fertilization involving a normal gamete and the one with the extra chromosome would yield a zygote with 47 chromosomes. Such is the case with certain abnormal human males who have an extra *X* chromosome (*XXY*). These individuals are sterile and sometimes mentally retarded.

On rare occasions a meiotic division may be so abnormal that all chromosomes are distributed to but one daughter cell. This would result in a diploid

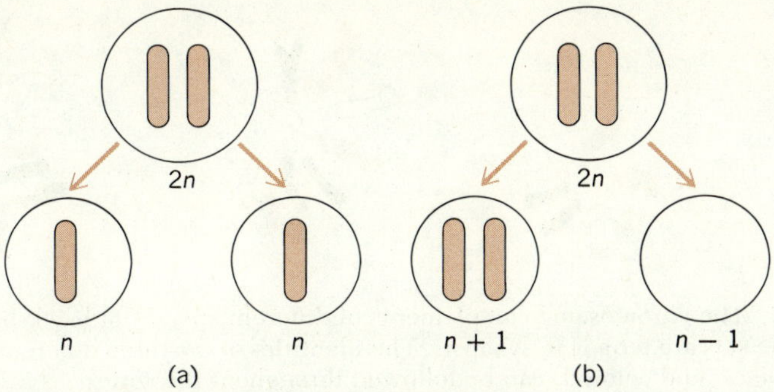

Figure 8-20 Schematic diagram of normal (a) and abnormal (b) meiosis. On rare occasions meiosis fails to apportion a chromosome set between daughter cells. This results in gametes half of which have one extra chromosome ($n + 1$) and half of which lack one chromosome ($n - 1$). This may also occur with two or more chromosome sets.

gamete rather than a haploid gamete.[5] A zygote formed by the union of such a gamete with a normal haploid sex cell would be triploid ($3n$). If two such diploid gametes were to unite in fertilization, the chromosome number in the zygote would be doubled ($4n$). Such cases in which the chromosome number is increased by some multiple of the haploid number are called *polyploid.*

	DIPLOID CHROMOSOME NUMBERS					DIFFER BY MULTIPLES OF
Four species of hawksbeard (*Crepis*)	6	12	18	30		3 pairs (6)
Three species groups of wheat (*Triticum*)	14 (1 species)	28 (5 species)	42 (3 species)			7 pairs (14)
Five species of chrysanthemum	18	36	54	72	90	9 pairs (18)

Figure 8-21 Polyploid series in three different groups of plants.

Polyploidy has played a significant role in the evolution of plants. Figure 8-21 depicts three different groups of plants (hawksbeards, wheats, and chrysanthemums) in which different species have differentiated on the basis of polyploidy. In the hawksbeards different species have chromosome numbers which differ from one another by multiples of three pairs. The identity of these three pairs can be followed throughout the entire series of species (Fig. 8-22). Polyploid plants

[5] Doubling of the chromosome number can also occur during cell division when the cytoplasm fails to divide.

Figure 8-22 The chromosome complements of different species of hawksbeard (*Crepis*). All chromosome sets are tetrads in synapsis. The identities of the three different chromosome types (white, black, and colored) can be followed throughout the series.

Figure 8-23 The Easter lily on the left was taken from a diploid plant; that on the right came from a plant in which the chromosome number had been doubled ($4n$) artificially by chemical treatment with colchicine. (U.S.D.A.)

are often large and vigorous (Fig. 8-23). Those produced artificially through chemical treatment (colchicine) are frequently less fertile than their diploid relatives.

GENE MUTATIONS

The occurrence of a gene mutation is clearly evident in the lineage and progeny of Queen Victoria (Fig. 8-13). There is no record of hemophilia in the ancestors of Victoria. Since she passed the gene to three of her children, it is probable that the hemophilic allele came from a new mutation supplied by either her father or her mother. It is not likely that the same mutation was produced on three separate occasions in the ova of the queen herself.

Changes in genes undoubtedly occur in many different ways. Altering the nucleotide sequence in DNA by the addition, deletion, or substitution of a different base would be one way. These modifications would result in corresponding changes in mRNA codons. Figure 8-24 illustrates such a case in which a viral mRNA was apparently altered by the deletion of one base and the insertion of another.

Gene mutations are responsible for the alternative alleles that complete a given gene set. A dominant, wild-type gene, for example, A, may mutate to a recessive a, which in turn may mutate to a second recessive allele a'. Reverse mutations can also take place; the mutant allele a, for instance, may revert to the wild-type state A.

The rate at which a particular gene mutates is generally constant, but this rate varies from gene to gene. The rate of forward (direct) mutation is usually

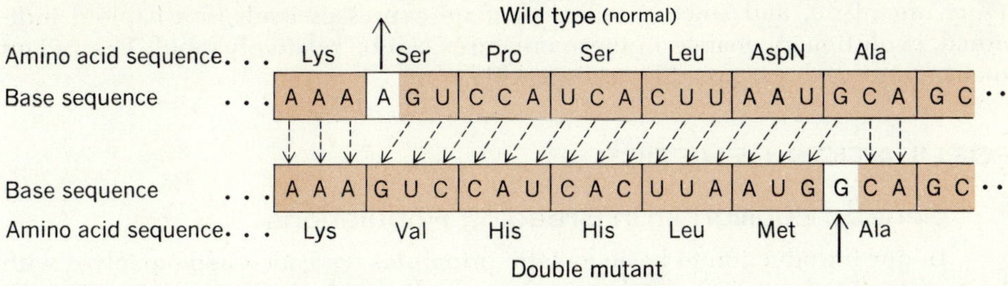

Figure 8-24 Gene mutations in a bacterial virus. It is believed that the removal of one base, A, and the insertion of another, G, about fifteen bases away, produced a double mutant mRNA specifying a new sequence of amino acids in the affected section of the macromolecule. Consult Figures 3-24 and 3-25. (Modified from F. H. C. Crick in *Scientific American*, October, 1966)

higher than that for reverse mutation. In the majority of organisms investigated, the natural mutation rates for specific genes usually fall between 1 and 100 mutations per million gametes.

Most gene mutations are recessive. They are also usually harmful. Since these mutations are haphazard changes in gene structure, the deleterious effects of most mutant alleles should not be surprising. Living organisms have been on this planet for more than 3 billion years. Over this vast period the elimination of unfit phenotypes through natural selection has tended to preserve only those genes that prove advantageous to their possessors. A random change in such genes will usually (but not always) be undesirable.

Over long periods of time advantageous mutations occur. Under certain conditions even nominally deleterious genes can prove beneficial. A case in point is the human mutant allele which specifies an anomalous type of hemoglobin. When homozygous, this allele is responsible for a generally fatal form of anemia (*sickle-cell anemia*). Heterozygous genotypes, however, are healthy and appear to suffer far less severely from malignant malaria than homozygous normal individuals who lack the allele. In West Africa where malignant malaria is rampant the frequency of sickle-cell anemia may approach 5 percent. In such populations persons who are homozygous for normal or for sickling hemoglobin run the risk of death or disability from either malaria or anemia, whereas heterozygous individuals are protected from both.

In organisms such as the "higher" plants and animals, a large investment has been made in elaborate structures and sophisticated design. Broad adaptive modifications have enabled these creatures to invade special environments such as dry, terrestrial habitats. Life in its conquest of energy is most frugal. It is probably not a coincidence that the dominant generation in these organisms is diploid. Were they haploid a single undesirable mutation could endanger the life of every individual who received it. In diploid populations many mutants of varying degrees of harmfulness can be carried safely in the heterozygous state. As these alleles spread through the population opportunity is afforded for their being tested in many environments and in many genetic combinations without adversely affecting the biologic vigor of the population as a whole. The "simpler" organisms, in which the dominant generation is commonly haploid, have less invested in "capital outlay." They compensate for the swift elimination of unfit individuals by high reproductive rates. Since natural selection operates only on the phenotype (the living organism), and since a recessive gene expresses itself in a haploid individual, evolutionary change in these creatures can be relatively rapid. They adapt to changing conditions readily and speedily.

POPULATION GENETICS

Genetic Equilibrium in Breeding Populations

In our introduction to basic genetic principles we concerned ourselves with the results of crosses between heterozygous individuals. In Figure 8-5, for example, the first-generation offspring of pink four-o'clocks were crossed ($RR' \times RR'$)

(a) Percent of plants in each phenotype	(b) Percent of gametes in each phenotype carrying		(c) Percent that each phenotype contributes to total alleles produced by all plants (column a × column b)	
	R	*R′*	*R*	*R′*
4 (*RR*)	100	0	4	0
32 (*RR′*)	50	50	16	16
64 (*R′R′*)	0	100	0	64
		Gene frequency =	20%	80%

Figure 8-25 In a population of 100 four-o'clock plants in which 4 have red, 32 have pink, and 64 have white flowers, the *gene pool* of the population consists of alleles 20 percent of which are for red flowers (R) and 80 percent of which are for white flowers (R'). The relative proportion that these alleles bear to one another in the gene pool is their *gene frequency*.

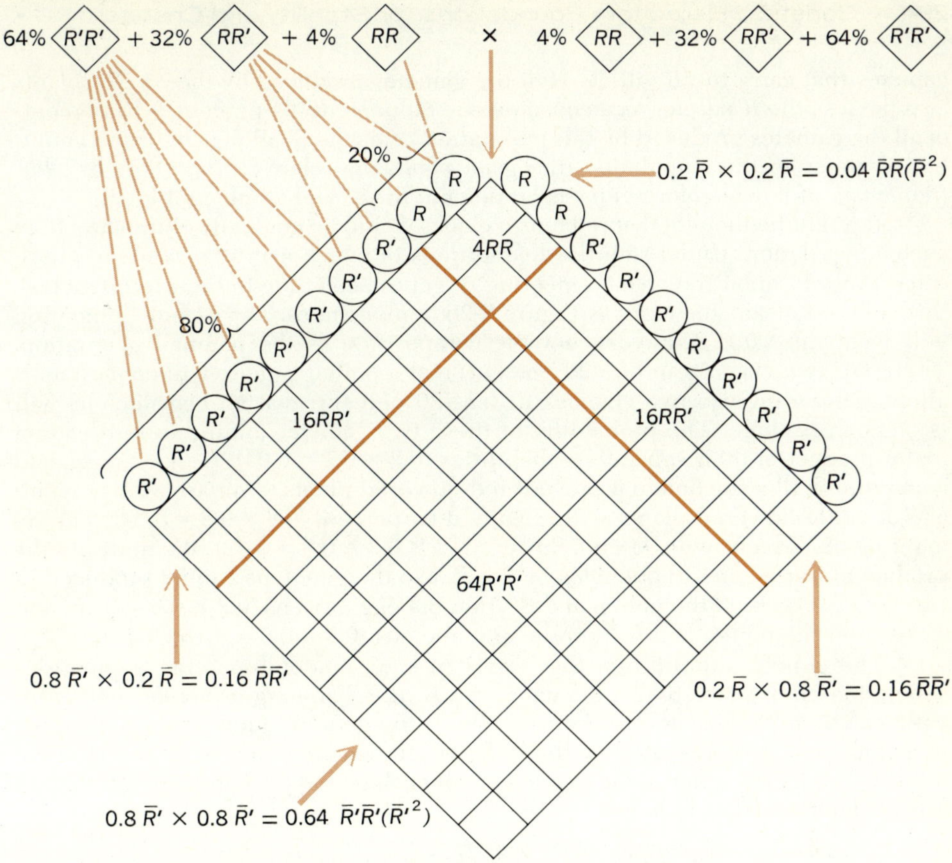

Proportion of *RR* in Population	Proportion of *RR'* in Population	Proportion of *R'R'* in Population
$\bar{R} \times \bar{R}$ (0.2 × 0.2 = 0.04 or 4%) $\bar{R}^2$	$2 \times \bar{R} \times \bar{R}'$ (2 × 0.2 × 0.8 = 0.32 or 32%) or simply $2\bar{R}\bar{R}'$	$\bar{R}' \times \bar{R}'$ (0.8 × 0.8 = 0.64 or 64%) $\bar{R}'^2$

Figure 8-26 Even though the red-flowered individuals in Figure 8-25 made up only 4 percent of the plants, their number remains unchanged in the next generation. This tendency for gene frequencies to remain constant from generation to generation is known as the Hardy-Weinberg principle. The symbols $\bar{R}$ and $\bar{R}'$ represent the gene frequencies of alleles *R* and *R'*, respectively.

and gave rise to a second generation, of which 25 percent bore red flowers, 50 percent pink, and 25 percent white flowers. The parents of this second generation produced gametes, one-half of which carried the allele for red (*R*) and one-half the allele for white (*R'*). Natural populations of organisms are seldom, if ever, so homogeneous. There is no reason why all the gametes of a breeding population should be equally divided between the alternative alleles of a gene set. Examine, for instance, the population of four-o'clocks shown in Figure 8-25. Out of 100 plants, 4 are red-flowered, 32 pink, and 64 white. The 64 *R'R'* plants can only produce gametes that carry the R' allele. Half the gametes produced by the 32 RR' plants also possess the R' allele. As a consequence 80 percent (64 percent + 16 percent) of all the gametes produced by this population bears the R' allele. Of the total number of genes in this population (the *gene pool*), the relative proportion, or *gene frequency*, of flower-color alleles is 80 percent for R' and 20 percent for R.

It might be thought that red four-o'clocks would be gradually eliminated from such a population. Under stabilized conditions in which mating is random (predicated entirely upon the chance meeting of genes according to their frequencies), this is by no means the case, as Figure 8-26 demonstrates. The second generation will have the same relative genotypic composition as the parental generation. There are two chances in ten (20 percent) that a pollen gamete will contain an R allele. There are also two chances in ten (20 percent) that an egg nucleus will possess an R allele. The probability of these two gametes joining in fertilization is the product of their separate probabilities ($0.2 \times 0.2 = 0.04$). Hence the second generation will also contain 4 percent red-flowered plants. Similarly, the percentage of white-flowered plants will remain at 64 percent ($0.8 \times 0.8 = 0.64$). The remaining 32 percent will be pink-flowered ($2 \times 0.2 \times 0.8 = 0.32$). We multiply the product of these gene frequencies by 2 because the genotype can be produced in two ways: (1) R from the pollen and R' from the egg nucleus ($0.2 \times 0.8 = 0.16$) and (2) R' from the pollen and R from the egg nucleus ($0.8 \times 0.2 = 0.16$).

The genetic equilibrium that exists in such randomly mating populations generation after generation is known as the *Hardy-Weinberg principle*. This principle can be broadly stated mathematically. If the gene frequency of a given gene in a gene pool is represented by the letter p and the frequency of its single allele represented by the letter q, the percentage of each genotype in the next generation can be determined as follows:

$$p^2 + 2pq + q^2 (p \times p + 2 \times p \times q + q \times q)^6$$

This mathematical approach has two obvious advantages: (1) it is not as cumbersome as the checkerboard technique and (2) it can be worked backward just as easily as forward.

For instance, suppose we wanted to determine the frequency of the sickle-cell allele in a West African population where 4 percent of the people had sickle-cell anemia (q^2). Assuming this population were in genetic equilibrium, the frequency of the sickling allele (q) would have to be that percentage which, when multiplied by itself, gave a product of 4 percent (q^2). That percentage is obviously 20 percent ($\sqrt{0.04} = 0.2$). Since $p + q$ must always equal 1.0, the frequency of the allele for normal hemoglobin (p) must be 80 percent. From these determinations we could make a significant deduction. In a population in which about 4 percent of the people die from sickling anemia, 32 percent (the heterozygotes) would be protected from both malaria and anemia ($2\,pq$ or $2 \times 0.8 \times 0.2 = 0.32$).

A state of genetic equilibrium exists, then, when changes do not occur in the gene pool of breeding populations. Multitudes of genes, many of which are present at low frequencies, are conserved in a stable state from generation to generation whether they are dominant, codominant, or recessive. As a matter of fact, however, we know that complete equilibrium is not likely to be achieved. No population is probably ever in an absolutely stable genetic state. Gene pools are subject

[6] Those who have had algebra will perhaps recall that the possible combinations of two different factors, p and q, multiplied by one another $(p + q)^2 = p^2 + 2pq + q^2$.

to change in their gene frequencies and genotypic composition. Let us see why this is so.

The Disruption of Genetic Equilibria: Organic Evolution

According to the Hardy-Weinberg principle, certain conditions must prevail in order for the gene pool of a population to be in genetic equilibrium.

1. The population must be large enough for the "laws" of probability to operate according to expectation.
2. The gene pool must not be added to by immigration from another gene pool or diluted by emigration.
3. Reverse mutations must occur at the same rate as direct (forward) mutations.
4. No phenotype can enjoy greater net reproductive success than another.

POPULATION SIZE AND GENETIC DRIFT

As observed early in this chapter when discussing Mendel's work, a small population is prone to significant deviations from expected probability ratios. For instance, in small breeding populations of 100 to 300 individuals only a few random and unpredictable genetic events may result in eliminating an allele from the gene pool. There is consequently a tendency in such populations to haphazardly reduce the frequency of some alleles while increasing the frequency of others. This, in turn, favors an increase in the number of homozygous individuals. Chance deviations of this nature that occur in small, isolated populations are known as *genetic drift.* Although genetic drift reduces the variability in any one population, it may augment variation in the species as a whole. A species that is segmented into a number of small, isolated populations commonly demonstrates very definite intergroup differences that surpass by a considerable margin the degree of variability found in one large, single population.

Genetic drift plays a unique role in the evolutionary process. It may facilitate the effectiveness of natural selection in some instances. If the percentage of individuals homozygous for a certain recessive defect happens to be momentarily augmented, more deleterious alleles are, in effect, chased from their hiding places in heterozygous individuals and made to "face the music." The phenomenon of genetic drift, however, generally involves those genes whose phenotypic expressions are, *at least for the time being,* neither particularly advantageous nor disadvantageous. Lacking for the most part any significant adaptive value, it results in indeterminant changes in temporarily "neutral" characters.

A good example of genetic drift was provided by a study of a religious sect in the eastern United States known as the Dunkers (German Baptist Brethren). These people have generally not married outside their own group since the sect was founded in West Germany between 1719 and 1729. Although 45 percent of the people in West Germany and 40 percent of the people in the United States have type A blood, 60 percent of the Dunkers studied in each of these countries had this blood type.

IMMIGRATION AND GENE FLOW

It scarcely need be said that a gene pool cannot remain stable if genes are added to it by an influx of individuals from a different population. The introduction of new genes in this fashion, called *gene flow,* serves to maintain variety even in small populations.

A clear illustration of gene flow was revealed in a study of the inhabitants of a small coastal village in southern Sardinia. Although this village lies close to the malaria-infested plains of Sardinia, the frequency of a gene for abnormal hemoglobin[7] was much lower here than in neighboring populations. On further inquiry it was learned that this particular village was established in 1700 by immigrants from a nonmalarial area in northern Italy. Until recently this community was reproductively isolated. The few carriers of the defective gene are of Sardinian ancestry.

MUTATION PRESSURE

As we have just seen, a population cannot be in genetic equilibrium if genes are being added to the gene pool. It makes little difference whether the genes are being added from without (by gene flow) or from within (by mutation). The mutation of genes (and chromosomes) is a slow but ceaseless process. If direct mutations occurred at the same rate as reverse mutations, they would cancel one another out. However, they do not. The rate for a direct mutation is seldom, if ever, the same as that for the reverse mutation; the direct mutation is usually more frequent. There is consequently a *mutation pressure* favoring an increase in the less mutable allele and a corresponding decrease in the more mutable allele. This mutation pressure is equivalent to the difference in rates between direct and reverse mutations and usually favors direct mutation.

If gene $A \rightarrow$ its allele a represents a direct mutation, mutation pressure usually tends to increase the frequency of allele a at the expense of gene A. Other compensatory factors such as natural selection can, of course, modify or completely nullify these effects. The type of diabetes caused by insulin deficiency is the result of a recessive allele in the homozygous state. In the last three decades the percentage incidence of this disease has more than doubled in the United States. It is thought that this relative increase in diabetes is due, in a significant degree, to mutation pressure. Natural selection against the recessive phenotype has been greatly reduced through effective insulin treatment.

VARYING REPRODUCTIVE FITNESS AND NATURAL SELECTION

One of the most critical requirements for maintaining stable gene frequencies is that all phenotypes in a breeding population have equal opportunities, commensurate with their relative numbers, for passing their genes on to succeeding generations. No genotype can, by reason of greater phenotypic fitness, have an advantage over another in its genetic contribution to the reproductive continuity of the population. Individuals with certain genotypes may give birth to disproportionately large numbers of offspring as long as there is also a disproportionately equivalent number of these offspring who fail to reproduce for one reason or another (higher infant mortality, general physical debility, sterility, and so on). The net reproductive fitness, generation in and generation out, must be the same for individuals of all genotypes.

Such criteria, of course, are never met in any natural breeding population. Different genotypes produce a spectrum of phenotypes with varying degrees of fitness for a variety of environmental conditions. Each generation receives most of its genes from those organisms which are reproductively fit. The reader will recognize this process as *natural selection.* Natural selection favors the fit and rejects

[7] Not the sickling gene in this case but another gene entirely which has quite similar phenotypic effects insofar as malarial resistance and anemia are concerned.

the unfit. Phenotypes lacking reproductive vigor tend to be eliminated from the evolutionary scene by the *selection pressure* which is exerted against them. Their genes contribute relatively little to the gene pool of succeeding generations.

An elegant example of these principles is provided by the African populations to which we have previously referred. Evidence suggests that the introduction of Malaysian agricultural practices into Africa about 1500 years ago allowed native African communities to penetrate the tropical rain forest and its organically rich soils. The augmented food supply accruing from these practices led to population growth which, presumably, resulted in an increased probability of malarial infection. Those heterozygous individuals carrying the sickle-cell gene would have a selective advantage over those phenotypes lacking the allele. As the sickling gene increased in frequency the correspondingly larger population of immune heterozygous individuals permitted further agricultural exploitation of the rain forest environment. The existing gene frequencies of the sickling allele in some cases exceeds 20 percent. It is believed that these high frequencies have developed in the manner suggested over the 50 to 60 generations since Malaysian agriculture was introduced into Africa.

Although these frequencies may have changed relatively little during the last 20 generations, the current use of insecticides in mosquito abatement together with recent improvements in the treatment of malaria will undoubtedly result in further evolutionary changes in the gene pool.

This study of human evolution is particularly intriguing because it points up so vividly the sensitive interrelationships that exist between man and his technology.

QUESTIONS

1. Why was Mendel's choice of garden peas for his original work a good one?
2. A cross between a red-flowered (dominant) and a white-flowered (recessive) pea plant yielded 69 red-flowered and 77 white-flowered offspring. What was the genotype of the red-flowered parent?
3. How can dominance, incomplete dominance, and recessiveness be distinguished at the level of enzyme action and protein structure?
4. A child with type O blood born to a type A mother could not have a natural father whose blood type is (a) A; (b) B; (c) AB; (d) O.
5. In chickens feathered legs are the result of either or both of two dominant genes located on different chromosomes. These genes and their recessive alleles may be indicated as follows:

   A = feathered legs B = feathered legs
   a = featherless legs b = featherless legs

   What phenotypic ratio would result in the offspring of $AaBb \times AaBb$ crosses?
6. In certain breeds of dogs two independently assorting gene sets are involved in the inheritance of coat color. These gene sets may be represented as follows:

   A = black color B = color inhibitor (white)
   a = brown color b = color expressor (black or brown depending on A-a gene set.)

   What phenotypic ratio would result in the offspring of $AaBb \times AaBb$ crosses?
7. Critically evaluate the following statement: The presence of a given characteristic or trait in an organism is the product of either heredity or of environment.

8. A normal-sighted woman whose father is red-green color blind marries a normal-sighted man. What is the probability of their daughters being color blind? Of their sons being color blind?
9. Had Mendel investigated eight rather than seven separate characteristics of pea plants in his studies, one of his conclusions might have been altered. Which conclusion would this have been and why?
10. Describe three kinds of chromosomal rearrangements (chromosomal mutations).
11. How may changes in chromosome number occur?
12. Describe what is meant by a population being in genetic equilibrium.
13. How can each of the following disrupt a genetic equilibrium:

    (a) Genetic drift (c) Mutation pressure
    (b) Gene flow (d) Selection pressure

Suggested Readings

Beadle, G. W., and M. Beadle. *The Language of Life.* Garden City, New York: Doubleday, 1966.

Beerman, W., and U. Clever. "Chromosome Puffs," *Scientific American,* April 1964.

Bonner, D. M., and S. E. Mills. *Heredity.* Foundations of Modern Biology. Englewood Cliffs, N.J.: Prentice-Hall, 1964.

Cavilli-Sforza, L. L. "Genetic Drift in an Italian Population," *Scientific American,* August 1969.

Crick, F. H. C. "The Genetic Code: III," *Scientific American,* October 1966.

Lerner, I. M. *Heredity, Evolution and Society.* San Francisco: Freeman, 1968.

Levine, R. P. *Genetics.* Modern Biology Series. New York: Holt, Rinehart and Winston, 1968.

Moore, J. A. *Heredity and Development.* New York: Oxford University Press, 1963.

Srb, A. M. *et al. General Genetics.* San Francisco: Freeman, 1965.

Tomasz, A. "Cellular Factors in Genetic Transformation," *Scientific American,* January 1969.

Wagner, R. P. "Genetics and Phenogenetics of Mitochondria," *Science,* **163,** March 7, 1969.

Wills, C. "Genetic Load," *Scientific American,* March 1970.

Yanofsky, C. "Gene Structure and Protein Structure," *Scientific American,* May 1967.

CHAPTER 9 Organic Evolution: the ebb and flow of life through time

We observed that populations do not achieve a fully stable condition in which the gene pool is in equilibrium. They are rather in a more or less continuous state of dynamic flux in which gene frequencies and genotypic composition are altered by genetic drift, gene flow, and/or mutation pressure. These essentially haphazard genetic changes confer on members of a population varying degrees of fitness and unfitness for a given set of environmental conditions. If the environmental circumstances are altered to a significant degree, the criteria for survival may be correspondingly modified and some of the fit and unfit may exchange places. The standards for biologic fitness are set by the environment. Those genotypes which are able or better able to exploit the energy resources about them will be favored over the unable or less able in their contribution to the gene pool of future generations. By differential reproduction there will be a natural selection of those genotypes which prevail through life's endless conquest of energy. The successful end-products of this self-perfecting regimen are adapted to the surroundings in which they live.

Adaptive changes in the gene pool of a population do occur. Such changes constitute a clear case of organic evolution. The delicate interplay between human technology and the gene pool in West Africa is an excellent example of *current* human evolution through natural selection at the population level.

We may well ask ourselves two opposing questions at this point. Is the phenomenon of adaptive change restricted to populations within a species? Or does the potential for biologic novelty transcend the species so that new species may arise through adaptive modifications in older ones? The evidence unequivocally supports the second alternative, as we are about to see.

CONFIRMATION OF A GENERAL PRINCIPLE OF ORGANIC EVOLUTION

There is no biological—or, for that matter, physical—principle that is more adequately or fully substantiated by scientific evidence than that of organic evolution. Nor is there any principle that more thoroughly pervades and unifies all biological phenomena within one all-embracing theme. There is no facet of biology that does not attest to the reality of organic evolution. We will restrict ourselves to a sampling of some of the evidence in six different biological sciences. The reader interested in plumbing the depths of the evidence may consult the references at the end of the chapter.

Paleontology

Paleontology is the study of the remains or traces (fossils) of prehistoric organisms preserved in the rocks. The history of life as revealed in the fossil record is perhaps its most basic aim. As such it affords a direct approach to the validity of organic evolution at and above the species level. Have existing species maintained

(a) (b)

Figure 9-1 *Archeopteryx* fossil and restoration. (a) Cast of fossil *Archeopteryx*. (b) Restoration based on fossil. Note the many reptilian features, such as the large number of tail bones (vertebrae), claws on (wing) "fingers," teeth, and poorly developed breast bone. None of these traits is found in modern birds. (American Museum of Natural History)

their identity throughout the span of life on earth? Has perhaps the history of life been marked by several special creations in which each group of creatures (except the last) has been wiped out by some cataclysmic event? Or has the progress of life through time been marked by gradual change and modification in which species adapt to changing conditions and give rise to new species or, failing in this, become extinct (or both)? Let us look at the record.

Fossils can take many forms. Sometimes the entire organism may be preserved, such as insects encased in amber (fossilized pine pitch) or mammoths which have been buried in ice (Siberia). On occasion feces are fossilized. This material may yield identifiable pollen grains or the bones of animals which were preyed upon. Footprints or trails (Fig. 1-3) as well as body impressions may be preserved. A beautiful example of body impressions is shown in Figure 9-1 in which a primitive bird (the toothed *Archeopteryx*) left clear impressions of its feathers that endured some 150 million years or more after its death. The petrified skeleton left unmistakable evidence of three claws on the ends of its wing "fingers" and the claws on its toes.

In most cases only the hard parts of organisms are durable enough to withstand destruction before their burial in sediment. Even woody stems, shells, and bone are usually consumed or weathered away before being covered by a protective layer of silt or sand. Once entombed they may still decompose. In short, the conditions which favor fossilization are not often met. Only a few of all the organisms that have ever inhabited the earth have left fossil remains or traces. Fossils of soft-bodied creatures are particularly rare. The scarcity of fossils is further compounded by the fact that many of them still lie deeply buried beneath overlying sedimentary rocks and lava flows beyond the reach of searching paleontologists.

Figure 9-2 The Grand Canyon of Arizona. Some of the rock formations over which the Colorado River flows are the altered remnants of ancient sediments laid down 2 billion years ago or more. Geologic formations at the canyon rim (a vertical mile above the river) are about 230 to 280 million years old and contain footprints of primitive reptiles which existed long before the first dinosaurs. (N. W. Carkhuff, U.S. Geological Survey)

Where the concealing overburden of rocks has been cut into or removed by erosion (or excavated by man), however, fossils are sometimes found, occasionally in great numbers. If nothing else this attests to the fact that rare events become commonplace over long periods of time involving many organisms. As one descends into the Grand Canyon of Arizona (Fig. 9-2) layer upon layer of increasingly older rocks come into view. Many of these rocks at one time were unconsolidated sediments laid down by wind, rivers, and coastal waters. Examining the fossils contained in these rocks is almost like scanning the pages of a great book backward. A most interesting story unfolds. Both the diversity and complexity of fossil organisms decrease as one approaches the bottom of the canyon (and the earlier "chapters" in the "book").

A brief sketch of some of the important landmarks in the history of life is outlined in Figure 9-3. The paleontological record is divided into eras, periods, and epochs based on major geologic changes that have occurred during the long history of our planet. The approximate ages of these geologic divisions have been determined by radioactive elements whose decay rates (see page 25) encompass billions of years (potassium-argon, uranium-lead, and so on).

A restoration of a Carboniferous forest based on collected fossils is shown in Figure 9-4. Trees up to 100 feet high were related not to any present trees but rather to our diminutive club mosses and horsetails (see plates in Chapter 3). Ferns that reproduced by seeds were also important components of these forests, but have long since vanished from the earth.

The Jurassic landscape depicted in Figure 9-5 was equally unique. Forests had become populated with cycads and conifers (cone-bearing trees). Although mammals and birds were both uncommon and primitive, reptiles existed in great

ERA	PERIOD	EPOCH (with approximate time to beginning of each in millions of years)	CHARACTERISTIC FORMS OF LIFE	
Cenozoic (63)	Quaternary (2–3)	Recent (0.025) Pleistocene (2–3)	Rapid proliferation of mammals and birds. Rise of grasses. Establishment of primate lineage leading to toolmaking types within the family of man about 2 million years ago.	The Age of Mammals
	Tertiary (63)	Pliocene (13) Miocene (25) Oligocene (36) Eocene (58) Paleocene (63)		
Mesozoic (230)	Cretaceous (135)	Progressive extinction of many reptilian types. First snakes. Rise of woody flowering plants.		The Age of Reptiles
	Jurassic (180)	Early mammals and birds. First flowering plants		
	Triassic (230)	Proliferation of many reptilian types, including ancestral dinosaurs. Extinction of forests of seed ferns, club mosses, and horsetails and their replacement by conifers and cycads.		
Paleozoic (600)	Permian (280)	Reptiles abundant. Extinction of many early Paleozoic organisms.		
	Carboniferous (345)	Dense forests of seed ferns and trees related to modern club mosses and horsetails. First mosses. Rise of insects and reptiles.		The Age of Amphibians
	Devonian (405)	First sharklike and bony fishes. Rise of amphibians, spiders, and ferns.		The Age of Fishes
	Silurian (425)	Appearance of land plants related to modern club mosses. First scorpions.		
	Ordovician (500)	First primitive vertebrates. No land animals or plants.		
	Cambrian (600)	All major groups of invertebrates represented including some no longer extant today. No land animals or plants.		
Pre-Cambrian ↓ Birth of Planet (4800)		Algae and/or bacteria up to possibly 3.5 billion years ago. Jellyfish and other relatively simple invertebrates approximately 0.8 to 1.0 billion years ago. The first billion years the planet was probably lifeless (prebiotic evolution).		

Figure 9-3 Some important landmarks in the history of life as revealed in the fossil record. For identification of the organisms alluded to, the reader should consult the illustrated classification of organisms in the Appendix.

Figure 9-4 A Carboniferous forest. Large trees related to modern horsetails (right foreground) and club mosses (left foreground and fallen logs) together with long-extinct seed ferns made up dense forests, the accumulated organic matter from which was a major contributor to coal formation in the United States. The dragonfly (right center) had a wingspread of 24 inches. (Field Museum of Natural History)

Figure 9-5 Life on a Jurassic landscape. Palmlike cycads and cone-bearing trees (conifers) commanded the uplands. Horsetails more like their modern counterparts abounded in swamps. Reptiles of all sizes and ecologic occupations were the dominant animal life. Birds (Archeopteryx) and mammals were small, uncommon, and primitive. The color markings of the animals are imaginary. (Peabody Museum, Yale University)

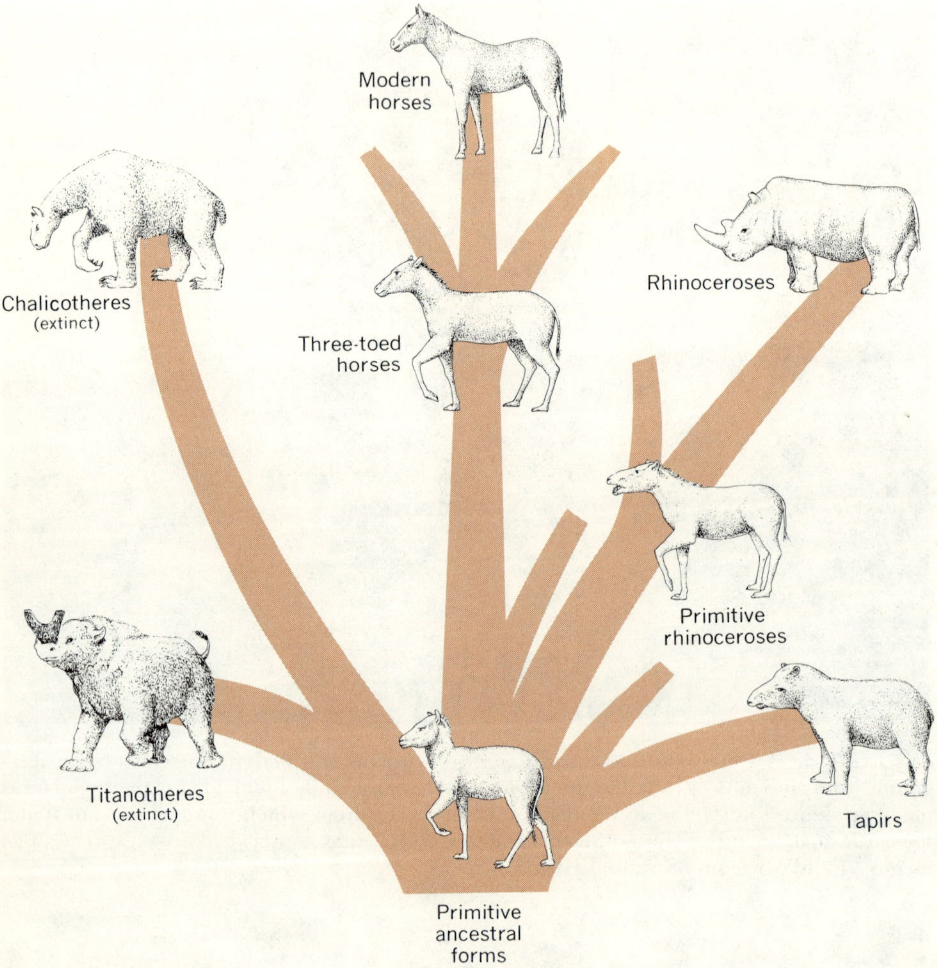

Figure 9-6 Evolution of odd-toed hoofed animals. The study of fossil mammals reveals that the tapirs of South America and Malaysia, the rhinoceroses of the Old World, horses and two extinct groups (Titanotheres and the horselike but clawed Chalicotheres) are all related because they descended from a common stock of primitive Paleocene mammals. Early horses and rhinoceroses were strikingly similar.

profusion and variety. Some species of terrestrial dinosaurs weighed as much as ten elephants. Others were no larger than crows. Most of this magnificent fauna of aerial, aquatic, amphibious, and terrestrial reptiles became extinct like the luxuriant seed fern and club moss forests before them.

In following the progress of a particular group of organisms through time, the fossil record amply supports the thesis of gradual change. Certain of the so-called mammal-like reptiles of Triassic time were very similar to the primitive mammals of the following Jurassic period. From a given group of ancestral mammals a series of fossil types diverged or *radiated* in several directions. The South American and Malaysian tapirs, the Old World rhinoceroses, horses, and two extinct groups appear to have originated in this fashion (Fig. 9-6). The evolution of the horse family is remarkably well documented by the paleontologic record (Fig. 9-7). The ancestral horses of 60 million years ago were the size of a fox. Their front feet bore four toes and their hind feet three. Their toes ended in tiny hooves, producing a foot nicely adapted for walking on spongy forest litter or swampy ground. Their teeth were simple, designed for browsing on vegetation that was soft or succulent. The radiation that this group and its descendants underwent was much more involved than Figure 9-7 indicates.

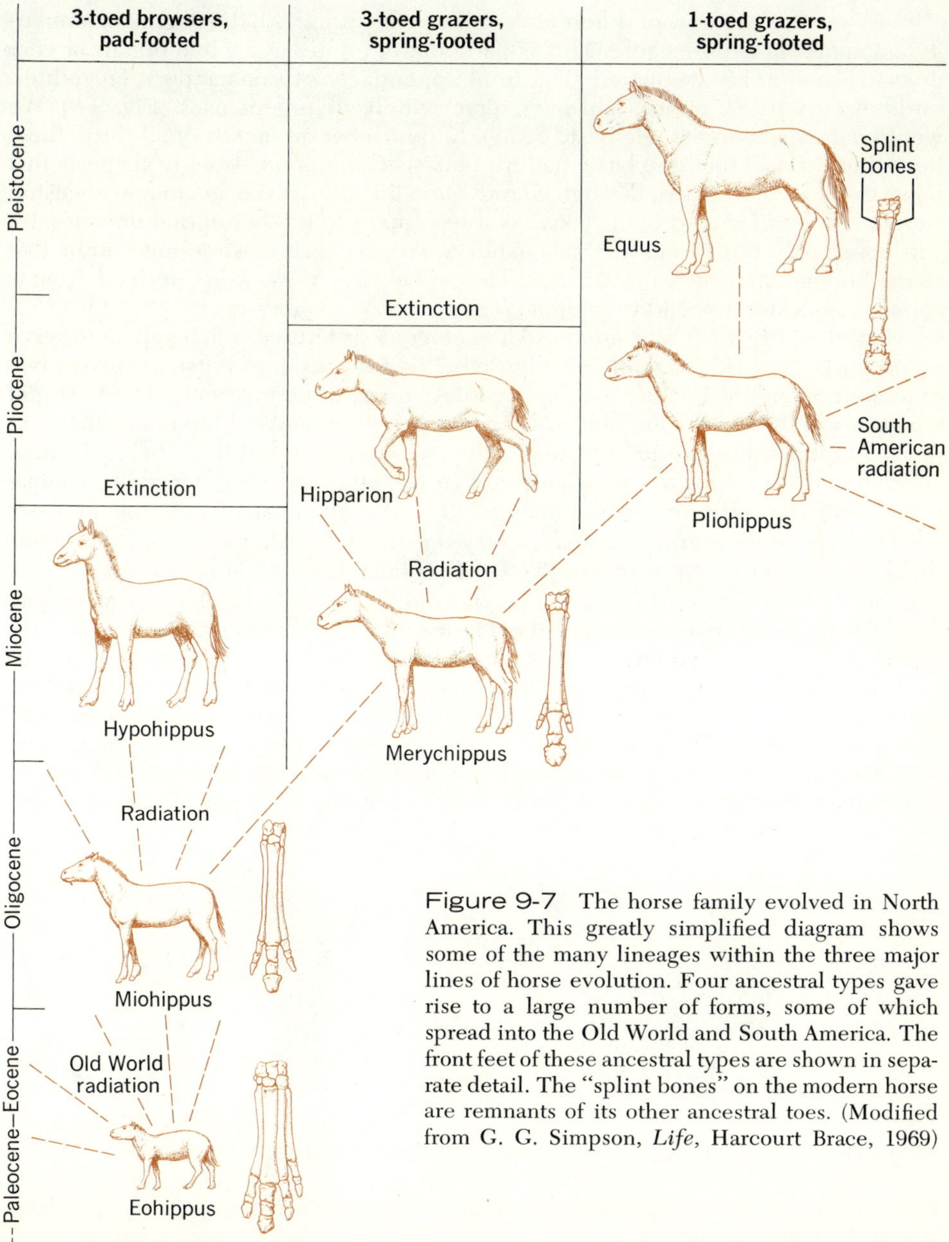

Figure 9-7 The horse family evolved in North America. This greatly simplified diagram shows some of the many lineages within the three major lines of horse evolution. Four ancestral types gave rise to a large number of forms, some of which spread into the Old World and South America. The front feet of these ancestral types are shown in separate detail. The "splint bones" on the modern horse are remnants of its other ancestral toes. (Modified from G. G. Simpson, *Life*, Harcourt Brace, 1969)

With the gradual development of extensive plains areas in the Miocene epoch, horse evolution yielded new types with teeth adapted to grazing on tough grasses. The weight on each foot was shifted to one toe which acted like a spring so that the animal could run over hard and often rough terrain. The intricate evolutionary history of the horse family as revealed through fossils is a convincing demonstration that organic evolution did indeed take place.

Comparative Anatomy and Embryology

When the anatomy of different vertebrates is examined, the degree of similarity among them is noticeable. This is particularly impressive when organs of very diverse function are compared. The front appendages of salamanders, crocodiles, birds, bats, whales, moles, and man serve widely divergent uses (Fig. 9-8). We would not expect any single basic design to best meet the needs of all these functions, but the fact that one basic pattern persists throughout strongly suggests that these types have evolved, through adaptive modification, from a common ancestor. In fact, the fossil record lends considerable support to the belief that the ancestor was not greatly unlike modern salamanders. Organs such as vertebrate limbs that share fundamental structural resemblances because they were derived from a common ancestor are said to be *homologous*.

Particularly intriguing are those homologous structures which appear to serve no purpose. Note, for example, in Figure 9-7 that the modern horse possesses two functionless "splint bones" that bear witness to its three-toed ancestry. *Vestigial* structures of this nature are not at all uncommon. Remnants of hip bones that normally support limbs are found in both the python snake and the whales. Certain crayfish that are native to caves have no eyes, yet they possess eye stalks. If organisms were specially created, the incorporation of useless parts would seem grossly inefficient. As an example the reader may recall the female gametophyte in flowering plants (Fig. 7-13). Why are five cells produced, for no apparent reason, that degenerate following fertilization? This question is logically resolved by the explanation, supported by paleontological evidence, that these cells represent relics of a former gametophyte generation possessed by the ancestors of flowering plants.

In following the embryonic development of related organisms one is further struck by the incongruities of the special creation hypothesis. All vertebrate embryos look astonishingly alike at corresponding stages in their development (Fig. 9-9). All of them possess embryonic gill pouches, yet none of the reptiles, birds, or mammals has aquatic larvae and all the adults breathe by lungs. These nonfunc-

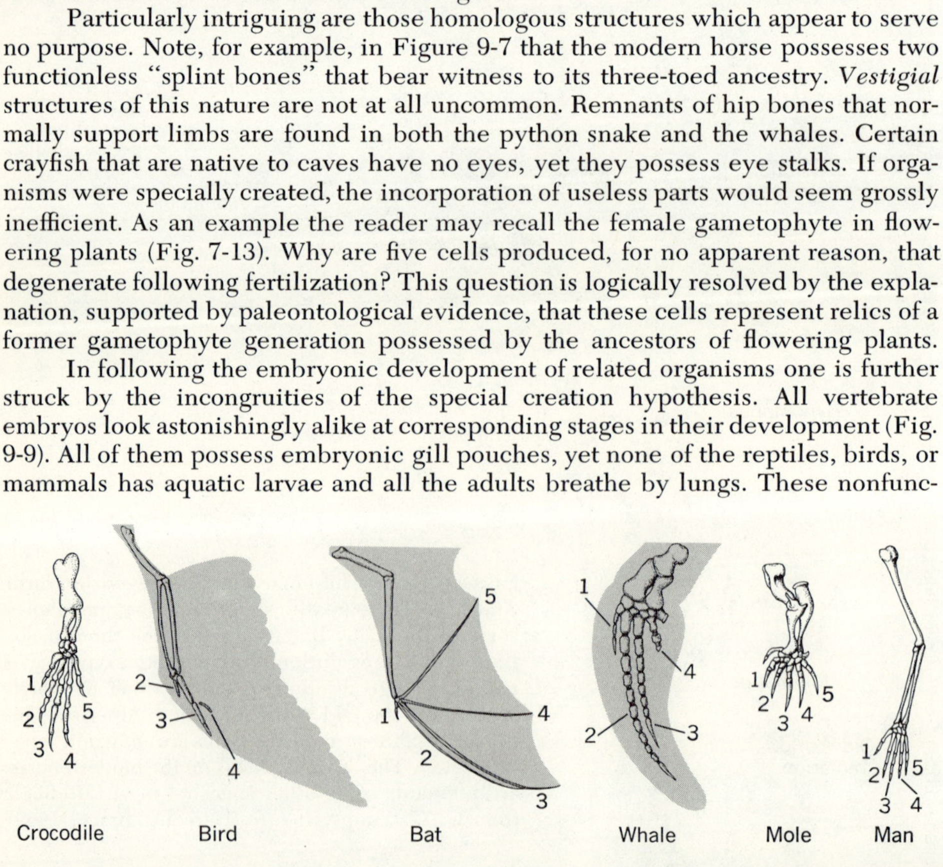

Figure 9-8 Left front leg bones of various vertebrates. Homologous bones have been adapted for different functions.

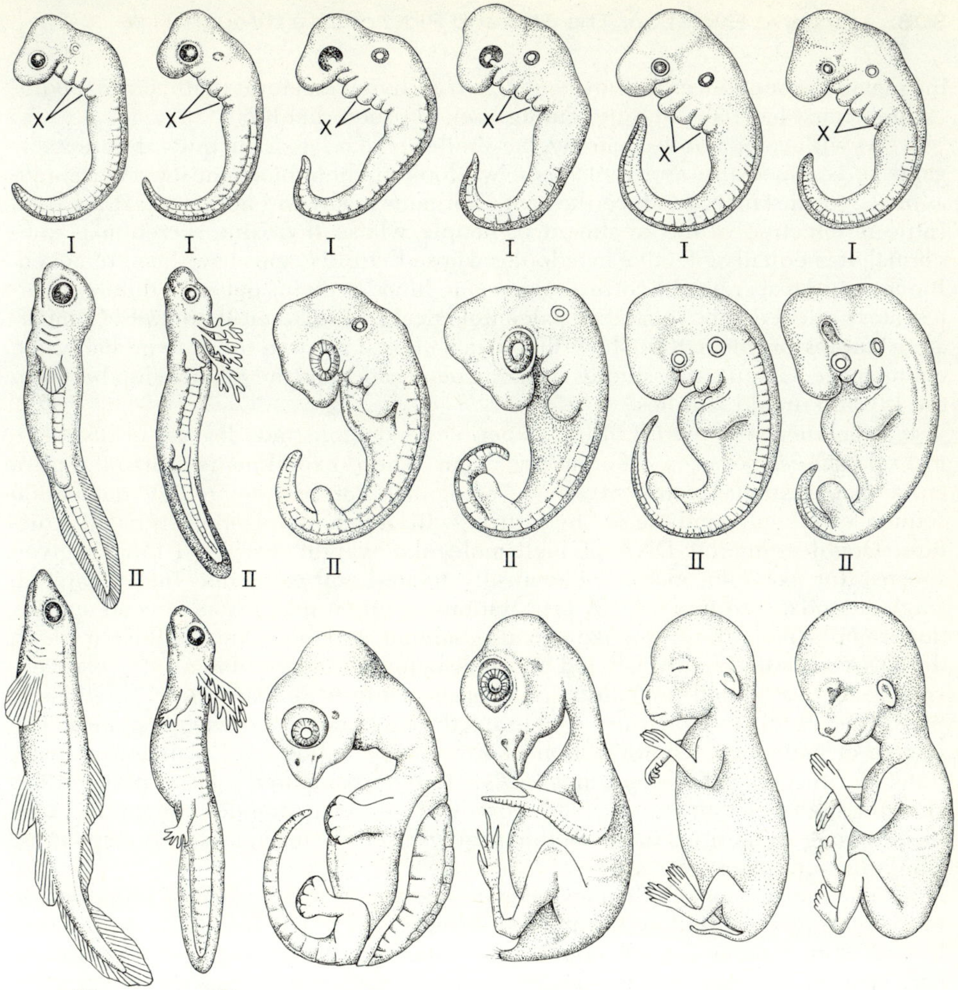

Figure 9-9 The embryos of representative vertebrates at corresponding stages in development. Gill pouches are indicated by the letter x. (From E. O. Dodson, *A Textbook of Evolution,* Saunders, 1952, after Haeckel.)

tional embryonic gill pouches take on a meaningful significance, however, when viewed as structures essential to embryological development in a diverse group of animals that share a common ancestry. Life does not abandon successful procedures; embryonic patterns and sequences that have proven to be efficient tend to be conserved. New forms appear to have arisen by diverging at some point from such established developmental patterns. The appearance of embryonic gill pouches represents a physiologically important step which, it would seem, has demonstrated its usefulness in the pharyngeal development of all vertebrates.

Comparative Biochemistry

If anything surpasses the architectural order that is woven through the diversity of living form, it is the biochemical uniformity that pervades all living processes. Chapters 3 and 4 certainly attest to this. It seems reasonable to assume that the metabolic sequences that are common to all living creatures (with minor modifications) must have an antiquity dating back to primordial life.

As with comparative anatomy, the similarity of related, but quite different, organisms is especially apparent when we look at their biochemistry. Vertebrate animals, for instance, produce the same hormones. Thyroid hormone extract from cattle is effective in the treatment of people whose thyroxine secretion is subnormal; it is equally effective in tadpoles whose thyroid glands have been removed. Biochemical tests reveal a correspondence in blood proteins between different vertebrates believed to be related on paleontological and anatomical grounds. Comparative studies of this nature show that man's closest relative today is probably the chimpanzee. Comparable studies utilizing plant sap show the relationship between the lily and orchid families.

Nucleic acid "hybridization" experiments demonstrate degrees of biochemical relatedness between different organisms. These experiments utilize the principle that a single strand of DNA "recognizes" a complementary nucleotide sequence and will combine or "hybridize" with it to form a double-stranded structure. Double-stranded DNA of high molecular weight, extracted from a given species (for example, man), is chemically treated and converted into separated, single strands. Additional DNA preparations are then made from several species (for example, man, rhesus monkey, mouse, salmon, and the colon bacillus) in which the DNA is radioactively labeled and mechanically broken down into relatively small, single-stranded fragments. Next each of these radioactive preparations is placed separately into mixtures containing the high-molecular-weight strands. The degree of molecular "hybridization" between the radioactively labeled preparations and the long-stranded human DNA is then compared. Those preparations showing high percentages of combination with the long-stranded human DNA indicate strong similarities in nucleotide sequences and consequently corresponding similarities in genes.

The rhesus monkey demonstrates this degree of genetic relationship to man; in decreasing order follow the mouse, salmon, and colon bacillus. Nucleic acid hybridization studies not only establish evolutionary relationships but also substantiate many of the same evolutionary relationships revealed through the comparative blood chemistry of vertebrates, comparative anatomy, and paleontology.

The evolutionary history of a group is even reflected in its embryonic biochemistry. Fishes excrete most of their nitrogenous waste as ammonia. Adult amphibians excrete urea for the most part (see pages 156–159). Although birds eliminate nitrogenous waste in the form of uric acid, a bird embryo first excretes ammonia. A period of urea excretion follows before the embryo finally starts to produce uric acid. Since the fossil record suggests that birds trace their ancestry from reptiles back through amphibians to primitive fishes, it would seem that the bird is biochemically recapitulating certain major events in its evolutionary history.

Animal and Plant Domestication

The great diversity of cultivated plants and domesticated animals which man has created, and continues to create, through selective breeding provides direct evidence of the continuing occurrence of organic evolution. Man has produced an impressive list of crop plants and farm animals, not to mention a profuse variety of ornamental plants and animal pets. It is, of course, true that only in a very few cases have any of these been considered to be new species. It is equally true, on the other hand, that in many instances these organisms have not been assigned to separate species only because (1) their history by domestication has been known and/or (2)

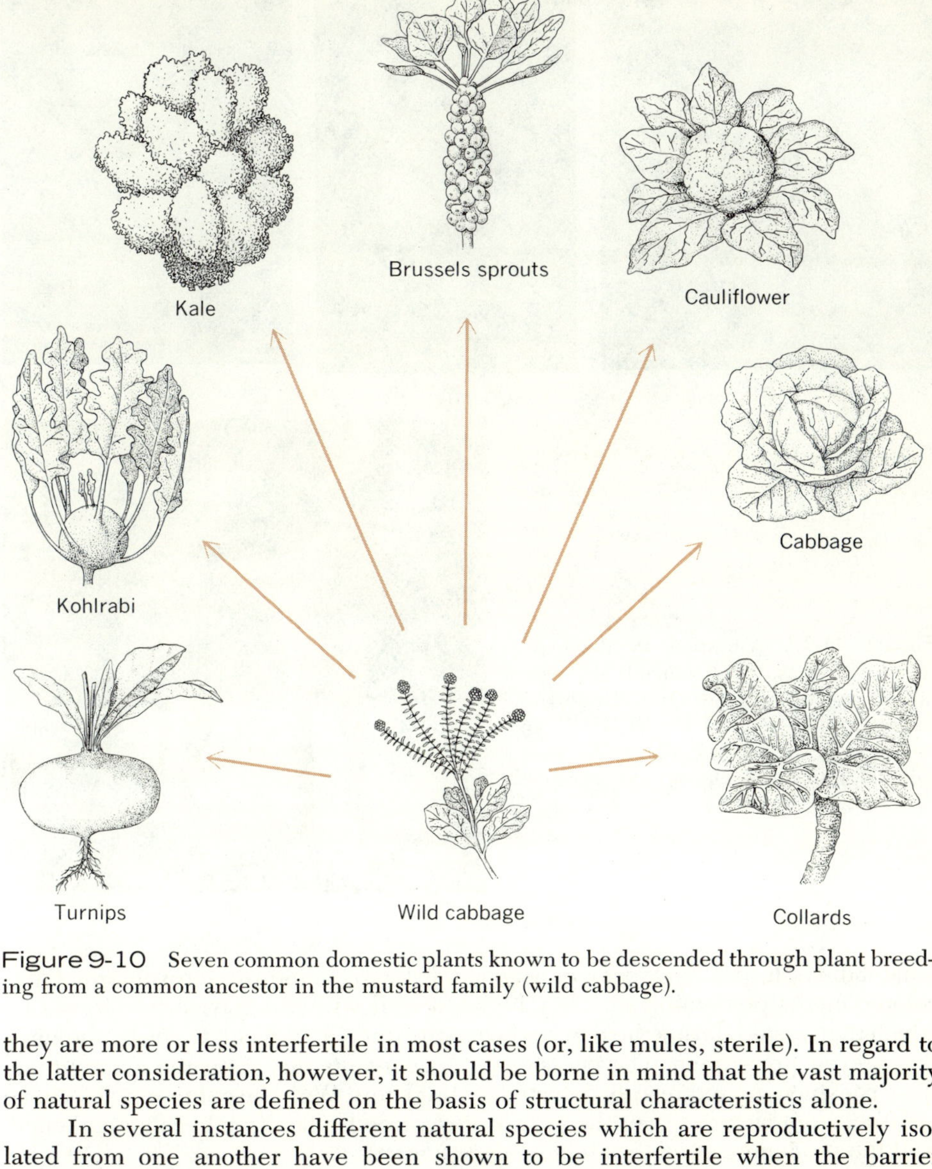

Figure 9-10 Seven common domestic plants known to be descended through plant breeding from a common ancestor in the mustard family (wild cabbage).

they are more or less interfertile in most cases (or, like mules, sterile). In regard to the latter consideration, however, it should be borne in mind that the vast majority of natural species are defined on the basis of structural characteristics alone.

In several instances different natural species which are reproductively isolated from one another have been shown to be interfertile when the barrier separating them was removed. It seems highly likely that if botanists had first encountered the crop plants shown in Figure 9-10 in a wild state, they would have described more than one species. As for the dogs illustrated in Figure 9-11, matings between Great Danes and Chihuahuas are highly unlikely under any sort of natural circumstances.

Ecological Genetics and Evolutionary Biology

The study of living populations of wild species offers further direct support of the validity of evolutionary processes. A classical example of this involves evolutionary changes that have been occurring in many different species of moths that

(a) (b) (c)

(d)

Figure 9-11 Variation in domestic breeds of dogs. (a) Samoyed, (b) Chihuahua, (c) Bulldog, (d) Great Dane [(a) Camilla Smith; (b) Sid Bernstein, Photo Researchers; (c) Black Star; (d) Russ Kinne, Photo Researchers]

inhabit the industrialized sector of England. Prior to the industrial revolution, light-colored moths predominated. They blended well with the grayish tree trunks on which they rested during the day, and were presumably difficult to see by the birds that preyed upon them. The occasional dark mutants that appeared among them were conspicuous and readily detected by predators. With the outpouring of smoke and soot from factories over the past century, tree trunks have become darkened. And the predominant moths are now dark colored.

An experimental study of one of these moth species (the peppered moth, Fig. 9-12) yielded very interesting results. Large numbers of marked individuals of both color phases were released in air polluted areas. Traps were then set out to recapture them. Of the moths recaptured the darks outnumbered the lights two to one. When this procedure was repeated in unpolluted areas, the recapture ratio was two lights to one dark. Motion pictures confirmed that predatory birds were largely responsible for the differential mortality of the two color phases.

In yet another investigation into evolutionary biology, a species was actually "synthesized." A group of plants known as the hemp nettles (*Galeopsis*) includes one species (*G. tetrahit*) with 16 pairs of chromosomes. Other species have 8 pairs.

Figure 9-12 The peppered moth and its dark form on a soot-covered tree trunk near Birmingham, England (left). The two color phases on a grayish lichen-covered trunk in a soot-free area (right). (H. B. D. Kettlewell)

Two species in the latter group are similar to *G. tetrahit* in a number of respects. An effort was consequently made to interbreed these two species in an attempt to produce the polyploid species. The effort was successful. The offspring were indistinguishable from natural *G. tetrahit* individuals. Crosses between the "synthetic" and natural species, moreover, proved to be interfertile, and the resulting seeds germinated, producing normal *G. tetrahit* plants. Furthermore neither the natural nor "synthetic" polyploid species could be made to interbreed with either of the diploid species from which *G. tetrahit* had apparently evolved.

Biogeography

If each species were the work of special creation, we should expect it to be found in all habitats for which it had been created. This, however, is usually not the case. The fauna and flora of the deserts of Africa and America are quite different despite the fact that the environments are comparable. On the other hand, the volcanic Galapagos Islands some 600 miles off the coast of Ecuador harbor 26 species of land birds, 23 of which are found nowhere else in the world. Yet all 23 appear to be related to birds on the mainland where the climates and environments are quite unlike any in the archipelago. Biogeographic examples of this sort are commonplace. They cannot be reasonably explained through special creation but are a natural consequence of organic evolution. A species has a place of origin from which it disperses into areas available to it. As it does so it is subject to modifications imposed by the environment.

As organisms disperse their distribution tends to be discontinuous. The evolution of the camel family is an excellent illustration of this (Fig. 9-13). The camels of Africa, the Near East, and Asia are halfway around the world from their closest relatives, the llamas of Andean South America. The paleontological record of fossil camels accounts for this discontinuity nicely. The distribution of fossil members of the family was continuous between the regions now occupied. The oldest of these fossil types lived in North America and was the common ancestor of modern camels and llamas. As this common ancestor dispersed from its place of origin it gradually diverged into two separate lines. One of these groups succeeded in moving southward over the land bridge between the Americas to become South American llamas. The other was successful in migrating over a Tertiary land bridge

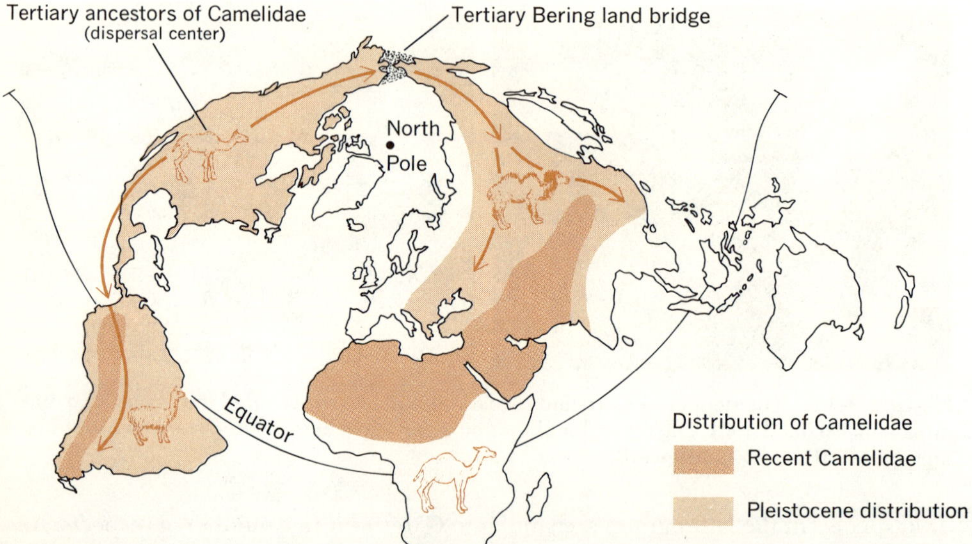

Figure 9-13 Comparison of past and present distributions of the Camelidae (camel family). Modern discontinuous distribution is a remnant of a once continuous distribution. (Modified from G. G. Simpson, *Life*, Harcourt Brace, 1969)

(now the Bering Straits) into Eurasia to ultimately become modern camels. Extinction in North and Central America and much of the Old World has left two widely separated remnants of a family that once spanned most of the globe.

Increasing interest has been aroused in the prehistoric connections between continents by a revival of the hypothesis of continental drift. An impressive body of geodetic, geologic and paleontologic evidence is accumulating which strongly suggests that the continents of today are actually separated parts of two Mesozoic supercontinents called Laurasia and Gondwanaland (Fig. 9-14). It is postulated that, possibly sometime about the middle of the Mesozoic era, these huge land masses started to break up into fragments (continents) that gradually began to drift away from each other. A primitive seed plant (the long extinct *Glossopteris* which bore unique reproductive structures on its leaves) is one of several especially convincing fossil discoveries that support the hypothesis of wandering continents. This plant (Fig. 9-15) was apparently the dominant element of a luxuriant flora that contributed to extensive coal deposits throughout Gondwanaland (South America, Southern Africa, Southern India, Antarctica, and Australia) from about 220 to 300 million years ago. At that time the floras on Laurasian continents were quite dissimilar. It is difficult to account for the distribution of *Glossopteris* and its associated species over such a vast area in the southern hemisphere unless indeed the continents of Gondwana were much closer together than they are now.

THE NATURE OF EVOLUTIONARY PROCESSES

The Directing Agency of Evolution

Any mechanism attempting to explain how evolution occurs must embody a directing agency of some kind to account for the skillful adaptations that living crea-

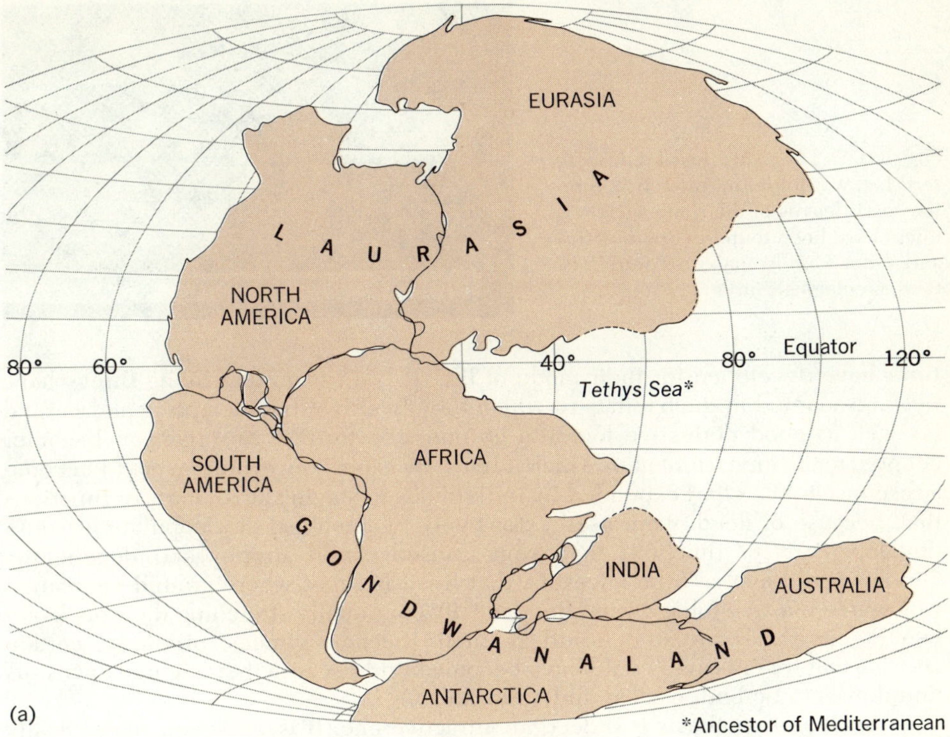

Figure 9-14 (a) The postulated supercontinents of Mesozoic time: Laurasia and Gondwanaland. (b) Gondwanaland reassembled. [(a) modified from R. S. Dietz and J. C. Holden in *Scientific American*, 1970; (b) modified from Hurley and Rand in *Science*, Vol. 164.]

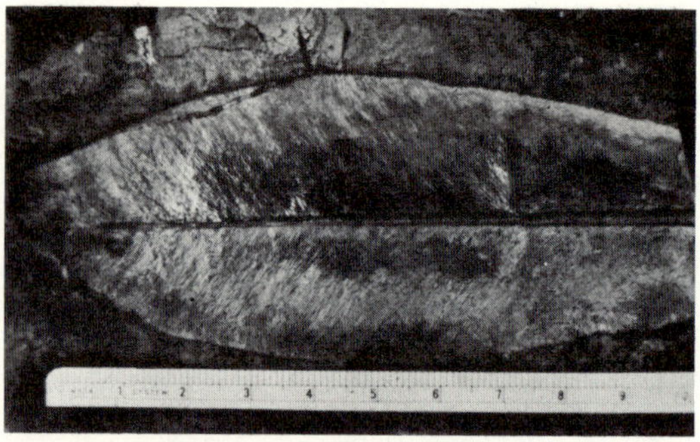

Figure 9-15 This fossil *Glossopteris* leaf was found in Antarctica. Similar fossil leaves of the same age (Permian) have been found in South Africa and other Gondwanaland continents. (U.S. Geological Survey)

tures have developed for their mode of life. It is a truism that living things have become adapted to the habitats in which they live. It is both tempting and natural (as well as good deductive logic) to go one step further. An organism becomes adapted to the environment in which it lives. By extension, evolution could become a response to the effort expended by individuals in staying alive. Parts or functions that, because of need, were used extensively by members of a population would develop. Parts or functions that went unused would atrophy and degenerate. Giraffes could evolve from ancestors that lived in areas where ground vegetation was scarce but tree foliage was abundant. By continually stretching their necks for generations these ancestors could gradually become giraffes. Similarly, certain crayfish that live in caves could have become blind because their sighted ancestors found eyes to be both useless and a nuisance.

Such an hypothesis is indeed an attractive one. It is, however, scientifically unsound for several reasons. First, we know that phenotype does not determine genotype; it is the other way around. Second, innumerable attempts to demonstrate the hereditary transmission of acquired characters have failed. The ancient practice of foot binding to which Chinese women of a former aristocracy subjected themselves from infancy produced a "club foot" symbolic of their caste. Although this custom was in vogue for many centuries, infants continued to be born with normal feet. Head-binding among certain African peoples has been similarly ineffective. Finally, there are many examples of evolution for which this hypothesis fails to account. How, for instance, did the ancestors of green grasshoppers practice being green? In view of the paleontologic evidence that all lineages ultimately end in extinction, why is it that all organisms are eventually unable to adapt? How (and for what purpose!) would they practice being biologic failures?

This hypothesis has been credited to the author of the first truly scientific treatise on organic evolution (Lamarck's *Zoological Philosophy*, 1809), but it is probably no more Lamarckian than it is Darwinian.[1] It was espoused by both men and claimed by neither. It was in 1859 with the appearance of Charles Darwin's *Origin of Species* that two monumental contributions were made which were to mark the beginning of the modern era in biology: (1) a great wealth of evidence was brought together for the first time in support of a general theory of organic evolution; and (2) among the hypotheses that were called upon to explain how evolution occurs, a well-reasoned and scientifically substantiated theory of natural selection emerged as the directing agency of evolutionary change.

[1] The use-disuse hypothesis (or the inheritance of acquired characters as it is now called) was even discussed in a hereditary context by ancient Greek scholars.

Darwin based his theory of natural selection on three well-established observations. He noted, first, that the reproductive potential of all organisms is such that more offspring are produced than could possibly survive if all checks against their increase were removed. He calculated that even a pair of elephants, which are very slow-breeding animals, could produce 19 million descendants in 750 years if nothing, except old age, eliminated elephants. He also observed, on the other hand, that populations of living things are relatively constant from year to year. From these two facts, Darwin reasoned, all organisms must be subject to a struggle for existence. At this point an obvious question arose. Is this struggle blind and senseless or is there a tendency for elimination to be systematic? The marshaling of fact number 3 resolved this question: there is a bewildering degree of hereditary variation among the individuals of any species. To Darwin the conclusion was clear. Those individuals whose hereditary makeup happened to be best attuned to survival would be most likely to pass their characteristics on to succeeding generations. The purpose of evolution's course is fitness and natural selection guides the way to that end.

The competition for survival and the contest for fitness should be thought of neither as a brutal nor a fearsome struggle for existence. On the contrary, natural selection is a slow and gradual process in which certain kinds of people, who are less susceptible to malaria, and certain kinds of moths, which are less liable to predation, increase their relative numbers in a population. The introduction of rabbits into Australia has posed a serious threat to many of the native pouch-bearing mammals. The competition there has hardly been marked by ferocity. The fact is that rabbits are highly efficient herbivores, very elusive prey, and notoriously prolific. The local herbivores and carnivores are simply outclassed. Natural selection, moreover, often results in establishing cooperative interrelationships between organisms, which is quite the antithesis of competition. As we shall see later in this chapter, there are many cases in which different organisms have evolved mutual dependencies on one another to further their common survival. This has occurred not only within species but also between species and between large segments of entire plant-animal communities.

In order for any living creature to survive it must sustain a flow of energy commensurate with the metabolic needs of its highly ordered state. This results in ramifications that extend far beyond the need for nutritive energy alone. The successful conquest of energy encompasses a multitude of factors including protection from predators, disease, and the vicissitudes of climate and fate. The specific and unbelievably numerous devices (Chapters 4 through 7 cataloged them) by which organisms accomplish these ends are called *adaptations*. And adaptations, biologists believe, are the products of natural selection.

No organism sustains the awesome pace of life forever. Before the metabolic machinery begins to "run down," reproduction occurs and a new generation comes into being. As we stressed in Chapter 3 life is a population-oriented phenomenon. It owes its nature to a long and continuous history of countless individuals distributed over a vast period of time. Individual organisms are like "booster pumps" providing the free flow of energy by which the continuity of life is perpetuated. In this capacity they also serve as test models whose biologic design must pass the test of contemporary acceptability. As such they are expendable and subject to natural selection. Those that pass the test of environmental suitability contribute heavily to the gene pool. They produce more mature offspring, who, in turn, produce more mature offspring as those biologic designs that pass inspection become increasingly incorporated into oncoming generations. The extent to which the varying genotypes in a population effectively harvest the resources about them is reflected by their contributions to the gene pool. Natural selection is differential reproduction which favors those individuals with the highest *net* reproductive efficiency.

Since natural selection operates only up to and through the reproductive period of an organism's life, survival beyond the breeding age is seldom selected for. Those exceptional cases where life span may extend, as with man, far beyond the procreative age are probably the result of (1) indirect or supplementary contributions of postreproductive individuals to the reproductive process (in the training or protection of young, serving as food for predators, thereby sparing the young from excessive predation, and so on) and/or (2) the possibility that some traits which favor survival early in life are also conducive to longevity. It must not be assumed, however, that factors which contribute to longevity are necessarily related to the perpetuation of species. In the interest of *population survival* the existence of noncontributing members may be a hindrance. It is doubtless no coincidence that the *maximum* life span of man has remained virtually unchanged throughout recorded history. The marked increase in degenerative diseases during old age is apparently not selected against, whether it be in aboriginal hunting societies or the civilized world of sophisticated medical arts.

Selection, Genetics, and Population Dynamics

It will be worthwhile at this time to review and summarize certain features of natural selection with respect to its operational characteristics in the evolutionary process. Three such characteristics are of special significance.

1. *Natural selection is creative as well as adaptive.* Moths adapted to the woods around industrial England were the result of selection pressure which increased the frequency of one allele at the expense of another. People adapted to malarial Africa have evolved through selection pressures which *adjusted* the frequencies of alternative alleles to produce an optimum number of malarial-resistant heterozygotes. Similar instances are known in a wide variety of organisms in which heterozygous individuals are fitter than either of the corresponding homozygotes. Natural populations are highly heterozygous. The reproductive superiority of such populations has been demonstrated experimentally (Fig. 9-16). Selection pressures on populations are incredibly complex since they are exerted against whole organisms and the countless multitudes of genes which determine their phenotypes. Most of these genes influence other genes and many if not all of them have several phenotypic effects. Some effects are advantageous and others may be disadvantageous. The adjustment of gene frequencies for overall fitness in the face of such complexity is truly a creative process.

2. *Natural selection exerts a stabilizing influence on evolution.* The phenotypic effects of varying combinations of genes differ. Some gene combinations are favorable, others are unfavorable. The natural selection of individuals with appropriate chromosome mutations can increase the likelihood of favorable (or decrease the likelihood of unfavorable) combinations. Two complementary genes located on different chromosomes could become part of the same linkage group through translocation (Fig. 8-18), thereby increasing the chance that the two genes would be transmitted together. Similarly, two complementary genes located at some distance on the same chromosome could be brought closer together by an inversion, decreasing the chance that they would be transmitted separately through crossover.

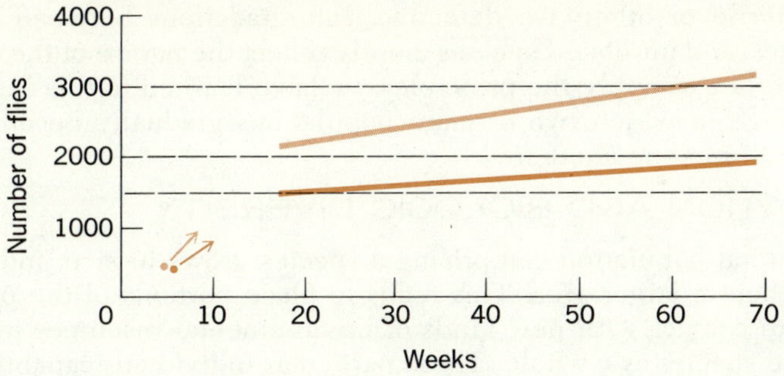

Figure 9-16 Evolutionary change and heterozygosity. Two fly populations (*Drosophila serrata*) of 300 individuals each were placed in separate cages under constant environmental conditions. One population was from a geographic strain collected in New Guinea (solid color). The other was a hybrid population derived from crossing the New Guinea with an Australian strain (light color). By the fifteenth week (three generations) competition in each cage for available food and space became intense. Although both populations demonstrated measurable adaptation to the crowded conditions, the rate of increase of the hybrid population was double that of the New Guinea population (an average increase of 20.4 flies per week versus 8.4 flies per week). (Adapted from F. J. Ayalo, *Science*, Vol. 162, pp. 1453–1459 (December 1969. Copyright 1969 by the American Association for the Advancement of Science)

3. *Natural selection requires genetic variability to effect continuing survival of a population.* Environments are unstable and subject to continuous oscillation and change. A population subject to natural selection cannot meet the demands of a fickle "market" unless it is supplied with a profuse variety of "experimental" phenotypes. Mutation, genetic reassortment, and recombination provide these by continually replenishing the gene pool with novelty and change. In the final analysis genetic variation is the most imperative adaptation of all. It is the only real protection a population can muster against extinction. The failure to generate adequate variation for the adaptive action of natural selection has probably been the major cause of extinction of many of the millions (billions?) of species that once populated the earth.

Isolation, Divergence, and Speciation

Now that we have identified the directing agency of evolution and explored some of the principles underlying adaptive changes within populations, we are prepared to examine the manner in which new species originate (*speciation*). In so doing we are really not examining a new issue but merely enlarging on one that has been before us. A species is a population that usually consists of several smaller populations. Although gene flow is possible between these subpopulations, the species as a whole is distinguished by being reproductively isolated from other species. In following the evolutionary development of a new species from some ancestral population one is confronted with a gradual accumulation of changes in the gene pool[2] over a long period of time. Specifying the point in history at which a descendant population became a new species is a somewhat arbitrary judgment. The degree of interbreeding between several subpopulations of a species may vary considerably. Closely related species may on occasion interbreed and produce fertile offspring, but this usually occurs infrequently and their gene pools are essentially isolated. Less closely related species do not interbreed or, if they do, their off-

[2] An exception to this is the occasional occurrence of speciation in plants through polyploidy, as in the hemp nettles referred to previously.

spring are sterile or otherwise defective. Intergradations between populations, related species, and unrelated species merely reflect the nature of the evolutionary process itself. Let us apply the principles we have learned to this higher level of adaptive change in which two or more populations gradually become reproductively isolated from one another.

ISOLATION AND BIOLOGIC DIVERSITY

As the total population comprising a species grows in size, individuals disperse throughout a larger area. This tends to place portions of the population in more intimate contact with new kinds of environmental resources until now untapped by the species as a whole. Those particular individuals capable of utilizing these unexploited resources will have an advantage. Their reproductive efficiency will be enhanced by lowered competition from other members of the species, and natural selection (plus genetic drift) will begin to produce a diversity of forms in response to those new environments lending themselves to exploitation.

An excellent example is provided by the leopard frog in the United States. This common amphibian species is distributed virtually over the entire country from deserts to high mountain tops. Several geographic races are readily recognized (Fig. 9-17). Frogs from populations (or, more properly, races) in adjoining regions can be crossed readily and produce normal young. When frogs from distant areas such as Vermont and Florida are interbred, however, the resulting embryos are abnormal. Very few survive, apparently because of a disparity in the rates of development between the northern and southern populations. The hybrid embryos in essence receive conflicting genetic directions and develop abnormally. These two populations could be considered separate species were it not for the fact, as shown in Figure 9-17, that they are bridged by a continuous chain of intervening populations between which gene flow can occur.

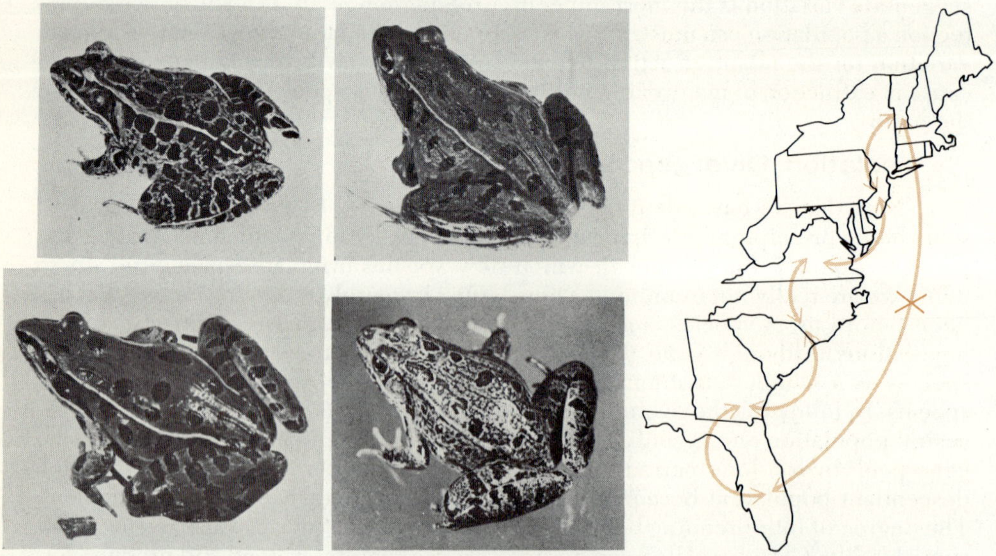

(a) (b)

Figure 9-17 Geographic variation and gene flow in the leopard frog. (a) Four different geographic races of the leopard frog (*Rana pipiens*). (b) Neighboring populations of this frog in the eastern United States can interbreed and produce viable young. Crosses between New England and Florida frogs yield very few young which survive. [(a) John A. Moore]

OPPORTUNITY FOR CONQUEST AND SPECIATION

The conquest of new environments and ways of life depends on certain prerequisite opportunities. George Gaylord Simpson, a highly respected student of evolution, identifies three kinds.

A species must first have physical access to a new environment. Monkeys in the rain forests of Central America lack the *physical opportunity* of entering the rain forests of Africa. Were these monkeys somehow able to reach the African forests the adaptations they had developed for living in New World forests would probably enable them to gain a foothold since the two environments are similar in many respects. New World monkeys have the *constitutional opportunity* to invade parts of tropical Africa. That they would be successful in doing so, however, is unlikely because *ecological opportunity* is wanting. The Old World monkeys are in control of all monkey habitats. New World invaders would probably not stand a chance.

The formation of new species is a response to the myriad *opportune* circumstances which have presented themselves in the long evolutionary history of life. We will examine two such circumstances, one involving ecological opportunity and the other constitutional opportunity.

EVOLUTION THROUGH ECOLOGICAL OPPORTUNITY

The Galapagos Islands mentioned previously in this chapter (Fig. 9-18) were formed over a million years ago by volcanic activity. They have never been connected to one another or to any other land mass. The physical opportunity to invade

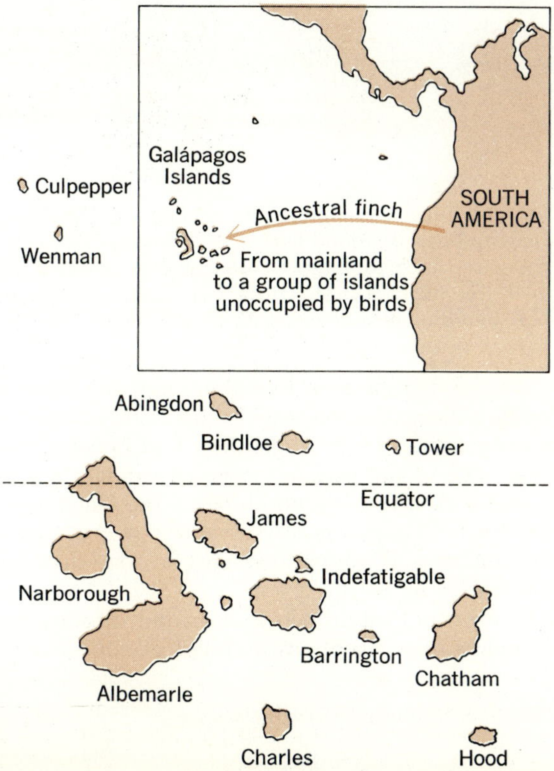

Figure 9-18 The Galapagos Archipelago, home of Darwin's finches. (Modified from G. G. Simpson, *Life*, Harcourt Brace, 1969)

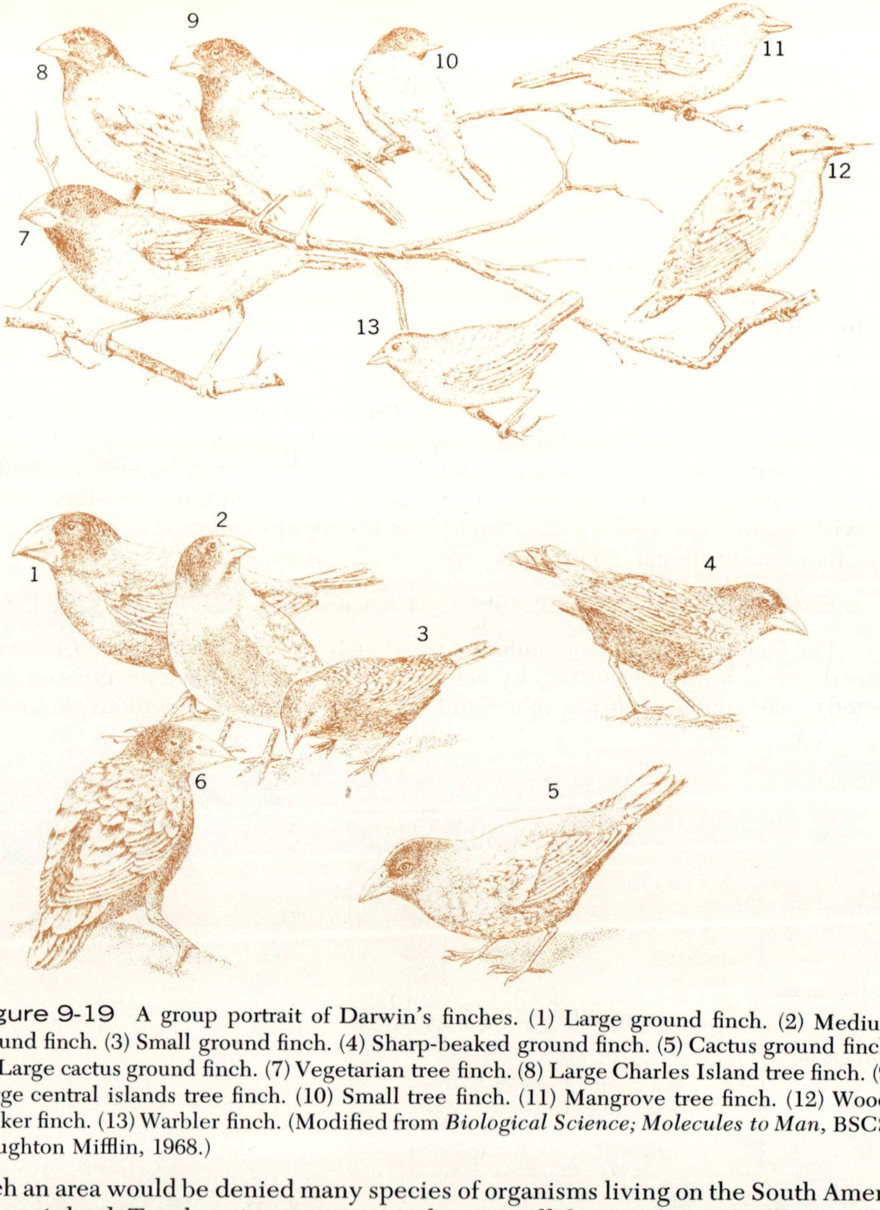

Figure 9-19 A group portrait of Darwin's finches. (1) Large ground finch. (2) Medium ground finch. (3) Small ground finch. (4) Sharp-beaked ground finch. (5) Cactus ground finch. (6) Large cactus ground finch. (7) Vegetarian tree finch. (8) Large Charles Island tree finch. (9) Large central islands tree finch. (10) Small tree finch. (11) Mangrove tree finch. (12) Woodpecker finch. (13) Warbler finch. (Modified from *Biological Science; Molecules to Man*, BSCS, Houghton Mifflin, 1968.)

such an area would be denied many species of organisms living on the South American mainland. To others immigration might range all the way from a highly improbable (but possible) event to a virtual certainty. Once vegetation had become established on the islands it is believed one of the original migrant animals was a species of vegetarian ground finch. Today all the small land birds of the Galapagos Islands trace their ancestry to these early finch pioneers. First studied by Darwin in 1832, they have been called Darwin's finches and comprise at least thirteen distinct species found nowhere else in the world[3] (Fig. 9-19).

Having once arrived, the possibilities for further colonization and subsequent speciation by the ancestral ground finches were excellent. Ecological opportunity

[3] A related finch is found some 600 miles to the northeast on Cocos Island.

was wide open with a variety of food resources and nesting sites and no competition to thwart their utilization. Even predators were scarce. Increase in numbers together with occasional island hopping gradually resulted in substantial populations on one and then another of the several islands in the archipelago. Gene flow between the island populations would have been negligible. The stage was set for speciation to occur.

The detailed histories of each of the 13 species, of course, is not clear. They are now spread over the archipelago in varying combinations, some islands having as many as ten species and some as few as three species. Most of the islands had about the same ecological opportunities,[4] however, and each species became spe-

[4] Most of them, for example, could accommodate large, medium, and small vegetarian ground finches.

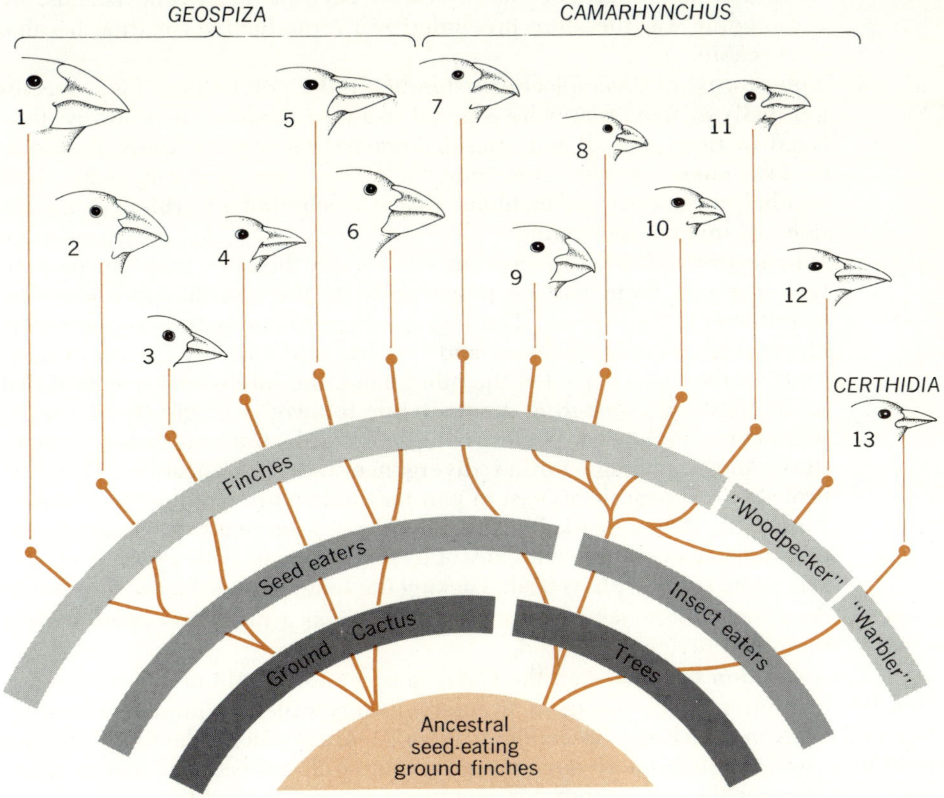

1. *G. magnirostris*
2. *G. fortis*
3. *G. fuliginosa*
4. *G. difficilis*
5. *G. scandens*
6. *G. conirostris*
7. *C. crassirostris*
8. *C. pauper*
9. *C. psittacula*
10. *C. parvulus*
11. *C. heliobates*
12. *C. pallidus*
13. *Certhidia olivacea*

Figure 9-20 Adaptive radiation in Darwin's finches. Related species are grouped within a unit called a *genus* (plural, *genera*). Three genera are shown: *Certhidia*, *Camarhynchus* and *Geospiza*. The scientific name of a species includes its generic group (capitalized) followed by its specific name (lower case). The scientific name of the large ground finch (1), for example, is *Geospiza magnirostris;* that of the woodpecker finch (12) is *Camarhynchus pallidus*. The numbers cross-refer to Figure 9-19. (Modified from G. G. Simpson, *Life*, Harcourt Brace, 1969)

cialized in response to the local conditions that led to its irrevocable reproductive isolation. It can be stated with some confidence that speciation followed three courses.

1. The relaxation of selection pressures on genes formerly selected against on the mainland led to increased variation and diversification. Geographic races, through reproductive isolation, became full-fledged species. Some ground finch species fed on large seeds, some on medium-sized seeds, and some on small seeds (1, 2, and 3, respectively in Figs. 9-19 and 20). Still others fed primarily on cactus flowers (5 and 6). In becoming adapted to their specialized food habits each species developed characteristic beak shapes and sizes. Members of one species recognized one another by their beaks (recall from Chapter 1 that gulls use their eye rings). As more than one species came to occupy each of the several islands, interbreeding was therefore precluded and reproductive isolation became irrevocable.
2. The process of diversification extended to the point where finch species gradually evolved that were adapted to a new mode of life entirely; they lived in trees and all but one of them fed on various kinds of insects (7–11). These species constituted a new genus (see Fig. 9-20). Still another became so "unfinchlike" that it resembled a warbler (13). It too belongs in a distinct genus.
3. Whenever two similar species come to occupy the same area the competition between them will be proportional to the specific environmental requirements they share. The less resourceful competitor is commonly eliminated in such situations and this has undoubtedly occurred many times in the Galapagos. On the other hand, the intensified selection that exists between competing groups tends to favor ecologically dissimilar phenotypes in the two species. Thus natural selection may reduce competition and encourage further divergence and specialization. This was probably responsible at least in part for the evolution of the Woodpecker Finch (12). This remarkable tree finch, by reason of a behavioral adaptation, does not compete with any of its insectivorous colleagues. It chisels into a tree trunk with its beak. Lacking the long tongue of a woodpecker, it then snaps off a cactus spine which it uses as a probe to extract insects from the crevice it exposed.

The evolution of finches on the Galapagos is an example of *adaptive radiation.* The evolution of life has been an interwoven complex of innumerable adaptive radiations in which the descendants of an ancestral stock, adapted in its own right to a particular habitat, diverge or radiate into unfilled ecologic niches wherever or whenever the opportunity for conquest has been favorable.

Let us now explore an adaptive radiation of far grander proportions and one to which we owe our very existence—the conquest of land by vertebrates. It began over 350 million years ago with an ancestral stock whose constitutional potential was not as restricted as that of a ground finch and it tapped ecological opportunities which encompassed the earth.

EVOLUTION THROUGH CONSTITUTIONAL OPPORTUNITY

There is good geologic evidence that large areas of the earth were subject to seasonal drought during the Age of Fishes and for millions of years thereafter (approximately 200 to 400 million years ago). Periods of abundant rainfall would be

Figure 9-21 *Left:* The lobe-fin fishes of 400 million years ago breathed air and used two pairs of their fleshy fins as body props for crawling. *Right:* Fifty million years later some of decendants of the early lobefins had modified their body-propping fins into legs and become primitive amphibians like those shown here of Carboniferous age. (American Museum of Natural History)

followed by rainless seasons in which streams would dwindle or dry up and ponds would become stagnant and relatively devoid of oxygen. In oxygen-depleted waters it would be highly advantageous for a gill-breathing fish to surface and somehow avail itself of atmospheric oxygen. Fossil evidence suggests that this was the case. Most early fresh water fishes were equipped with both gills and lungs. (All the fishes today which have true lungs also live in regions of seasonal drought.) Many of these fishes also bore two pairs of fleshy fins. It is believed these fins acted as body props by which the fish could wiggle or crawl. This would enable fish to move along a stream bed to a new pond before the one they were in dried up completely. A pair of lungs and two pairs of appendages which served as both fins and "legs" were fish adaptations which allowed fishes to continue to function as fishes.

They also just happened to produce a vertebrate with a new-found constitutional opportunity to invade a different environment, the land. It is a singular fact, however, that this invasion did not really get underway in earnest for many millions of years. Although land plants had become common in Devonian time (Fig. 9-3), the ecologic resources of terrestrial environments were vastly inferior to those that were to come. Primitive amphibians gradually evolved from a certain group of fleshy-finned fishes, called the lobe-fins, as body-propping fins became modified into legs (Fig. 9-21). Some of the descendants of these ancient amphibians, in turn, gave rise to early reptiles whose chief claim to fame was an egg that would develop away from water. But even the first land eggs were probably an aquatic adaptation because they were laid by amphibious fish-eating reptiles. These creatures produced an egg that was no longer at the mercy of transitory ponds and streams. A vertebrate had evolved whose semiaquatic existence was protected during its developmental as well as its adult stages.

During the period of early reptilian evolution, land plants were undergoing a series of impressive adaptive radiations (Fig. 9-4). Ecologic opportunities for a conquest of the land began to open up on a lavish scale. Reptiles happened to be constitutionally capable of exploiting them. The result was an adaptive radiation of majestic proportions in which dinosaurs (and many other reptiles), birds, and mammals have been produced in unbelievable variety. All of these, including man himself, trace their origins back through a chain of circumstances which began with some fishes facing the vicissitudes of a Devonian climate. Had the circumstances been altered to a significant degree the results would not have been the same. Our familiar world of life today could have been one of a million others—and we might have been in none of them.

THE BROAD SWEEP OF EVOLUTION —ITS SALIENT FEATURES

From the two adaptive radiations just described we could conclude that the course of evolutionary history has been marked by two features: (1) a great diversity of organisms has been produced from one or, at the most, a very few related ancestral forms and (2) these evolutionary products have been remarkably well adapted to harvest the diverse resources of varying unstable environments. A moment's reflection, however, reveals that these are but different faces of a common coin. The evolution of a great variety of organisms *has been* an adaptation to a multitude of environments. Stated simply, the salient feature of evolution's course has been the effective harvest of all available resources that would sustain life. It *is* life's conquest of energy.

Although organic diversity and adaptation are just different aspects of the same issue, it is convenient to discuss them separately. We will treat certain general considerations pertinent to the broad sweep of evolution's course under the topic of organic divergence. Those fascinating particulars by which organisms have solved the problems of survival and conquest will be reserved for a discussion on adaptation.

The Divergent Aspects of Evolution

THE INCREASING DIVERSITY OF ORGANIC FUNCTION AND FORM

Figure 9-20 illustrates the divergent pathways taken by the descendants of one ancestral species of finch in the Galapagos Islands. Thirteen species of birds evolved, seven of which embarked on new modes of life and led to the formation of three distinct genera (species groupings). These related and unique genera plus a 14th species (which is in a separate genus) on Cocos Island constitute what biologists call a *family*.

Some idea of the adaptive radiation of birds since their origin about 150 million years ago can be gained by reference to Figure 9-22. Darwin's finches are one of many different but relatively similar families which, because of their general similarity, are placed within a common *order* (the perching birds). The order of perching birds in turn is one of many other orders which belong in the *class* of birds. And, to complete the classification, birds are finally one of several classes of organisms belonging to a great group or *phylum* of animals that embraces the vertebrates and their allies. Even at this topmost level of the hierarchy we are dealing with related types. Birds themselves are a part of a dramatic series of radiations that began with fishes.

The fossil record substantiates a general tendency throughout the history of

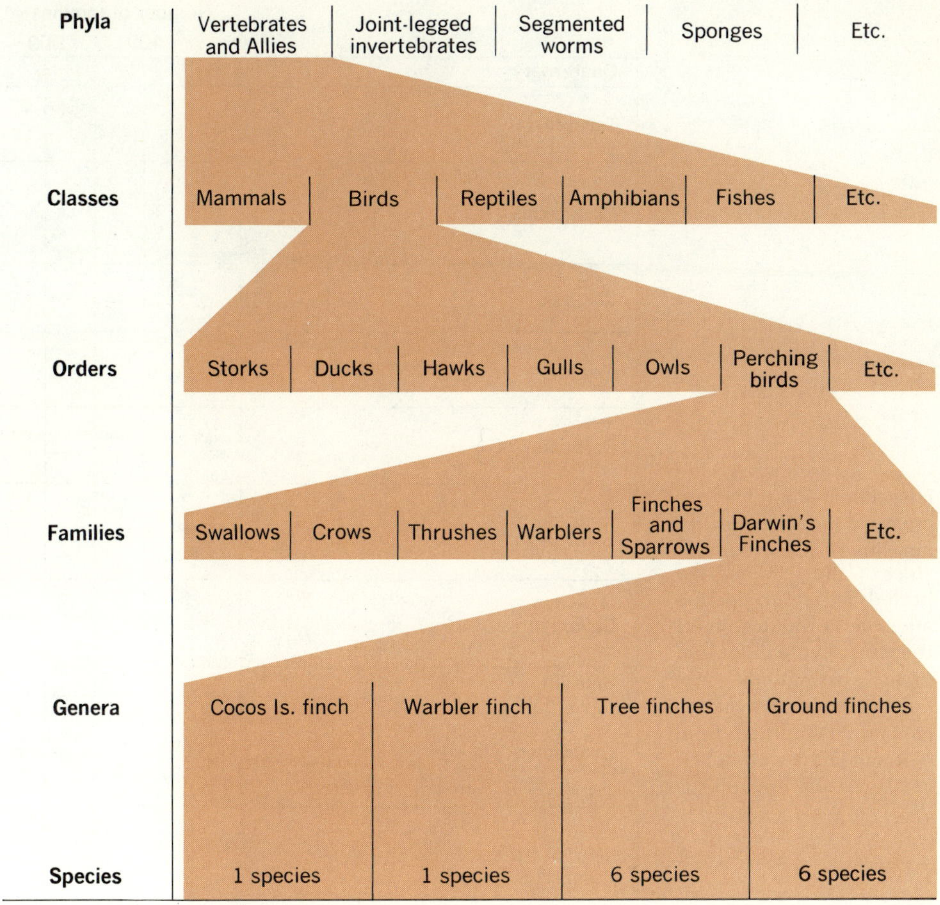

Figure 9-22 The family of Darwin's finches is here placed in the hierarchy of categories used by biologists to express levels of kinship. Illustrated in this fashion the immensity of avian adaptive radiation is apparent. (Refer to Fig. 9-20.)

life toward increasing the kinds of organisms (Figs. 9-23 and 24). This proliferation of variety has produced creatures of greater and greater complexity. Complexity, however, is not an evolutionary end in itself. Just as often evolution has steered a course toward simplification. Parasites, for instance, have become highly efficient reapers of life's harvest by abdicating entire organ systems. A tapeworm has dispensed with a digestive system. It is supplied with one by its host. The evolution of mammals from reptiles was accompanied by great reductions in skeletal complexity. Even the elegant transformations in function and form undergone by the adaptive radiation in vertebrate limbs (Fig. 9-8) often entailed the loss or fusion of skeletal elements.

The tendency of evolution to radiate in many directions champions neither complexity nor simplicity. The extension of organic diversity through time reflects an ever-growing proficiency on the part of living creatures to utilize an ever-enlarging variety of this planet's resources.

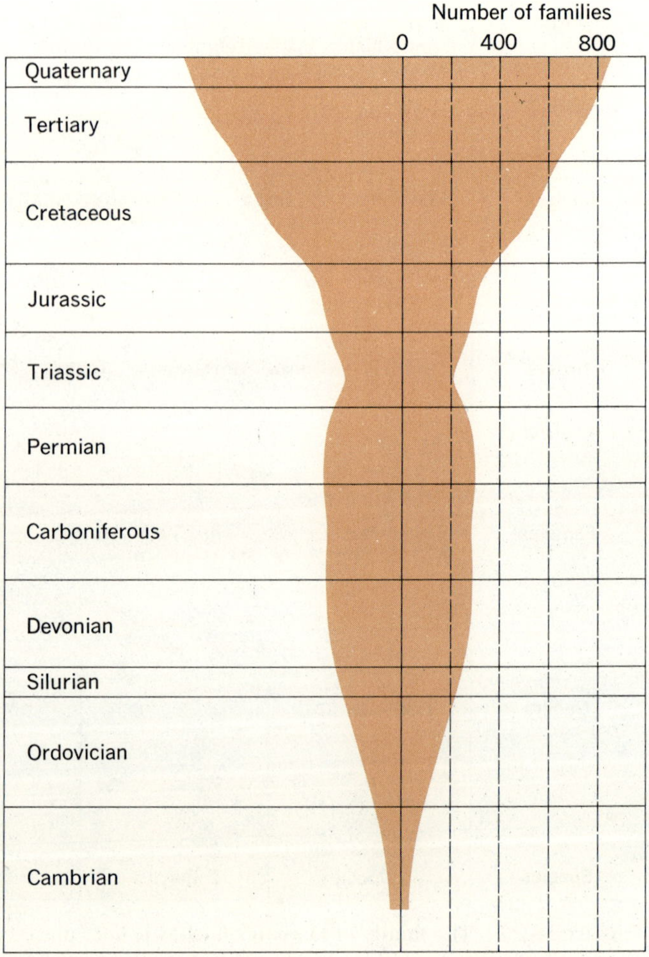

Figure 9-23 The increase in number of major families of fossil animals since the Cambrian period. The sharp decline after the Permian reflects the most severe of several great extinctions involving a variety of organisms. (Modified from Norman D. Newell in *Scientific American*, 1963)

THE INEXORABLE TOLL OF ENVIRONMENTAL CHALLENGE

Most of the millions or possibly billions of species that have ever lived failed to maintain their adaptiveness indefinitely. Although a few organisms still exist that represent groups of great antiquity (Figs. 9-25 and 26), their exceptional status but lends emphasis to this general rule. Certainly on the basis of fossil evidence it would seem that extinction is the eventual lot of most organisms. Only a third of about 2500 families of animals that have left a fossil record are still living. A few of the extinct groups evolved into new families, but the majority passed into oblivion without issue after an average life span of approximately 75 million years.

The high incidence of extinction is probably the result of one of two situations. A species may actually lack sufficient genetic variability to meet the demands of environmental change. This is a possible plight faced by organisms adapted to very specialized ecologic niches. The great sabertooth cat of Pleistocene time was

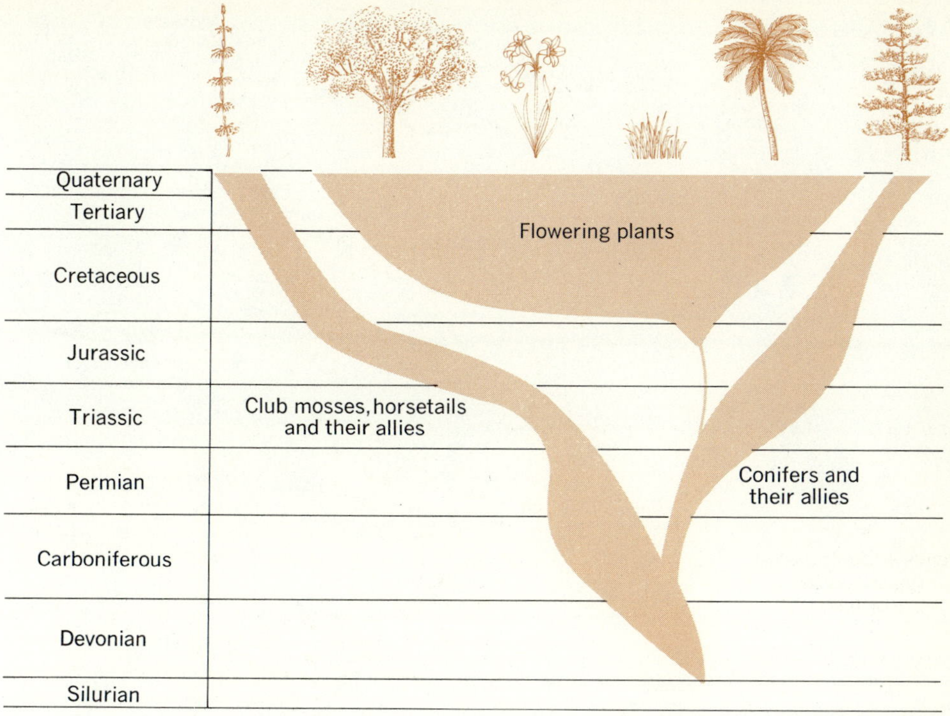

Figure 9-24 The history of land plants reveals a spectacular radiation of flowering plants. Note that our diminutive horsetails and clubmosses had their heyday in the Carboniferous (Fig. 9-4) while conifers reached their peak in the late Triassic and Jurassic (Fig. 9-5). The width of the bands is proportional to the number of genera in each group. (Modified from Norman D. Newell in *Scientific American*, 1963)

Figure 9-25 The ginkgo or maidenhair tree (*Ginkgo biloba*) is the sole member of an order dating back to Permian time. This species, which grows to 100 feet or more in height, produces motile sperms like the ferns but is related to conifers. Tree on the left is without leaves. Leaves and seeds are shown on the right. (Eugene Memmler)

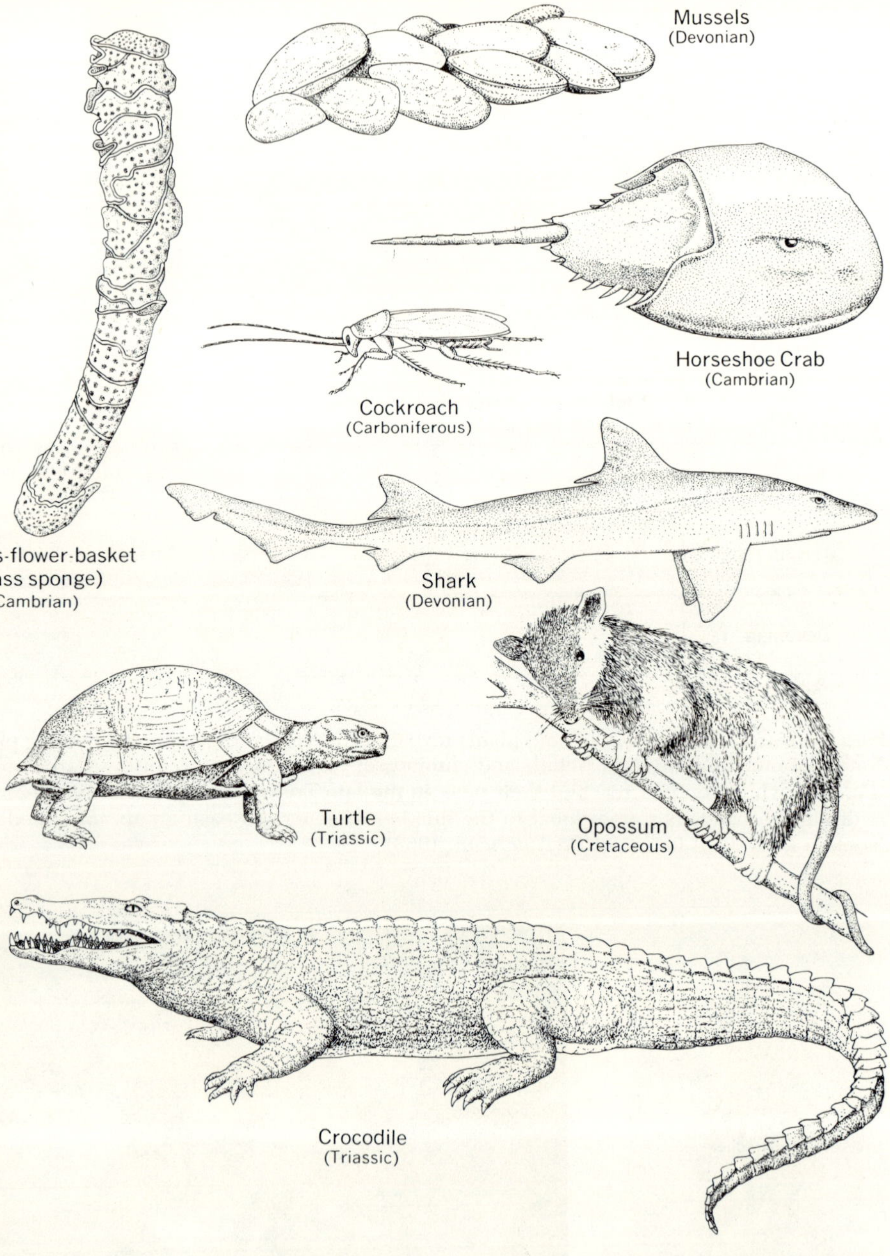

Figure 9-26 These hardy animals represent groups of ancient lineage. The geologic period in parentheses below the name is their age of origin. (Modified from Norman D. Newell in *Scientific American*, 1963)

apparently adapted for preying particularly on thick-skinned elephantlike mastodons. When the mastodons became extinct, the fleet-footed, wary herbivores that remained (deer, antelope, and so on) were too elusive for the lumbering cats. In some instances, it is also possible that, even in highly adaptable species, natural selection could not keep pace with rapid environmental change.

In view of the high rate of extinction, one may wonder how a persistent gain in species diversification is maintained. Reference to Figure 9-27 should help to clarify this point. As a group or groups of organisms begin to decline in numbers, the habitats and ecologic niches they vacate are left open for conquest by other groups. The radiation of bony fishes more than compensated for the attrition in their invertebrate counterparts. Following the period of the "Great Dying" during late Cretaceous when dinosaurs and several other reptile groups became extinct, mammals underwent a tremendous radiation, utilizing ecologic niches even more efficiently than the previous occupants.

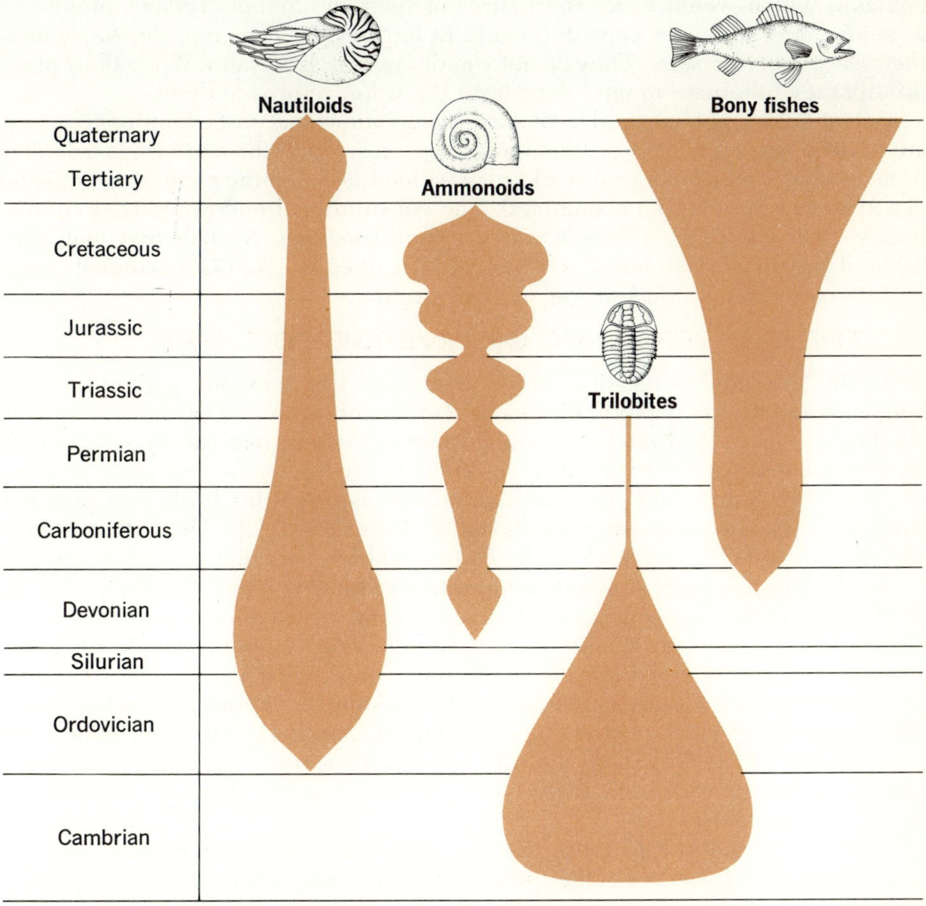

Figure 9-27 The varying destinies of four groups of marine swimmers and bottom-dwellers. Three great groups of invertebrates were at one time dominant animals in the seas of the world. Two are now extinct and one (nautiloids) is represented by only one genus of three species. The ammonoids and nautiloids (both related molluscs) gradually replaced trilobites (related to horseshoe crabs) in the habitats vacated by the latter's decline. The ammonoids and nautiloids in turn were ecologically replaced by the adaptive radiation of bony fishes. Note that the ammonoids suffered near extinction twice before their final demise. Band width represents relative numbers of families. (Modified from Norman D. Newell in *Scientific American*, 1963)

THE UNIQUENESS OF EVOLUTIONARY CHANGE

As pointed out evolution has occurred when certain ecological opportunities have become accessible to organisms constitutionally capable of exploiting them. Each of these interdependent opportunities—physical, constitutional, and ecological—is a singular circumstance. Each derives its distinctive character from a specific time and place. That each of the opportunities responsible for a particular evolutionary result should ever be duplicated at the same time and in the same place is an incomprehensible event. The lung-breathing fish today would not be up to a repeat performance of their Devonian relatives. The land now teems with denizens which would make short shrift of such an attempt. Today's lungfishes, moreover, are no longer constitutionally fit for the job. *They are not the same as their Devonian forebears.* They do not wander in search of water. When their ponds dry up, they hibernate in mud dens until the water returns to them.

It should not be inferred from these observations that past evolutionary events are of little consequence to contemporary species. Quite the contrary is true. The form and function of ancestral stocks may be modified, but the evolutionary history of a species can never be abandoned. The constitutional endowments of contemporary creatures are the altered features of their ancestors. A whale may look like a fish and live like a fish, but it acts like and breathes like and *is* a mammal. Major evolutionary changes, once made, are irrevocable.

THE CONVERGENCE OF DIVERSIFIED TYPES

The enormous diversity which evolution has wrought within the four kingdoms and the many phyla, classes, and orders of organisms has led to extensive overlapping in the utilization of available resources. A great variety of relatively unrelated creatures has evolved in response to environments and ways of living which are very much alike. The adaptive radiations of reptiles, birds, and mammals have included the conquest of land, sea, and air (Fig. 9-28). Those that became adapted to life in the sea, such as the extinct ichthyosaurs, the penguins, and porpoises, look superficially similar to other active swimmers which live there also, the sharks and squids. This general similarity of distantly related forms is known as *convergence* and is a consequence of natural selection. In our example all active swimmers, regardless of their constitutional nature, must meet certain common requirements. One of these is streamlining. The squid (a mollusc) as well as its vertebrate cohabitors must meet the qualifications demanded by the environment and general mode of life they share. This does not make a porpoise a fish any more than it makes a penguin a squid. It does add, however, to the diversity of molluscs, mammals, and birds.

There are many excellent examples of evolutionary convergence similar to that just described. Species of the spurge family of plants in the deserts of Africa and Australia show a striking resemblance to the true cactuses of American deserts (Fig. 9-29). Both plant families contain many species whose shoot systems are capable of storing large quantities of water and whose transpirational surface has been minimized either through the loss of leaves or their reduction (in cacti) to spines. (The spines of spurge "cactuses" are thought to be branches.) Cactus spines also protect the plants against many large herbivores.

The classic demonstration of convergence appears in the primitive pouch-bearing mammals of Australia and the placental mammals which occupy the rest of the earth. When the island continent of Australia was separated from its mainland (probably in Cretaceous times), the only kinds of mammals in that part of the world

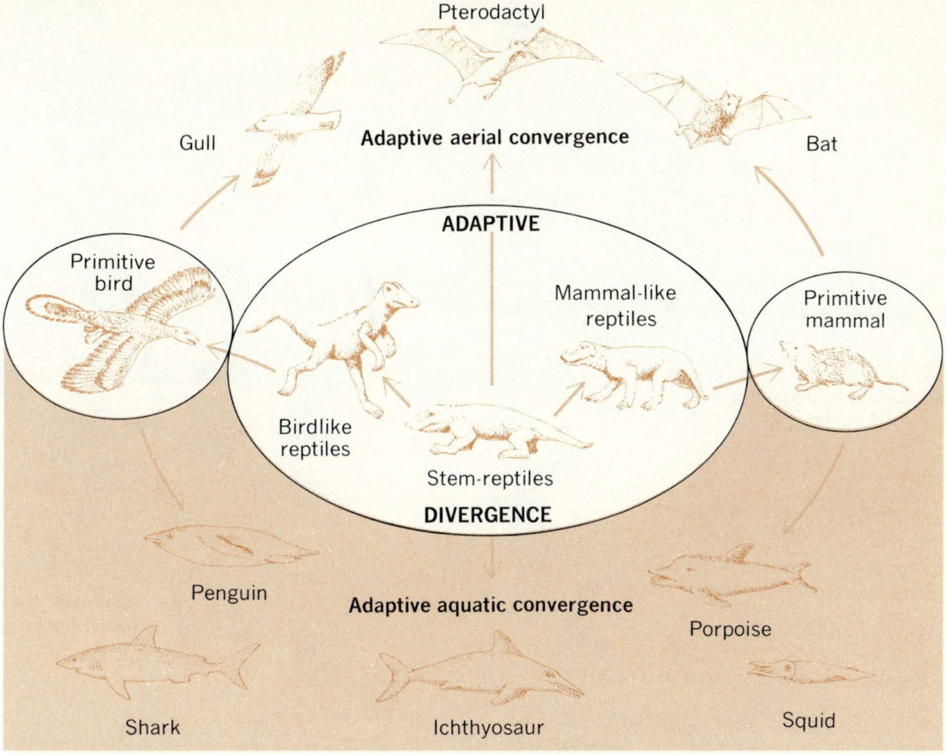

Figure 9-28 Evolutionary divergence and convergence compared. The ancient "stem reptiles" were semi-aquatic animals from which the ancestors of birds as well as the ancestors of mammals evolved through divergence. Further divergence within reptiles, birds, and mammals led to a variety of animals adapted to land, sea, and air. Organisms, even though distantly related, tend to resemble one another when they have evolved in response to a common environment. Such adaptive convergence is shown here for aerial forms (gull, pterodactyl, and bat) and aquatic forms (penguin, ichthyosaur, porpoise, shark, and squid).

Figure 9-29 Convergent evolution in plants. The plant on the right is a member of the cactus family; that on the left belongs to the unrelated spurge family. (Photographs by author)

Figure 9-30 Convergent evolution in animals; some placental mammals (left) and their marsupial equivalents from Australia (right). (From G. G. Simpson, *Life,* Harcourt Brace, 1969, after A. E. Brehms, 1912)

were pouched marsupials related to present-day kangaroos and a few reptilelike egg-laying types similar to the modern duckbill. Placental mammals evolved later on the continental land masses after Australia had become isolated; this evolution did not include the subcontinent. The marsupial fauna consequently underwent an adaptive radiation independent of placentals elsewhere. The correspondence in phenotypes between the placental and marsupial forms which evolved in response to similar ecologic opportunities is most impressive (Fig. 9-30). Even more extraordinary is the realization that each of the animals shown on the left is more closely related to a sperm whale (a placental) than to its ecologic equivalent on the right.

The Adaptive Aspects of Evolution

Everything that has ever been written in the broad field of biology (and much outside of it) describes in some fashion the adaptations of living things. In Chapters 4 through 7 we dealt exclusively with them: devices and processes for food getting or making, digestion, internal transport, gas exchange, waste elimination, and integrative mechanisms, adaptive behavior, procreation, and development.

One aspect of adaptation which we have not specifically covered in any detail is that which characterizes the interactions between the many and varied species composing a single plant-animal community. If the community is a terrestrial one, trees, shrubs and/or smaller autotrophs compete with one another for sunlight, water, or both. A multitude of animals are dependent on these plants for shelter, breeding sites, and food. The herbivores, both large and small, who prey on these plants are in turn preyed on by carnivores. Parasites live on all of these. Dead carcasses support scavengers and, added to by leaf fall, dead vegetation, and animal excrement, a host of microscopic decomposers that return organic matter to the soil. On this rich refuse thrive heterotrophic fungi and animals from microscopic to toadstool and earthworm size. And finally, bringing the chain of interdepencies full sweep, a vast multitude of soil bacteria converts organic matter and soil nitrogen into inorganic nitrates. These nitrates support the growth and reproduction of green plants around which this enormous complex is built—a complex in which every ecologic niche is occupied and utilized by a separate specialist we have come to know as a species.

A plant–animal community such as that just briefly outlined is obviously an integrated system of diverse species that have evolved together. These species are in varying degrees dependent on one another; they interact in bringing about the harvest of every available source of energy. These complex interactions not only tend to assure full and continuing utilization of the environment but they also favor, in broad perspective, the survival of every species in the community. In this respect they are certainly adaptive. In fact, some of the most fascinating examples of adaptation are found in this area of *interspecific interaction.* We will turn our attention briefly to some of them.

SYMBIOSIS

Symbiosis is a term used for all kinds of close associations between two species. Three types of symbiosis are usually recognized. *Commensalism* refers to those relationships in which one species benefits, whereas the other is neither benefited nor harmed. *Mutualism* includes those in which both species benefit. *Parasitism*, on the other hand, involves associations in which one organism benefits at the expense of the other. Although these categories appear to be quite distinct, they are not. They commonly intergrade and are difficult to distinguish. A parasite that lives in such harmony with its host that the host is only rarely harmed is more like a commensal than a parasite. A commensal that takes advantage of its associate on occasion is a part-time parasite, but if the associate is sometimes benefited, the relationship becomes partially mutualistic. The intergradation between these categories is, as a matter of fact, a most significant feature because it throws light on their evolutionary origins.

COMMENSALISM AND MUTUALISM

The advantages derived by commensal and mutualistic species are varied but commonly involve food, protection, and reproduction. Indian pipe (a flowering

(a)

(b)

Figure 9-31 Commensalism. (a) Indian pipe, a heterotrophic flowering plant believed to live off organic matter digested by soil fungi. (b) the remoras attach to sharks temporarily by a sucker on their backs. They eat "table scraps" the shark misses but do not seem to harm the shark. [(a) T. H. Everett; (b) George Whitley, Photo Researchers]

plant) and remoras (fish) obtain their nutrition from commensal relationships with soil fungi and sharks, respectively (Fig. 9-31). The pea family of plants has a mutualistic association with soil bacteria that reduce soil nitrogen to ammonia in exchange for a home and a source of energy (Fig. 9-32). Both plant and bacteria convert the ammonia to protein. A large number of marine fish species either make a living by cleaning the parasites from other fishes or benefit by being cleaned (Fig. 9-33). The cleaners are usually brightly colored, advertising their presence to prospective customers. In many cases the immunity of cleaners from predators has become so firmly established that other species have come to mimic their color and shape to share in this immunity. Some of these have even reversed the relationship and prey on the fish that confuse them with cleaners.

In the plant world lichens are classic examples of mutualism in which algae and fungi are associated (Fig. 9-34). In any one type of lichen a species of alga provides the food. The fungus produces a protective covering around the alga as well as rootlike processes that secure water and minerals.

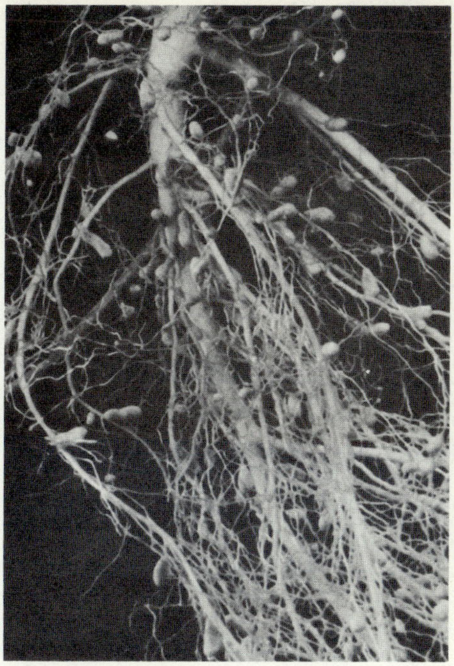

Figure 9-32 Many members of the pea family of plants have nodules on their roots which harbor *nitrogen-fixing* bacteria. Using sugar supplied by the plant, these bacteria reduce soil nitrogen to ammonia in excess of their needs for protein synthesis. The plant converts the excess into its own protein. The plant shown here is a clover. (U.S.D.A.)

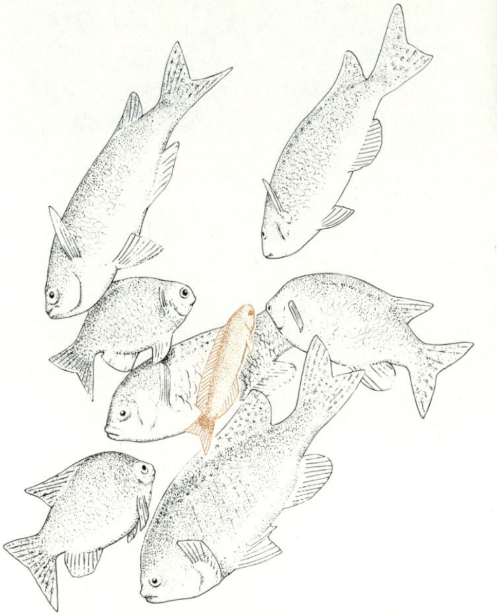

Figure 9-33 Cleaning symbiosis. Many species of marine fish make a living by eating the parasites on the bodies of other fish species. Here several different blacksmith fish are shown waiting their turn to be groomed by a senorita fish (color). (Modified from Conrad Limbaugh in *Scientific American*, 1961)

Remarkable cases of mutualism are often encountered between a plant and animal species. One of these is illustrated in Figure 9-35. The flowers of certain sages are so designed as to bring about cross-pollination by bees. A sugary nectar, secreted near the base of the ovary, attracts the insect. A lever arrangement at the base of the two stamens assures that the backs of visiting bees will be dusted with pollen. The stigma matures later than the anthers and, as it does, grows downward and is then in position to receive pollen from a bee that has previously entered a younger flower.

PARASITISM

The kinds of organisms that have become parasites are those which have had easy access into or onto the bodies of other creatures. Protists and worms, occurring commonly in the water animals drink, consequently make up a large proportion of

Figure 9-34 Types of lichens. (a, b) Two types of leaflike lichens growing on rock; (c) a shrublike form. [(a) Mary M. Thacher, Photo Researchers; (b) Russ Kinne, Photo Researchers; (c) Gerard from Monkmeyer]

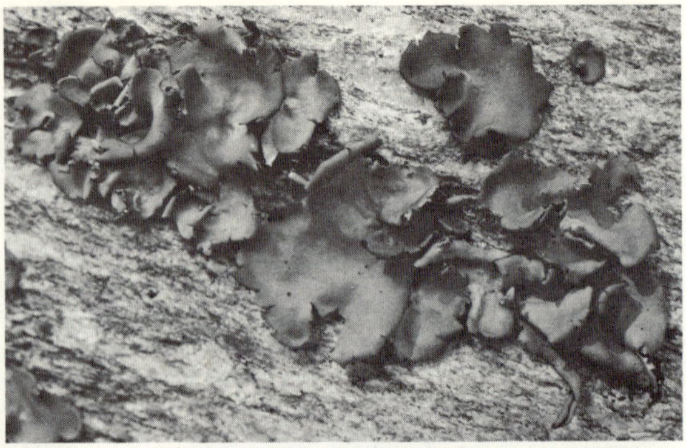

(a)

(b)

(c)

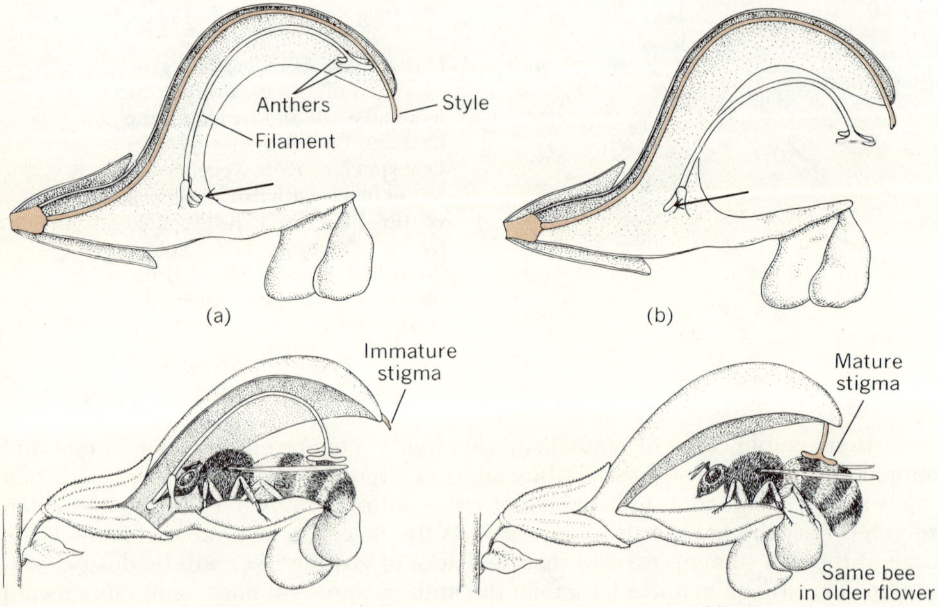

Figure 9-35 Mutualism in the cross-pollination of a sage (*Salvia pratensis*). In this species the stigma matures later than the anthers (stigma, style and ovary are colored). (a) Arrow points to plate formed by fusion of bases of the two stamens. (b) Pressure against the plate (indicated by arrow) causes anthers to tip forward. (c) Bee pushes against plate and receives pollen on back. (d) Bee visits older flower. The stigma has now grown downward, the lobes open, and the bee transfers pollen to stigma.

animal parasites. In the early stages of their evolution such organisms may have existed as commensals, assuming a parasitic role later in their development.

In any case parasitism is a product of evolutionary adaptation in both parasite and host. Most parasites are usually restricted to one or a few related species of hosts. Great specialization on the part of the parasite together with high tolerance of the host to its presence are characteristic of host-parasite associations that have been in business a long time. It is as profitable for a host to produce a body that can tolerate parasites as it is for a parasite to continue living in a home without destroying it. The mutuality of host-parasite relationships is beautifully illustrated today in both agriculture and medicine. A new fungi-resistant strain of some grain crop or a new antibiotic are soon followed by mutants of both fungus and bacteria capable of infecting the grain or tolerating the drug.

In becoming specialized to a parasitic mode of life, organisms are often said to be degenerate because of the loss of structures associated with the free-living habit like roots and leaves (Fig. 9-36) or digestive tracts (Fig. 9-37). Although this is true, it should be borne in mind that a tapeworm, for instance, has no need for a digestive system. Why, then, support one? The animal is nicely adapted to living in the intestines as Figure 9-37 shows. Its outermost layer protects it from digestive enzymes but allows the products of digestion to be absorbed by the worm. Of particular note is the highly efficient reproductive system. Each segment becomes a complete reproductive unit. Enormous fecundity is characteristic of parasites. It helps assure their transfer from one host to the other.

The problem of finding new hosts is a formidable one. The eggs of many tapeworm species pass out in the feces. How do these eggs and the larvae they contain find their way into a new host? The chances of their being eaten are not at all good. To ensure reaching their appropriate host, parasites have adopted a sophisticated variety of complex life histories. In the fish tapeworm of man, for instance, about

Figure 9-36 Dodder is a flowering plant which parasitizes other plants by twining around them. Outgrowths of the parasite penetrate into the phloem and absorb food from the host. (Russ Kinne, Photo Researchers)

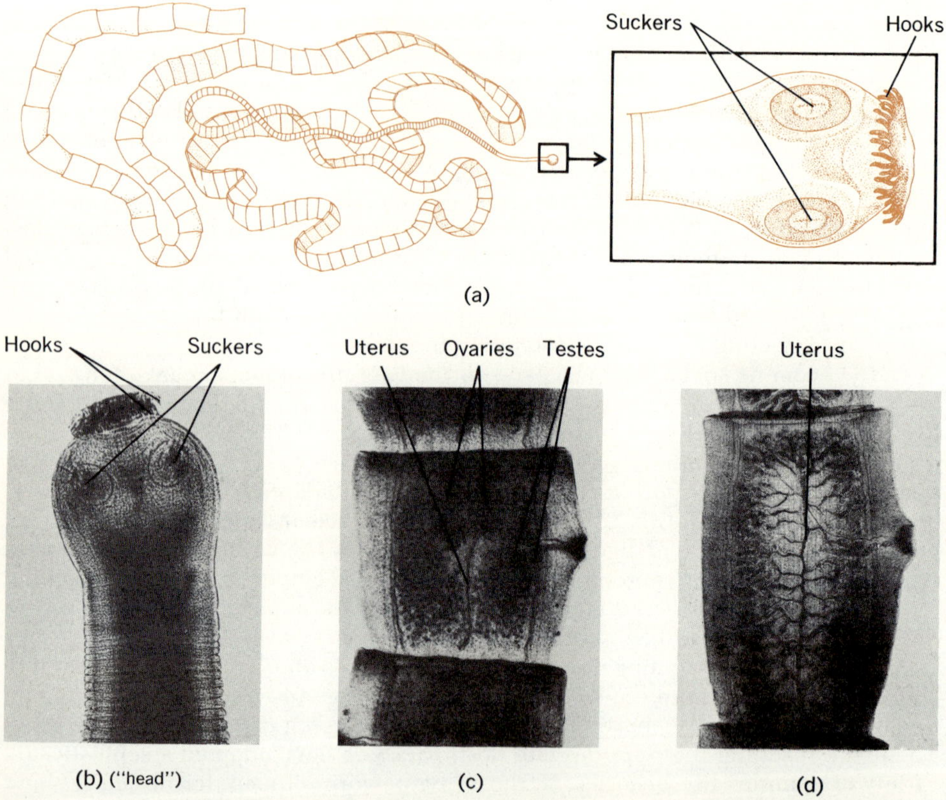

Figure 9-37 A typical tapeworm (*Taenia serrata*). (a) General sketch of entire worm. (b, c, d) Sections from anterior, middle, and terminal portions, respectively. Each segment of worm becomes a complete reproductive unit. Terminal segments, little more than distended uteri filled with eggs, break off and pass out with feces. "Head" of worm is imbedded in intestinal wall and secured by hooks and suckers. Segments are budded off asexually below "head." [(b, c, d) General Biological Supply House]

one million eggs are discharged daily by each worm into the feces. These hatch into minute swimming larvae which are carried via sewage systems into fresh water lakes and streams. Here they are eaten by tiny invertebrates that become parasitized by the larvae. These invertebrate *intermediate hosts* are then eaten by fishes (perch, trout, whitefish, and so on). The parasitic larvae move into the muscles of the fish, the second intermediate host. Man now eats raw or improperly cooked fish and the parasite's life cycle is finally completed.

Parasite life histories are without a doubt one of the most extraordinarily elaborate adaptations to be found in all of nature. It is not at all uncommon for parasites to use the services of other parasites in order to ensure their transmission from host to host.[5] In this instance, the intermediate host is known as a *vector*. Mosquitoes are vectors for the small protists that parasitize the red blood cells of many vertebrates

[5] Often parasitic species are used simply as hosts by other species, resulting in *hyperparasitism:* parasites which parasitize the parasites of a parasite.

(and the gut walls of mosquitoes), causing malaria. There is considerable merit in the old rhyme:

The many fleas which dogs do tease
Have other fleas to bite 'em
And these, in turn, have other fleas
And so *ad infinitum.*

PREDATION

The distinction between parasites and predators cannot always be clearly made. For example, how small must a herbivore (plant predator) be before it is considered a plant parasite? Predation can, nevertheless, ordinarily be distinguished from parasitism in four ways: (1) predators do not live with their prey in an intimate association[6] (parasite and host bodies are intimately associated); (2) with the exception of herbivores, predators usually kill outright and devour a goodly portion, if not all, of their prey (few parasites kill their hosts and if they do, it is not immediate and host consumption is relatively light); (3) predators are in the approximate size range of their prey or larger (parasites are much smaller than their hosts); and (4) predators are outnumbered by their prey (parasites usually outnumber their hosts).

The devices used by predators to catch and hold their prey as well as those employed by the prey to avoid capture are highly varied. Sharp teeth, claws, beaks, and mandibles together with poisonous or paralyzing venoms and trapping mechanisms like spider webs assist a host of predators. On the other hand, the turtle's armor, the porcupine's quills, cactus spines, the antelope's speed, and a multitude of poisonous or noxious substances produced by a great number of both plants and animals serve the survival of the preyed-upon.

[6] The relationship is not symbiotic.

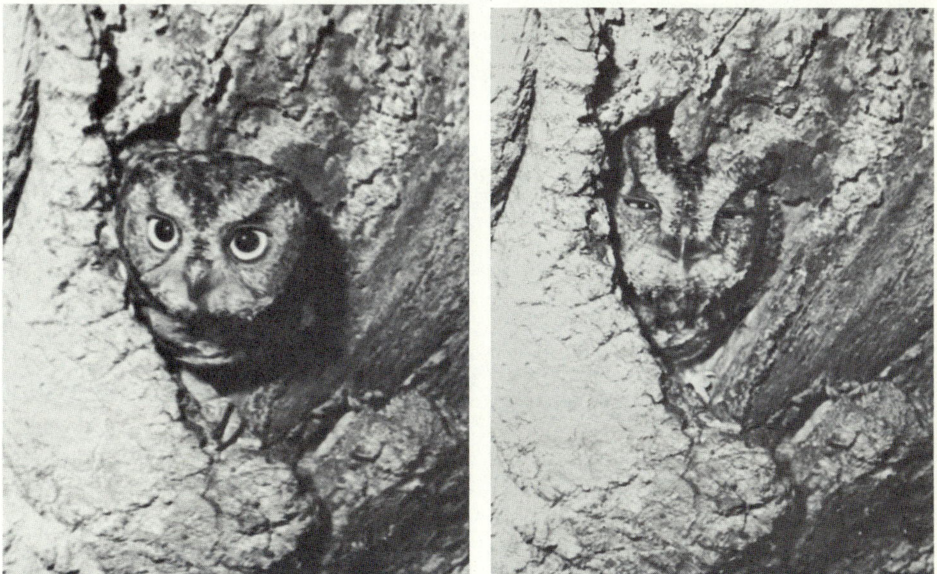

Figure 9-38 Concealing coloration can be as useful to a predator like this screech owl as to its prey. The photograph on the right was taken 10 seconds after the one on the left. (Lewis W. Walker)

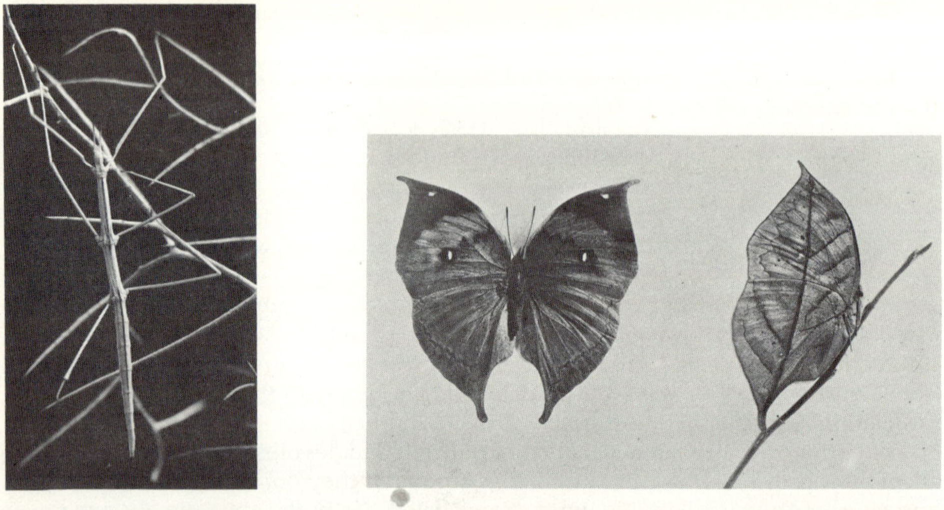

Figure 9-39 Mimicry of inanimate objects. The walking stick insect on the left looks like twigs. The leaf butterfly on the right resembles a leaf when its wings are folded. (Lewis W. Walker; American Museum of Natural History)

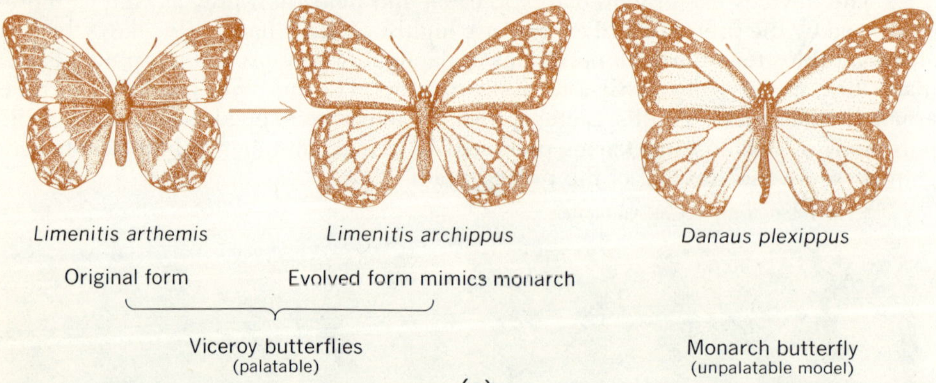

(a)

(b)

Figure 9-40 Mimicry and warning coloration. (a) Monarch butterfly on the right advertises its unpalatability by conspicuous warning coloration. The palatable viceroy butterflies have evolved a form that avoids undue predation by mimicking the monarch. (b) The two photographs on the right indicate the powerful emetic effects of eating a monarch butterfly (left). This blue jay will hereafter avoid both monarchs and viceroy mimics on sight. (L. P. Brower, Amherst College)

Adaptive behavior is also employed successfully. The stealthy approach of a bobcat is matched against the keen perception of the sentinel quail or goose whose job it is to detect danger and spread the alarm to others. The killdeer who feigns a broken wing can often lead a predator away from its shallow nest in the sand.

Adaptive modifications in color and form are particularly striking and quite effective. *Concealing coloration* enables an organism to blend with its surroundings and is obviously advantageous to both predator and prey (Figs. 5-22 and 9-38). The survival of a variety of insects is enhanced by *mimicking* the form and color of inanimate objects (Fig. 9-39). Some organisms, on the other hand (many insects, frogs, reptiles, and so on), advertise their presence to predators by bright or contrasting color patterns. This *warning coloration* is associated with formidable defense mechanisms such as irritating or toxic secretions (Fig. 9-40). A predator usually makes only one mistake in capturing such a creature and thereafter avoids it on sight.

Predatory immunity of this nature has led to interesting consequences. Generally related species, each with effective defense mechanisms (for example, two or more species of butterflies), have come to mimic one another. This becomes a form of multiple predation insurance. Even defenseless species have evolved that mimic these color patterns. An excellent example of evolutionary adaptation involves such a case.

Many of the plants in the milkweed family produce substances that induce violent vomiting in the vertebrates that eat them. They are thereby protected from vertebrate predators. The larvae of certain insects, however, are able to feed on these plants. These insects, one of whom is the Monarch butterfly, are consequently unpalatable to vertebrates because of their larval diet. The Monarch butterfly exhibits warning coloration, advertising its unsavory nature to potential predators. Among the palatable Viceroy butterflies (they do not eat milkweed) a Monarch mimic has evolved (Fig. 9-40). As suggested in the illustration, a bird which has eaten a Monarch will avoid both Monarchs and Viceroys thereafter with nothing more than a glance.

An interesting chain of evolutionary events was triggered by an insect species with the constitutional opportunity to exploit the predation-resistance conferred on members of a plant family through natural selection.

QUESTIONS

1. Describe the manner in which five different biological sciences substantiate the principle of organic evolution. Use a specific example in each case.
2. What is the use/disuse hypothesis? What are some of its shortcomings?
3. Discuss at some length the adaptive, creative, and stabilizing roles that natural selection plays in directing the course of evolution.
4. In a growing population how does increasing proximity to new environments and isolation favor biologic diversity and the formation of new species?
5. Elaborate upon the following thesis: "Speciation is a response to opportunity."
6. In the evolution of Darwin's finches speciation followed three main courses. Describe each and illustrate.
7. What significance can be placed on the fact that evolution results in ever-increasing biologic diversity?
8. Since species extinction is the eventual fate of most organisms, how can evolution show a persistent gain in species diversification?

9. Life may well be a common phenomenon in the universe. Is it probable that some other world (planet) will have species similar to ours? Why?
10. When the island continent of Australia was separated from its mainland, the highest form of mammals were presumably marsupial-like forms. A new kind of mammal, the placental, ultimately evolved on the continental land masses. Can you offer a reasonable hypothesis to explain why mammalian evolution in Australia never went beyond a marsupial stage?
11. Compare and contrast commensalism, mutualism, and parasitism, indicating how they may intergrade with one another.
12. Discuss the evolutionary development of parasitism and host-parasite relationships.
13. In Chapter 8 mention was made of a number of plant species which produce cyanide. How may this serve as an adaptive advantage to the plant?
14. Distinguish between concealing coloration, mimicry, and warning coloration.
15. How can mimicry in combination with warning coloration become a kind of multiple predation insurance?

Suggested Readings

Brace, C. L. *The Stages of Human Evolution.* Foundations of Modern Anthropology Series. Englewood Cliffs, N.J.: Prentice-Hall, 1967.

Brower, L. P. "Ecological Chemistry," *Scientific American,* February 1969.

Burnett, A. L., and T. Eisner. *Animal Adaptation.* Modern Biology Series. New York: Holt, Rinehart and Winston, 1964.

Clark, Sir W. E. LeGros. *Man-Apes or Ape-Men?* New York: Holt, Rinehart and Winston, 1967.

Colbert, E. H. *Evolution of the Vertebrates.* New York: Wiley, 1969.

Dayhoff, M. O. "Computer Analysis of Protein Evolution," *Scientific American,* July 1969.

Delevoryas, T. *Plant Diversification.* Modern Biology Series. New York: Holt, Rinehart and Winston 1966.

Dobzhansky, T. *Mankind Evolving.* New Haven, Conn.: Yale University Press, 1962.

Fingerman, M. *Animal Diversity.* Modern Biology Series. New York: Holt, Rinehart and Winston, 1969.

Hamilton, T. H. *Process and Pattern in Evolution.* Current Concepts in Biology. New York: Macmillan, 1967.

Hanson, E. D. *Animal Diversity.* Foundations of Modern Biology Series. Englewood Cliffs, N.J.: Prentice-Hall, 1964.

Lack, D. L. *Darwin's Finches.* Harper Torchbooks. New York: Harper and Row, 1961.

Lerner, I. M. *Heredity, Evolution and Society.* San Francisco: Freeman, 1968.

Lewis, H. "Speciation in Flowering Plants," *Science,* **152,** April 8, 1966.

Mayr, E. *Animal Species and Evolution.* Cambridge, Mass.: Harvard University Press, 1963.

McAlester, A. L. *The History of Life.* Foundations of Earth Science Series. Englewood Cliffs, N.J.: Prentice-Hall, 1968.

Moody, P. A. *Introduction to Evolution.* New York: Harper and Row, 1968.

Newell, N. D. "Crises in the History of Life," *Scientific American,* February 1963.

Romer, A. S. "Major Steps in Vertebrate Evolution," *Science,* **158,** December 29, 1967.

Romer, A. S. *The Vertebrate Story.* Chicago: University of Chicago Press, 1959.
Savage, J. M. *Evolution.* Modern Biology Series. New York: Holt, Rinehart and Winston, 1969.
Simpson, G. G. *The Meaning of Evolution.* New Haven, Conn.: Yale University Press, 1967.
Spiegelman, S. "Hybrid Nucleic Acids," *Scientific American,* May 1964.
Wallace, B., and A. M. Srb. *Adaptation.* Foundations of Modern Biology Series. Englewood Cliffs, N.J.: Prentice-Hall, 1964.
Washburn, S. L., and P. C. Jay (eds.). *Perspectives on Human Evolution 1.* The Society for the Study of Human Evolution. New York: Holt, Rinehart and Winston, 1968.

CHAPTER 10 Organic Interaction: the ebb and flow of life in space

It is clear from the previous discussion as well as numerous others throughout this book that evolution takes place as a response of a species to its environment. We have gone to some lengths elaborating on the long-range nature of this response. We have traced it through time to its inevitable consequence in either adaptation or extinction. Except for occasional references to rather vague and broadly defined "selection pressures" and "environmental changes," however, we have neglected a cardinal element in the evolutionary process, namely that which evolution is a response to. The environment metes out the qualifications for continued survival by levying a multitude of selection pressures against its citizenry. These pressures are neither broad nor vague but quite specific and often exacting. It is therefore no coincidence that the life of every organism is inextricably associated with its surroundings. So intimate is this association that the physical environment itself may be profoundly influenced by the life it supports. In a terrestrial plant-animal community green plants alter the amount of sunlight that reaches the ground surface and the moisture in the earth below. Their roots assist in converting rock into soil. Bacteria increase the amount of soil nitrate.

The list could be extended almost without end. The physical environment is as much influenced by the life it supports as a community of living things is affected by its physical surroundings. It is therefore with good reason that biologists speak of an ecological system or *ecosystem* as the unit of organic interaction. A lake, a meadow, a forest, or a desert are ecosystems, each one characterized by certain physical conditions that affect and are effected by a biological community of the many populations (species) which live there.

It is within an ecosystem where life goes on and the challenge for change is met. Diverse species interact and evolve together and with the physical surroundings about them as part of a common feedback system. The environment, in short, is an interdependent complex of physical and biological components.

As we direct our attention to the scene where "the action is"—to the day-to-day setting in which life moves and evolves—we will be entering the lively and alluring field of environmental biology or ecology. We will first attempt to make the term "environment" more meaningful before considering the dynamic interactions that characterize the life of a species within the environment of which it is a part.

THE NATURE OF ENVIRONMENT

Some Physical Components

SOLAR RADIATION

The singular importance of solar radiation can hardly be overstated since it is the ultimate source of life's energy. The green plants and protists on which all living things depend are not found where light fails to penetrate. Organisms inhab-

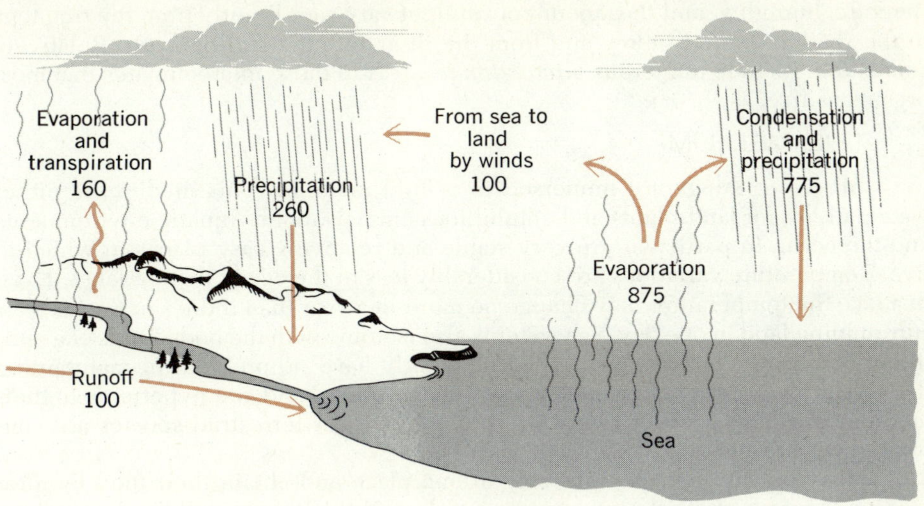

Figure 10-1 The water cycle. Fresh water flowing from land to sea is replaced by water evaporating from the sea and falling back onto the land. Evaporation from inland bodies of water and from plants and animals follows a parallel course of condensation and is similarly cycled back. The numbers refer to the volume of water in cubic kilometers.

iting the darkness of the ocean deeps live on organic matter that drifts down from above, dead bodies, or on one another. A similar situation prevails among soil organisms. In both instances, as in all others, life relies on the autotrophic productivity of lighted regions.

Radiation is of further significance because of its roles in water cycling and in maintaining environmental temperatures. A great deal of the energy received from the sun is absorbed in the evaporation of water from oceans, lakes, and streams and from the bodies of organisms (transpiration, respiration, and so on), as shown in Figure 10-1. The water vapor so formed eventually condenses, falling back to earth as precipitation to replenish fresh-water habitats and to supply the needs of terrestrial organisms. Without this cycling of life's most precious ingredient there would be no usable environment beyond the ocean's shore. The deserts of the world come the closest to that dire eventuality. Even they, however, receive enough moisture from dew and scattered rainfall to support a diverse, although sparse, biota (fauna and flora).

A very large amount of solar energy is transformed into heat as it penetrates the earth's atmosphere. Variations in temperature (as well as precipitation) accompany changes in both latitude and altitude. As one moves toward the equator or to a lower elevation, temperatures tend to increase. Corresponding changes in the distribution of many organisms usually follow such variations.

Strictly local conditions superimpose additional modifications on those of a more regional nature. The winter temperatures over valley floors at night may be colder than along surrounding slopes as dense cold air formed close to the ground flows to lower levels, displacing warmer and lighter air upwards. In the northern hemisphere, mountain or valley slopes that face the south (and the sun) are drier and warmer than north slopes. On an even more restricted scale one will find climatic variation within an orchard, a woodlot, or a city park. Air movement, temperature, humidity, and the amount of sunlight vary significantly from the tree tops to the shaded ground below and from the litter into the soil beneath. Relatively small areas include numerous *microclimates.* It is in these microclimates that most organisms live.

THE MEDIUM

All living things are immersed in a fluid medium. This medium is either water, air, or, for land plants and amphibious animals, both. Aquatic environments and the ocean in particular are very stable and relatively easy places in which to live. Temperature variations are considerably less in the ocean than in lakes. Even in a lake the temperature may change no more in a year than it does in the air over surrounding land in one day. Sea water is also isotonic with the body fluids of a multitude of marine creatures, which consequently have no problem in maintaining their water balance. Fresh water species, on the other hand, are hypertonic to their medium and have a water removal problem, whereas terrestrial species are confronted with evaporation and water loss.

Many aquatic heterotrophs stay in one place and obtain their food by filter feeding. Their terrestrial counterparts are denied this luxury and must somehow pursue their food. Aquatic species can practice external fertilization. This easy way out is similarly denied to creatures living on the land. Because of the buoyancy of water, aquatic organisms are not concerned with problems of body support. Delicate jellyfishes can float and swim about quite effectively. Algae lack woody supportive tissue, yet huge marine algae known as kelp are commonly up to 100 feet long. Giant clams 5 feet long can support massive protective shells 6 inches thick. The Blue Whale, the largest animal that has ever lived, weighs up to 170 tons. When it becomes beached, its rib cage is literally crushed by its own weight.

Aquatic media hold by no means all the trump cards. Air is 21 percent oxygen. Even well-aerated water contains less that 1 percent. It is no coincidence that the animals with the highest metabolic rates (insects, birds, and mammals) are air breathers. Water, moreover, absorbs light rather effectively. Aquatic autotrophs are consequently restricted to a relatively shallow illuminated zone near the surface where water turbulence and the risk of mechanical damage are greatest. Creatures that live in water are additionally limited in their distribution by the sheer density of their medium. Although a diversity of species inhabit the ocean deeps, the whales are presumably the only group of organisms that can tolerate the enormous range of water pressures between the ocean bottom and its surface.

Land plants are unique organisms in that their shoots are surrounded by air and their roots are immersed in both soil air and water (Fig. 4-14). The nature of the soil itself is of the greatest importance in determining the functional relationships of the root system to its surrounding media.

Unless soil has been transported from someplace else by wind or water it rests on its parent material of limestone, lava, granite, and so on. Above this basement rock a layer of weathered and fragmented parent material grades upward to the subsoil and the topsoil with its thin covering of plant litter. Plant litter gradually becomes incorporated into the topsoil, along with the decomposing bodies of dead soil organisms, as organic matter or humus.

Root systems are confined largely to the topsoil. Its depth, degree of aeration, water-holding capacity, and chemical constitution are very important to vegetative growth. Sandy soils have excellent aeration, but retain very little water. Clay soils hold on to water tenaciously, but aeration is poor. Intermediate soils (loams) are the most productive.

Areas with moderate to heavy rainfall (mountains, eastern United States, tropics, and so on) are often characterized by leached soils. Water percolates through the soil to the ground water below, carrying important elements like calcium with it. This also lowers the pH of the soil. Bog soils are extreme examples of this. Here leaching not only has removed soil nitrates but also has depressed the pH to the point where nitrogen-fixing bacteria can no longer live. Plants like Venus's flytrap (Fig. 3-18b) are found only in such soils and seem to be specifically adapted to them. These seemingly incongruous creatures are carnivorous autotrophs that apparently obtain their nitrate by digesting captured insects.

Regions with light rainfall (deserts, western United States), on the other hand, commonly produce alkaline soils because of high evaporative losses. As water evaporates near the soil surface it is replaced through capillary action by water from below. This tends to concentrate calcium salts and their ions in the uppermost topsoil layers. The resulting increase in pH often has an interesting effect. It tends to make a variety of soil ions unavailable to plants by increasing the number that are adsorbed onto colloidal soil particles at the expense of those that are free in the soil water. The osmotic problem of increased alkalinity plus the availability of only a portion of the total soil nutrients pose special challenges for plant survival. Under desert conditions these problems are further compounded by extreme summer heat, winter cold, and excessive aridity.

Some Biological Components

The biological environment which any species faces is obviously all the other organisms with which it is associated either directly or indirectly in the ecosystem. A great variety of species is engaged in equally numerous interspecific interactions. These interrelationships are by no means random, but are highly organized. The life of a biologic community, like that of an individual organism, depends on the use of nutrient materials and the maintenance of energy flow. The fundamental principles of metabolism pertain no less to communities than to individuals, and are indeed the characteristic feature of the biotic environment. We can properly consider, therefore, community metabolism: the flow of materials and energy through the ecosystem as a result of the integrated interactions between the diverse species in that system.

INTERSPECIFIC INTERACTIONS AND COMPETITION

We discussed several kinds of interspecific interaction in Chapter 9: (1) commensalism and mutualism in which one or both of two associated species are benefited with harm done to neither and (2) parasitism and predation in which one of two interacting species is benefited at the expense of the other. A third category of interaction that we have not yet covered is *competition*, in which two (or more) associated species are harmed by their relationship. Pronghorn antelope and deer compete for forage where their ranges overlap. The mesquite bush competes with other desert plants for the soil moisture around it (usually with high success).

Among autotrophs competition is for sunlight, water, and chemical nutrients. Heterotrophs may compete for shelter, food, and breeding or nesting sites. The intensity of competition between the two species is in direct proportion to the degree of similarity in their ecologic requirements. That is, the intensity of interspecific competition is dependent on the amount of overlap in the *niches* of the two species.

The niche of an organism relates to the specific role it plays in the ecosystem, especially as a user of energy (solar or chemical). Niche has been likened to the biologic profession of a species. It has a certain location or "address" that we call its *habitat* (a sandy beach, in treetops, and so on). It involves the kind of food an organism eats and whether feeding occurs at night or during the day. If the food is insects, what kind of insects, at what stage of development (eggs, larvae, adults), and how are they captured? On the Galapagos Islands we saw that three of Darwin's finches occupied separate niches as ground-feeding seed-eaters (1, 2, and 3, Figs. 9-19 and 9-20) based essentially on the difference in the size of seeds eaten. Although there is doubtless some overlap in the size of seeds consumed by these species in their ground-feeding habitat, competition is relatively minimal because that overlap is minimal.

It is assumed by most ecologists that two species cannot occupy the same niche in the same ecosystem for any length of time. To do so would mean that the two species are so alike in their biological constitutions and ecologic needs that they would be one, not two species. Two species, however, may attempt to occupy similar niches, in which case intense competition may develop for those factors in the ecosystem that they share. Some of the consequences of such an eventuality we discussed on page 322. They are threefold.

First, one species may be incapable of withstanding the competition of another. Such a case is shown in Figure 10-2 where a species of animal-like protist (*Paramecium caudatum*) is unable to coexist with a closely related form (*P. aurelia*). There are instances, on the other hand, where the successful competitor varies according to environmental circumstances. An example of this nature was inves-

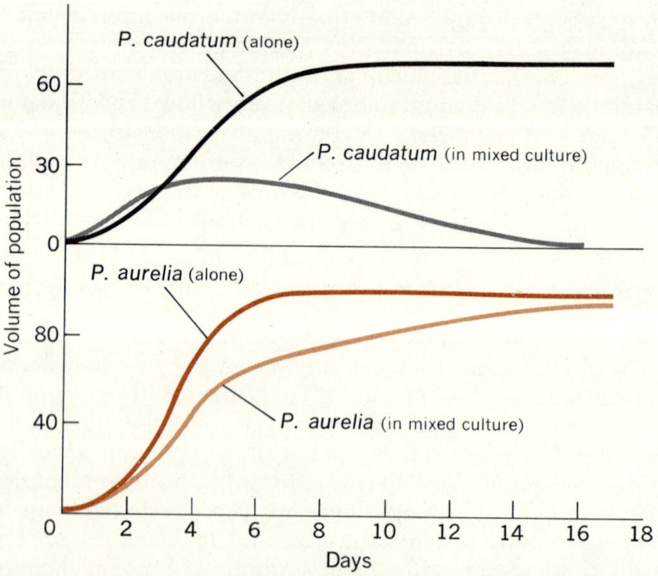

Figure 10-2 The effects of competition between two species of *Paramecium*. The population growth of two species of *Paramecium* grown in separate cultures under standardized conditions is compared with their growth in a mixture culture. Because of the competitive pressures in the mixed culture, one species (*P. caudatum*) was eliminated. (G. G. Gause, *Science*, Vol. 79, 1934)

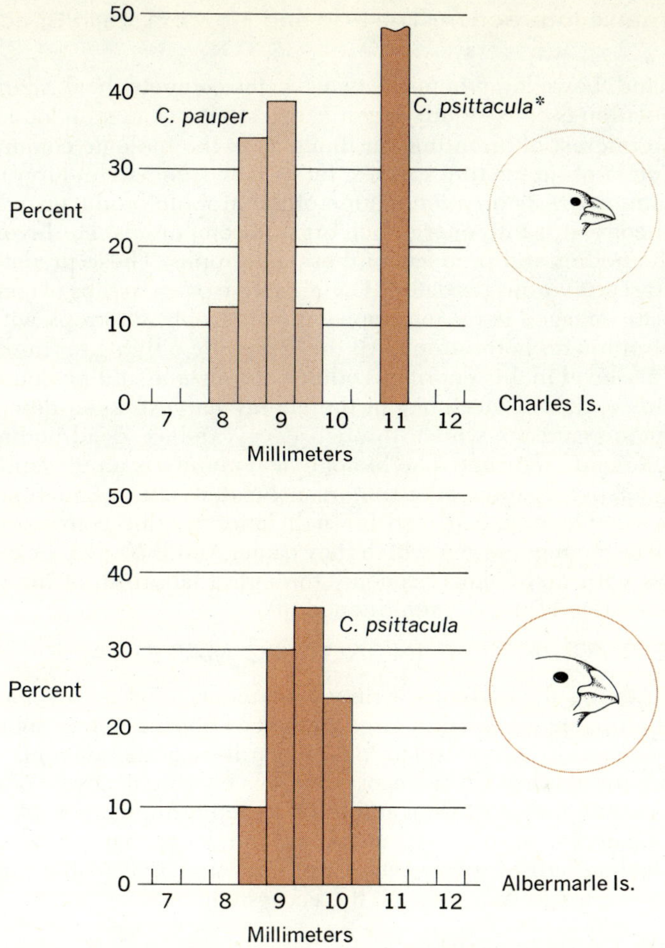

* The number of collected specimens—16—was too small to compute reliable percentages, but all fell within the size range indicated.

Figure 10-3 The effects of competition in two of Darwin's finches. The large tree finch found on Albemarle Island, *C. psittacula* (dark-colored histogram), has about the same range in beak depth as the large tree finch confined to Charles Island, *C. pauper* (light-colored histogram). The *C. psittacula* population on Charles Island, however, is significantly different, suggesting a change in food requirements where the two species compete. (Adapted from David Lack, *Darwin's Finches*, Cambridge University Press)

tigated in which two species of flour beetles were made to compete in different microclimates. One species always won out under hot-moist conditions, but was invariably ousted from the scene under cool-arid situations. The outcomes varied considerably between the two contestants over a range of microclimates between these two extremes.

The third possible effect is a reduction in competition because of the intensified selection that accompanies strong ecologic rivalry. Selection tends to foster survival in both competing species by encouraging ecological divergence between them. This is well demonstrated by Darwin's finches (Fig. 10-3). The large central islands tree finch (*Camarhynchus psittacula*) has about the same range in beak sizes as the large tree finch that is restricted to Charles Island (*C. pauper*). On

Charles Island, however, where the two species compete, beak size in the *C. psittacula* population is consistently larger, suggesting a change in food requirements.

In the interest of orienting our thoughts to the biologic community as an integrated system of interacting species, let us now summarize our remarks thus far. Every community rests on a foundation of autotrophic producers which trap solar energy and convert it into energy-rich organic compounds. Herbivorous predators consume the bodies and products of these autotrophs. These predators in turn are preyed on by carnivorous predators. Collaborative partnerships of commensalism or mutualism are engaged in by autotroph with autotroph, autotroph with heterotroph, and heterotroph with heterotroph. All these provide a living for innumerable parasites. At every level in this enormous edifice of consume-and-be-consumed, competition assures optimum efficiency of the energy harvest, even down to the multitudes of microorganisms which, through decay, reduce dead bodies and organic wastes to inorganic nutrient. The biologic community is an elegantly coordinated system of interspecific interactions. Inorganic materials are taken from the media by autotrophic producers and cycled through heterotrophic consumers and decomposers back to the media from which they came. And the energy captured by autotrophs flows with meticulous economy through a labyrinth of interdependencies until every vestige of it has been drained off.

THE CYCLING OF INORGANIC MATERIALS

Living things depend on a variety of elements (Fig. 2-1). Some of these, like carbon and nitrogen, are used in large amounts. Others (such as cobalt and manganese) are required in minute quantities. Regardless of the amounts utilized, these materials are routed through the ecosystem in a continuous cycle of use and reuse. From the medium they pass from autotroph to heterotroph and back to the medium by way of metabolic wastes, excrement, and the decay that follows death.

The carbon and nitrogen cycles are representative of the kind of itinerary followed by such materials through the ecosystem.

THE CARBON CYCLE

In the carbon cycle (Fig. 10-4) the source of inorganic carbon is largely CO_2. The atmospheric supply is bolstered by a vast reservoir of CO_2 dissolved in the oceans as bicarbonate. Some of this carbon is reduced to organic compounds by autotrophs. These compounds serve not only the autotrophs but the heterotrophs that feed on them and on one another. To these we must also add the decomposers that utilize the organic carbon still tied up in animal wastes and in dead bodies. The cellular respiration of all these creatures recycles the carbon back to its source.

Organisms whose dead bodies escape complete decay remove carbon from the cycle in the form of peat, coal, oil, gas, limestone, and so on. Some of the limestone, of course, is subject to gradual weathering and decomposition. Man, on the other hand, is returning the carbon of fossil fuels to the atmosphere by burning in an amount which is currently equal to more than 10 percent of that respired by all living things on the planet. This will undoubtedly increase the percentage of CO_2 in the atmosphere, if indeed, as many believe, it has not done so already. Since atmospheric CO_2, like glass, is transparent to light but not to heat, this might exert a greenhouse effect by warming the climates over the earth to a significant degree if existing practices continue.[1]

[1] This is a complicated issue, however, and reliable predictions cannot be made at this time.

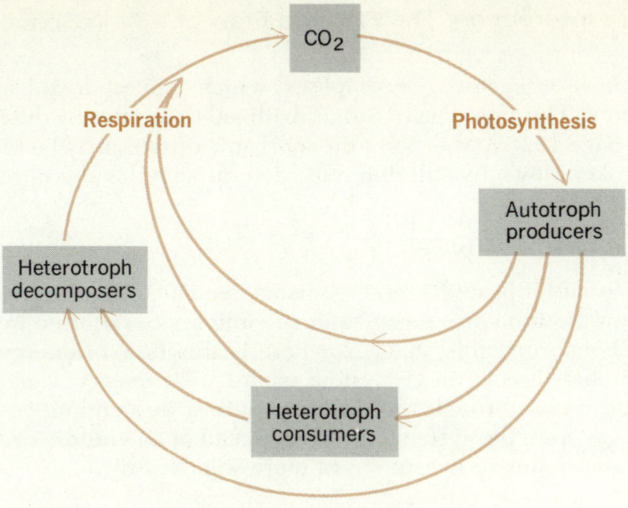

Figure 10-4 The carbon cycle.

THE NITROGEN CYCLE

The atmosphere, which is 79 percent nitrogen, constitutes a more than ample supply of this element for the biosynthesis of proteins, nitrogen-containing lipids, and nucleic acids. Nitrogen, however, is relatively inactive chemically and very few organisms are capable of converting or "fixing" it into compounds that lend themselves to further biosynthesis. The mutualistic bacteria which we described in Chapter 9 (Fig. 9-32) perform this function along with other nitrogen-fixing bacteria and blue-green algae which live freely in the soil. Biological nitrogen-fixation is supplemented by the production of nitrogen oxides in the atmosphere by lightning. Rain carries these oxides into the soil where they are transformed into nitrates.

From the raw materials provided by nitrogen-fixation (ammonia, nitrates, and so on) autotrophic biosynthesis yields organic compounds which pass along routes comparable to the carbon cycle (Fig. 10-5). Autotrophs are consumed by heterotrophs. The dead bodies of both, together with the wastes of the heterotrophs, are

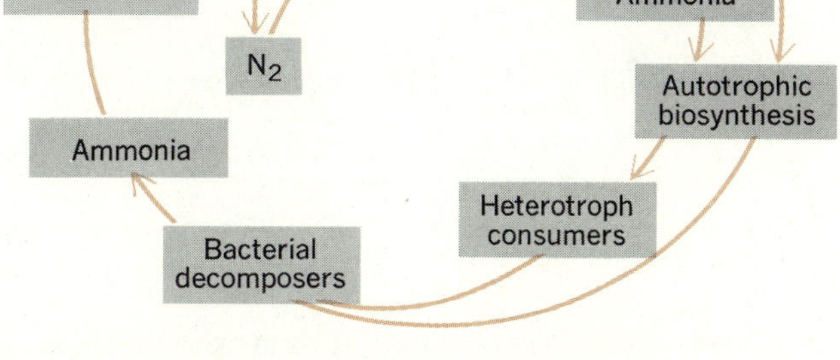

Figure 10-5 The nitrogen cycle.

cycled through hosts of bacterial decomposers which convert organically bound nitrogen to ammonia. The ammonia in turn is oxidized to nitrates by different bacteria (see footnote 1, page 122). At this point the inorganic nitrates may be recycled to the producers or broken down by still different bacteria to molecular nitrogen and end the cycle where it began.

THE FLOW OF ENERGY

As we have said repeatedly, all organisms need a constant input of energy in order to maintain the immense organizational complexity required by their highly energized and hence unstable state. Nowhere is this flow of energy more beautifully demonstrated than in an ecosystem where solar energy, captured by autotrophs, is passed down through the community in a descending cascade until it finally drains away from the system entirely. Instead of an endless cycle of use and reuse there is continuous reinvestment in more solar energy.

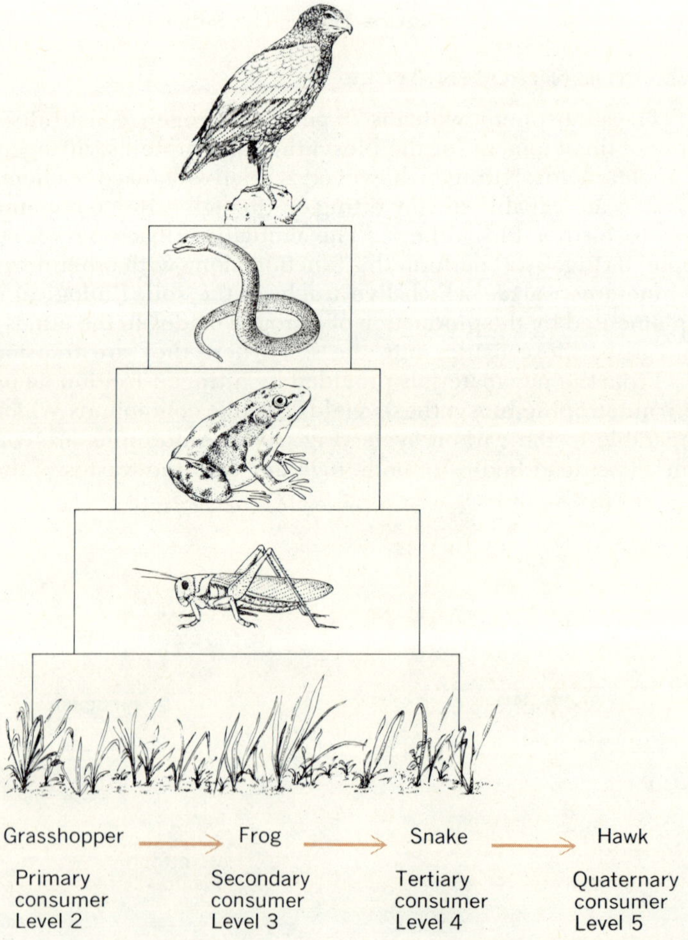

Figure 10-6 A single food chain from grass to hawk (parasitic consumers and decomposers not shown). Note the general pyramid shape due to the loss of energy at each successive level of energy conversion.

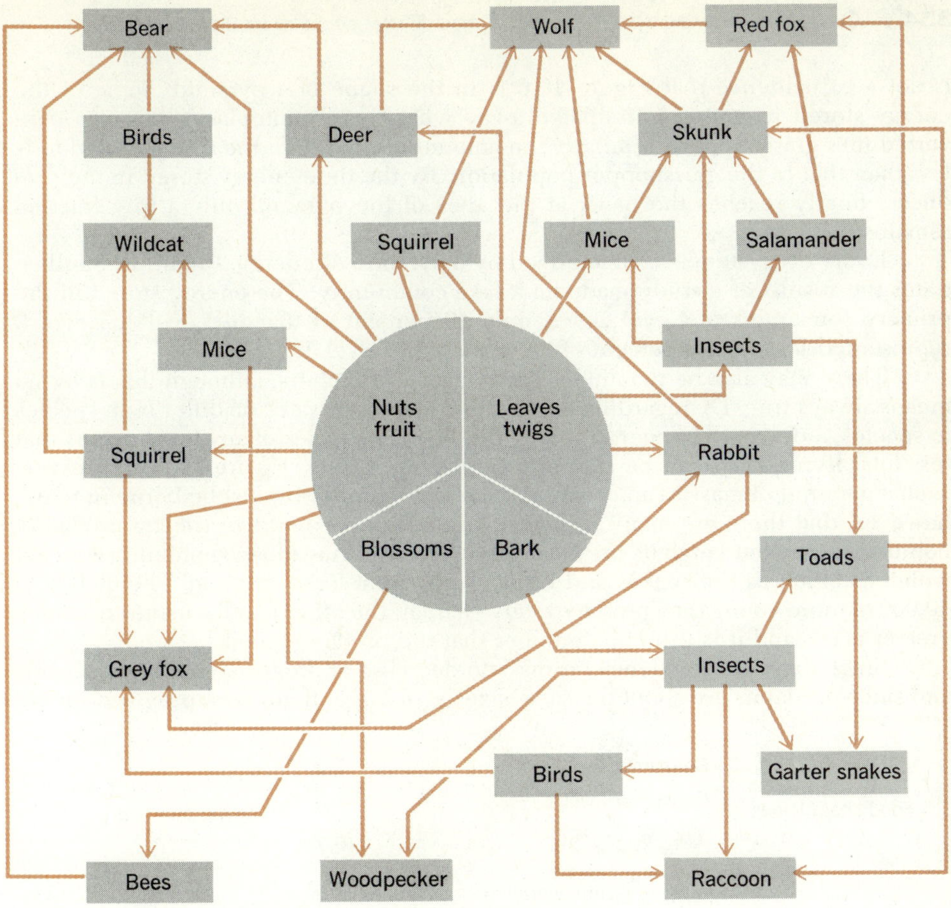

Figure 10-7 Food chains in a deciduous forest in Illinois. (From V. E. Shelford)

To take a simple and classical example (Fig. 10-6), grass, the *producer*, is fed on by grasshoppers, the primary *consumers*. Grasshoppers are eaten by frogs, the secondary consumers, which, in turn, are preyed on by snakes, the tertiary consumers. At the end of this *food chain* stands the hawk, which, as the fourth consumer, dines on snakes. Many parasitic consumers, of course, are found at all levels in the chain as well as multitudes of bacteria and fungi responsible for decay. These *decomposers*, as we have seen, reduce dead bodies and organic wastes to inorganic nutrients that are recycled back to the producers.

The sequence of species from grass to hawk in Figure 10-6 represents a single food chain. The flow of energy through a community usually involves a large number of interrelated food chains and may be exceedingly complex (Fig. 10-7). Regardless of the complexity, however, community food chains can be broken down into their respective levels of energy flow. Level 1 consists of all the producers; Level 2, all the herbivores (primary consumers); Level 3, the predators that prey on herbivores (secondary consumers); Level 4, the predators which live on Level 3 predators (tertiary consumers), and so on, depending on the nature of the community. The significant point is that at each successive level there is a progressive loss of energy from the community. It will be recalled from Chapter 4 (page 124) that every energy transformation is accompanied by a loss of usable energy. It is not a coincidence that Figure 10-6 is in the shape of a pyramid. Some of the energy stored in grass is dissipated away when grass protoplasm becomes converted into grasshoppers. Similarly the amount of energy in the frog population is less than that in the grasshopper population. By the time energy stored in the producers finally reaches the hawk at the apex of the pyramid only a tiny fraction remains.

Every ecosystem is characterized by a pyramid of energy. Figure 10-8a illustrates the results of a study made on a lake community. The energy stored in the primary consumers of Level 2 was only 10 percent of that in Level 1. Level 3 represented less than 2 percent of the energy in Level 1.

There may also be pyramids of mass and of numbers, although this is by no means always true. Disregarding the fact that metabolic rates do differ from species to species, a decrease in energy among the diverse species of any level means that less total living mass can be supported per unit of time. Figure 10-8b represents such a pyramid of mass in an ocean eelgrass community. It must be borne in mind, however, that there are many instances where, *at one point in time*, a predatory population may outweigh its prey. The 1000 mice in a meadow community may not weigh as much as the snakes and birds of prey that feast on them. The 30,000 to 40,000 or more young they produce every year, on the other hand, constitute a much greater mass, and it is from this number that the predators feed.

Since there is a tendency for mass to decrease at progressively higher levels and since predators are about the same size as (or larger than) their prey, we should

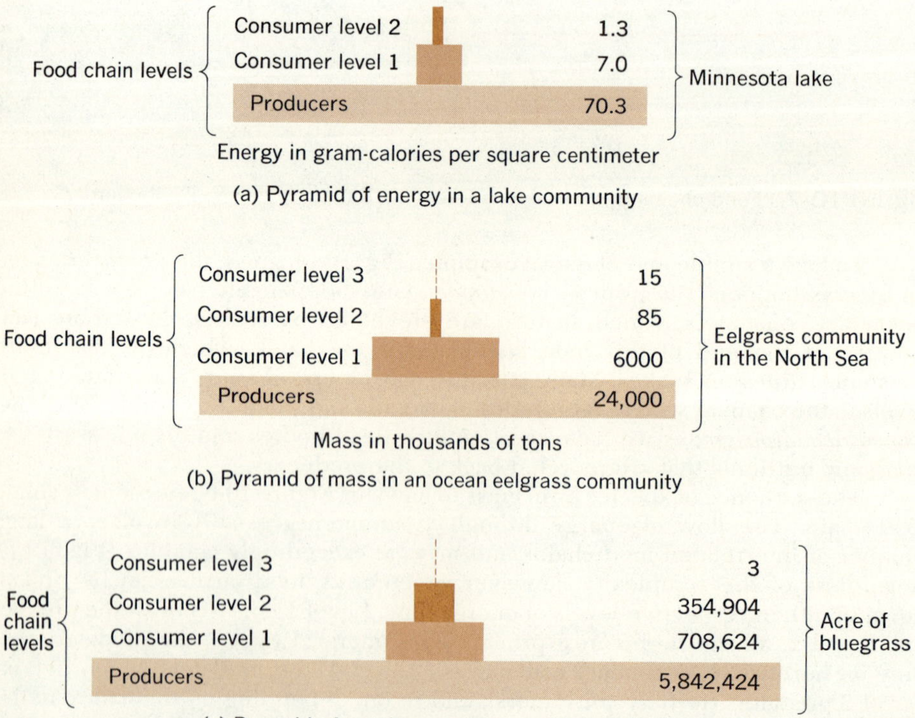

(a) Pyramid of energy in a lake community

(b) Pyramid of mass in an ocean eelgrass community

(c) Pyramid of numbers on an acre of grassland

Figure 10-8 Pyramids of energy, mass, and numbers. (Adapted from E. P. Odum, *Fundamentals of Ecology*, Saunders, 1959)

expect to find communities in which there is also a pyramid of numbers. Figure 10-8c illustrates such a case on an acre of grassland. For almost 6 million producers there were only three consumers at the fourth level in the community. Here again there may be many exceptions. Parasites and plant-eating insects have less mass than their hosts and usually outnumber them.

THE LIFE OF A SPECIES IN ITS ENVIRONMENT

Now that we have some understanding of what environment is, we can intelligently discuss how a species responds to it. We will therefore switch our emphasis from the ecosystem to one of its constituent species, bearing in mind that that species is not only affected by the system but is, at the same time, a part of it. As long as the ecosystem does not undergo significant change, the life of that species will be relatively stable. Nothing, of course, is changeless and ecosystems are no exception to the rule. The contingency of community change is taken up later. We will first consider the course of events that characterize comparatively short periods of time during which environmental conditions are reasonably constant.

The Paradox of Populations: Collaboration versus Competition

It is characteristic of most species that individual members tend to cooperate with one another, even if it is solely restricted to the act of mating. Aggregations of organisms (schools of fish, flocks of birds, swarms of insects, clumps of trees, and so on) not only enhance reproductive efficiency by bringing prospective mates together but commonly aid survival under adverse conditions. A clump of trees exposes its total mass to less surface than a single tree. Through decreased transpiration and better shading of the ground, the clump might survive better under arid conditions than a lone tree. Sowbugs (pillbugs) are commonly found clustered together under logs as a means of conserving moisture for their delicate gills. Six to ten ducks will maintain a small area of open water in a frozen pond by their activity and the warmth of their bodies. Mice even engage in the communal nursing of their young. Several mothers combine their litters and share their milk supply. The added warmth of many bodies in close contact and the more equable apportionment of total milk production (plus other factors) result in growth rates which exceed that of litters raised separately.

Individuals in groups also collaborate in food gathering and in protection against predators. Each individual in a pack of wolves enjoys better hunting than a lone wolf. The combined eyes of a herd of antelope are better protection than a single pair. There are, however, levels of collaboration that go beyond mere aggregation and involve social organization which often assumes a high level of sophistication. From Chapter 5 the reader will recall social regulation among insects through the use of pheromones as well as the elaborate society of the honeybee (and other social insects) where food-getting, protection, and even the size and structure of the population itself are under social control. The structure of a chimpanzee troop, on the other hand, is much more informal, but nonetheless characterized by an extreme degree of collaborative effort.

In the face of forces that bind members of a population into cooperative organizations there is an opposing circumstance that may become intensely competitive in nature. The members of a species occupy the same niche in an ecosystem. Since their requirements are essentially identical, the contest for ecologic resources may become extremely vigorous if a population grows to the point where one or more of these resources are in short supply. The specter of intraspecific competition is a formidable one because it may carry the seeds for self-extinction. Interspecific competition, predation, and parasitism help to stave off this dire possibility. The species itself, in many instances, has evolved inborn mechanisms that prevent excessive concentrations of individuals. It is no coincidence among certain rodents (chipmunks, gophers, and so on) that the young at a particular stage of their development begin to fight among themselves. Their mother also joins in the familial friction. This ostensibly disruptive behavior results in dispersing the young, thereby preventing population concentration in one area. It is to such mechanisms which control population size that man has coined the phrase "balance of nature."

Population Size and Environmental Capacity

Although the reproductive potentials of different species vary, the rate of reproductive increase per unit of time always exceeds the death rate from old age alone. When mature organisms reproduce, they tend to replace more than their number. Their offspring repeat the process when they mature. Since the product of population increase also increases, population growth is potentially exponential (for example, 2, 4, 8, 16, 32, etc.). If this potential is allowed to express itself without restriction, a growth curve similar to that on the left in Figure 10-9 is possible for any species. Fortunately, there are various factors that limit unrestrained growth. In our illustration this *limiting factor* is available nutrient. It gradually decelerates the rate of growth in the yeast population until the number of cells levels off at a constant figure. The S-shaped curve on the right that is produced through the interaction of reproductive potential and limiting factor is typical of population growth within an ecosystem. It is through this interaction that population size is adjusted to the maximum number the environment can support indefinitely. This number is the *environmental capacity* of the ecosystem.

The limiting factors which regulate populations and keep them within the environmental capacity of the community may be grouped into three categories: (1) predation and parasitism, (2) competition, and (3) self-regulation. Let us consider each one of these in turn.

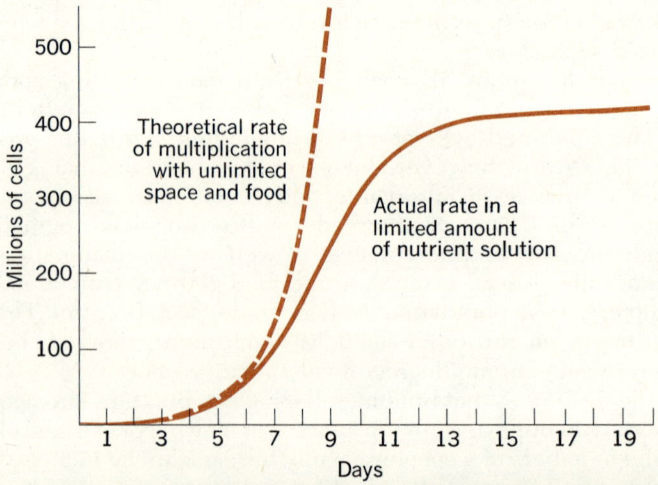

Figure 10-9 Multiplication of yeast cells.

PREDATION AND PARASITISM

The relationship between predator and prey is a very sensitive one. When the prey is abundant, predation is relatively heavy because detection and capture are comparatively easy. As the numbers of prey decline from this predation (and other factors), there is not only less food for predators, but it becomes more difficult to find and catch it. The food shortage results in a drop in the predator population. With predation pressure reduced, the numbers of prey begin to increase. As they do so the food supply of the predators is augmented and they ultimately respond with corresponding increases. Such a cycle of oscillating population peaks is shown in Figure 10-10 between the snowshoe hare and its predator, the lynx.

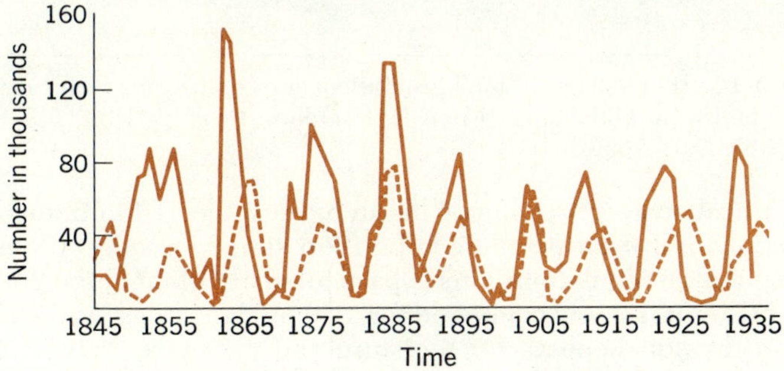

Figure 10-10 Corresponding cycles of population density between a predator, the lynx (dashed line), and its prey, the snowshoe hare (solid line), based on pelts received by Hudson's Bay Company from 1845 to 1935. (University of Toronto Studies, University of Toronto Press)

Similar cycles are commonly reflected between parasite and host. Diseases of epidemic proportions usually strike during periods of high population density. Not only is it easier for parasites to be transferred to new hosts at this time, but it also appears, in certain animals at least, that stress due to overcrowding somehow increases the susceptibility to infection.

There is also a high degree of sensitivity between parasite and host. This is beautifully demonstrated in the attempt that was made to reduce the population of European rabbits introduced into Australia by infecting them with a virus obtained from South American rabbits. The first epidemic killed about 98 percent of the rabbit population. The second was fatal to some 85 percent. By the sixth epidemic mortality had been reduced to 25 percent. Natural selection had (1) produced a more resistant population of rabbits and (2) produced a less virulent strain of virus that had the advantage of being able to "live" in many more rabbits.

COMPETITION

The hazards of competition and overexploitation of the environment are avoided in great numbers of species by the simple expedient of rarity. Of the 29 plant species comprising the vegetation of a tall-grass prairie in Oklahoma, 20 were found to make up only 16 percent of the total stand (Fig. 10-11). Comparable results were obtained in another study involving the random collection of nocturnal moths in one locality (Fig. 10-12). Of the 240 species which were collected, 115 were represented by 10 or fewer individuals. Only 35 species were represented by more than 100 individuals.

SPECIES	PERCENT OF COVER
Sorghastrum nutans	24
Panicum virgatum	12
Andropogon gerardi	9
Silphium laciniatum	9
Desmanthus illinoensis	6
Bouteloua curtipendula	6
Andropogon scoparius	6
Helianthus maximiliana	6
Schrankia nuttallii	6
Twenty additional species (less than 1 percent each)	16
Total	100

Figure 10-11 Percentage of total ground cover of the plant species making up a tall-grass prairie in Oklahoma. (From E. P. Odum, *Ecology*, Holt, Rinehart and Winston, 1963. Data from Rice.)

When critical ecologic resources begin to be subjected to intensified utilization, interspecific and intraspecific competition start to assume increasing effectiveness in population regulation. This is particularly evident in the density of trees in a forest where light is being competed for. Although not visible, the same competition exists in the soil beneath for moisture and minerals. Natural selection, of course, has produced shade-tolerant species as well as deep-rooted and shallow-rooted plants. This diversification not only alleviates competition but makes for fuller utilization of ecologic resources.

One of the most notable (and tragic) examples of the effectiveness of intraspecific competition involved the deer herd on the Kaibab plateau in Arizona. Before 1907 the environmental capacity of this range for deer was estimated to have been about 30,000 animals (assuming none of the feed would have been used by range livestock with which deer compete). The deer herd, numbering about 4000, was kept well below this limit and in healthy vigorous condition by predators. Through the misguided belief that predators are the enemies of their prey, man stepped in to "help" the deer. Between 1907 and 1923[2] 3000 coyotes, 674 pumas ("mountain

[2] Predatory animal control actually continued to 1939 and resulted in the extinction of the wolf.

NO. OF SPECIES	NO. OF INDIVIDUALS COLLECTED IN EACH SPECIES
35	1 {35 moths / 35 species}
50	2-5
30	6-10
90	11-100
35	101 or more
240 species	

Figure 10-12 Population density of 240 species of nocturnal moths as reflected by the number of individuals collected in one locality. [Data extracted from Fisher, Corbet and Williams, *J. Anim. Ecol.*, Vol. 12, 42 (1943)]

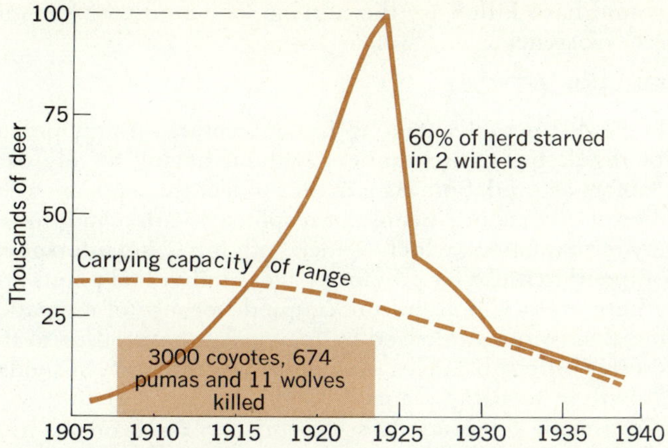

(a)

(b)

Figure 10-13 (a) Man's systematic killing of predators on the Kaibab plateau of Arizona from 1907 to 1923 caused an explosive increase in the deer population far in excess of available feed (carrying capacity). The end result was wholesale starvation, range damage and decreased forage production, and the loss of many more deer than the predators would have killed. (b) Deer do not normally browse on pine trees unless forced to do so by starvation, at which point the trees become denuded up to a conspicuous "browse line." [(a) modified from Aldo Leopold in Wisconsin Conservation Dept. Publ., 1943; (b) Courtesy of U.S. Forest Service.]

lions"), and 11 wolves were killed (Fig. 10-13). The deer responded with a population explosion that resulted in 60,000 animals starving to death in two winters alone (1924–1925). The damage of this exercise in ignorance continues to accrue. So far wholesale winter starvation, range damage which has reduced carrying capacity to less than one-third of its former value, and the loss of many, many more deer than the predators would have killed are the heritage of a herd which even today has a precarious winter existence.

SELF-REGULATION

It would be a distinct advantage to living creatures if they possessed inborn mechanisms for regulating their numbers without having to rely exclusively on such extrinsic factors as predation. As a matter of fact this appears to be true in an imposing number of diverse organisms. In response to arid conditions plants have evolved a variety of adaptations which reduce both intra- and interspecific competition. We have alluded to these on previous pages. Individual plants are commonly surrounded by bare areas which are unoccupied because of aromatic substances and other chemical secretions released by leaves that have fallen to the ground or by roots in the soil. These substances prevent the germination of seeds or interfere with the root growth of seedling plants.

We have also made reference to self-regulatory mechanisms in animals. On page 194 we noted that the scent of a strange male (whose proximity might be due to overcrowding) may terminate the pregnancy of mated female mice. We observed that the honeybee queen cannot only control the sex of her offspring but their number as well. In the current discussion we mentioned the disruptive behavior in chipmunk and gopher families that encourages dispersal of the young. Also mentioned was the possibility that the psychological and physiological stress that accompanies crowded conditions somehow lowers the resistance of individuals to disease. In fact, there is good evidence, at least in laboratory mammals, that population stress evokes endocrine and behavioral responses that adversely affect reproduction itself in many ways. These effects include abnormal estrus, small, defective litters, and inadequate milk secretion.

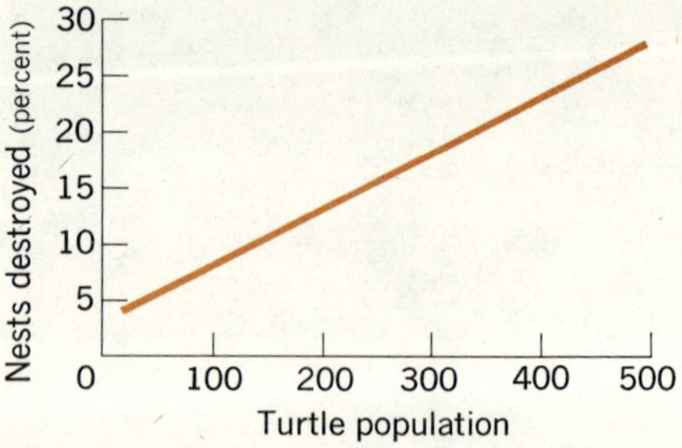

Figure 10-14 The diagonal line indicates the approximate correlation between population density and nest destruction in the green sea turtle. As population increases, turtles destroy a higher percentage of their nests. (Adapted from H. R. Bustard and K. P. Tognetti, *Science*, Vol. 163, pp. 939–941, February 1969. Copyright 1969 by the American Association for the Advancement of Science.

Green sea turtles on the Great Barrier Reef of Australia appear to regulate their numbers through egg destruction. A given population excavates its nests in the sand above the spring high tide mark of one particular beach. Since this nesting territory is restricted in size, increased population density results in more and more nests being dug into and destroyed by successive groups of nest-digging females (Fig. 10-14).

A great variety of vertebrates establish breeding territories. One individual (usually a male) or a pair defends a certain area against intrusion by other members of the same species (especially other males). The familiar sight and lovely sound of a male bird vocalizing in the spring may be a dulcet love song, but it is more often a vociferous warning to similar males that this territory has been claimed. Territoriality is an effective means of controlling population density. It does so by limiting the number of breeding individuals. Adults who do not have access to a breeding territory simply do not breed that year.

CHANGING BIOLOGICAL DESTINIES: ECOLOGIC SUCCESSION

We have been considering the ecosystem as a fairly stable entity (which it is during short intervals of time) so that we could concentrate on the more immediate problems that species face in order to survive in their environment. It is as if we had stopped a motion picture projector to study one frame of the film more intently. The ecosystem, of course, is in a constant state of flux for the simple reason that the biologic community is continuously altering its physical environment. When sun-loving trees invade an open area, the shade they cast may interfere with the growth of seedlings they reproduce. This may permit the gradual intrusion of other trees which, when mature, cast such heavy shade that the pioneering species is eliminated for want of enough light. In most cases the biological interactions within one community so modify the physical surroundings that the community itself becomes transformed. One community tends to pave the way for another and the end result is an *ecologic succession* of communities.

We have all watched one crop of weeds replaced by another of a different kind on a vacant city lot or the succession of plant species that follow one another on fallowed or abandoned farmland. In the southeastern United States the general pattern of ecologic succession on abandoned farmlands is shown in Figure 10-15. The first year bare fields are invaded by annual weeds, followed in the second year by grass and a different crop of weeds. Grass becomes the dominant plant type for several years. After about ten years shrubs begin to invade the area and become the dominant vegetation. In the meantime the soil is becoming richer by the accumulation of plant litter and increasing populations of soil microorganisms. Pine seedlings start to make their appearance and a pine forest eventually replaces the shrubs. But pine seedlings fail to grow well in deep forest shade. Shade-tolerant, deciduous trees, particularly hickories and oaks, begin to intrude and the pines gradually give way to the hardwoods. The hardwood forest so formed is relatively stable and will normally maintain itself for a long time. Because of this comparative stability it is said to represent the *climax* community for this area and these conditions. Throughout the succession of dominant plants there are, of course, corresponding changes in the animal community. This is reflected in the illustration by some of the alterations made in the perching bird population.

Even a swampy lake, like abandoned farmland, may be converted into a forest climax (Fig. 10-16). Sediment and the remains of dead aquatic vegetation gradually accumulate in such lakes. As the lake begins to fill, the shallow perimeter slowly becomes transformed into a meadowlike bog (b). Rank growths of bulrushes and cattails around the lake as well as pond lilies and pondweeds in the lake contribute a large amount of organic residue to the growing bog at the expense of the shrinking lake (c). The lake finally disappears altogether, leaving a meadow which, in due time, will be completely taken over by sun-loving forest trees (d).

The process of ecologic succession begins as soon as a new potential habitat becomes available. If this happens to be a massive slab of rock which has freshly tumbled from a mountain face in California, the first pioneers to colonize it are usually lichens (Fig. 9-34). The lichens secrete acids which etch tiny crevices in the rock and provide bits of soil. Weathered rock particles, wind-blown soil, and

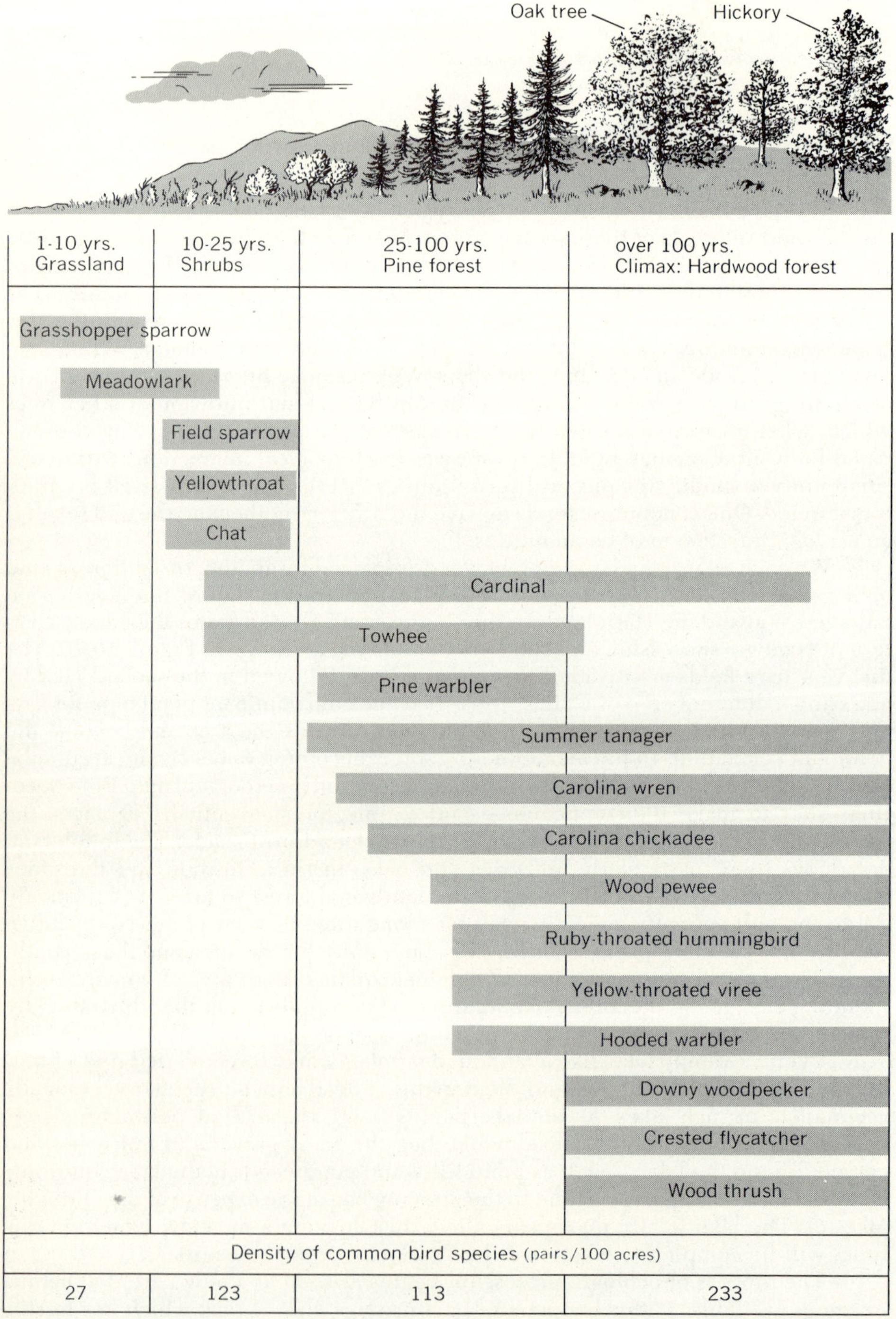

Figure 10-15 The general pattern of plant succession (top) on abandoned farmland in the southeastern United States, showing some of the accompanying changes in the bird population (bottom). (Modified from E. P. Odum, *Ecology*, Holt, Rinehart and Winston, 1963)

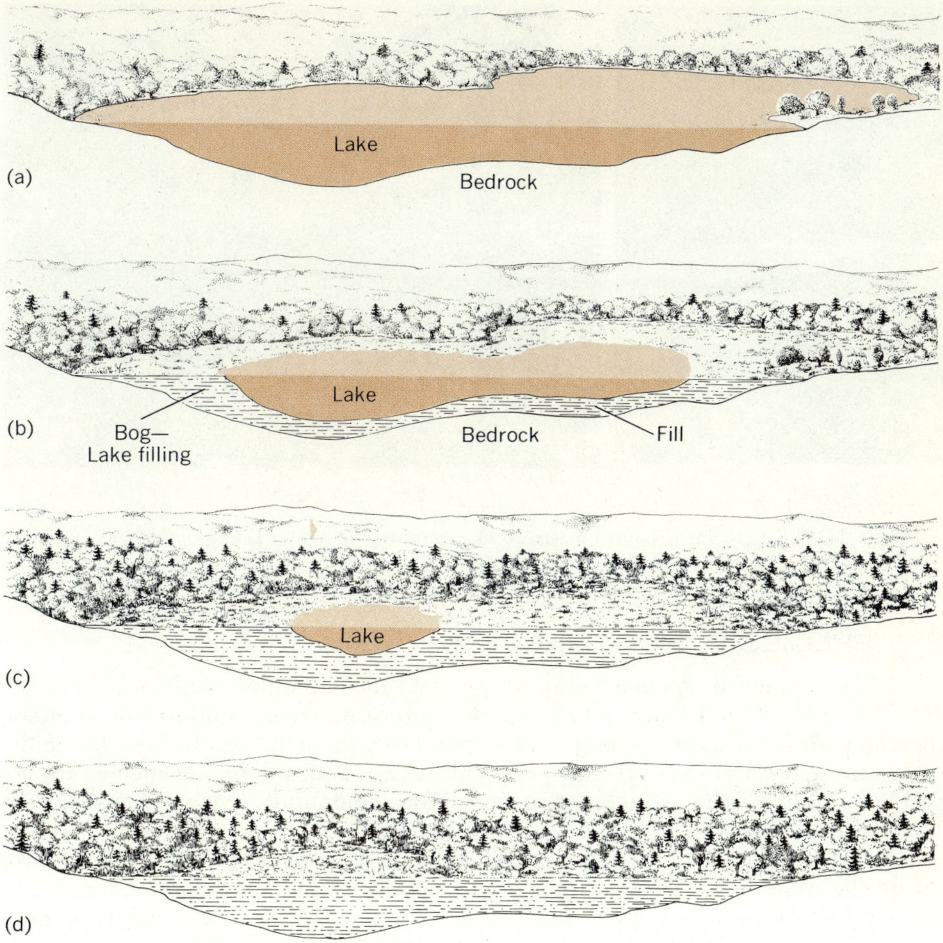

Figure 10-16 The conversion of a shallow lake to a forest through succession (see text). Compare (b) with Figure 10-20. (Modified from A. K. Lubeck, *Geomorphology*, McGraw-Hill, after Dachnowsky.)

pieces of dead lichen gather in these crevices. Mosses often appear shortly after the lichens. Accumulating bits of dead vegetation slowly build up a mulchlike layer capable of supporting grasses and herbs (Fig. 10-17). Small woody plants gradually gain a foothold in the larger rock crevices. Even pine trees or junipers may become established in these cracks. The combined work of physical weathering and biological activity ultimately converts the rock into soil and the biological community moves on to its climax stage. In so doing the original capacity of the rock slab for supporting life has been augmented enormously.

The Ecological Consequences of Succession

A number of ecologists believe that in the majority of cases certain fundamental features characterize the change from pioneering to climax community in any given succession. These features concern species structure, total biological mass, and energy flow.

(a) (b)

Figure 10-17 (a) Succession on bare rock; (b) close-up view. Lichens, mosses, grasses, herbs, and small shrubs are early invaders. Even pine trees or junipers may become established in the crevices of the rock. [(a) author; (b) Sol Karlin, Los Angeles Pierce College]

SPECIES STRUCTURE

The change in species composition is in the direction of greater diversity. This is evident in Figure 10-15 where a representative sampling of common perching birds shows two species in the grassland, six in the shrub stage, ten in the pine forest, and 12 in the climax hardwoods. In general, colonizing species are not likely to be important participants in the climax community. Many are absent entirely. Some species in the succession have broader niches than others and will persist for longer periods of time (for example, the cardinal in Fig. 10-15). The increase in species number reflects additions to the food web (chains) of the community. As succession gets underway there is a general trend toward greater diversity in autotrophs. The diversification in microorganisms and heterotrophic plants and animals becomes particularly striking during the later stages of succession.

TOTAL BIOLOGICAL MASS

Referring again to Figure 10-15 note that the number of perching birds increased from 27 pairs per 100 acres in the pioneering community to 233 pairs per 100 acres in the climax community. There is a tendency for biological mass to increase at every level in the pyramid of energy until the climax is reached. Decomposing organic matter appears to continue growing in bulk with the passage of time.

ENERGY FLOW

Autotrophic production steadily grows during the early stages of succession but levels off at a more or less constant volume in the climax community. Paralleling this growth in autotrophic productivity, food chains become progressively more elaborate, refined, and specialized. More and more of the energy captured by producers is utilized by consumers. The mature climax community achieves stability because it has arrived at a dynamic equilibrium in which a certain optimal biological mass is being supported by minimal energy input. Such a community operates essentially on its detritus (dead organic matter) through complex and little understood food chains which have microscopic decomposers at one end and animals at the other. Investigation has revealed that less than 10 percent of the annual autotrophic production of a mature forest is consumed in the living state by herbivores and parasites.

The Convergence of Climax Communities

The complex species structure and balanced metabolism which give the mature community its biologic stability also buffer and control physical factors such as temperature, light, moisture, and wind. As long as no major environmental changes occur, such a community can be self-perpetuating and last for centuries.

Over the earth, areas with similar climatic and physiographic features tend to support climax stages which, although composed of different species, correspond remarkably well in their dominant life forms. The mature grasslands of Oregon, Wyoming, the Argentine pampas, and northern Australia represent the culmination of widely separated successions which have converged on a common ecologic configuration in which grasses of some kind are the predominant vegetation. These common forms toward which ecologic succession tends to converge are called *biomes*. They are characteristic of all ecosystems whether they be tide-pool, estuary, marsh, lake, river, or desert. In order to gain some understanding of the nature of biomes we will very briefly consider six common terrestrial types: (1) tundra, (2) coniferous forest, (3) temperate deciduous forest, (4) tropical rain forest, (5) grassland, and (6) desert (Fig. 10-18).

TUNDRA

The land of tundra is a gently rolling plain dotted with many bogs, lakes, and ponds (Fig. 10-19). It is the northernmost biome where temperatures are low enough that the subsoil is in a continuously frozen state. This, together with strong drying winds and a growing season of two months or less, renders the region largely treeless. Mosses, lichens, grasses, and dwarf shrubs make up most of the vegetation. Great herds of caribou, vast multitudes of flies and mosquitoes, and enormous numbers of shore birds and waterfowl populate the area. Most of the bird population is migratory, however, and leaves after the summer nesting season. Although the number of organisms is astronomical, the number of resident species is less than in any other type of mature community except the most barren deserts. Mature communities in this biome are also less stable than those in any other terrestrial ecosystem.

CONIFEROUS FOREST

The coniferous forests lie in a broad zone south of the tundra. In those latitudes of the world characterized by milder climates they are confined to mountain ranges. Although the winters are just as cold as those of the tundra, the summers are longer and warmer. The subsoil does not remain frozen and the growing season lasts for three months or more. Many lakes, bogs, and meadows dot a landscape commanded by conifers (Fig. 10-20). The characteristic evergreen trees are spruces and firs. In the western United States pines and junipers are also common. Birch and aspen are typical deciduous species. Deer, moose, bears, squirrels, and a variety of birds live in this region. During the winter many of the mammals hibernate and most of the birds migrate. Although this biome boasts of more species than the tundra, it claims less than other types of forest communities.

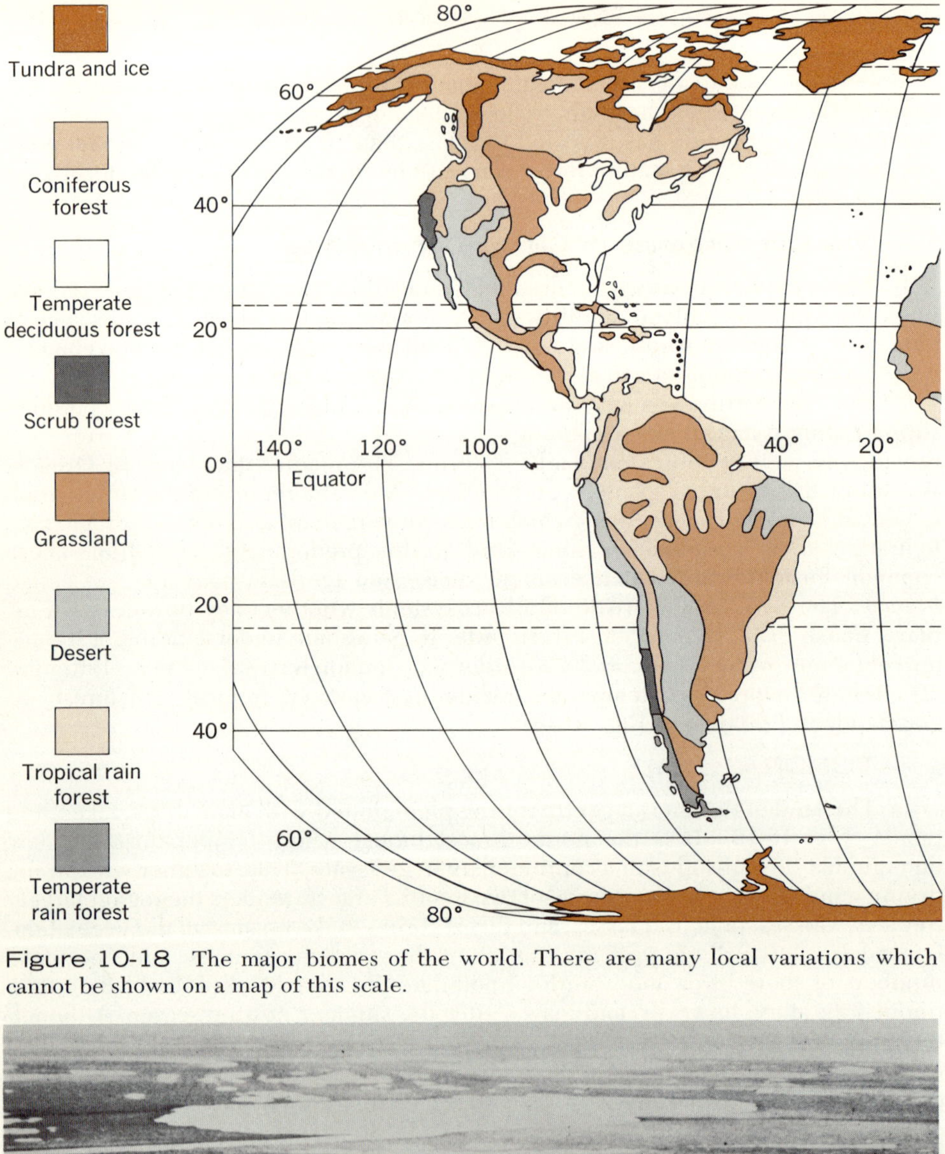

Figure 10-18 The major biomes of the world. There are many local variations which cannot be shown on a map of this scale.

Figure 10-19 Tundra in aerial view. (G. Dallas and M. H. Hanna, California Academy of Sciences)

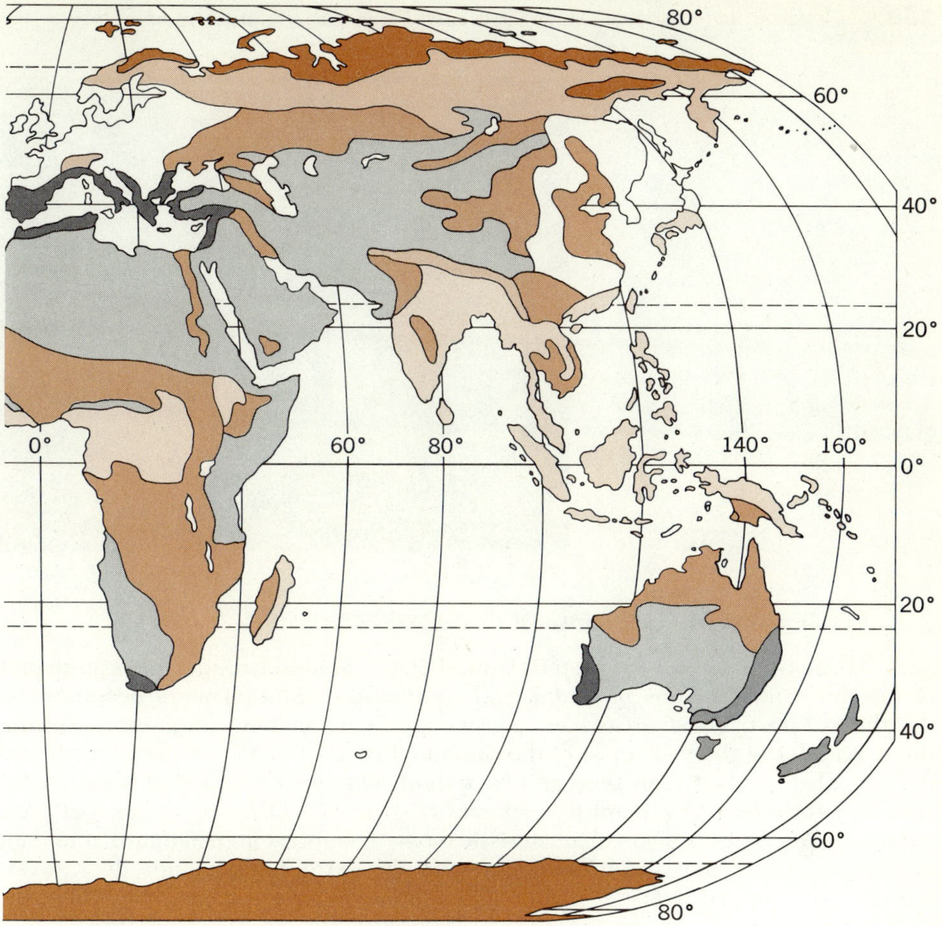

Figure 10-20 A coniferous forest. The lake in the foreground is being encroached upon by a bog succession (middle ground). Compare with Figure 10-16b. (C. Wilson and W. Loomis, *Botany*, Holt, Rinehart and Winston, 1967)

Figure 10-21 Temperate deciduous forest. (U.S. Forest Service)

TEMPERATE DECIDUOUS FOREST

At one time most of the eastern United States, central Europe, and a large part of the temperate regions of China and southeastern Siberia were occupied by deciduous forest. These areas have a growing season of about six months. During the remainder of the year most of the perennial plants have shed their leaves and the annuals have cast their seed and have died. The deciduous forest usually consists of one to three dominant tree species (Fig. 10-21). Oaks, hickories, beeches, maples, ashes, and so on, are characteristic types. The forest floor contains abundant organic detritus, fungi, and invertebrate animals. Vertebrates include deer, foxes, raccoons, tree squirrels, and a considerable number of tree-nesting species of birds.

TROPICAL RAIN FOREST

The tropical rain forest receives abundant rainfall continuously throughout the year. The growing season is often yearlong. Such conditions favor vegetative growth of utter profusion. One to several hundred species of evergreen trees form a dense canopy of interlacing branches which breaks the incessant fall of rain and filters out almost all the sunlight before it reaches the forest floor. Despite the high population density, two trees of the same species rarely grow close to one another and are commonly hundreds of yards apart. The forest floor is so dimly illuminated that autotrophs are often lacking entirely. Here water drips constantly, the humidity is high, and the temperature is almost unvarying. If plants are present, they are usually ferns, mosses, and low herbs. As one proceeds vertically the vegetation is stratified into levels or stories. Low palms, tree ferns, and evergreen shrubs above the forest floor are in turn below trees of moderate height.

The uppermost story of emergent trees have trunks that are unbranched up to 100 to 150 feet (Fig. 10-22a). Within such a multistoried structure one can imagine the variation in microclimates from full exposure at the top to the dank, dimly lit region far below. Climbing woody plants called *lianas* use the trunks of trees for support (Fig. 5-34c) so they may spread their leaves into shafts of light that penetrate the canopy. Plants known as *epiphytes* (Fig. 10-22b) gain access to light by attaching to the branches of trees and shrubs, forming penthouse gardens at various

(a)

Figure 10-22 (a) Tropical rain forest in Hawaii National Park. (b) An epiphyte supported in the crotch of a limb. Such plants are common in rain forests where they use their position high in the trees to gain access to light. They are not parasitic but send their roots into "soil" composed of their own humus. [(a) First National Bank of Hawaii; (b) Sol Karlin, Los Angeles Pierce College]

(b)

levels from the canopy down. Arboreal animals, including monkeys, squirrels, birds, and frogs (not to mention a great diversity of insects), follow a similar vertical distribution. The particular species composing various rain forests may differ sharply, but there is nevertheless a remarkable uniformity in species structure among them. These communities are without peer among terrestrial biomes, both for the complexity of their structure and for their stability.

All the four biomes we have discussed receive adequate precipitation, both in amount and distribution, to support an abundant and diversified biota. The distinguishing feature among them is the length of the vegetative growing season. The major climatic influence regulating this is temperature, since moisture is not a restrictive factor. As we have observed, temperatures increase as one progresses from the polar regions toward the equator, growing seasons lengthen, and biome type is modified accordingly. Temperature, however, is just as sensitive to changing altitude as it is to changing latitude. As one moves up the slopes of the Appalachians, the Rockies, the Sierra Nevadas, or the Coast Range mountains a series of biomes is encountered which corresponds quite well with those that could be seen by moving northward (Fig. 10-23). Patches along the north-facing slopes of many peaks in the Rockies and the Sierras are perpetually blanketed in snow. Below these peaks are communities dominated by lichens, mosses, and small herbaceous plants which overlook vast expanses of coniferous forest.

The two remaining biomes differ fundamentally from the previous four in that the critical variable is moisture rather than temperature. The moisture-laden masses of air that move eastward through the Pacific states dump most of their precious burden as they rise over the western slopes of the Coast Ranges, Sierras, and finally the Rockies. As a result the southwestern and Great Plains regions of this country are relatively arid. The biologic communities found here are unique. They are nev-

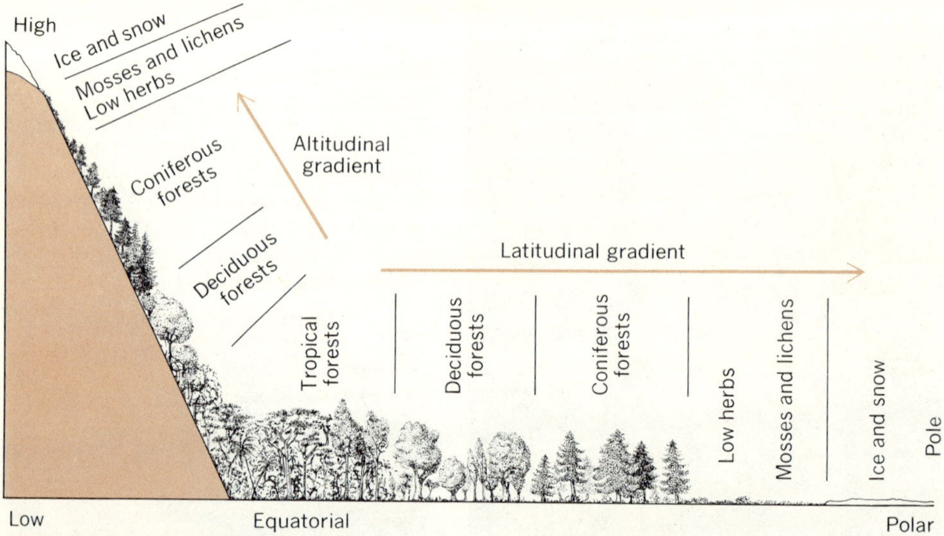

Figure 10-23 Correspondence between the horizontal and vertical distribution of biomes.

ertheless remarkably similar to those in other parts of the world where species that are utterly different are living under corresponding conditions.

GRASSLAND

In the temperate and tropical parts of the world there are regions where the annual rainfall is such that trees will not grow but grasses flourish in great abundance (Fig. 10-24). Here the limited rainfall is largely restricted to the early part of the growing season.[3] A large part of this region in the United States is now devoted to corn, wheat, and other grains. The exceptionally high volume of good quality forage which grasslands produce, plus the lack of conspicuous shelter, are in great

[3] The western grasslands of the United States receive most of their moisture from melting snow.

Figure 10-24 A grassland biome in Colorado. (U.S. Forest Service)

measure responsible for the two kinds of mammals that are found in greatest abundance: (1) large, fleet-footed herbivores like the zebras and antelopes in Africa (or the bison and pronghorn antelope in this country prior to settlement by whites) and (2) burrowing rodents and rabbits (gophers, prairie dogs, and so on, in the United States). The fauna of the grassland community (mammals, birds, reptiles, and insects) is certainly more conspicuous than that of any other terrestrial climax and is probably second only to the rain forest in absolute numbers.

DESERT

The deserts of the world lack enough rain to support large stands of grasses. Temperature fluctuations are also extreme. During the day air temperatures often exceed 110° F, especially during the summer. Because of the scarcity of vegetation ground surface temperatures commonly reach 160° F or more. As nightfall approaches, the loss of this heat through radiation is unimpeded so that evening temperatures may become cool to subfreezing cold. Shrubs such as mesquite, creosote bush, and sagebrush are common and possess a variety of drought-resistant adaptations. Among these are (1) water storage in roots, (2) reduced leaf size, and (3) leaflessness except for short periods following a rain. Succulent plants such as cactus (Figs. 9-29 and 10-25a) store water in their tissues. Some deserts have a variety of short-lived annual herbs that develop from seeds that germinate after early spring rains. They blossom, produce seed, and die in just a few days to a couple of weeks. At this time the desert is a riot of color, but throughout the rest of the year it appears as in Figure 10-25b. Animal life includes rodents, a few birds, lizards, snakes, spiders and their allies, and insects. Most of them are active only at night,

(a) (b)

Figure 10-25 The desert biome. (a) Cactus in the Sonoran desert of southern Arizona. (b) Shrub vegetation in the Mojave desert of southern California. [(a) Ward's Natural History Establishment; (b) Sol Karlin, Los Angeles Pierce College]

secreted by day under rocks, buried in the sand, in underground burrows, or in recesses excavated into cactus trunks. It is in the desert perhaps more than in any other type of community that one is impressed with the majestic proportions of life's adaptability. Here, although challenged by the severest of climatic extremes, living creatures carry on in a world which, to them, is a friendly and familiar one. In the face of physical adversity which is frequently brutal the conquest of energy continues to be successfully waged.

QUESTIONS

1. Describe the role of solar radiation in water cycling.
2. Compare the advantages and disadvantages of water and air as biologic media.
3. What determines the degree of competition that may develop between two species?
4. Discuss the various ecological consequences of intense competition.
5. In what ways do bacteria participate in the nitrogen cycle?
6. What is a food chain and how is it related to energy flow in a biologic community?
7. Why is every biologic community characterized by a pyramid of energy?
8. Discuss what is meant by the expression "the paradox of populations."
9. What limiting factors keep populations within the environmental capacity of the community?
10. The interaction between a predatory species and its prey is distinguished by a cycle of oscillating population peaks in both species. Why is this?
11. Describe five different ways in which species may regulate their own numbers, giving an example of each.
12. Why does ecologic succession take place?
13. Certain fundamental changes are believed to characterize the transition from pioneering to climax community in any given succession. Discuss these changes and their ecological consequences.
14. What exactly is meant by the term "biome"?
15. Describe a terrestrial biome that is found in regions in which
    (a) Adequate precipitation is received but the critical variable is temperature.
    (b) Precipitation is the critical variable rather than temperature.

Suggested Readings

Allee, W. C. *et al. Principles of Animal Ecology.* Philadelphia: Saunders, 1949.

Boughey, A. S. *Ecology of Populations.* Current Concepts in Biology. New York: Macmillan, 1968.

Bresler, J. B. (ed.). *Environments of Man.* Life Sciences series. Reading, Mass.: Addison-Wesley, 1968.

Carson, R. L. *The Sea Around Us.* New York: Oxford University Press, 1951.

Clarke, G. L. *Elements of Ecology.* New York: Wiley, 1966.

Croll, N. A. *Ecology of Parasites.* Cambridge, Mass.: Harvard University Press, 1969.

Hazen, W. E. *Readings in Population and Community Ecology.* Philadelphia: Saunders, 1964.

Hutchinson, G. E. *The Ecological Theater and the Evolutionary Play.* New Haven, Conn.: Yale University Press, 1965.

Isaacs, J. D. "The Nature of Ocean Life," *Scientific American*, September 1969.
Kormondy, E. J. *Readings in Ecology.* Englewood Cliffs, N.J.: Prentice-Hall, 1965.
Odum, E. P. *Ecology.* Modern Biology Series. New York: Holt, Rinehart and Winston, 1963.
Odum, E. P., and H. T. Odum. *Fundamentals of Ecology.* Philadelphia: Saunders, 1959.
Park, T. "Beetles, Competition, and Populations," *Science*, **138**, December 28, 1962.
Phillipson, J. *Ecological Energetics.* Studies in Biology. New York: St. Martin's Press, 1966.
Pimentel, D. "Population Regulation and Genetic Feedback," *Science*, **159**, March 29, 1968.
Slobodkin, L. B. *Growth and Regulation of Animal Populations.* New York: Holt, Rinehart and Winston, 1961.
Wynne-Edwards, V. C. "Population Control in Animals," *Scientific American*, August 1964.

CHAPTER 11 Organic Revolution: man and his culture versus life and its future

So far we have considered biologic communities in their pristine state, unfettered by human intervention so that we might gain some comprehension of the dynamic forces that mold and stabilize them under natural conditions. This position, of course, is a very unrealistic one, for there are probably but relatively few creatures living anywhere on earth that are not affected directly or indirectly by human activity. In the vast majority of cases, moreover, these effects are significant and have far-reaching consequences. Man's presence has altered, in some instances drastically, the faunas and floras of most of this planet for more than 10,000 years.

Man as a species (*Homo sapiens*) has existed for about 150,000 years, perhaps more. From his origin in the Old World he later spread to the Western Hemisphere. Paleontological evidence leaves little doubt that prehistoric man was a predator without peer. His bill of fare ran the gamut from fish and shellfish to birds, rabbits, horses, bears, and mammoths. Being omnivorous, his diet also included fruits, nuts, berries, tubers, and roots. Filling such a wide niche he could find food in a great diversity of natural environments and he was probably included in nearly all food chains that could support a large animal. Because of the breadth of his omnivorous appetite his position in most communities was complex since he could play the roles of primary, secondary, or tertiary consumer (or higher) within the same ecosystem. His aptitude as a predator was formidable. There is good evidence that intensive hunting by early man in North America was probably in large measure responsible for the extinction of the mammoths on this continent about 11,000 years ago. (These animals suffered a similar fate in Asia as well.)

Between 8000 and 10,000 years ago man began to domesticate animals and plants. He had learned that the nomadic lot of the hunter-gatherer could be greatly improved by controlling and protecting the growth and harvest of useful organisms. This resulted in more productive yields. It also guaranteed man a place at the two most advantageous levels in the pyramid of energy, namely as a primary and secondary consumer.[1] It allowed for plant food storage over the winter season and periods of drought. Exhausting migrations, which often proved hazardous to the young of both man and his domestic animals, were no longer necessary. The increased environmental capacity, together with the relative stability that this agricultural revolution made possible, led to a very significant increase in population and the gradual development of villages and ultimately cities. Population density gradually climbed to a new level, at which point it was held in check to an important degree by the recurring incidence of epidemic diseases in urban centers. As a case in point, bubonic plague periodically decimated the urban populations of Europe and the Middle East from the fifth century B.C. into the nineteenth century A.D. In one year alone (1348) one-third to one-half of the population of England was wiped out by "The Black Plague."

[1] Why are these the most advantageous levels for a consumer in the pyramid of energy?

The coming of the Industrial Revolution brought with it the elimination of one of the last effective checks against population increase. The work of Pasteur in the nineteenth century and many others since that time has produced a medical and public health technology which has shorn disease of its power to keep human populations in check. Over the last century this technology has spread throughout the world and is in effective use over most of the globe. The result has been a population explosion of unbelievable proportions with all its dreaded complications.

Starting with prehistoric man as a hunter-gatherer and his probable role in big game extinctions, our species has a long history in the disruption of stable climax communities. Domestication, urbanization, and disease prevention have compounded man's assault on his environment. We are demonstrably a pioneering, colonizing species that has shown no propensity to live as a collaborator in a well-ordered mature ecosystem. We have finally run out of virgin environments to exploit. The consequences were inevitable. A cataclysmic chain of events has been set in motion, slowly and insidiously at first, but as man spread throughout the world these events have grown to an avalanche of destructive fury. The mistakes we have made are of unparalleled magnitude. As a result of increasing the short-term productivity of the earth's resources we have abused and reduced the long-term productive capacity of the world's ecosystems while at the same time creating populations that are neither adequately fed nor cared for.

In directing our attention to these problems we will first consider the nature and extent of the damage done. Having accomplished that we will be in a position to address ourselves to what must become mankind's foremost commitment the world over: well-planned and fully supported corrective measures aimed at *long-range* human survival.

THE DECLINE IN ENVIRONMENTAL PRODUCTIVITY

Soil and Water Loss

Two of the many indispensable functions served by mature forest and grassland communities are the perpetuation of soil and fresh water resources. The forest canopy breaks the impact of rainfall on steep slopes. Grass and other ground vegetation as well as organic litter protect the soil from the erosive action of raindrops and impede the flow of runoff waters. Without such protective cover both surface water and wind are free to remove the topsoil and transport it great distances. Organic matter, plant stems, and roots also contribute greatly to the porosity of soil, allowing rainfall to infiltrate so that minimal amounts of surface water remain for runoff. The capacity of soils to store such water is remarkable. This is particularly true in meadows, swamps, and forests where the soils are spongy because of abundant organic matter and a maze of interlacing roots. The litter alone of a mature forest is capable of absorbing 40 to 50 million cubic feet of water per square mile in 10 minutes. Rather than rushing to the ocean in intermittent streams that become raging torrents after every storm, this water is released slowly to groundwater storage and to springs and streams which flow yearlong. It is small wonder that we refer to the more than 281,000 square miles of National Forests in this country as priceless watersheds.

The damage which man has inflicted on these communities has been devastating. Clearing forests and plowing prairie sod to raise domesticated crops resulted

Figure 11-1 Severe gully erosion on farmland. (U.S. Forest Service)

in quick returns at much higher yields than could have been realized from the native vegetation in the same length of time. But the soils were left unprotected between the fall harvest and spring planting. Wind and runoff removed the topsoil (Fig. 11-1). Often the transported soil came to rest on another farmer's field and buried his crops or filled streams with silt, killing the food of native fish. Pasturing wild range lands with excessive numbers of livestock, improper logging, and strip mining have had corresponding effects (Figs. 11-2 to 4). The streams that drain these lands are like conveyor belts to the sea, ironically moving their loads of silt by the very water which a well-managed watershed would issue gradually and sediment free. Damming these streams to prevent flood damage after heavy storms is

Figure 11-2 This Arizona range land normally supports a stand of grasses (left of fence). Overgrazing by livestock (right of fence) has converted it into a minimal vegetative cover dominated by snakeweed. (U.S. Forest Service)

Figure 11-3 Excessive logging on private land (left side of picture) left soil unprotected and susceptible to erosion. Selective logging on U.S. Forest Service land (right side of picture) not only leaves protective cover of trees but a forest that can continue to be harvested in the future on a sustained yield basis. (U.S. Forest Service)

Figure 11-4 Strip mining removes vegetation and the entire mantle of soil from extensive areas, promoting erosion and leaving the area unfit for future vegetative growth. (*Louisville Courier Journal*)

Figure 11-5 The reservoir above this dam in California has been completely filled by the silt which has been eroded from the watershed above. The photograph was taken three years after the dam was built. (U.S. Forest Service)

Figure 11-6 Badly eroded land may be stabilized if adequate subsoil remains to sustain the planting of a soil-binding crop. The planting of pines here completely checked gully erosion by providing a canopy of pine forest and a protective cover of ground litter. (U.S. Forest Service)

not only a costly measure but a temporary one as well. It has been estimated that most if not all of the flood control dams in this country will have suffered the same fate as that shown in Figure 11-5 within the next 30 years. Such dams wind up impounding a reservoir of silt rather than water and consequently no longer serve the purpose for which they were built.

The land surface of the continental United States, excluding mountains and deserts, encompasses more than 2.7 million square miles. The United States Soil Conservation Service has estimated that about 468,750 square miles of this has been either ruined or severely damaged by accelerated (man-induced) soil erosion. Abandoned farmland includes another 78,125 square miles. An additional 43 percent (1.2 million square miles) has lost between one-quarter and three-quarters of its topsoil. Every year this nation's croplands and range lands lose 3 billion tons of soil through accelerated erosion. Nor is this country by any means the sole victim. Large areas in South America, Africa, the Mediterranean region, the Middle East, India, and China are subject to the same menace.

Losses of this magnitude are unrecoverable. Thousands of years are required to produce fertile soils. Steps, of course, can be taken to reduce further losses from damaged lands and to curb soil depletion in productive areas. The planting of hardy, well-adapted plants has in many instances reestablished a semblance of soil stability on badly eroded soils (Fig. 11-6). Although such lands must be kept in agricultural retirement, their value as a home for wild life and a pleasant place to stroll are becoming of increased worth to our cramped society. Soils that are still agriculturally productive can be protected by a variety of techniques such as contour farming and terracing (Fig. 11-7). The success of these methods depends to a large ex-

Figure 11-7 Growing crops in strips that follow the contour of the land increases infiltration of rain and decreases runoff and erosion. Strips of corn and hay shown on left, grain and hay on right. (U.S.D.A. Soil Conservation Service)

tent on the degree of cooperation that can be elicited from adjacent landowners in a given area.

Biologic Damage from Agricultural Development

Agricultural development by its very nature amounts to changing or destroying biologic communities and replacing them with those of human design. Swamps are drained, forests are cut, and grasslands are plowed, altering if not removing the natural habitats of diverse and numerous organisms. Many native species have been affected adversely by these activities. Replacing a stable community composed of many species with farmland devoted to a single crop also creates additional problems which arise from the biologic imbalance such practices force on the environment. It must be granted, of course, that man, like any other creature, must use his environment to the fullest advantage. That fullest advantage, however, calls for long-range survival based on permanent productivity rather than a short-term existence dependent on overutilization of natural resources.

SPECIES EXTINCTION

The promiscuous draining of swamps has not only obliterated excellent water storage basins but has erased extensive waterfowl nesting areas. Additional habitats for a great variety of species have been seriously damaged or destroyed through overgrazing by livestock and the clearing of lands which, because of steepness of slope or shallowness of soil, should never have been farmed in the first place. Excessive trapping, hunting, and fishing decimated wild populations. Since 1850 about 140 species of warm-blooded animals have become extinct (Fig. 11-8). All

PERIOD	NUMBER OF SPECIES BECOMING EXTINCT IN PERIOD	AVERAGE RATE OF EXTINCTION PER PERIOD
1650–1850	40	1 every 5 years
1850–1900	64	1 every 9+ months
1900–	76	1 every 11 months
	180 species total	

(a)

(b)

(c)

Figure 11-8 Extinction in warm-blooded vertebrates. (a) Record of extinctions of warm-blooded species according to period in which extinction occurred. Decrease in extinction rate in the twentieth century reflects beginning of conservation measures. (b) The passenger pigeon was once one of the most abundant birds in the United States. The last individual died in a zoo in 1914. (c) The flightless dodo of Mauritius Island in the Indian Ocean was exterminated in 1693 by man and by domestic hogs which ate its eggs and young. [(a) Modified from data in Milne and Milne, *The Biotic World of Man*, Englewood Cliffs, N.J.: Prentice Hall; (b) National Audubon Society; (c) British Musuem of Natural History]

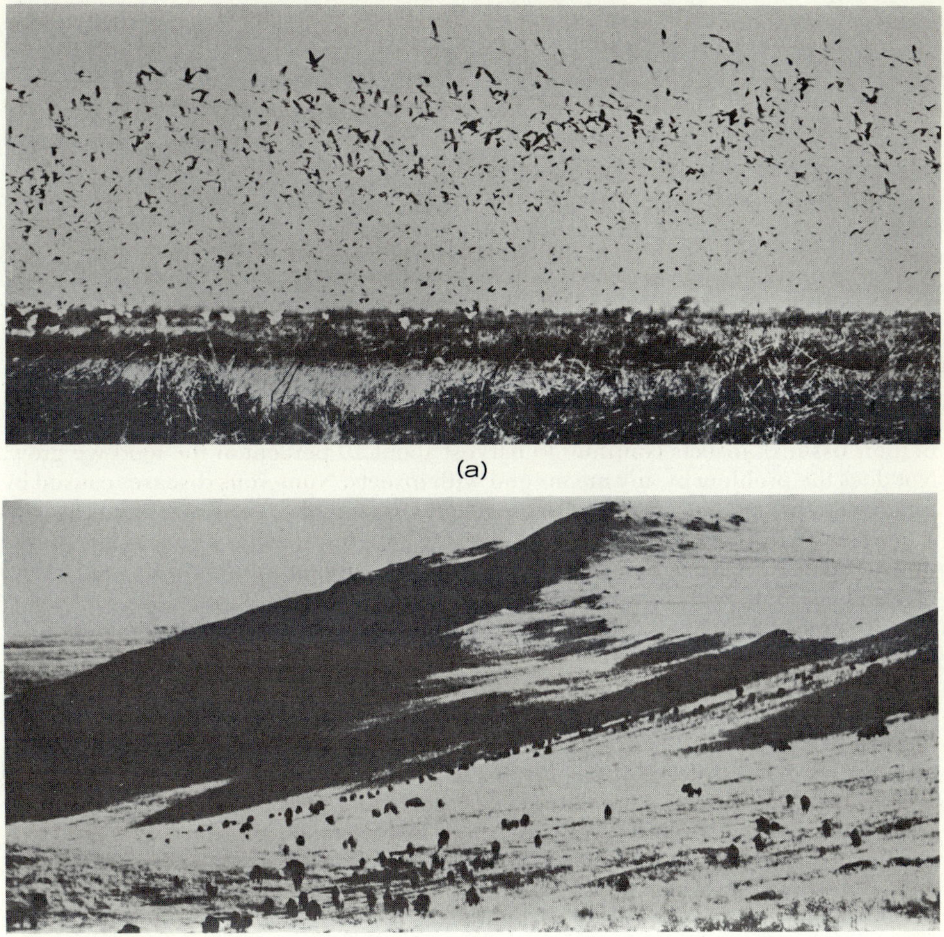

(a)

(b)

Figure 11-9 State and federal parks and wildlife refuges help to conserve our nation's wildlife. The U.S. Fish and Wildlife Service administers many refuges throughout the 50 states, two of which are shown. (a) Sacramento National Wildlife Refuge in California. (b) National Bison Range in Montana. (U.S. Dept. of the Interior)

these creatures were contributing agents in the balance of nature and an integral part of the great out-of-doors that millions of Americans spend billions of dollars to visit and enjoy.

Irresponsible practices still go on in many parts of the world, and a number of species, like the Orangutan, are at the brink of oblivion, to be lost like the others forever. Conservation measures, initiated during this century, have become increasingly effective, at least in this country. Sounder regulations governing the killing and capture of animals (including treaties with Canada and Mexico relative to migratory waterfowl), better law enforcement, and a system of federal and state refuges and parks over the country (Fig. 11-9) have had salutory effects. Several formerly endangered species, including the trumpeter swan and the American bison, have been saved by these efforts, sometimes narrowly. The National Bison Range was stocked in 1906 from a small private herd. There were no wild bison to be found anywhere at that time. When the Red Rock Lake refuge was established in the 1930s for the trumpeter swan there were just 33 birds left to carry on.

THE PROBLEM OF MAINTAINING UNBALANCED COMMUNITIES

It is no more possible to maintain a community in which there is only one species of plant, say wheat, than it is to maintain one with a single species of animal, such as cattle. The simple food chains of wheat→man or forage→steer→man are unattainable goals which we must expend great effort and vast sums of money to approach. The harder we press the attack the more imminent the prospects of disaster become. Although man has contaminated himself and the environments of the world with his insecticides (even seals and penguins of Antarctica contain DDT in their tissues), insects continue to harvest about 10 percent of the food we grow. Nor does the problem by any means end with insects. Numerous diseases caused by fungi and other protists, worms, bacteria, and viruses afflict domestic plants as well as animals. Vegetative crops receive fierce competition from weeds while being preyed upon by rodents and rabbits. As herbivores rodents and rabbits play a dual role and compete with livestock.

In trying to manage these complications man is beset with the perennial problem of interfering with involved food chains which include farmlands as a part of a larger biologic community. Attempts have been made, for example, to eliminate coyotes because they occasionally kill sheep. Since coyotes feed largely on rabbits, the cost of this practice has been uncommonly great when runaway rabbit populations drove sheep from their pasture for lack of feed. The use of insecticides has produced similar results since beneficial insects as well as other forms of life are killed along with destructive species. Many insect sprays and dusts are toxic to a great diversity of organisms, such as honeybees, earthworms, fish, a variety of birds (many of which eat insects), and mammals, including man.

The indiscriminate killing of beneficial insects which prey on agricultural pests has been responsible for extensive damage on innumerable occasions. Destructive species which had formerly been held in check by their natural insect predators are suddenly released from predation pressure. The farmer who has just rid his crops of one kind of pest finds himself victimized anew by another, or several. In order to control pest populations man has, in many instances, introduced predators from foreign countries. The ecologic nature of an alien species *of any kind* must be thoroughly appraised before it is introduced.[2] Otherwise such measures can have serious consequences. In an attempt to protect the sugar cane crop of the West Indies from rat damage a predatory mammal of India known as the mongoose was introduced. The mongoose proceeded to reduce the rat population and then turned to native wildlife and domestic animals for its food supply. The mongoose is now a major pest in this area.

As a result of an impressive record of what not to do, man is finally beginning to turn his attention to more enlightened procedures in the handling of these agricultural predicaments. Welcome programs of planned biologic regulation rather than chemical control are starting to gain momentum. Large populations of preda-

[2] The intentional introduction of the English sparrow and the accidental introductions of the European corn borer insect and the Russian thistle are excellent cases in point.

tory insects, like the native ladybird beetles in this country, are reared and then released in areas infested by insects on which the predators normally feed. In one case this technique has even proved to be effective in weed control. A native stock-poisoning plant of Europe called Klamath weed was inadvertently introduced into the western United States and Australia and proceeded over a period of about 50 years to overrun many hundreds of square miles of range lands, rendering them unfit for grazing. Two species of beetles which live exclusively on this weed in Europe were introduced (after thorough testing) onto the range lands in question and have reduced the weed population to the point where cattle can now graze safely.

Another biologic method that has been shown to limit specific populations of insects entails the release of large numbers of males made sterile by exposures to ionizing radiation (page 24). Normal females which mate with sterile males lay eggs (in most species) which do not hatch. Through the systematic release of treated males the population gradually dies out for want of an adequate proportion of normal males. This technique totally eradicated a serious pest of livestock, the screwworm or horn fly, over a vast area of Texas and the southeastern United States. A modification of the sterile male procedure which holds great promise and is also species-specific makes use of pheromones. If males can be lured into traps baited with natural or synthetic sex-attractant pheromones (page 194), a local population could be eliminated quickly.

Environmental Pollution by the Urban-Industrial Complex

Environmental pollution by man has begun to assume fearsome proportions. Technological agencies and products which were originally intended to improve man's lot now threaten to contaminate our soil, our water, our atmosphere, and ourselves to a degree which can no longer be tolerated.

SOIL POLLUTION

The production of solid waste (trash) in this country has reached 8 pounds per person per day. As one drives through the countryside the offensive sight of great numbers of bottles and aluminum cans is surpassed only by the stark realization that we are condoning a policy of waste that makes throwaway containers out of durable materials. The metals that wind up in littered cans and junked automobiles are nonrenewable resources. Their supply is by no means inexhaustible.

Of much more immediate concern, however, is the contamination of soil by toxic materials. Such substances can reach the soil from the air, from water, or by direct application. Dramatic examples like the following are not particularly uncommon. In southeastern Tennessee copper smelting in bygone days used to liberate sulfur dioxide fumes into the atmosphere. This noxious gas killed vegetation over an area of about 36 square miles. Even though the emission of SO_2 stopped in 1907 (with its industrial conversion into salable sulfuric acid) the area still remains devoid of vegetation (Fig. 11-10). Of growing concern today is the widespread and increasingly dangerous concentration of nitrates in both soil (and consequently foods) and water. Intestinal bacteria convert nitrate into a related compound which, if present in sufficient quantity, seriously interferes with oxygen transport by hemoglobin. Nitrate food poisoning of young children, who are especially vulnerable, has already been reported. The insidious aspect of this problem—and of all pollution today—is that it was created and is now perpetuated by overpopulation. During the past 25 years in the United States, oxides of nitrogen produced by the exhausts

Figure 11-10 Copper Basin, Tennessee, where SO_2 emissions from a copper smelter at the turn of the century denuded the area for miles around. Attempts at reclamation still remain largely fruitless. (U.S. Forest Service)

of automobiles and power plants have increased 300 percent. In a still-losing battle to feed hungry mouths the use of nitrogen fertilizers has increased 1400 percent. These two sources release 10 million tons of nitrogen compounds to the environment of our country every year, which is more than the nitrogen cycled by all the processes of nature combined.

WATER POLLUTION

The maximum permissible concentration of nitrates in drinking water, according to public health standards, is 10 parts per million. This limit has now been exceeded in many wells in Southern California, Illinois, and Minnesota. One town in Minnesota has already found it necessary to replace its entire water system. Sewage, of course, contributes its share of nitrogen[3] to rivers, lakes, and ground water. Aquatic ecosystems are being overloaded with human organic wastes. Rivers, lakes, and seacoasts are becoming less and less fit for aquatic life through contamination and reduced oxygen supply. At the same time the necessity for reuse of our inland waters is becoming more and more urgent as the demands of an ever-burgeoning population exceed the supply of uncontaminated sources.

We have the technical skill in sewage processing to reclaim water for redistribution. In many instances this reclaimed water has a lower bacterial count than it had when it was consumed in the first place. It is also heartening to note that man can build facilities to correct the growing problem of thermal pollution. The return of water used for industrial cooling back to its source can result in the heating of natural waters to a point where aquatic life can be endangered. Electric power production (particularly nuclear-powered plants) is the chief contributor of this heat. Considering that the 30 year projection for electric power expansion in this country would require one-third of the average daily runoff in the United States for cooling purposes, the presence of massive cooling towers at power plants is an encouraging sign of ecologic awareness (Fig. 11-11).

The problem that man has as yet shown little inclination to solve is the foreboding one of industrial contamination. The chemical industries in particular pose special problems in water pollution because they produce such great quantities of substances, such as detergents, which contain ingredients that cannot be removed by most sewage treatment systems (Fig. 11-12). The growth of these industries since World War II has been spectacular (Fig. 11-13). Hundreds of new products (and new waste products) are produced each year. The task of checking their

[3] Increase in sewage nitrogen over the past 25 years has been 70 percent, reflecting population growth.

Figure 11-11 Three massive cooling towers in a Kentucky power plant being built by the TVA will cool 282,000 gallons per minute through a range of 27.5°F. Water is cooled by evaporation. (Tennessee Valley Authority)

Figure 11-12 Detergents, which pass through sewage treatment plants, damage the aquatic community in this stream and contaminate the water for human consumption downstream. (Hays, Monkmeyer)

TYPE OF PRODUCTS MANUFACTURED	1940 PRODUCTION	1959 PRODUCTION	INCREASE OF 1959 OVER 1940 PRODUCTION
Plastics	75,000 tons	2,500,000 tons	33.3×
Detergents	7,500 tons	650,000 tons	86.7×
Insecticides and agricultural chemicals	4,000 tons	270,000 tons	67.5×

Figure 11-13 Increase in production of chemical industries in the United States since World War II. These industries pose special problems for pollution.

toxicity or possible harmfulness as they wend their way into the waters of the world is a monumental one. According to a study made by the National Academy of Sciences on the current rate of pollution increase, there will be enough waterborne wastes by 1980 to consume or make unavailable all the oxygen in all the 22 river systems of the United States during dry-weather flow. Continuing manufacture, consumption, and disposal by present standards can create a world environment no longer suitable for the existence of many living things by merely extending a phenomenon that is already well underway.

AIR POLLUTION

Air pollution is by no means a new problem. England enacted legislation 200 years ago to regulate the burning of coal. Today the major source of air pollution is still combustion. Automobiles, trucks, aircraft, power plants, steel mills, lumberyards, and burning dumps are some of the chief producers. Air pollution occurs in two distinct forms: (1) smoke, including gaseous contaminants such as oxides of sulfur which act as reducing agents and (2) smog, a product of the internal combustion engine and composed of oxidized hydrocarbons[4] which, because of their ozone and peroxide content, act as oxidizing agents. Both smoke and smog are human nuisances as well as irritants, affecting the respiratory passages and eyes. They are also capable of much more serious damage. London's black fog of 1952 caused about 4000 deaths, many of which were the result of suffocation. Deaths have occurred in Pennsylvania, Mexico, and Belgium from similar causes. Crop damage from smog in California has been considerable. Alfalfa, oats, sugar beets, and spinach are the most sensitive. The list of crops, as well as other plants, that are adversely affected is a long one and even includes pine trees in the National Forests.

There is no longer any excuse for the kind of air pollution shown in Figure 11-14a. Devices are now in extensive use whereby the solid visible particles in smoke or dust can be precipitated out. The gaseous components (oxides of sulfur and nitrogen, and so on) still constitute a problem. Smog is becoming a serious threat to every major city (Fig. 11-14b). The problem is particularly acute because the primary offender, the gasoline-powered automobile, has become a pivotal center in American life. Major changes in, if not the abandonment of, the internal combustion engine are clearly necessary now. The two industrial giants directly involved, the automobile and petroleum interests, will resist such changes unless they are subject to extreme social pressure as well as offered some sort of economic assistance to effect such revolutionary measures. It would indeed seem that a government which can subsidize an industry to produce tanks in time of war could find it economically, socially, and morally justifiable to render assistance here in the interest of human survival. In the interim, pollutants from automobile exhausts are being dumped into the air over Los Angeles at the rate of 1200 tons per day as the burning of fossil fuels and trash over the country at large brings the total atmospheric contamination to the staggering total of 142 million tons per year, or three-fourths of a ton for every American.

GENERAL POLLUTION AND SELF-CONTAMINATION

Some forms of contamination either do not fall properly within the categories of soil, water, and air pollution or, because of their general nature, pertain equally well to all three. The former group includes such things as the bewildering hodgepodge of highly touted patent medicines, drugs, and other pharmaceutical and med-

[4] Lead, found in many gasolines, is also a component of smog.

(a)

(b)

Figure 11-14 Air pollution. (a) One industrial plant like this is capable of contaminating a vast area around it. (b) Most of the smog in the Los Angeles basin is the result of 1200 tons of hydrocarbons emitted daily from automobile exhausts and crankcases. A portion of the city on a clear day (left) and a smoggy day (right) is contrasted. [(a) Grant Heilman; (b) Air Pollution Control District, County of Los Angeles]

ical products and agents which, because of inadequate regulation, often do more harm than good. For want of a better term we will refer to these as self-contaminants. In the area of general pollution we will confine our remarks largely to what has probably become man's most dangerous global pollutant, those pesticides known as chlorinated hydrocarbons,[5] especially DDT.

DDT is known to be toxic to a great variety of invertebrates and vertebrates, including man. It is very doubtful if any animal is immune to the danger of this substance and its chemical allies. It is effective, moreover, in minute dosages. Several species of fishes, known to have been killed by DDT, contained whole body concentrations averaging from 4 to 7 parts per million. Several species of birds that died for the same reason had concentrations averaging 112 parts per million. In

[5] Mercury-containing fungicides used in seed treatment also appear to present serious problems.

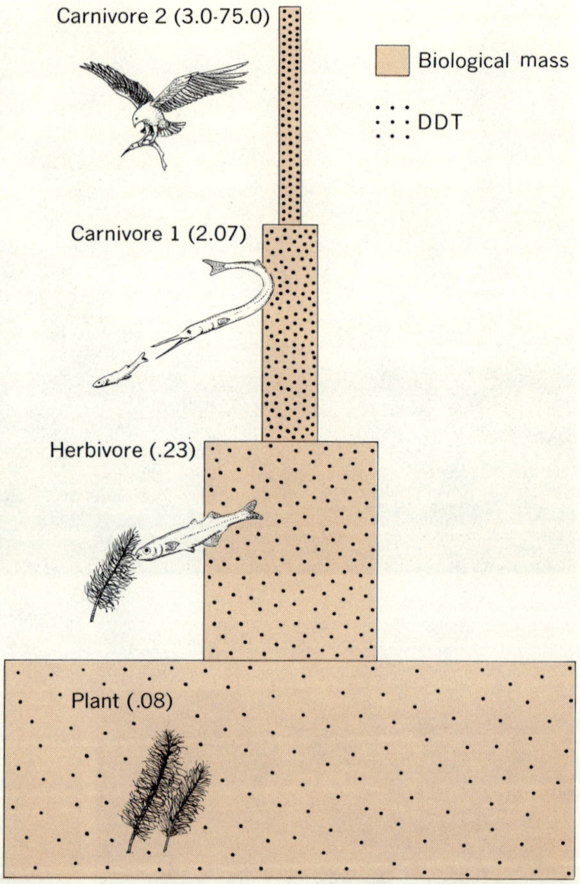

Figure 11-15 A greatly simplified pyramid of mass of a Long Island estuary that has been sprayed with DDT to control mosquitoes. The number in parentheses is parts of DDT per million parts of body weight. (Modified from George M. Woodwell, "Toxic Substances and Ecological Cycles," in *Scientific American*, March 1967)

sublethal doses the pesticide has been shown to affect the reproductive systems in both birds and mammals and to seriously interfere with calcium metabolism in birds. This has resulted in excessive nest mortality in a number of species because of eggshells that have become too thin and fragile to resist breaking. The nesting population of peregrine falcons in the eastern United States has been wiped out and the species faces a similar fate, along with the brown pelican and perhaps others, in California.

DDT is usually applied as a spray (sometimes a dust) and consequently pollutes both soil and air. Commercial applications utilize low-flying aircraft where about 50 percent of the pesticide has been found to remain dispersed in the atmosphere, to be circulated by air currents and deposited on the ground by rainfall. Surface runoff carries the pesticide to rivers, lakes, estuaries, and the open sea where it is further transported by ocean currents and the movement of predatory animals that have preyed on contaminated organisms. It is a persistent substance, resisting biologic degradation to relatively harmless residues for 30 years or more. As it enters food chains its concentration increases with each ascending level in the pyramid of energy since it takes no part in normal metabolism and becomes stored in fatty tissue. A marsh sprayed with DDT to kill mosquitoes showed concentrations in scavenging and carnivorous birds 1000 times higher than that found in the producers of the community (Fig. 11-15). The concentration in carnivorous fishes and birds approached that in animals dying from DDT. Concentrations such as this are known to occur widely throughout this country and the world.

The blubber of Florida dolphins contains 220 parts per million. The edible flesh of California bass has 4 to 138 parts per million. The gulls, penguins, and seals of Antartica contain DDT in their fat stores. And of especial concern is the fact that people all over the world have been found to contain it in their fatty tissues (Fig. 11-16). The total ecological consequences of this situation are only beginning to be understood, but it is probably safe to say that they may prove to be little short of disastrous for many forms of life.

LOCATION	CONCENTRATION (PARTS PER MILLION)
U.S. (Average)	11
Canada	5.3
Alaska (Eskimo)	2.8
England	2.2
West Germany	2.3
France	5.2
Hungary	12.4
Israel	19.2
India	12.8-31.0

Figure 11-16 Known concentrations of DDT residues in the fatty tissues of man in various parts of the world. (From George M. Woodwell, "Toxic Substances and Ecological Cycles," in *Scientific American,* March 1967)

Another group of general pollutants include fission products, like strontium 90 and cesium 137 which are portions of uranium atoms left over after atomic fission, and nonfission isotopes, particularly iodine 131, which have been made radioactive by high-energy particles. These substances have been injected into the atmosphere by nuclear weapons testing and dispersed throughout the world. Since the huge macromolecules of living systems are extremely vulnerable to high-energy radiation damage, this, together with possible additional contamination by radioactive wastes from nuclear reactors, consistutes a clear hazard to man and the ecosystems of the world. Ionizing radiation not only produces certain types of cancer but it is also a powerful mutagenic agent (page 286). In both instances the amount of fission or nonfission product required is infinitesimal. As with DDT the concentration of many of these substances progressively increases along the course of a food chain. Both strontium 90 and cesium 137 have slow decay rates (28 and 30 years, respectively). Strontium 90 is chemically similar to calcium in behavior and becomes incorporated in bone. Cesium 137 is similar to potassium and becomes widely distributed in the body. Although iodine 131 is short-lived (decay rate of about 8 days), it is quickly localized in the thyroid gland. As a result of our 1954 atomic tests in the Pacific, thyroid abnormalities in children who were exposed to fallout on one atoll began to appear about ten years later.

Man's greatest threat from radiation damage, ironic though it may seem, is from an instrument designed to save lives by the use of a high-energy form of ionizing radiation called X rays. According to the United States Public Health Service, 232 million medical X rays and 227 million dental X rays were received by Americans in 1964 alone. It need hardly be said, of course, that use of the X-ray machine is essential in the detection of many human afflictions. It is equally true that it is capable of inflicting serious bodily and genetic damage when its use is not subject to regulation by local, state, or federal public health agencies or when it is administered by persons with insufficient training. Under favorable circumstances the possible benefits of X-ray exposure must always be weighed against the possible harm of radiation damage. In a variety of cases the benefits far outweigh the risks. In a few instances the reverse is true.

Routine biannual or even annual X-ray examinations for dental caries (cavities) seem to be unjustifiable for three reasons:

1. Caries can be adequately diagnosed in the early stages by other techniques which are harmless to the patient.
2. The average dental X ray exposes the patient, according to the United States Public Health Service, to more than 25 times the radiation received by the average chest X ray.
3. Repeated X-radiation in the general area of the mouth, throat, and neck in children, at dosages which had been thought to be safe, has been shown to be positively correlated with the growth of thyroid tumors a number of years later.

Another area in which X-radiation appears to be ill advised is its use in determining the fetal position in pregnant women. Again, a competent obstetrician can usually determine the position of the fetus by techniques that do no harm to either the mother or the child. An impressive body of data is accumulating, on the other hand, indicating that the incidence of leukemia and certain other forms of cancer runs from 10 to 40 percent higher among children who were exposed to X rays before birth.

The importance of genetic damage should also be stressed. X-ray beams, like light-ray beams, cannot be focused exclusively on a single spot. They scatter in all directions as they are deflected by atomic nuclei. X-ray exposures, whether of the mouth or foot, may also bombard the gonads with high-energy radiation. The genetic effect of this bombardment is cumulative. Mutations increase in a linear fashion both with increased X-ray dosage and the repetition of exposures from year to year. The evolutionary consequences of this still lie in the area of speculation and controversy, but most assuredly are a cause for serious concern.

We must guard against the aforementioned hazards of X-radiation or ionizing radiation in any form. However, in those cases regulated by public health agencies and administered by competent and responsible medical authority, when the possible benefits of X-ray exposure are weighed against possible harm from radiation damage, the benefits usually outweigh the risks.

THE POPULATION EXPLOSION

Starvation and disease were neutralized as the last remaining checks against human population increase by the agricultural and industrial revolutions. Decreased mortality coupled with higher birth rates have produced a growth in world population that is characteristic of unrestrained reproductive increase (Fig. 11-17). The spectacular dimensions of this growth are reflected in the fact that the population has become exponentially larger as the time interval required to double the

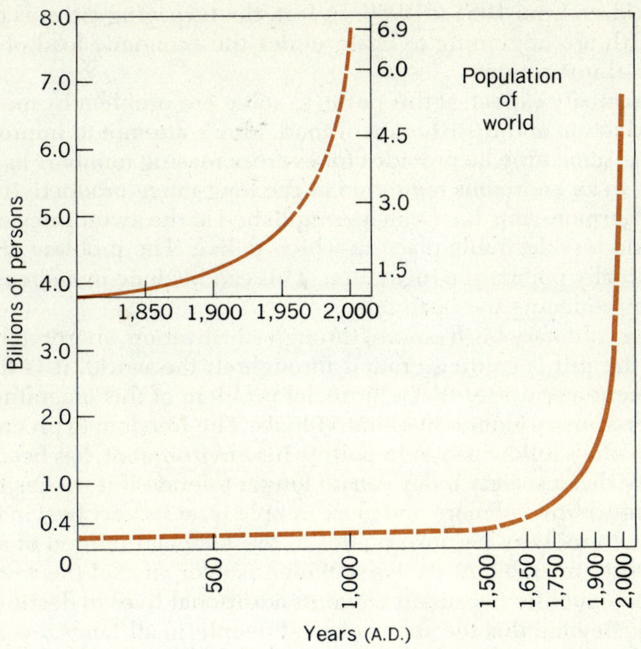

Figure 11-17 Estimated population of the world, A.D. 1 to A.D. 2000.

population has become progressively shorter (Fig. 11-18). At the current rate of increase, a world population of 4 billion in 1975 will have reached 8 billion 35 years later in 2010.

Two-thirds of the world's people are underfed and one-half of these are starving. The great majority of this famine-ridden population lives in technologically underdeveloped countries with low living standards. During the decade 1950 to 1960, 80 percent of the population increase was in Asia, Africa, and Latin America. Nor is malnutrition the only stress visited upon such peoples. The economic burden that high population density imposes on newly emerging nations is stifling. The five-year plan of India required facilities and teachers for 20 million

YEAR (A.D.)	POPULATION (BILLIONS)	NUMBER OF YEARS TO DOUBLE
1	0.25(?)	1650(?)
1650	0.50	200
1850	1.1	80
1930	2.0	45
1975	4.0	35
2010	8.0	?

Figure 11-18 The number of years required to double the population of the world at various times during the Christian Era. [From "World Population Growth." Dorn, *Science*, Vol. 135, Jan. 26, 1962 and United Nations Publications No. ST/SOA/Ser. A/28(1958) and No. ST/SOA/Ser. A/17(1953).]

additional children from 1961 to 1966. In fact, the taxpaying citizens of the richest country on earth are beginning to chafe under the economic load of their educational and social institutions.

We can scarcely expect, at this point, to solve our problem by merely manipulating the production and distribution of food. Man's attempt to improve his living standards at the same time he provides for ever-increasing numbers has resulted, as we have seen, in an enormous *reduction* in the long-range productivity of his environment. Furthermore, this has been accomplished at the awful price of making the world a less and less desirable place in which to live. The problem of overpopulation must be met by population regulation. This can be done by increasing the mortality rate or by reducing the birth rate.

Although voluntary birth control through sterilization, abortion, contraceptive devices, and the pill is gaining ground throughout the world, it is the consensus among competent authorities that a biosocial problem of this magnitude cannot be left to the discretionary judgment of individuals. The freedom to procreate, like the freedom to rob one's fellowman or to pollute his environment, has become a human irresponsibility that a society today can no longer tolerate if it wishes to avoid ruin. The continuing addition of more and more people is, at its very best in the richest of countries, akin to robbery because it adds to the financial burden of all and to the despair of poverty in many. At its worst, which is over most of the world, it is little better than manslaughter because it commits additional lives to destinies of squalor and starvation. Beyond this the sheer mass of people in all lands, even our own, is endangering the environments of life everywhere and threatening to overrun and multilate vast areas of wild lands reserved for public use and enjoyment.

A number of programs and policies have been suggested in recent years aimed at achieving some degree of effective population control. The opponents of many of these measures feel they are much too drastic. The proponents reply that, considering the urgency of the problem, some of them may not be drastic enough. Some critics go further and state that government has no place meddling in such affairs. In answer supporters point out that the government already meddles in these affairs when it subsidizes population increase by income tax deductions and social welfare payments based on family size and when it legislates against abortion. In any event some of the suggestions that have been made are listed for the reader's consideration. They have been extracted from a compilation prepared by Bernard Berelson, President of the Population Council in New York.[6]

I. Extensions of Voluntary Fertility Control

A. Institutionalized maternal care should be made available to the rural areas of developing countries, providing competent family-planning education and services.

B. Restrictions against induced abortion should be liberalized everywhere.

II. Establishment of Involuntary Fertility Control

A. The addition of a chemical fertility-control agent to food or water supplies. This agent would be developed through research and used in amounts adequate to reduce fertility to some predetermined level.

B. The issuance of licenses to have children, given to women in such numbers as to ensure a reproductive rate of one (zero growth).

[6] "Beyond Family Planning" *Science*, **163**, 533–542, February 7, 1969.

C. The requirement that all illegitimate pregnancies be subject to induced abortion.

III. Intensified Educational Campaigns

A. Incorporation of educational materials on population into both elementary and secondary school curriculums.

B. The promotion of enlightened attitudes on population and family-planning through extensive, well-coordinated use of national satellite television systems.

IV. Incentive Programs

A. A payment to couples who effectively practice contraception or who accept sterilization.

B. A bonus for each 12-month period in which no child is born, with a balloon payment for every 5-year interval.

V. Social Service Systems That Discourage Child-Bearing

A. Levy a tax on all births beyond a certain number and give tax relief to the parents of small families and to the unmarried.

B. Limit governmentally sponsored medical care, housing, scholarships, loans, etc., to small families.

C. Special pension payments to poor parents of small families during their old age.

VI. Broad Changes in Social and Economic Institutions

A. Increase the minimal age at marriage by legislation, by imposing a costly marriage license fee or by the payment of bonuses to couples who delay the event.

B. Promote measures that improve the status of women and encourage them to obtain employment as an alternative or supplementary role to marriage.

VII. Political Methods and Organizations

A. U.S. restriction of food aid to only those governments sponsoring population control, and the application of political pressure to those governments or religious institutions which impede progress in this area.

B. The establishment of worldwide national agencies together with one international body empowered to take necessary steps to regulate population size, and whose ultimate goal is the promotion of zero growth.

VIII. Augmented Research Efforts to Implement All of the Above Areas

It is probable that whatever degree of control man can bring to bear on reducing his numbers will come from a combination of medical, legal, economic, and social efforts. As Berelson has stated, the value of any particular measure or technique will be predicated on whether it is "scientifically available, politically acceptable, administratively feasible, economically justifiable, and morally tolerated." And all of these in turn hinge on how well mankind perceives and appreciates the enormous gravity of his circumstances.

LOOKING TOWARD LONG-RANGE SURVIVAL

In Chapter 10 we discussed some of the ecological consequences of succession (pages 363–365). It will be helpful to review and elaborate on these at this point because they are most germane to man's continued survival on this planet. It is the early or young stages of ecologic succession that agriculture emulates. A wheat crop or pastured cattle utilizes species that grow rapidly and have high reproductive potentials. The aim is to achieve an economic harvest of maximum productivity with as little standing crop (breeding stock, crop plant residues, and so on) remaining as possible. So it is with young nature in the early stages of succession. A relatively few plant species characterized by high fecundity and vigorous growth sustain a rate of photosynthesis that exceeds by a considerable margin the respiratory needs of the community. The biosynthesis of energy-rich organic compounds surpasses the community's need for them. The result is an increase in biological mass. Young nature, too, is productive. But it is also, like its agricultural counterpart, unstable. Its food chains are simple, symbiotic relationships are minimal, and its primary producers cannot withstand competition. Like weeds overwhelming unattended cropland, young nature passes through successional stages to the relatively stable climax of mature nature. As it does so the total amount of biological mass grows. Among primary producers, species featuring rapid growth and reproduction are exchanged for those with slow growth and high competitiveness. Food chains become vast webs of rich symbiotic interaction. The community reaches a stable state of dynamic equilibrium imposed through exquisite biologic controls. The ratio of autotrophic biosynthesis to community respiration approaches 1 : 1. An optimal biologic mass is reached that is supported by minimal energy input.

A study of a mature forest showed, as we have mentioned, that less than 10 percent of net photosynthetic production is grazed in the living state. More than 90 percent of all energy flow goes into the maintenance of community stability. This flow passes from community detritus through a maze of food chain networks involving decomposition, scavenging, predation, parasitism, commensalism, and mutualism, and includes a great variety and abundance of microorganisms, plants, and animals. It is in these symbiotic interactions that nutrients such as calcium, phosphorus, and nitrogen are cycled and conserved by the community. The accumulation of a rich litter serves to hold and store water and the entire community protects its soil and buffers its climatic extremes.

Mature nature, unlike young nature, is protective and stable. These are the very qualities which man has been destroying in his exploitations and for which he now has such a desperate need. Man requires the productivity and growth that are exemplified in young nature and agriculture. But he also needs pure, that is, *unproductive* water. He must breathe *unproductive*, that is, uncontaminated air. He must produce but he must also protect. His long-range wants demand an atmosphere in which CO_2 and O_2 are in balance and in which climatic extremes are held in check. He will realize these wants only when great expanses of natural vegetation and perhaps all of the open seas are held as protected reserves to continue their function in regulating the terrestrial environments of life. In so doing he will reinforce the recreational and esthetic dimensions of human experience by which man also lives. Our naive attempts to conquer nature have overlooked the rather obvious fact that we are an integral part of nature. *To control her we must first learn how to control ourselves.*

In this regard the positive recommendations of Eugene P. Odum are most welcome and timely. Pointing to the parallels between ecologic succession and the development of human society, he states: "In the pioneer society, as in the pioneer ecosystem, high birth rates, rapid growth, high economic profits, and exploitation of accessible and unused resources are advantageous, but, as the saturation level is approached, these drives must be shifted to considerations of symbiosis (that is, "civil rights," "law and order," "education," and "culture"), birth control, and the recycling of resources. A balance between youth and maturity in the socio-environmental system is therefore the really basic goal that must be achieved. . . ."[7]

The administration of land areas according to a compromise system of multiple use (such as national forests that permit limited use by some combination of grazing, lumbering, or mining together with recreation, wildlife conservation, and watershed protection) is often a successful practice. It is not applicable to all environments, however, and has its limitations in any environment since one use commonly interferes with others. In view of this, Odum recommends that the entire landscape be zoned in accordance with restrictions establishing and delimiting four kinds of ecosystems that would serve one another: (1) the urban-industrial system commingled with (2) a productive system of intensive cropping, (3) a compromise, multiple-use system, and (4) a protective mature system of wilderness areas (Fig. 11-19). A largescale program of ecological study would be set up to determine what proportions these ecosystems should bear to one another in order to end man's conflict with nature. The United States with a large share of its total area held in public domain (national forests, parks, and so on) and states like California and New Jersey that recently set out to increase their holdings in protected lands would provide fertile ecopolitical opportunities for such an endeavor and, indeed, constitute a beginning.

Resource utilization under such a program would be contingent upon a policy that man participates in biologic communities, he does not conquer them. Working in collaboration with natural ecosystems and patterning our food chains after those in mature nature can be highly rewarding. Opportunities abound throughout the world to *share* in the productivity of young nature. Communities that have been kept youthful because of some more or less regular disturbance lend themselves beautifully to compromise systems of use. Intertidal zones such as estuaries and shallow bays are excellent examples of this in which nutrient cycling brought about by alternating tides, together with the natural fertilization that comes from rich terrestrial runoff, produce communities capable of exceptional yields. Japan uses such areas to raise oysters (which, as filter-feeders, live on microorganisms). This activity

[7] "The Strategy of Ecosystem Development," *Science*, **164**, 262–270, April 18, 1969.

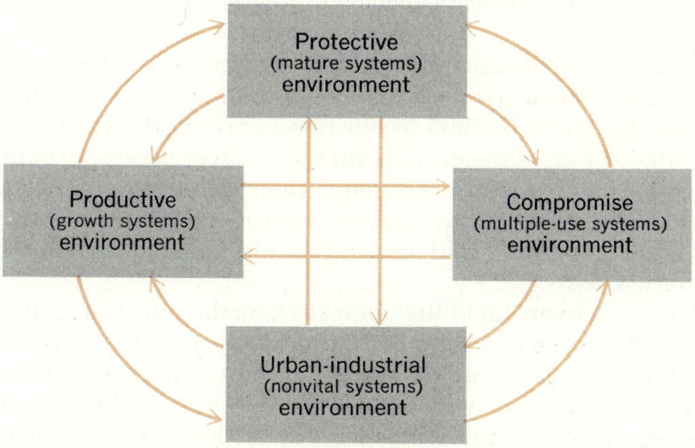

Figure 11-19 Basic kinds of ecosystem development required by man. (Modified from E. P. Odum, *Science*, Vol. 164, p. 269, (1969). Copyright 1969 by the Association for the Advancement of Science)

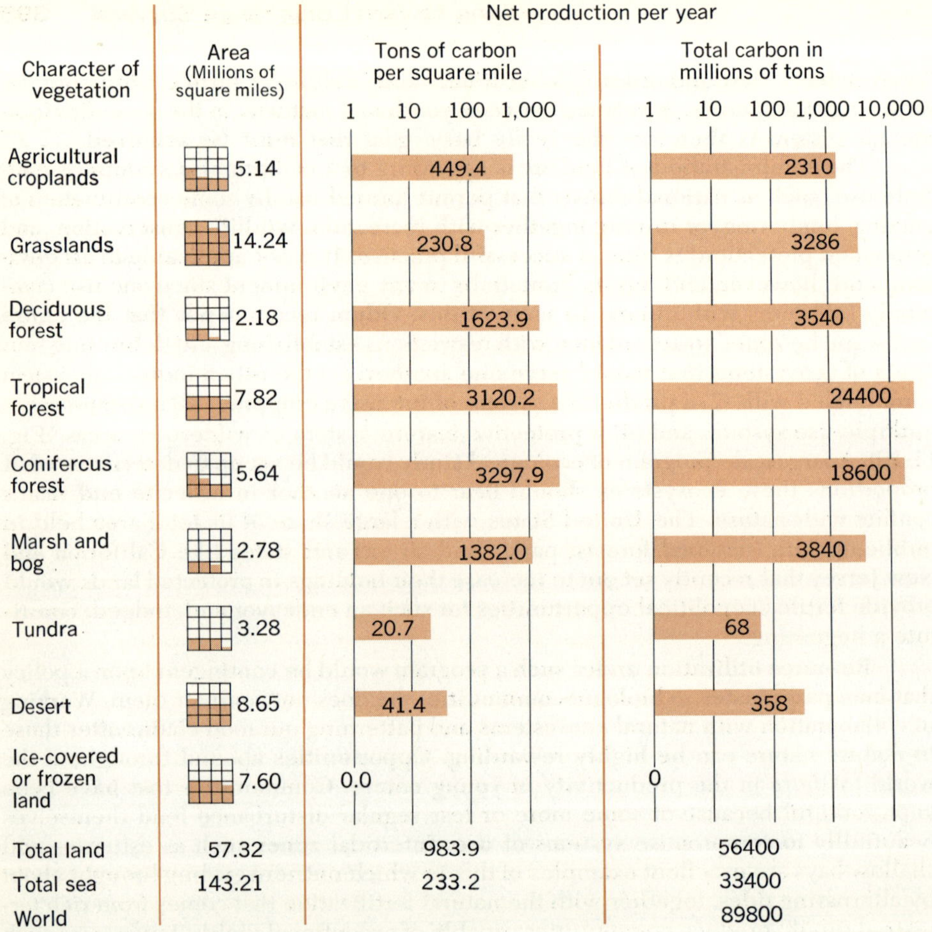

Figure 11-20 Total net production of organic matter per year by the vegetation of the world in terms of amount of carbon incorporated into organic compounds. (Modified from Edward S. Deevey, Jr., "The Human Population," in *Scientific American*, 1960)

neither mars the natural beauty nor otherwise adversely affects the ecosystem. Using about 25 percent of a given estuary for cultivation, the oyster harvest averages 15¾ tons per acre per year. In 1965 oyster production in the Hiroshima area alone amounted to 240,000 tons of meat. In Spain the per acre harvest of cultured mussels (also edible, filter-feeding mollusks) more than quintuples the Japanese oyster yield. In this country oyster culture could be practiced along the Atlantic and Gulf Coasts and mussel culture would be practicable along the shores of either the Pacific or Atlantic Coasts.

Any intensive harvesting of the open seas, on the other hand, does not appear to be advisable. The ocean expanses of the world are suspected of playing the major role in maintaining the atmospheric oxygen supply so necessary to terrestrial life. Intense cultivation in the open seas could have detrimental effects on the earth's reservoir of oxygen. Since almost 63 percent of the world's total production of organic matter comes from the land anyway (Fig. 11-20), it would seem prudent to reserve the open seas as part of the protective environment.

Our productive systems can be greatly improved by an agricultural economy that places increasing emphasis on detritus food chains. The standard policy of planting large areas in a single crop that grows rapidly and is both palatable and nutritious results in attracting great multitudes of pests. Protecting this investment by the widespread use of pesticides has not only not solved the problem but has resulted in contaminating ourselves. As Odum suggests, "Why not also practice the reverse strategy—that is, select plants which are essentially unpalatable, or which produce their own systemic insecticides while they are growing, and then convert the net production into edible products by microbial and chemical enrichment in food factories? We could then devote our biochemical genius to the enrichment process instead of fouling up our living space with chemical poisons!" For example, yeast not only has a high protein content but is also rich in B vitamins, making a good food base or food supplement.

It has been conservatively estimated that a fermenting tank seeded with 500 pounds of yeast and a nutrient medium composed of acid-digested plant fiber and ammonium salts will produce a ton of yeast per day (equivalent in protein content to 20 pigs). Certain kinds of algae are also highly nutritious, containing large amounts of protein, lipids, and vitamins (especially A and C). Some of these have been grown in sewage disposal ponds where they provide the oxygen necessary for sewage decomposition. Properly prepared and cooked, these algae are edible and could be used for consumption by either humans or livestock. Perhaps the most dramatic demonstration of the prolific harvests possible in a detritus agriculture (or, more properly, aquaculture) is found in the fish culture program of Indonesia. Bottom-feeding fish known as carp are confined in bamboo cages placed in fast-flowing streams of high sewage content. The fish, feeding on the dense carpet of worms and larval insects which covers the stream bed, grow rapidly and yield an annual harvest in West Java of more than 220 tons per acre. A somewhat similar technique is in operation in Munich, Germany, utilizing sewage ponds.

The corrective measures we have been discussing are rational approaches to a solution of the major problems that beset mankind today. They and others like them might solve these crises. They *will* do so only if they are effectively implemented and wholeheartedly supported everywhere. Herein lies the real issue. Mankind's biological crisis today is rooted in morality not technology. Human culture and codes of moral conduct have served man, the colonizer, exceptionally well. We are the most powerful species of all time and certainly the most unique. Man, however, is no longer a pioneer in new lands on this planet. He cannot continue to assume his make-believe role of nature's "conqueror" if he wishes to participate in her conquest of energy much longer. We have finally come to rest as settled residents. The freedom of abandon that is so admirable in exuberant youth must be replaced by the responsible freedom of maturity. New codes of moral conduct are needed befitting our new station in life so that man may live in symbiotic peace with his world. Human society, as the generator and guardian of all human rights and freedoms, must be protected from promiscuous acts of free will which weaken social structures and threaten the survival of all. The liberty to procreate at will in young nature is the license to destroy in mature nature. It seems incomprehensible that a species whose capabilities fill a spectrum from the genius of gods to the folly of fools should end its reign by the ultimate imbecility of seeking life in space as life is wantonly destroyed in a world of its birth through moral immaturity.

Man is a unique creature because he alone understands that he and other living things evolved. This knowledge can put him in the enviable position of being able to control his future once he learns how to control himself. If the value system governing this control is based, like nature's, on long-range survival, man's future can be secure and bright. To those who say that value judgments are nebulous evaluations at best, we can point to natural selection, which has made billions of them accurately, although amorally. The intelligent and sympathetic self-direction of human destiny could do the same, morally and with humanity.

In drawing this book to a close, we have come full circle. We began with man and we end with man. It perhaps can be said that everything that has ever been written is, in the last analysis, a study in man himself. This is as it should be. Nothing is so worthy or so in need of study.

QUESTIONS

1. Substantiate the following statement: "Historically man (*Homo sapiens*) has played the role of a pioneering species in young nature rather than that of a collaborating species in mature nature."
2. Discuss the extent to which man's exploitation of biologic communities has resulted in a decline in environmental productivity through
   (a) Soil erosion
   (b) Nitrate contamination of soil and water
   (c) Air pollution
   (d) General pollution by pesticides such as the herbicide 2,4,5-T, the insecticide DDT, and other chlorinated hydrocarbons
3. Describe the nature of the growth in the human population over the past 2000 years.
4. What are some of the major consequences of overpopulation?
5. What kinds of measures are being suggested today to achieve some degree of effective population control?
6. Discuss the Odum proposal of ecosystem development as a means toward long-range human survival.
7. Describe some of the ways in which man, acting in collaboration with natural ecosystems, shares in the bountiful productivity of young nature.
8. What are some of the ways in which man, patterning his agricultural economy after the detritus food chains of mature nature, can achieve high productivity without endangering the environment?
9. Discuss the meaning, implications, and veracity of the following statement: "Human culture and codes of moral conduct have served man the colonizer exceptionally well. But man is no longer a pioneer in new lands on this planet; he is a settled resident. New codes of moral conduct are needed befitting our new station in life so that man may live in symbiotic peace with his world."
10. What basis is there for the belief that once man learns how to control himself his prospects for long-range survival will be secure and bright?
11. Relative to the self-direction of human destiny, what are some of the ways in which computer analyses might be of great assistance in decision and prediction making?

Suggested Readings

Bardach, J. E. "Aquaculture," *Science*, **161,** September 13, 1968.
Bates, M. *Man In Nature.* Englewood Cliffs, N.J.: Prentice-Hall, 1964.
Bates, M. *The Forest and the Sea.* Vintage Books. New York: Random House, 1965.
Berelson, B. "Beyond Family Planning," *Science*, **163,** February 7, 1969.
Bresler, J. B. (ed.). *Environments of Man.* Life Sciences Series. Reading, Mass.: Addison-Wesley, 1968.

Carson, R. L. *Silent Spring*. Boston: Houghton Mifflin, 1962.
Carter, L. J. "Galveston Bay: Test Case of an Estuary in Crisis," *Science*, **167**, February 20, 1970.
Clark, J. R. "Thermal Pollution and Aquatic Life," *Scientific American*, March 1969.
Commoner, B. *Science and Survival*. New York: Viking, 1966.
Cox, G. W. (ed.). *Readings in Conservation Ecology*. Educational Division Meredith Corporation. New York: Appleton-Century-Crofts, 1969.
Davis, K. "Population Policy: Will Current Programs Succeed?" *Science*, **158**, November 10, 1967.
Deevey, E. S., Jr. "Human Population," *Scientific American*, September 1960.
Dobzhansky, T. "The Present Evolution of Man," *Scientific American*, September 1960.
Dobzhansky, T. *Mankind Evolving*. New Haven, Conn.: Yale University Press, 1962.
Dobzhansky, T. "Changing Man," *Science*, **155**, January 27, 1967.
Dorn, H. F. "World Population Growth: An International Dilemma," *Science*, **135**, January 26, 1962.
Ehrenfeld, D. W. *Biological Conservation*. Modern Biology Series. New York: Holt, Rinehart and Winston, 1970.
Enke, S. "Birth Control for Economic Development," *Science*, **164**, May 16, 1969.
Hardin, G. "The Tragedy of the Commons," *Science*, **162**, December 13, 1968.
Hardin, G. (ed.). *Population, Evolution and Birth Control: A Collage of Controversial Ideas*. San Francisco: Freeman, 1969.
Hoagland, H. "Science and the New Humanism," *Science*, **143**, January 10, 1964.
Holt, S. J. "The Food Resources of the Ocean," *Scientific American*, September 1969.
Johnson, C. E. (ed.). *Social and Natural Biology*. Princeton, N.J.: Van Nostrand, 1968.
Lerner, I. M. *Heredity, Evolution and Society*. San Francisco: Freeman, 1968.
Love, R. M. "The Rangelands of the Western U.S.," *Scientific American*, February 1970.
Odum, E. P. "The Strategy of Ecosystem Development," *Science*, **164**, April 18, 1969.
Peakall, D. B. "Pesticides and the Reproduction of Birds," *Scientific American*, April 1970.
Seaborg, G. T. "Science and the Humanities: A New Level of Symbiosis," *Science*, **144**, May 5, 1964.
Shepard, P., and D. McKinley (eds.). *The Subversive Science: Essays Toward an Ecology of Man*. Boston: Houghton Mifflin, 1969.
Simpson, G. G. *The Meaning of Evolution*. New Haven, Conn.: Yale University Press, 1967.
Stycos, J. M. "The Outlook For World Population," *Science*, **146**, December 11, 1964.
Wallace, B., and T. Dobzhansky. *Radiation, Genes and Man*. New York: Holt, Rinehart and Winston, 1959.
Woodwell, G. M. "Toxic Substances and Ecological Cycles," *Scientific American*, March 1967.

Carson, R. *Silent Spring.* Boston: Houghton Mifflin, 1962.
Carter, L. J. "[illegible]." *Science,* 167, [illegible] 1970.
Clark, J. R. "Thermal Pollution and Aquatic Life." *Scientific American,* March 1969.
Commoner, B. *Science and Survival.* New York: Viking, 1966.
Cox, G. W., ed. *Readings in Conservation Ecology.* Educational Division, Meredith Corporation. New York: Appleton-Century-Crofts, 1969.
Davis, K. "Population Policy: Will Current Programs Succeed?" *Science,* 158, November 10, 1967.
Deevey, E. S., Jr. "The Human Population." *Scientific American,* September 1960.
Dobzhansky, T. "The Present Evolution of Man." *Scientific American,* September 1960.
Dobzhansky, T. *Mankind Evolving.* New Haven, Conn.: Yale University Press, 1962.
Dobzhansky, T. "Changing Man." *Science,* 155, January 27, 1967.
Dorn, H. F. "World Population Growth: An International Dilemma." *Science,* 135, January 26, 1962.
Ehrenfeld, D. W. *Biological Conservation.* Modern Biology Series. New York: Holt, Rinehart and Winston, 1970.
Enke, S. "Birth Control for Economic Development." *Science,* 164, May 16, 1969.
Hardin, G. "The Tragedy of the Commons." *Science,* 162, December 13, 1968.
Hardin, G., ed. *Population, Evolution, and Birth Control: A Collage of Controversial Ideas.* San Francisco: Freeman, 1969.
Hoagland, H. "Science and the New Humanism." *Science,* 143, January 10, 1964.
Holt, S. J. "The Food Resources of the Ocean." *Scientific American,* September 1969.
Johnson, C. E., ed. *Social and Natural Biology.* Princeton, N.J.: Van Nostrand, 1968.
Lerner, I. M. *Heredity, Evolution, and Society.* San Francisco: Freeman, 1968.
Love, R. M. "The Rangelands of the Western U.S." *Scientific American,* February 1970.
Odum, E. P. "The Strategy of Ecosystem Development." *Science,* 164, April 18, 1969.
Peakall, D. B. "Pesticides and the Reproduction of Birds." *Scientific American,* April 1970.
Roberts, W. O. "Science and the Humanities: A New Level of Symbiosis." *Science,* 144, May 8, 1964.
Shepard, P., and D. McKinley, eds. *The Subversive Science: Essays Toward an Ecology of Man.* Boston: Houghton Mifflin, 1969.
Simpson, G. G. *The Meaning of Evolution.* New Haven, Conn.: Yale University Press, 1967.
Stycos, J. M. "The Outlook for World Population." *Science,* 146, December 11, 1964.
Wallace, B., and T. Dobzhansky. *Radiation, Genes, and Man.* New York: Holt, Rinehart and Winston, 1959.
Woodwell, G. M. "Toxic Substances and Ecological Cycles." *Scientific American,* March 1967.

APPENDIX: An Illustrated Classification of Organisms*

Kingdom *Monera* (Representatives unicellular, some colonial; microscopic; cells lack membrane-bound nuclei and such organelles as plastids, mitochondria, and Golgi apparatus; heterotrophic or autotrophic)

Phylum *Schizophyta* Bacteria (heterotrophic or autotrophic)

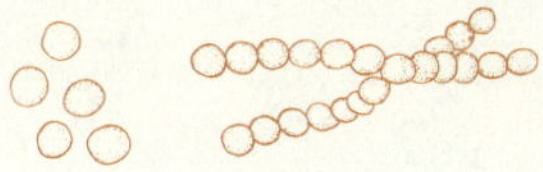

Coccus

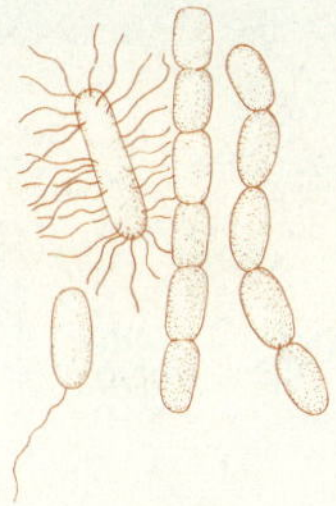

Bacillus

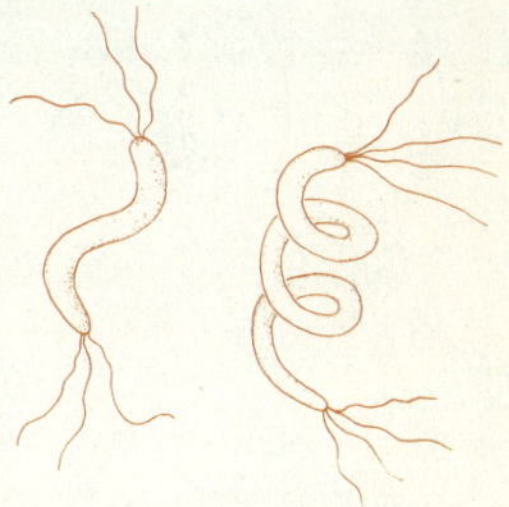

Spirillum

Phylum *Cyanophyta* Blue-green algae (autotrophic)

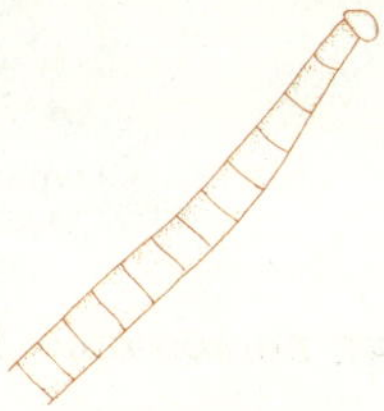

Oscillatoria

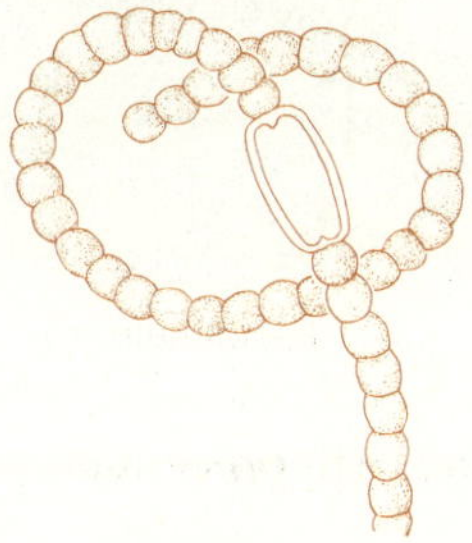

Nostoc

Kingdom *Protista* (Representatives unicellular to multicellular; microscopic to large; if multicellular, relatively little specialization of internal structure; heterotrophic or autotrophic)

* A few phyla of animals that are relatively unimportant in the world fauna today have been omitted. These include the Ctenophora, Mesozoa, Nemertina, Acanthocephala, Phoronidea, and Chaetognatha.

Phylum *Mastigophora* Flagellates (unicellular or colonial; microscopic; heterotrophic or autotrophic)

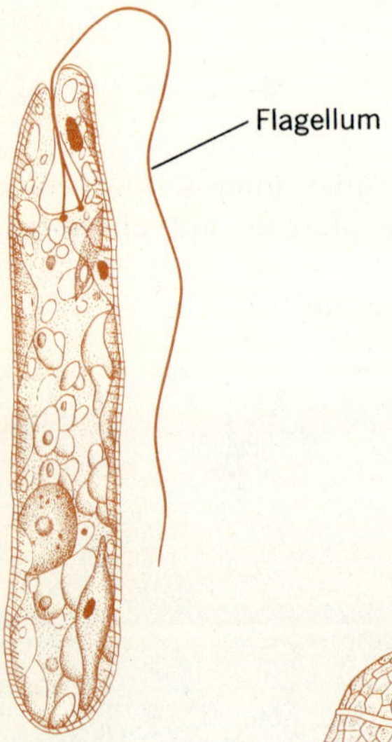

Euglena

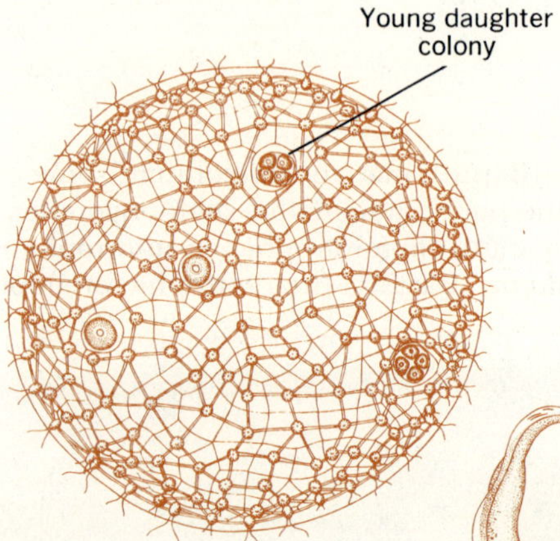

Volvox

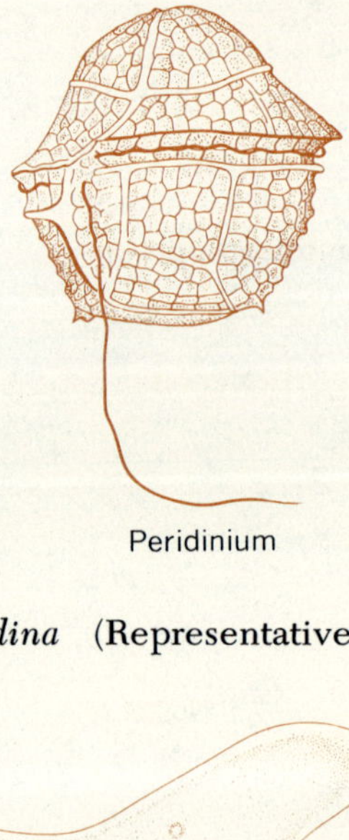
Peridinium

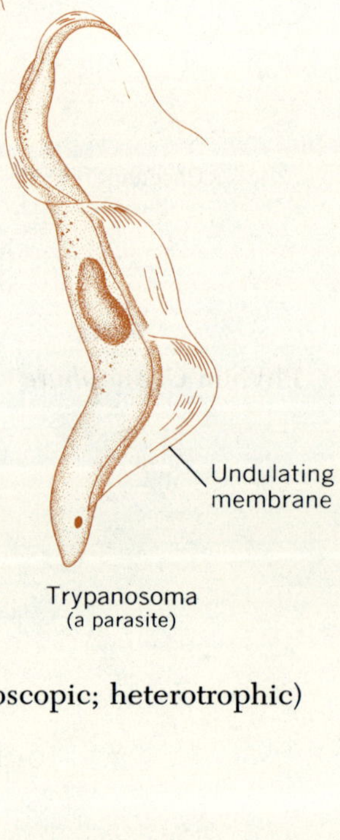

Trypanosoma
(a parasite)

Phylum *Sarcodina* (Representatives unicellular; microscopic; heterotrophic)

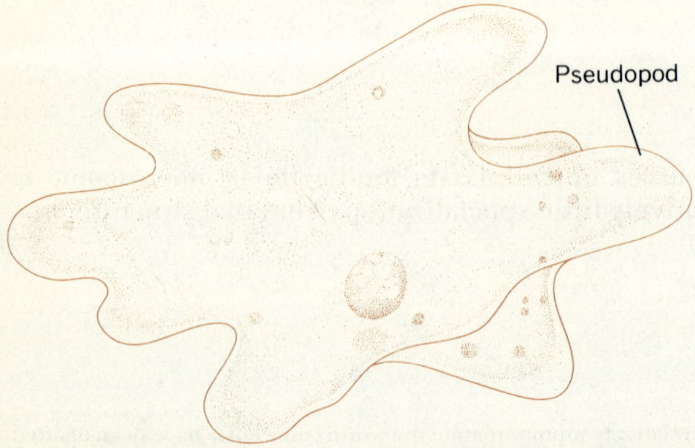

Ameba

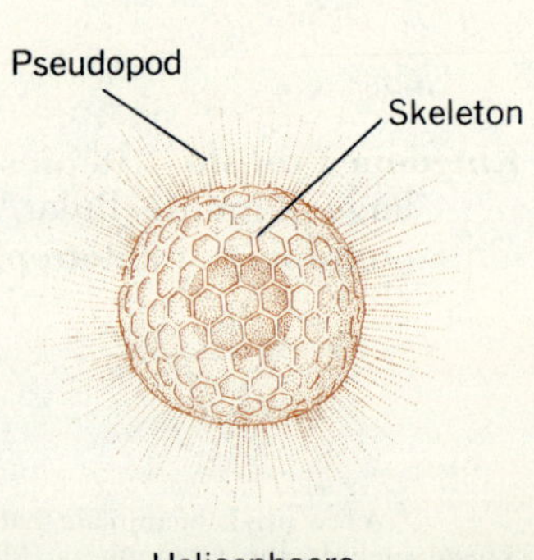

Heliosphaera

Phylum *Ciliophora* Ciliates (Representatives unicellular; microscopic; heterotrophic)

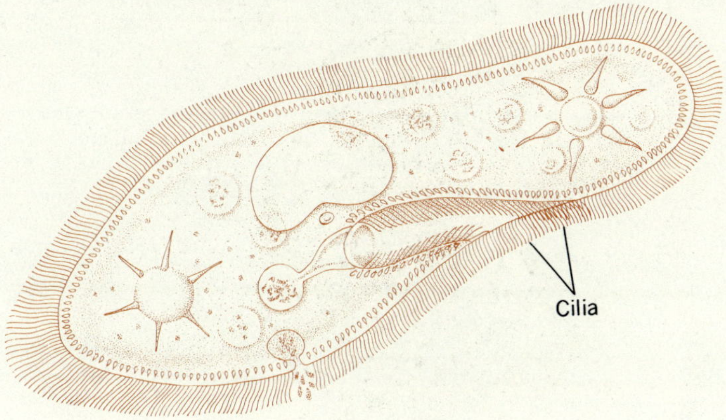

Paramecium

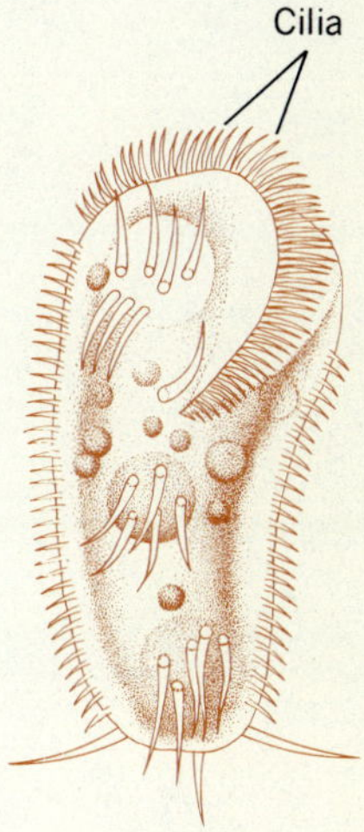

Stylonchia

Phylum *Sporozoa* (Representatives unicellular; microscopic; parasitic)

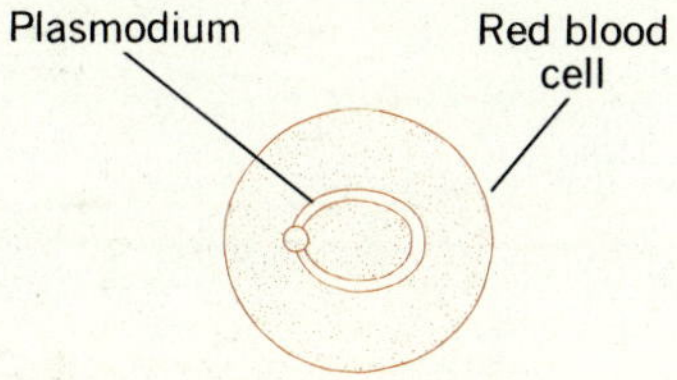

Phylum *Myxomycetes* Slime molds (Unicellular or cellular aggregates; microscopic to small; heterotrophic)

Dictyostelium

Phylum *Chlorophyta* Green algae (Unicellular to multicellular; microscopic to small; mostly autotrophic)

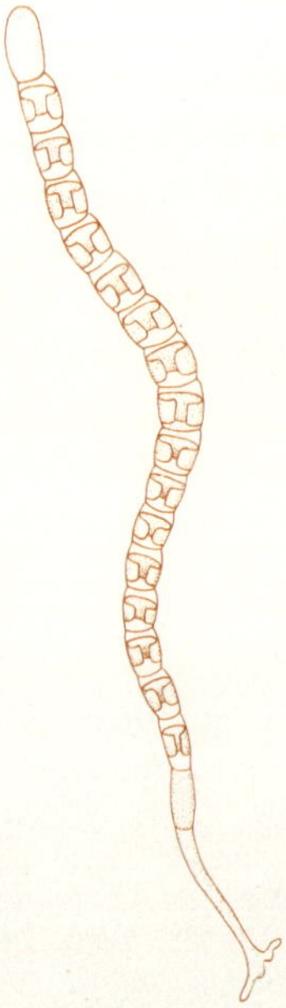

Ulothrix

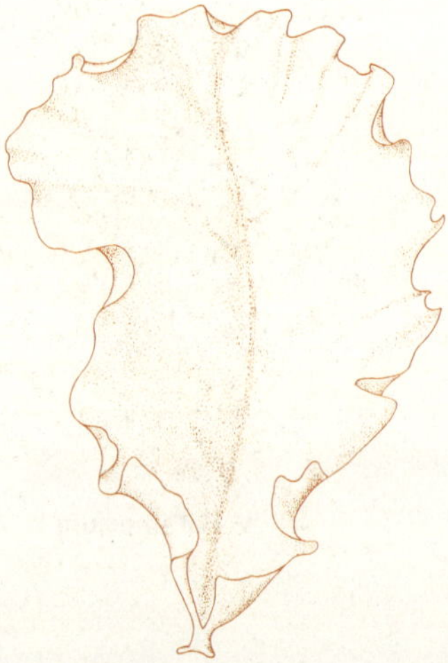

Ulva

Phylum *Chrysophyta* Yellow-green algae (Unicellular or colonial; microscopic; autotrophic)

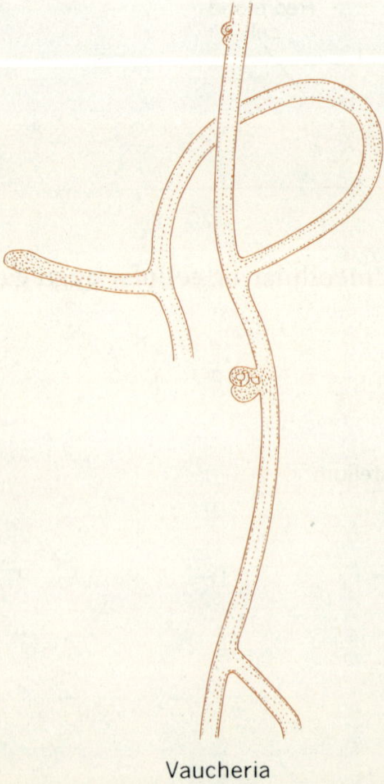

Vaucheria

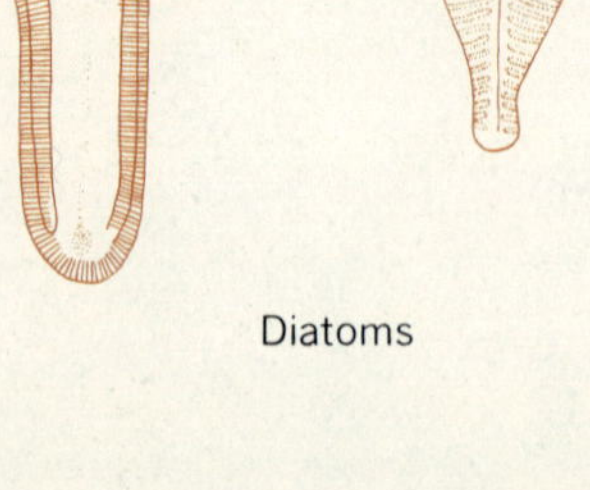

Diatoms

Phylum *Phaeophyta* Brown algae (kelp) (Multicellular; small to large; autotrophic)

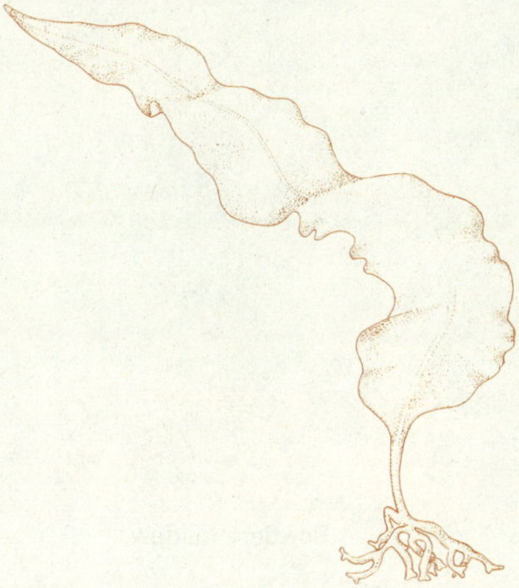

Laminaria

Phylum *Rhodophyta* Red algae (Multicellular; usually small; autotrophic)

Gigartina

Phylum *Mycophyta* Fungi (Unicellular or multicellular; microscopic to small; heterotrophic)

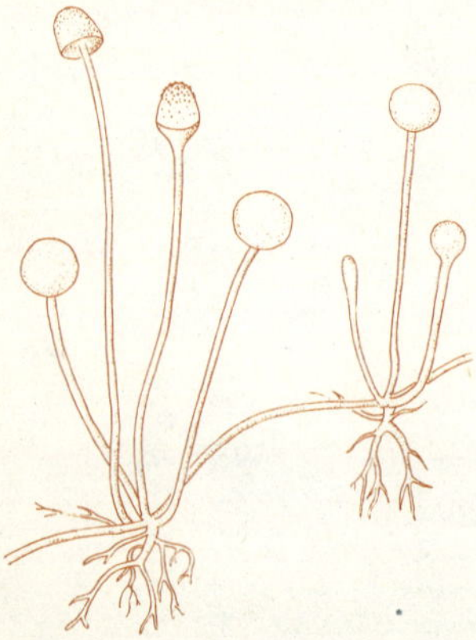

Rhizopus
(bread mold)

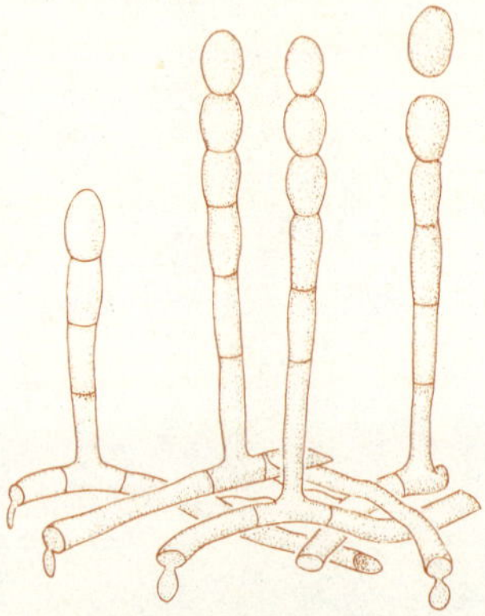

Powdery mildew

Amanita
(mushroom)

Kingdom of *Plants* (Multicellular, largely autotrophic organisms with considerable specialization of internal structure)

Phylum *Bryophyta* Liverworts and mosses (small autotrophs)

Marchantia
(liverwort)

Polytrichium
(moss)

Phylum *Tracheophyta* Vascular plants (Small to large plants with specialized supporting and conducting tissues)

Subphylum *Lycopsida* Lycopods (Small autotrophs. Carboniferous species reached the size of large trees)

Lycopodium

Subphylum *Sphenopsida* Scouring rushes or horse-tails (Small autotrophs. Carboniferous species reached the size of trees)

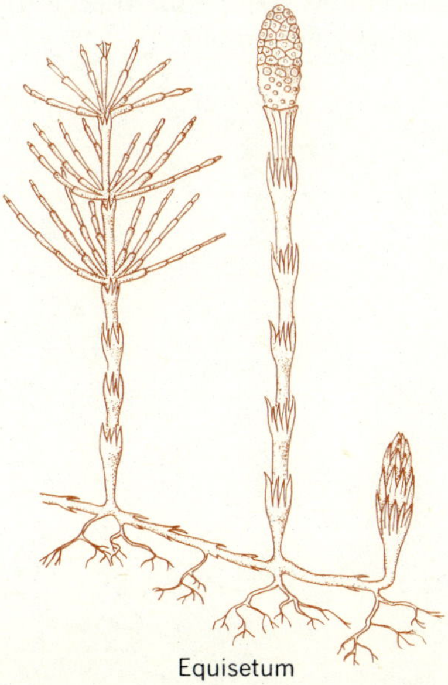

Equisetum

Subphylum *Pteropsida* Ferns and seed plants

Class *Filicineae* Ferns (Small to tree-sized plants)

Polypodium

Class *Gymnospermae* Gymnosperms (Usually large trees)

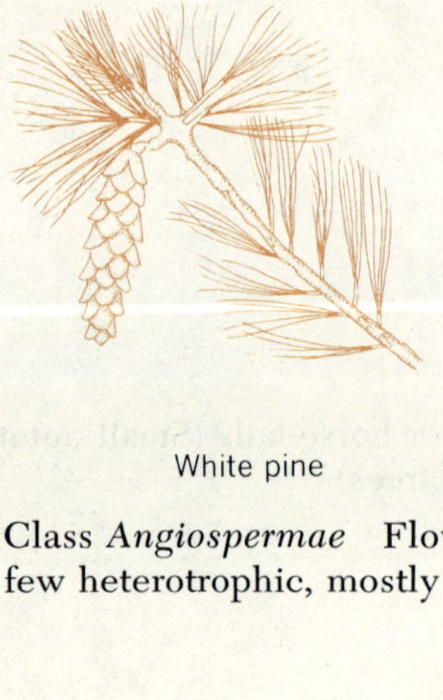

White pine

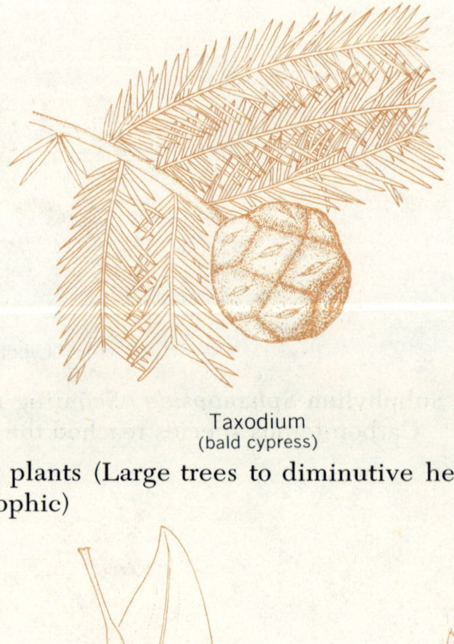

Taxodium
(bald cypress)

Class *Angiospermae* Flowering plants (Large trees to diminutive herbs; a few heterotrophic, mostly autotrophic)

Sugar maple

Clintonia
(a lily)

Wheat
(a grass)

Kingdom of *Animals* (Multicellular, heterotrophic organisms with considerable specialization of internal structure)

Phylum *Porifera* Sponges (Small animals)

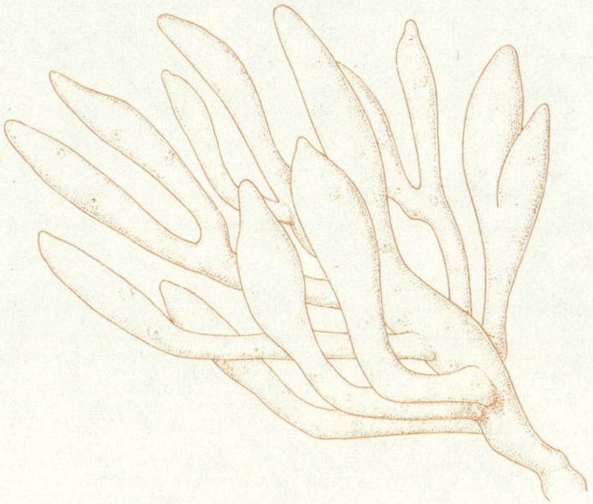

Haliclona

Scypha

Phylum *Coelenterata* Coelenterates (Small animals, although colonial forms such as the corals may form huge masses)

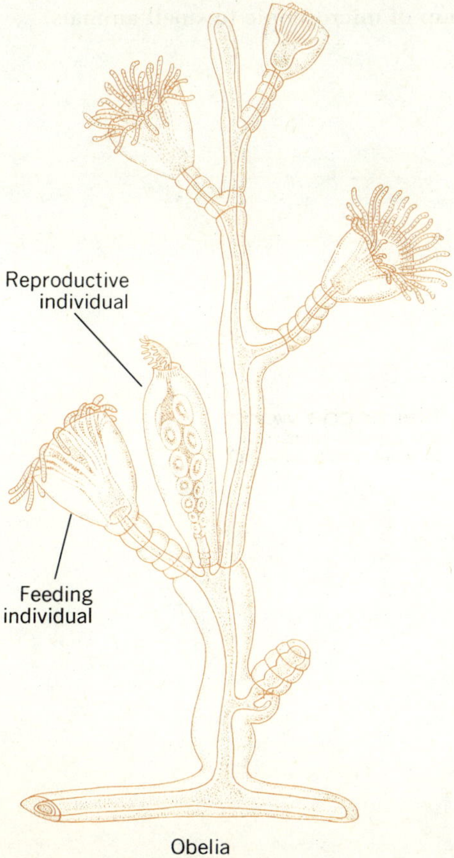

Obelia
(a colonial form)

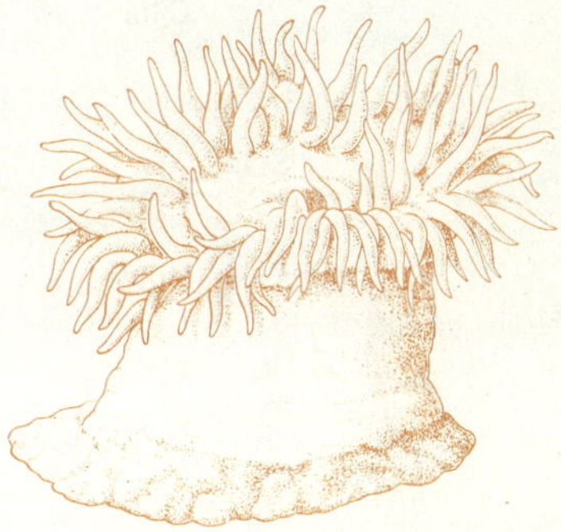

Sea anemone

Jellyfish

Phylum *Platyhelminthes* Flatworms (Microscopic to small worms, although some tapeworms are over 20 feet in length; a few free-living species, mostly parasitic)

Dugesia
(planaria)

Sheep liver fluke

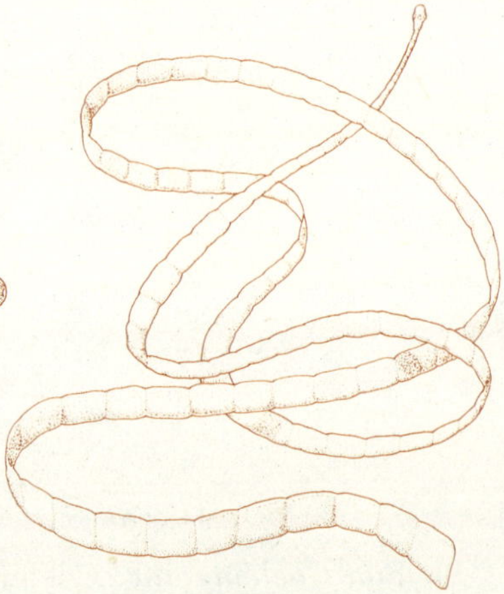

Beef tapeworm

Phylum *Aschelminthes* (A diverse group of microscopic to small animals)

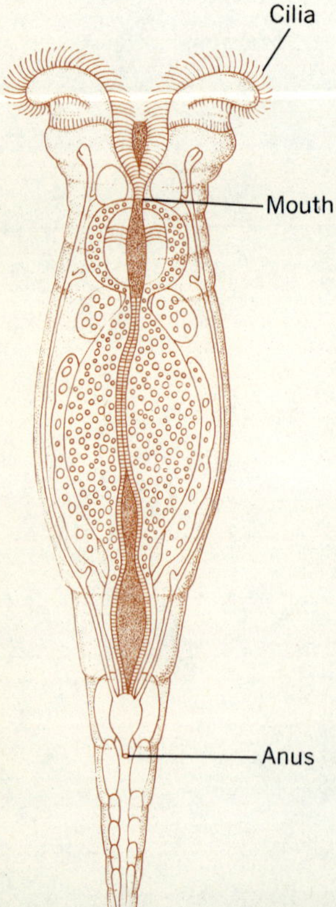

A rotifer
(microscopic)

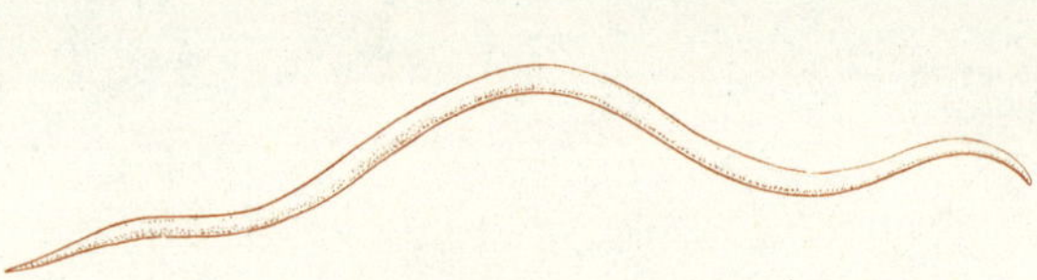

Nematode worm
(as a group microscopic to small; mostly parasitic)

Phylum *Bryozoa* Moss animals (Microscopic to very small colonial organisms)

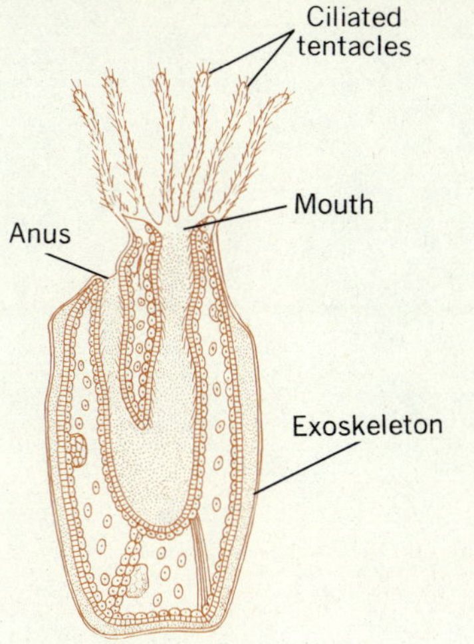

Structure of individual bryozoam organism

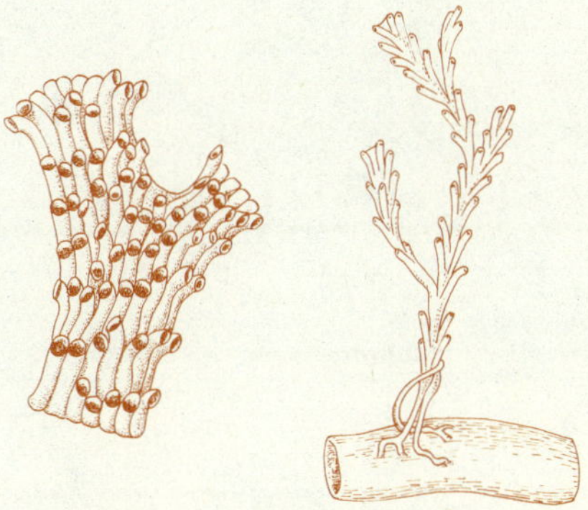

Idmonea Crisulipora

Two forms of branched bryozoam colonies

Phylum *Brachiopoda* Lampshells (Small animals which resemble clams only superficially)

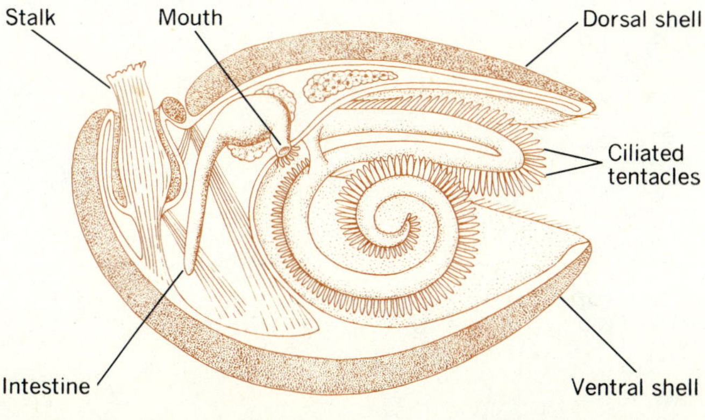

Phylum *Mollusca* Mollusks

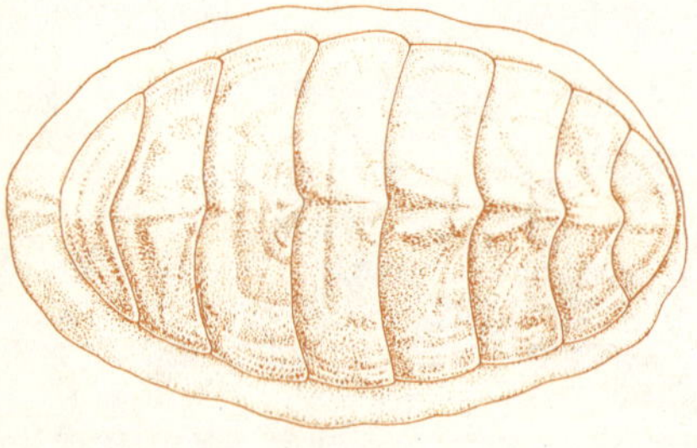
Chiton

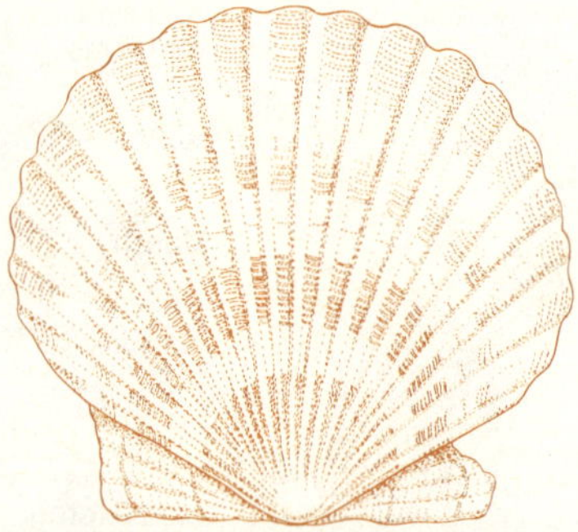
Scallop

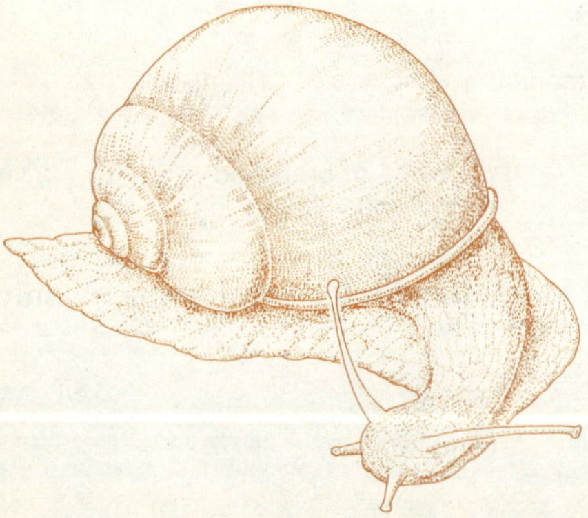
Snail

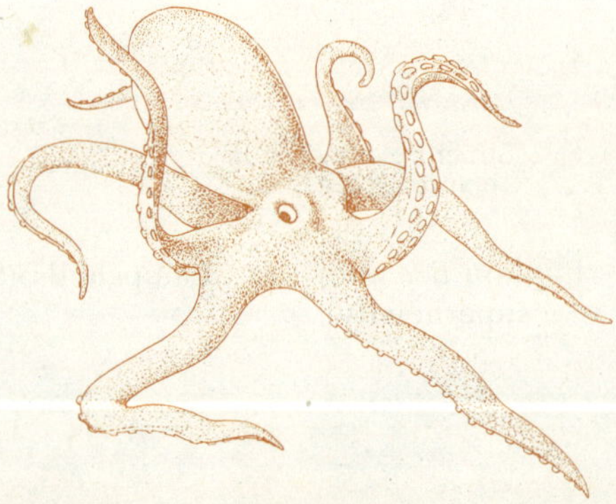
Octopus

Phylum *Annelida* The segmented worms

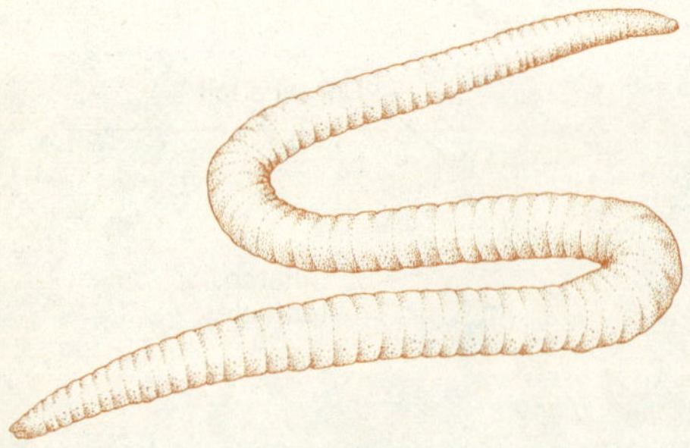
Earthworm

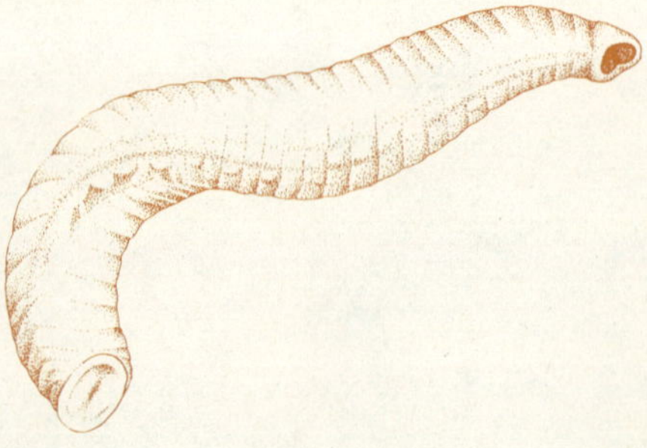
Leech

Phylum *Arthropoda* The joint-legged animals

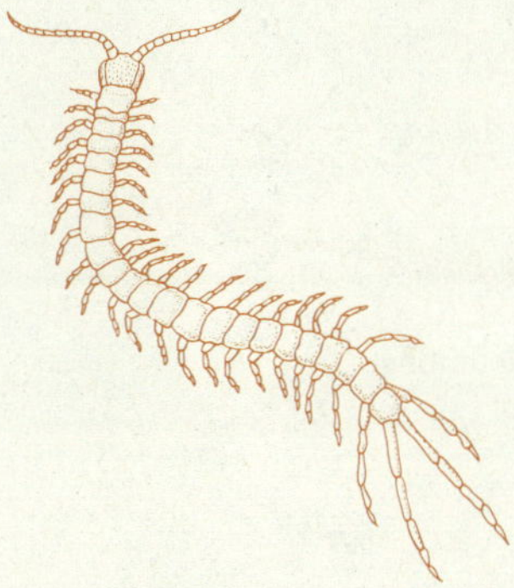

Centipede

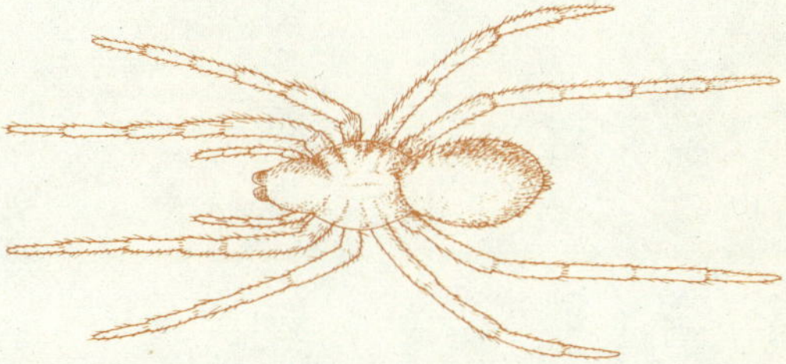

Wolf spider

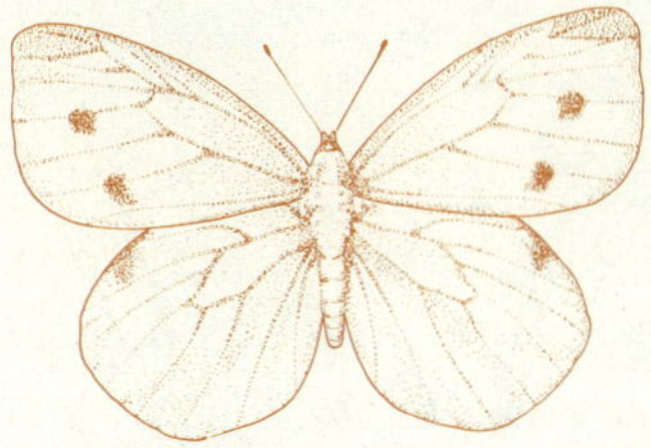

Cabbage butterfly

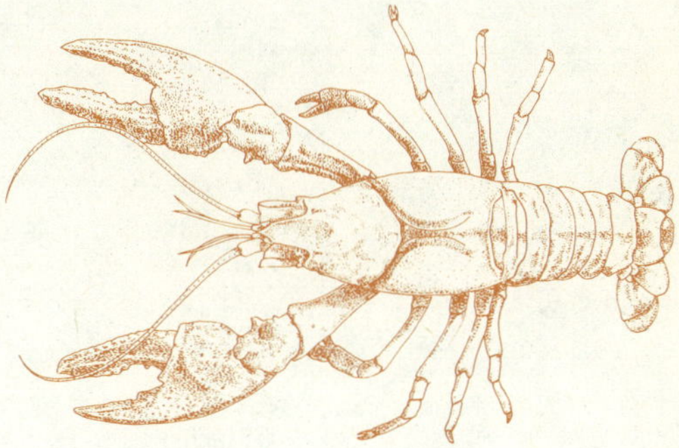

Crayfish

Phylum *Echinodermata* Echinoderms

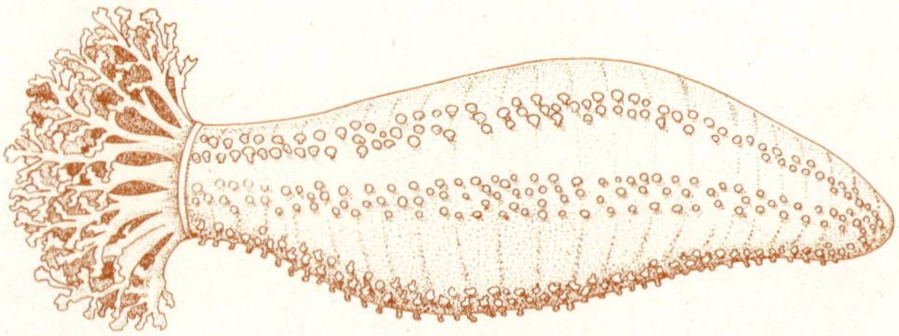

Sea cucumber

Starfish

Sea lily

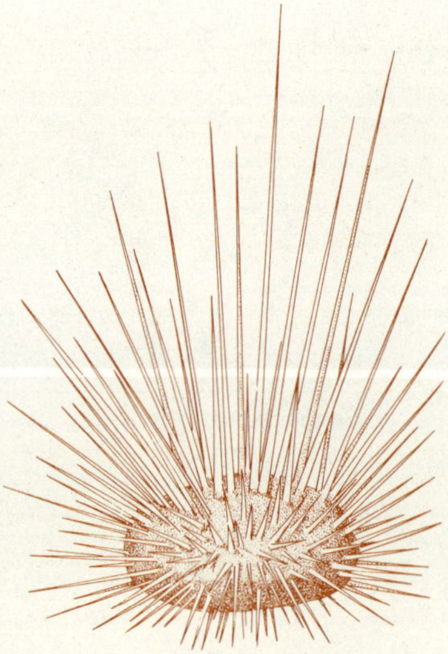

Sea urchin

Phylum *Chordata* Chordates

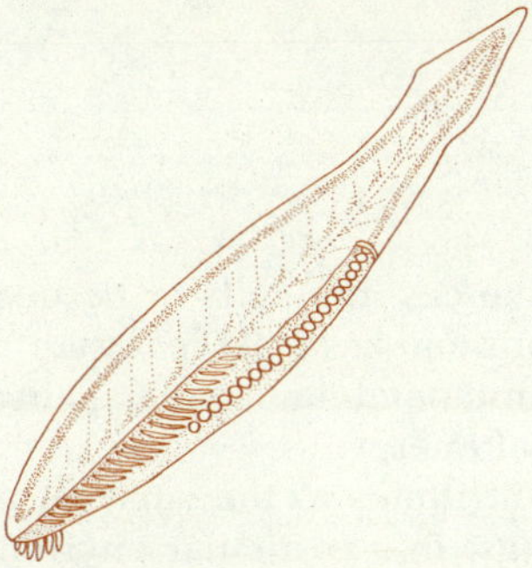

Amphioxus

Fish

Frog

Turtle

Bird

Rabbit

Acknowledgments

The following drawings in the appendix have been adapted from Adams, Baker, and Allen, *The Study of Botany*, 1970, Addison-Wesley, Reading, Mass.

Bacteria and algae, page 401; *Ulothrix* and diatoms, page 404; *Gigartina*, page 405; *Rhizopus, Amanita,* and *Polytrichium,* page 406; *Equisetum* and *Lycopodium,* page 407; *Polypodium,* sugar maple, and wheat, page 408.

APPENDIX: Glossary

Absorption The passage of fluids and dissolved substances into cells or tissues.

Acid A chemical compound that liberates hydrogen ions and has a pH less than 7.

Active transport The movement of materials across a unit membrane, often against the diffusion gradient, requiring an expenditure of energy.

Adaptation Any characteristic of an organism that contributes to the survival of that organism, and the species to which it belongs, in a particular environment.

Adaptive radiation The divergence or splitting up of a group of organisms into a number of species with different ways of life (ecologic niches).

Adenosine triphosphate (ATP) The major energy-carrying molecule in cell metabolism.

Adsorption The attachment of molecules of one substance to the surface of some other substance.

Allele One of two or more differing forms of a gene which occupy the same position in homologous chromosomes.

Alternation of generations Alternation of a haploid, gamete-producing generation and a dipoid, spore-producing generation in the life history of plants and certain protists.

Amino acids Any of about 20 organic acids out of which proteins are synthesized.

Amphibia A class of vertebrates characterized by a life cycle, a portion of which takes place in the water and a portion of which occurs on land.

Amphibious Any organism that is able to live both in the water and on land.

Anabolism That phase of metabolism involving the synthesis of organic molecules.

Anaerobic Pertaining to organisms or biologic processes that do not require the presence of free oxygen.

Anaphase That stage in mitosis and meiosis in which the chromosome sets begin to migrate toward the poles of the spindle figure.

Anatomy The study of the structure of organisms.

Anemia A deficiency of hemoglobin or red blood cells in the blood.

Annual A plant that completes its entire life cycle in one growing season and then dies.

Anterior The head or front end of an organism.

Anther That part of the stamen in plants that bears pollen.

Arboreal Living in trees.

Artery A blood vessel in vertebrates that carries blood away from the heart.

Asexual reproduction Any reproductive process that does not involve a reorganization of nuclear material; without the union of gametes or nuclear material.

Atom The most minute unit into which a chemical element can be divided without losing its characteristic properties.

Autotroph Any organism that can manufacture its own food.

Axon The process of a nerve cell (neuron) conducting impulses away from the cell body.

Base A compound that liberates hydroxyl ions and has a pH greater than 7.
Biologic clocks Mechanisms by which organisms regulate biologic rhythms (the photoperiodism of plants, etc.)
Biome A characteristic form which climax communities assume, determined by certain climatic conditions, such as grasslands and coniferous forest.
Biota All of the living organisms in a specified geographic region or in a given geologic age.
Blastula An early embryonic stage which usually consists of a single layer of cells arranged in the form of a hollow sphere.
Botany The biology of plants.

Carbohydrate An organic compound containing carbon, hydrogen, and oxygen in which the hydrogen to oxygen ratio is usually 2:1. Includes sugars and starches.
Catabolism That phase of metabolism in which organic molecules are broken down into smaller ones.
Catalyst A substance that facilitates a chemical reaction without becoming altered permanently itself.
Centromere The point on a chromosome to which a spindle fiber attaches during mitosis or meiosis.
Chlorophyll Any of several green pigments that absorb the light used in photosynthesis.
Chloroplast An organelle (found in plants and protists) containing chlorophyll in which photosynthesis takes place.
Chromosomes Elongated bodies in the cell nucleus consisting of DNA and protein and bearing the genes.
Cilia Short, hairlike organelles that protrude from the surface of many types of cells and beat in a coordinated fashion; usually numerous.
Class A usually diverse group of generally similar organisms, consisting of one or more orders and which, together with other related classes, make up a phylum.
Climax The final community in an ecologic succession; more or less stable unless the environment is changed.
Coenzyme A nonprotein organic compound that combines temporarily with an enzyme and thereby makes the enzyme active.
Cohesion The attractive force exhibited between similar molecules.
Colloid A mixture in which the particulate matter is in permanent suspension, but the particles are too large to form a solution.
Commensalism An intimate interspecific association (interaction) of two species in which one is benefited and the other is neither benefited nor harmed.
Community All of the organisms living in a common environment; the living portion of an ecosystem.
Compound A substance that is composed of chemical elements that are present in definite proportions by weight; its smallest units are molecules (molecular compounds) or ions (ionic compounds).
Conjugation A type of sexual reproduction in which nuclear material is exchanged during the temporary union of two cells; common in protists (*Paramecium*, *Spirogyra*) and some bacteria.
Convergence The resemblance between relatively unrelated organisms as a result of adaptation to a common environment (for example, a porpoise and a fish).

Copulation A temporary physical contact between two animals by means of special male and female organs whereby sperm cells are transferred from one to the other.

Cross-over The exchange of corresponding segments of homologous chromatids during the first meiotic division.

Cytoplasm That portion of the cell outside the nucleus and inside the cell membrane.

Deciduous Pertaining to trees and shrubs which shed all their leaves in one season of the year.

Decomposers Organisms like many bacteria and fungi which convert organic matter and dead organisms into plant nutrients.

Deletion A chromosomal mutation involving the loss of part of a chromosome.

Dendrites The process of a nerve cell (neuron) conducting impulses to the cell body.

Differentiation The developmental process by which an unspecialized cell becomes structurally and functionally modified into a specialized one.

Diffusion The migration of molecules and ions from a region of higher to a region of lower concentration as a consequence of their kinetic energy and random motion.

Digestion The process by which insoluble organic macromolecules are broken down by hydrolysis into simpler, soluble components.

Diploid Having chromosomes in homologous pairs.

DNA (*deoxyribonucleic acid*) A kind of nucleic acid found mostly in chromosomes (also occurring in chloroplasts and mitochondria) which embodies the hereditary information in a cell.

Dominant Pertaining to a gene that expresses its full phenotypic effect whether in the homozygous or heterozygous state.

Ecologic niche The specific role that a species plays in an ecosystem, especially in regard to the way it obtains food.

Ecologic succession The sequence of biologic communities that occupy a given area.

Ecology Study of the relationships of organisms to one another and to their environment; environmental biology.

Ecosystem All of the organisms in a given area and the associated physical factors with which they interact.

Effector A structure by means of which an organism responds (for example, a muscle or gland).

Electron A negatively charged particle outside the nucleus of an atom.

Electronegative Possessing an affinity for electrons.

Electropositive Showing a tendency to release electrons.

Element A chemical substance composed of atoms of a single kind (about 100 different kinds).

Embryonic induction Developmental process whereby one tissue in an embryo directs the differentiation of adjacent tissue.

Endergonic Pertaining to chemical reactions that require energy in the synthesis of energy-rich compounds.

Endocrine gland An organ producing a secretion (hormone) that is emptied directly into the blood.

Energy That which has the capacity to do work.
Environment The sum of all the physical and biological factors to which an organism is exposed.
Enzymes Catalysts produced by living organisms, protein in nature, or related to proteins (polypeptides).
Evolution The genetic changes that occur in populations of organisms through time.
Excretion The elimination of metabolic wastes from an organism.
Exergonic Pertaining to chemical reactions which release energy through the oxidation of energy-rich compounds.
External fertilization The union of gametes outside the body of a parent.

Family The next full category below order in the hierarchy of classification, and consisting of one or more genera.
Fauna All of the animal life in a given geographic area or in a given geologic age.
Fertilization The union of egg cell and sperm nuclei to form a zygote.
Fission Asexual reproduction by division of an organism into two or more equal parts.
Flagellum A long, whiplike organelle that protrudes from the surface of many kinds of cells; used for locomotion and feeding.
Flora All of the plant life in a given geographic area or in a given geologic age.
Food chain A chain of organisms in a biologic community in which each link feeds on the next lower member and is eaten by the next one above.
Fossil The remains or evidence of the presence of ancient organisms preserved in the earth.
Fruit The ripened pistil(s) of one or more flowers with or without associated flower parts.

Gamete A mature, haploid reproductive cell which, after uniting with another gamete, produces a cell (zygote) that develops into a new individual.
Gametophyte The haploid, gamete-producing generation in the life cycle of a plant.
Gastrula An embryonic stage of development following the blastula in which cell movements usually give rise to the formation of a cavity (the archenteron).
Gel The semisolid or jellylike phase of a colloid.
Gene The functional unit of heredity carried in a chromosome.
Gene frequency The percentage frequency (occurrence) in a population of any one of the alleles in a gene set (for example, $A = 0.2$, $a = 0.8$).
Gene pool All of the genes in a given population.
Genetic drift The random fluctuations in gene frequencies due to sampling errors, particularly evident in very small populations.
Genotype The genetic constitution of an individual.
Genus (*plural genera*) The next full category below family in the hierarchy of classification, and consisting of one or more species.
Germination The resumption of growth of a seed, spore, or other reproductive structure.
Gland An organ producing one or more secretions.

Habitat The location where an organism lives.
Haploid Having a single set of chromosomes, as in gametes.

Herb A plant with very little woody material.

Heterotroph Any organism that cannot manufacture its own food.

Heterozygous Having two different alleles of the same gene set in a homologous pair of chromosomes (for example, *Aa*).

Homeostasis The maintenance of a relatively steady state in the internal environment of an organism (may also be applied to a steady state between interacting organisms in an ecosystem).

Homologous Having a fundamental similarity of structure or a common origin.

Homozygous Having identical alleles of the same gene set in an homologous pair of chromosomes (for example, *AA* or *aa*).

Hormone A cell product which exerts specific effects on the activity of other cells to which it is transported in the same organism.

Hydrolysis Decomposition of a compound by the insertion of water molecules between certain bonds.

Hypertonic A solution having an osmotic pressure exceeding that of the cell contents.

Hypothesis A tentative explanation of a scientific problem.

Hypotonic A solution having an osmotic pressure less than that of the cell contents.

Imprinting A form of learning which occurs rapidly during a very short period in the early life of an animal.

Ingestion Taking food or water into the body.

Instinct A complex, genetically determined pattern of behavior which is essentially unlearned.

Intermediate host A host normally occupied by a parasite during some larval stage in the life cycle of the parasite.

Interphase That period during which a cell is neither dividing nor visibly preparing to divide.

Inversion A chromosomal mutation in which a segment of a chromosome is reversed.

Invertebrates Animals without backbones.

Ion An atom or group of atoms bearing an electric charge.

Isolating mechanism Any mechanism that prevents gene flow between different populations.

Isotonic A solution having the same osmotic pressure as that of the cell contents.

Isotope Atom that differs from other atoms of the same element because it has a different number of neutrons in its nucleus.

Kinetic energy Energy associated with molecular motion.

Kingdom Highest category in the hierarchy of classification, consisting of phyla.

Krebs cycle That portion of cellular respiration in which citric acid is continuously reformed.

Learning The processes by which adaptive changes in individual behavior occur through experience.

Lichen A mutualistic association between an alga and a fungus.

Life cycle The chain of events in the growth and development of an organism from zygote formation to death.

Linkage The tendency of genes located in the same chromosome to be inherited together.

Lipid Any of the many fatlike compounds, including waxes and steroids.

Macromolecule Any molecule of very high molecular weight (nucleic acids, proteins, polysaccharides).

Meiosis Two successive cell divisions in which the chromosome number is reduced from diploid to haploid.

Metabolism All of the chemical reactions occurring within a living system.

Metaphase That stage in mitosis or meiosis when the chromosomes become aligned along the equator of the spindle.

Micron One-thousandth of a millimeter; abbreviated μ.

Millimicron One-thousandth of a micron; abbreviated $m\mu$.

Mimicry Imitation in form, color, or behavior of certain organisms to others.

Mitosis Cell division characterized by exact chromosome replication whereby each daughter cell has the same chromosome number as the parent cell.

Molecule The smallest unit of a molecular compound that still retains the properties of that compound.

Monomer A molecular unit that can be linked with others like it to produce a polymer (for example, glucose is the monomer of starch).

Mutualism An intimate interspecific association (interaction) between two species in which each is benefited by the relationship.

Mutation A change in the genetic makeup of a cell (may involve a single gene or entire chromosomes).

Natural selection The differential survival of genetically different individuals in a population through successive generations.

Nerve A bundle of nerve fibers (nerve cell processes) with accompanying connective tissue.

Neuron A nerve cell.

Nucleic acids Macromolecules composed of nucleotide units (monomers); principal types are DNA and RNA.

Nucleus A more or less centrally located body in a cell, usually enclosed in a membrane, which contains the chromosomes.

Olfactory Pertaining to smell.

Operator gene The gene which regulates the activity of those structural genes closely linked to it.

Operon The operator gene and the structural genes under its control.

Order The next full category below class in the hierarchy of classification, and consisting of one or more families.

Organelle A part of a cell that is structurally and functionally differentiated.

Osmosis The diffusion of water (or some other solvent) through a semipermeable membrane.

Ovary The structure that produces female gametes (eggs or ova) in plants and animals.

Ovulation The discharge of one or more eggs from the ovary of an animal.

Ovum (*plural ova*) Female gamete.

Oxidation Chemical reaction in which electrons are removed from a substance.

Paleontology The study of the history of organisms through their fossil remains.

Parasitism An intimate interspecific association (interaction) between two species in which one species is benefited and the other harmed.

Parthenogenesis The development of an unfertilized egg into a new individual.

Penis The male copulatory organ.

pH A measure of the hydrogen ion concentration of a solution on a scale from 0 (very high) to 14 (very low); pH of 7 is neutral.

Phagocytosis The engulfing of solid particles by a cell.

Phenotype The physical appearance or performance of an organism in contrast to its genetic constitution (genotype).

Pheromone Substance externally secreted by an animal that influences the behavior of other animals of the same species.

Phloem Conducting tissue composed of living cells (sieve tube elements) that transport food throughout the body of certain vascular plants.

Photosynthesis The manufacture of simple carbohydrates from carbon dioxide and water using light as the energy source.

Phylum (*plural phyla*) The next full category below kingdom in the hierarchy of classification, and consisting of one or more classes.

Physiology The study of process and function in living organisms.

Pinocytosis The taking in of fluid droplets by a cell.

Pistil The female organ of a flower; consists of ovary, style, and stigma.

Pollination The transfer of pollen from anther to stigma by wind, insects, and so on.

Polymer A molecule consisting of many repeating units (monomers) linked together.

Polyploid A cell having three or more sets of chromosomes.

Posterior Pertaining to the rear end (the opposite of anterior).

Predation The act of securing food by the killing and devouring of other organisms.

Producers The autotrophs in an ecosystem.

Prophase The first stage in a mitotic or meiotic division when the chromosomes condense and become visible under the light microscope.

Protein A macromolecule composed mostly of a chain of many amino acid units.

Pyramid of energy Refers to the progressive decrease in stored energy in successive levels of energy flow in community food chains; producers represent the greatest amount of stored energy, the terminal consumers the least.

Queen A fertile female among the social insects (bees, ants, termites) whose function is to lay eggs.

Radioactive isotope An isotope with an unstable nucleus that emits ionizing radiation.

Receptor Sensory cell(s) or structure that responds to environmental changes (stimuli) usually by generating nerve impulses.

Recessive Pertaining to a gene that expresses its phenotypic effect only in the homozygous state.

Recombination New combination of genes as a result of cross-over during meiosis.

Redox reactions Oxidation-reduction reactions.

Reduction Chemical reaction in which electrons are added to a substance.

Reflex A simple, involuntary response to a stimulus in those animals possessing a central nervous system.

Regeneration The regrowth or repair of lost or damaged parts in an organism.

Regulator gene A gene which, through the production of a repressor substance, inactivates an operator gene.

Repressor A substance produced by a regulator gene which inactivates an operator gene and all of its structural genes.

Respiration The intake of oxygen and the release of carbon dioxide in breathing; also the oxidative chemical reactions of metabolism, called cellular respiration.

RNA (*ribonucleic acid*) A class of nucleic acids found both in the nucleus and in the cytoplasm that functions in protein synthesis; the hereditary material of many viruses.

Scavenger An animal that feeds on dead animals it has not killed.

Seed An embryo plant which has developed from a fertilized ovule and which, along with food-storage tissue, is enclosed within protective seed coats; found in gymnosperms and angiosperms (see Appendix).

Sex-linkage Genes located on the *X* or *Y* chromosomes.

Sexual reproduction Reproduction involving a reorganization of nuclear material; usually entails the union of gametes or nuclear material from two individuals.

Sol The semifluid, free-flowing phase of a colloid.

Solution A mixture of molecules or ions less than 1 mμ in diameter suspended in a fluid (usually water in living systems).

Species (*plural species*) A population or group of populations of potentially interbreeding individuals which ordinarily do not interbreed with other species; the basic unit of classification.

Sperm A male gamete; a spermatozoan.

Spindle The spindle-shaped structure composed of fine fibers which appears during cell division.

Spore An asexual reproductive structure, usually unicellular, capable of developing into a new organism.

Sporophyte The diploid, spore-producing generation in the life cycle of a plant.

Stamen The male organ of a flower, consisting of a pollen-producing anther and its stalk (filament).

Stigma That part of the pistil which receives pollen grains, and on which the latter germinate.

Stimulus (*plural stimuli*) An environmental change that elicits a response in an organism.

Structural gene A gene that embodies the genetic code for the transcription of a specific polypeptide or protein.

Substrate The substance on which an enzyme acts.

Symbiosis An intimate association of two or more species; includes parasitism, commensalism, and mutualism.

Synapse The region between two neurons, across which a nerve impulse is transmitted.

Telophase The final stage in cell division when the daughter cells begin to separate as the chromosomes reorganize into new nuclei at the two poles of the spindle.

Territory An area occupied by a pair of animals or a group; the area is usually defended by one or more of the occupants against intrusion by other members of the same species.
Testis The structure that produces male gametes (sperm) in animals.
Theory An hypothesis that has been thoroughly tested and which warrants a high degree of confidence.
Threshold The minimal stimulus that will elicit a response in a given structure.
Tissue An association of usually similar cells that perform a common function in multicellular organisms.
Transfer RNA (tRNA) A small RNA molecule that carries amino acids to the sites of protein synthesis; each type of amino acid has its own kind of tRNA molecule.
Translocation A chromosomal mutation in which chromosome segments are interchanged between nonhomologous chromosomes.
Transpiration The loss of water from plants by evaporation.
Turgor Distention of the plant cell wall by the accumulation of water inside the cell.

Unit membrane A membrane composed of an outer and an inner layer of protein molecules with two layers of lipid molecules sandwiched between; most of the membranous organelles of protists, plants, and animals share this basic design.

Vagina The female copulatory organ.
Valence The number of electrons gained, lost, or shared by an atom in its chemical bonds with other atoms.
Vascular Possessing vessels that conduct fluid.
Vein A blood vessel in a vertebrate carrying blood to the heart.
Vessel A tube composed of a series of dead cells aligned end to end for conducting water and minerals through a plant; found in the xylem of most flowering plants.
Viable Capable of remaining alive, especially after exposure to unfavorable conditions.
Virus An ultramicroscopic, noncellular particle composed of a nucleic acid core with a protein shell; capable of replication only within a living cell.
Viscera All of the soft internal organs of an animal.

Warm-blooded Maintaining a relatively uniform body temperature.
Warning coloration Conspicuous coloration possessed by certain animals that are distasteful or harmful.
Worker A sterile female among the social insects (bees, ants, termites).

Xylem Conducting tissue of plants that transports water and minerals from the roots upward; often provides mechanical support (wood consists of xylem).

Yolk The food material stored in the egg cells of animals.

Zoology The biology of animals.
Zygote A single cell formed by the fusion of two gametes; develops into a new individual.

INDEX

Page numbers in italics refer to illustrations. Page numbers followed by asterisks refer to footnoted material.